T0336195

ADVANCED CONTROL OF AIRCRAFT, SPACECRAFT AND ROCKETS

Aerospace Series List

Basic Helicopter Aerodynamics: Third Edition	Seddon and Newman	July 2011
Cooperative Path Planning of Unmanned Aerial Vehicles	Tsourdos et al	November 2010
Principles of Flight for Pilots	Swatton	October 2010
Air Travel and Health: A Systems Perspective	Seabridge et al	September 2010
Design and Analysis of Composite Structures: With applications to aerospace Structures	Kassapoglou	September 2010
Unmanned Aircraft Systems: UAVS Design, Development and Deployment	Austin	April 2010
Introduction to Antenna Placement and Installations	Macnamara	April 2010
Principles of Flight Simulation	Allerton	October 2009
Aircraft Fuel Systems	Langton et al	May 2009
The Global Airline Industry	Belobaba	April 2009
Computational Modelling and Simulation of Aircraft and the Environment: Volume 1 - Platform Kinematics and Synthetic Environment	Diston	April 2009
Handbook of Space Technology	Ley, Wittmann and Hallmann	April 2009
Aircraft Performance Theory and Practice for Pilots	Swatton	August 2008
Surrogate Modelling in Engineering Design: A Practical Guide	Forrester, Sobester and Keane	August 2008
Aircraft Systems, Third Edition	Moir and Seabridge	March 2008
Introduction to Aircraft Aeroelasticity And Loads	Wright and Cooper	December 2007
Stability and Control of Aircraft Systems	Langton	September 2006
Military Avionics Systems	Moir and Seabridge	February 2006
Design and Development of Aircraft Systems	Moir and Seabridge	June 2004
Aircraft Loading and Structural Layout	Howe	May 2004
Aircraft Display Systems	Jukes	December 2003
Civil Avionics Systems	Moir and Seabridge	December 2002

ADVANCED CONTROL OF AIRCRAFT, SPACECRAFT AND ROCKETS

Ashish Tewari

Department of Aerospace Engineering
Indian Institute of Technology, Kanpur, India

A John Wiley & Sons, Ltd., Publication

Registered office
John Wiley & Sons Ltd, The Atrium, Southern Gate, Chichester, West Sussex, PO19 8SQ, United Kingdom

For details of our global editorial offices, for customer services and for information about how to apply for permission to reuse the copyright material in this book please see our website at www.wiley.com.

Library of Congress Cataloguing-in-Publication Data
Tewari, Ashish.
Advanced control of aircraft, rockets, and spacecraft / Ashish Tewari.
p. cm.
Includes bibliographical references and index.
ISBN 978-0-470-74563-2 (hardback)
1. Flight control. 2. Airplanes–Control systems. 3. Space vehicles–Control systems. 4. Rockets (Aeronautics)–Control systems. I. Title.
TL589.4.T49 2011
629.1'1–dc22

2011009365

A catalogue record for this book is available from the British Library.

Print ISBN: 9780470745632
ePDF ISBN: 9781119971207
oBook ISBN: 9781119971191
ePub ISBN: 9781119972747
Mobi ISBN: 9781119972754

Set in 9/11pt Times by Thomson Digital, Noida, India.

The intellectual or logical man, rather than the understanding or observant man, set himself to imagine designs – to dictate purposes to God.

– Edgar Allan Poe

Contents

Series Preface

The field of aerospace is wide ranging and multi-disciplinary, covering a large variety of products, disciplines and domains, not merely in engineering but in many related supporting activities. These combine to enable the aerospace industry to produce exciting and technologically advanced products. The wealth of knowledge and experience that has been gained by expert practitioners in the various aerospace fields needs to be passed onto others working in the industry, including those just entering from University.

The *Aerospace Series* aims to be practical and topical series of books aimed at engineering professionals, operators, users and allied professions such as commercial and legal executives in the aerospace industry. The range of topics is intended to be wide ranging, covering design and development, manufacture, operation and support of aircraft as well as topics such as infrastructure operations and developments in research and technology. The intention is to provide a source of relevant information that will be of interest and benefit to all those people working in aerospace.

Flight Control Systems are an essential element of all modern aerospace vehicles, for instance: civil aircraft use a flight management system to track optimal flight trajectories, autopilots are used to follow a set course or to land the aircraft, gust and manoeuvre load alleviation systems are often employed, military aircraft are often designed for enhanced manoeuvrability with carefree handling in flight regimes that have statically unstable open loop behaviour, Unmanned Autonomous Vehicles (UAVs) are able to follow a pre-defined task without any human intervention, rockets require guidance and control strategies that are able to deal with rapidly time varying parameters whilst ensuring that rocket's orientation follows the desired flight path curvature, and spacecraft need to be guided accurately on missions that may last many years whilst also maintaining the desired attitude throughout.

This book, *Advanced Control of Aircraft, Spacecraft and Rockets*, considers the application of optimal control theory for the guidance, navigation and control of aircraft, spacecraft and rockets. A uniform approach is taken in which a range of modern control techniques are described and then applied to a number of non-trivial multi-variable problems relevant to each type of vehicle. The book is seen as a valuable addition to the *Aerospace Series.*

Peter Belobaba, Jonathan Cooper, Roy Langton and Allan Seabridge

Preface

There are many good textbooks on design of flight control systems for aircraft, spacecraft, and rockets, but there is none that covers all of them in a similar manner. The objective of this book is to show that modern control techniques are applicable to all flight vehicles without distinction. The main focus of the presentation is on applications of optimal control theory for guidance, navigation, and control of aircraft, spacecraft, and rockets. Emphasis is placed upon non-trivial, multi-variable applications, rather than those that are handled by single-variable methods. While the treatment can be considered advanced and requires a basic course in control systems, the topics are covered in a self-contained manner, without the need frequently to refer to control theory textbooks. The spectrum of flight control topics covered in the book is rather broad, ranging from the optimal roll control of aircraft and missiles to multi-variable, adaptive guidance of gravity-turn rockets, and two-point boundary value, optimal terminal control of aircraft, spacecraft, and rockets. However, rotorcraft and flexible dynamics have been excluded for want of space.

The task of designing an optimal flight controller is best explained by the parable of a pilgrim (cf. *The Pilgrim's Progress* by John Bunyan), who, in order to attain a certain religious state, $\mathbf{x}_f$, beginning from an initial state, $\mathbf{x}_i$, must not stray too far from the righteous (optimal) path. In other words, the pilgrim must be aware of the desired trajectory to be followed, $\mathbf{x}_d(t),\ t_i \leq t \leq t_f$, and must also have an in-built tracking system that detects deviations therefrom and applies corrective control inputs, $\mathbf{u}(t)$, in order to return to the optimal path. A successful tracking system is the one which keeps deviations from the desired trajectory,

$$\| \mathbf{x}(t) - \mathbf{x}_d(t) \|, \quad t_i \leq t \leq t_f,$$

to small values. A linear tracking system can be designed for maintaining small deviations by the use of linear optimal feedback control, and is guaranteed to be successful provided the initial deviation is not too large. The tracking system must only require control inputs, $\mathbf{u}(t)$, that are within the capability of a pilgrim to generate. Furthermore, the tracking system must be robust with respect to small (but random) external disturbances (like the various enticements and drawbacks experienced on a pilgrimage), which would otherwise cause instability, that is, a total departure from the righteous path.

The job of generating a desired trajectory, $\mathbf{x}_d(t),\ t_i \leq t \leq t_f$, which will be followed by the tracking system from a given initial state, $\mathbf{x}_i$, to the desired terminal state, $\mathbf{x}_f$, is by no means an easy one. It requires an understanding of the pilgrim's behavior with time, $\mathbf{x}(t)$, which is modeled for example by a nonlinear, ordinary differential equation,

$$\dot{\mathbf{x}}(t) = \mathbf{f}[\mathbf{x}(t), \mathbf{u}(t), t],$$

and should be solved subject to the boundary conditions at the two points,

$$\mathbf{x}(t_i) = \mathbf{x}_i, \quad \mathbf{x}(t_f) = \mathbf{x}_f.$$

Now, it is quite possible that there exist several solutions to such a boundary value problem, each depending upon the applied control inputs, $\mathbf{u}(t)$. One must, however, select the solution which requires the minimization of an objective (e.g., effort on the part of the pilgrim, $\| \mathbf{u}(t) \|$). Thus, one has to solve an optimal two-point boundary value problem (2PBVP) for creating a reference path to be followed by the tracking system. The part of the control system that generates the desired trajectory is called a terminal controller. It could be argued that different religious philosophies (control strategies) emphasize different objective functions, and thus produce different trajectories for a given set of conditions.

Much like a righteous pilgrim, the combined terminal and tracking control system applied to a flight vehicle enables it to automatically reach a desired state in a given time, despite initial deviations and small but random external disturbances.

This book is primarily designed as a textbook for senior undergraduates as well as graduate students in aerospace engineering/aeronautics and astronautics departments. It may also be used as a reference for practicing engineers and researchers in aerospace engineering, whose primary interest lies in flight mechanics and control. Instead of detailing the control theory in a theorem-like environment, only those mathematical concepts that have a bearing on optimal control are highlighted. Emphasis is laid on practical examples and exercises that require some programming, rather than on those of the pen-and-paper type.

At the end of each chapter, short summaries and a limited number of exercises are provided in order to help readers consolidate their grasp of the material. A primary feature of the book is the use of MATLAB® and Simulink®,[1] as practical computational tools to solve problems across the spectrum of modern flight control systems. The MATLAB/Simulink examples are integrated within the text in order to provide ready illustration. Without such a software package, the numerical examples and problems in a text of this kind are difficult to understand and solve. Such codes can also be useful for a practicing engineer or a researcher. All the codes used in the book are available for free download at the following website: http://home.iitk.ac.in/~ashtew.

The reader is assumed to have taken a basic undergraduate course in control systems that covers the transfer function and frequency response methods applied to single-input, single-output systems. Furthermore, a fundamental knowledge of calculus and linear algebra, as well as basic dynamics, is required. A familiarity with the basic linear systems concepts highlighted in Chapters 1 and 2 and Appendix A is also desirable. It is, however, suggested that the introductory material be supplemented by basic examples and exercises from a textbook on linear control systems, especially if the reader has not had a fundamental course on linear systems theory.

I would now like to suggest a possible coverage of the book by course instructors. The curriculum in an aerospace engineering department traditionally includes separate courses on automatic flight control of aircraft and spacecraft, depending upon the students' interest in pursuing either atmospheric or space flight in further study (or research). These basic courses are supplemented by advanced elective courses in the same respective areas. There is a modern trend to move away from this tradition, and to allow courses that cover aircraft, spacecraft, and rocket control systems at both introductory and advanced levels. Having taken a first course on flight vehicle control, the student is prepared to handle such a unified coverage at the next level. The suggested coverage of topics is in two parts, namely (i) optimal terminal control, followed by (ii) tracking (feedback) control. To this end, a single-semester course can be taught covering Chapter 1, Sections 2.1–2.8, 3.1, 3.2, 3.10, 4.1–4.8, 6.1–6.3, and 7.1–7.3. After the mid-semester recess, the second part can begin with Sections 2.9–2.12, and then take the students through various tracking control topics in Sections 3.3–3.9, 4.9–4.13, Chapter 5, Sections 6.4–6.6, and 7.4–7.7. The instructor can easily tailor the topics to suit the requirements of a particular course. Even those teachers who do not entirely agree with the unified coverage are free to choose the specific topics they consider relevant to a particular type of flight vehicle.

[1] MATLAB®/Simulink® are registered products of The MathWorks, Inc., 3 Apple Hill Drive, Natick, MA 01760-2098, USA. http://www.mathworks.com

While the coverage of flight dynamics is well within reach of an average senior undergraduate student in aerospace engineering, it may help to read a unified flight dynamics text, such as my *Atmospheric and Space Flight Dynamics* (Birkhäuser, 2006).

I would like to thank the editorial and production staff of John Wiley & Sons, Ltd, Chichester, especially Nicola Skinner, Debbie Cox, Eric Willner, Jonathan Cooper, and Genna Manaog, for their constructive suggestions and valuable insights during the preparation of this book. I am also grateful to Richard Leigh for copyediting the manuscript. This work would not have been possible without the constant support of my wife, Prachi, and daughter, Manya.

Ashish Tewari
December 2010

1

Introduction

1.1 Notation and Basic Definitions

Throughout this book, we shall represent scalar quantities by italic letters and symbols (a, α), vectors in boldface ($\mathbf{a}$, $\boldsymbol{\alpha}$), and matrices in boldface capitals ($\mathbf{A}$). Unless otherwise mentioned, the axes of a frame are specified by unit vectors in the same case as that of respective axis labels, for example, the axis ox would be represented by the unit vector, $\mathbf{i}$, while OX is given by $\mathbf{I}$. The axes are labeled in order to constitute a right-handed triad (e.g., $\mathbf{i} \times \mathbf{j} = \mathbf{k}$). Components of a vector have the same subscripts as the axis labels, for example, a vector $\mathbf{a}$ resolved in the frame $oxyz$ is written as

$$\mathbf{a} = a_x\mathbf{i} + a_y\mathbf{j} + a_z\mathbf{k}, \tag{1.1}$$

or alternatively as

$$\mathbf{a} = \begin{Bmatrix} a_x \\ a_y \\ a_z \end{Bmatrix}. \tag{1.2}$$

An overdot represents a vector (or matrix) derived by taking the time derivative of the components in a frame of reference, for example,

$$\dot{\mathbf{a}} \doteq \begin{Bmatrix} \frac{da_x}{dt} \\ \frac{da_y}{dt} \\ \frac{da_z}{dt} \end{Bmatrix} = \begin{Bmatrix} \dot{a}_x \\ \dot{a}_y \\ \dot{a}_z \end{Bmatrix}. \tag{1.3}$$

The vector product of two vectors $\mathbf{a}$, $\mathbf{b}$ is often expressed as the matrix product $\mathbf{a} \times \mathbf{b} = \mathbf{S}(\mathbf{a})\mathbf{b}$, where $\mathbf{S}(\mathbf{a})$ is the following skew-symmetric matrix of the components of $\mathbf{a}$:

$$\mathbf{S}(\mathbf{a}) = \begin{pmatrix} 0 & -a_z & a_y \\ a_z & 0 & -a_x \\ -a_y & a_x & 0 \end{pmatrix}. \tag{1.4}$$

Table 1.1 Control system variables

Symbol	Variable	Dimension
$\mathbf{u}(t)$	control input vector	$m \times 1$
$\hat{\mathbf{u}}(t)$	optimal control input vector	$m \times 1$
$\mathbf{w}(t)$	measurement noise vector	$l \times 1$
$\mathbf{x}(t)$	state vector	$n \times 1$
$\hat{\mathbf{x}}(t)$	optimal state vector	$n \times 1$
$\mathbf{y}(t)$	output vector	$l \times 1$
$\mathbf{z}(t)$	state vector for augmentation	$q \times 1$
$\boldsymbol{\nu}(t)$	process noise vector	$p \times 1$

The generic term *planet* is used for any celestial body about which the flight is referenced (Earth, Moon, Sun, Jupiter, etc.). The Euclidean (or L_2) norm of a vector, $\mathbf{a} = (a_x, a_y, a_z)^T$, is written as

$$\| \mathbf{a} \| = \sqrt{a_x^2 + a_y^2 + a_z^2}. \tag{1.5}$$

Standard aerospace symbols define relevant flight parameters and variables as and when used. The nomenclature for control system variables is given in Table 1.1. Any departure from this labeling scheme, if necessary, will be noted.

Control is the name given to the general task of achieving a desired result by appropriate adjustments (or manipulations). The object to be controlled (a flight vehicle) is referred to as the *plant*, while the process that exercises the control is called the *controller*. Both the plant and the controller are *systems*, defined as self-contained sets of physical processes under study. A system has variables applied to it externally, called the *input* vector, and produces certain variables internally, called the *output* vector, which can be measured. In modeling a system, one must account for the relationship between the input and output vectors. This relationship generally takes the form of a set of differential and algebraic equations, if the system is governed by known physical laws. A system having known physical laws is said to be *deterministic*, whereas a system with unknown (or partially known) physical laws is called *non-deterministic* or *stochastic*. Every system has certain unwanted external input variables – called *disturbance inputs* – that cannot be modeled physically and are thus treated as stochastic disturbances. The disturbances are generally of two types: (i) *process noise* that can arise either due to unwanted external inputs, or internally due to uncertainty in modeling the system, and (ii) *measurement noise* that results from uncertainty in measuring the output vector. The presence of these external and internal imperfections renders all practical systems stochastic.

The condition, or *state*, of a system at a given time is specified by a set of scalar variables, called *state variables*, or, in vector form, the *state vector*. The vector space spanned by the state vector is called a *state space*. The state of a system is defined as a collection of the smallest number of variables necessary to completely specify the system's evolution in time, in absence of external inputs. The number of state variables required to represent a system is called *order* of the system, because it is equal to the net order of differential equations governing the system.

While the size of the state space (i.e., the order of the system) is unique, any given system can be described by infinitely many alternative state-space representations. For example, a flight vehicle's state can be described by the position, $\mathbf{r}(t)$, velocity, $\mathbf{v}(t)$, angular velocity, $\boldsymbol{\omega}(t)$, and orientation, $\boldsymbol{\xi}(t)$, relative to a frame of reference. Thus, the *state vector* of a flight vehicle's motion is $\mathbf{x}(t) = \{\mathbf{r}(t), \mathbf{v}(t), \boldsymbol{\omega}(t), \boldsymbol{\xi}(t)\}^T$. However, $\mathbf{x}(t)$ can be transformed into any number of different state vectors depending upon the choice of the reference frame.

A system consisting of the plant and the controller is called a *control system*. The controller manipulates the plant through a *control input* vector, which is actually an input vector to the plant but an output of the controller. In physical terms, this output can take the form of either a force or a torque (or both) applied to a flight vehicle. Often, only electrical (or mechanical) signals are generated by the controller through wires (cables, hydraulic lines), which must be converted into physical inputs for the plant by a separate subsystem called an *actuator*. Also, controllers generally require measurement of the output variables of the plant. Whenever a measurement of a variable is involved, it is necessary to model the dynamics of the measurement process as a separate subsystem called a *sensor*. Generally, there are as many sensors and actuators as there are measured scalar variables and scalar control inputs, respectively. The sensors and actuators can be modeled as part of either the plant or the controller. For our purposes, we shall model them as part of the plant.

The design of a control system requires an accurate mathematical model for the plant. A plant is generally modeled by nonlinear differential equations that can be expressed as a set of first-order ordinary differential equations called the *state equations* such as

$$\frac{\mathrm{d}\mathbf{x}}{\mathrm{d}t} = \mathbf{f}[\mathbf{x}(t), \mathbf{u}(t), \boldsymbol{\nu}(t), t], \tag{1.6}$$

where t denotes time, $\mathbf{x}(t)$ is the state vector (of size $n \times 1$), $\mathbf{u}(t)$ is the control input vector ($m \times 1$), and $\boldsymbol{\nu}(t)$ is the process noise vector ($p \times 1$). The dimension n of the state vector is the order of the system. The nonlinear vector functional, $\mathbf{f}(.)$, is assumed to possess partial derivatives with respect to $\mathbf{x}(t)$, $\mathbf{u}(t)$, and $\boldsymbol{\nu}(t)$ in the neighborhood of $\mathbf{x}_\mathrm{d}(t)$ that constitute a special solution of the state equation (called the *nominal trajectory*). The nominal trajectory usually satisfies equation (1.6) for the unforced case, that is, for $\mathbf{u}(t) = \mathbf{0}$, $\boldsymbol{\nu}(t) = \mathbf{0}$:

$$\frac{\mathrm{d}\mathbf{x}_\mathrm{d}}{\mathrm{d}t} = \mathbf{f}[\mathbf{x}_\mathrm{d}(t), \mathbf{0}, \mathbf{0}, t], \quad t_\mathrm{i} \le t \le t_\mathrm{f}, \tag{1.7}$$

where ($t_\mathrm{i} \le t \le t_\mathrm{f}$) is called the *control interval* with initial time, t_i, and final time, t_f.

The output variables of a plant result from either direct or indirect measurements related to the state variables and control inputs through sensors. Certain errors due to sensor imperfections, called *measurement noise*, are invariably introduced in the measurement process. Therefore, the output vector, $\mathbf{y}(t)$, is related to the state vector, $\mathbf{x}(t)$, the control input vector, $\mathbf{u}(t)$, and the measurement noise vector, $\mathbf{w}(t)$, by an *output equation* given by

$$\mathbf{y}(t) = \mathbf{h}[\mathbf{x}(t), \mathbf{u}(t), \mathbf{w}(t), t], \tag{1.8}$$

where $\mathbf{h}(.)$ is a vector functional and $\mathbf{w}(t)$ is generally of the same size as $\mathbf{y}(t)$.

The most common task of a control system is to bring the plant to a desired state in the presence of disturbances, which can be achieved by either an *open-loop* or a *closed-loop* control system. In an open-loop control system, the controller has no knowledge of the actual state of the plant at a given time, and the control is exercised based upon a model of the plant dynamics, as well as an estimate of its state at a previous time instant, called the *initial condition*. Obviously, such a blind application of control can be successful in driving the plant to a desired state if and only if the plant model is exact, and the external disturbances are absent, which is seldom possible in practice. Therefore, a closed-loop control system is the more practical alternative, in which the actual state of the plant is provided to the controller through a *feedback loop*, so that the control input, $\mathbf{u}(t)$, can be appropriately adjusted. In practice, the feedback consists of measurements of an output vector, $\mathbf{y}(t)$, through which an estimate of the plant's state can be obtained by the controller. If the feedback loop is removed, the control system becomes an open-loop type.

1.2 Control Systems

Our principal task in this book is to design and analyze automatic controllers that perform their duties without human intervention. Generally, a control system can be designed for a plant that is *controllable*.

Controllability is a property of the plant whereby it is possible to take the plant from an initial state, $\mathbf{x}_i(t_i)$, to any desired final state, $\mathbf{x}_f(t_f)$, in a finite time, $t_f - t_i$, solely by the application of the control inputs, $\mathbf{u}(t)$, $t_i \le t \le t_f$. Controllability of the plant is a sufficient (but not necessary) condition for the ability to design a successful control system, as discussed in Chapter 2.

For achieving a given control task, a controller must obey well-defined mathematical relationships between the plant's state variables and control inputs, called *control laws*. Based upon the nature of the control laws, we can classify control systems into two broad categories: *terminal control* and *tracking control*. A terminal control system aims to change the plant's state from an initial state, $\mathbf{x}(t_i)$, to a *terminal* (or final) state, $\mathbf{x}(t_f)$, in a specified time, t_f, by applying a control input, $\mathbf{u}(t)$, in the fixed control interval, $(t_i \le t \le t_f)$. Examples of terminal control include guidance of spacecraft and rockets. The objective of a tracking control system is to maintain the plant's state, $\mathbf{x}(t)$, quite close to a nominal, reference state, $\mathbf{x}_d(t)$, that is available as a solution to the unforced plant state equation (1.7) by the application of the control input, $\mathbf{u}(t)$. Most flight control problems – such as aircraft guidance, orbital control of spacecraft, and attitude control of all aerospace vehicles – fall in this category.

While the design of a terminal controller is typically based on a nonlinear plant (equation (1.6)) and involves iterative solution of a *two-point boundary value problem*, the design of a tracking controller can be carried out by linearizing the plant about a nominal trajectory, $\mathbf{x}_d(t)$, which satisfies equation (1.6).

A tracking control system can be further classified into *state feedback* and *output feedback* systems. While a state feedback system involves measurement and feedback of all the state variables of the plant (which is rarely practical), an output feedback system is based upon measurement and feedback of some output variables that form the plant's output vector, $\mathbf{y}(t)$. The tracking controller continually compares the plant's state, $\mathbf{x}(t)$, with the nominal (desired) state, $\mathbf{x}_d(t)$, and generates a control signal that depends upon the error vector,

$$\mathbf{e}(t) = \mathbf{x}_d(t) - \mathbf{x}(t). \tag{1.9}$$

Clearly, the controller must be able to estimate the plant's state from the measured outputs and any control input vector, $\mathbf{u}(t)$, it applies to the plant. The estimation of plant's state vector from the available outputs and applied inputs is called *observation* (or *state estimation*), and the part of the controller that performs this essential task is called an *observer*. An observer can only be designed for *observable* plants (Appendix A). Since the observation is never exact, one only has the state estimate, $\mathbf{x}_o(t)$, in lieu of the actual state, $\mathbf{x}(t)$. Apart from the observer, the controller has a separate subsystem called the *regulator* for driving the error vector, $\mathbf{e}(t)$, to zero over a reasonable time interval. The regulator is thus the heart of the tracking control system and generates a control input based upon the detected error. Hence, the control input, $\mathbf{u}(t)$, depends upon $\mathbf{e}(t)$. Moreover, $\mathbf{u}(t)$ may also depend explicitly upon the nominal, reference state, $\mathbf{x}_d(t)$, which must be fed forward in order to contribute to the total control input. Therefore, there must be a third subsystem of the tracking controller, called a *feedforward controller*, which generates part of the control input depending upon the desired state.

A schematic block diagram of the tracking control system with an observer is shown in Figure 1.1. Evidently, the controller represents mathematical relationships between the plant's estimated state, $\mathbf{x}_o(t)$, the reference state, $\mathbf{x}_d(t)$, the control input, $\mathbf{u}(t)$, and time, t. Such relationships constitute a *control law*. For example, a *linear* control law can be expressed as follows:

$$\mathbf{u}(t) = \mathbf{K}_d(t)\mathbf{x}_d(t) + \mathbf{K}(t)\left[\mathbf{x}_d(t) - \mathbf{x}_o(t)\right], \tag{1.10}$$

where $\mathbf{K}_d$ is called the *feedforward gain matrix* and $\mathbf{K}$ the *feedback gain matrix*. Both $(\mathbf{K}, \mathbf{K}_d)$ could be time-varying. Note that in Figure 1.1 we have adopted the convention of including sensors and actuators into the model for the plant.

Example 1.1 Consider the problem of guiding a missile to intercept a moving aerial target as shown in Figure 1.2. The centers of mass of the missile, o, and the target, T, are instantaneously located at $\mathbf{R}(t)$ and

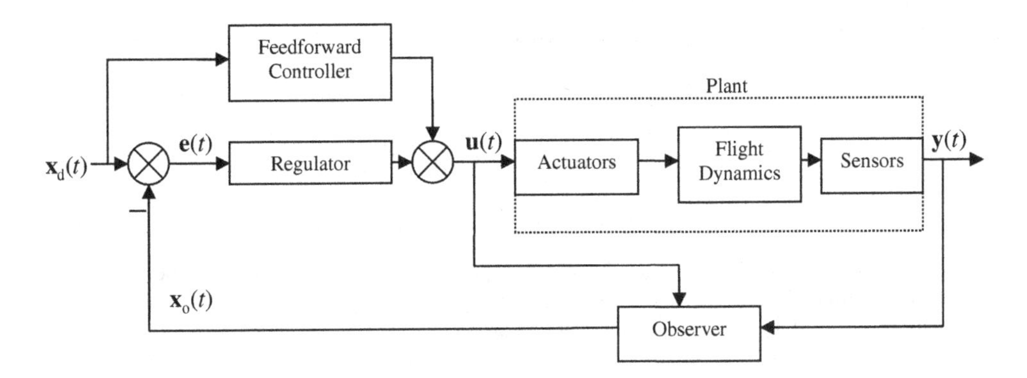

Figure 1.1 Tracking control system with an observer

$\mathbf{R}_T(t)$, respectively, with respective velocities $\mathbf{V}(t)$ and $\mathbf{V}_T(t)$ relative to a stationary frame of reference, ($SXYZ$). The instantaneous position of the target relative to the missile is given by (the vector triangle Soo' in Figure 1.2)

$$\mathbf{r}(t) = \mathbf{R}_T(t) - \mathbf{R}(t), \tag{1.11}$$

while the target's relative velocity is obtained by differentiation as follows:

$$\mathbf{v}(t) = \frac{\mathrm{d}\mathbf{r}}{\mathrm{d}t} = \frac{\mathrm{d}\mathbf{R}_T}{\mathrm{d}t} - \frac{\mathrm{d}\mathbf{R}}{\mathrm{d}t} = \mathbf{V}_T(t) - \mathbf{V}(t). \tag{1.12}$$

Without considering the equations of motion of the missile and the target (to be derived in Chapter 4), we propose the following control law for missile guidance:

$$\mathbf{V}(t) = \mathbf{K}(t)\,[\mathbf{R}_T(t) - \mathbf{R}(t)] = \mathbf{K}(t)\mathbf{r}(t), \tag{1.13}$$

where $\mathbf{K}(t)$ is a time-varying gain matrix. A linear feedback control law of the form given by equation (1.13) is called a *proportional navigation guidance law* (PNG), whose time derivative gives the required

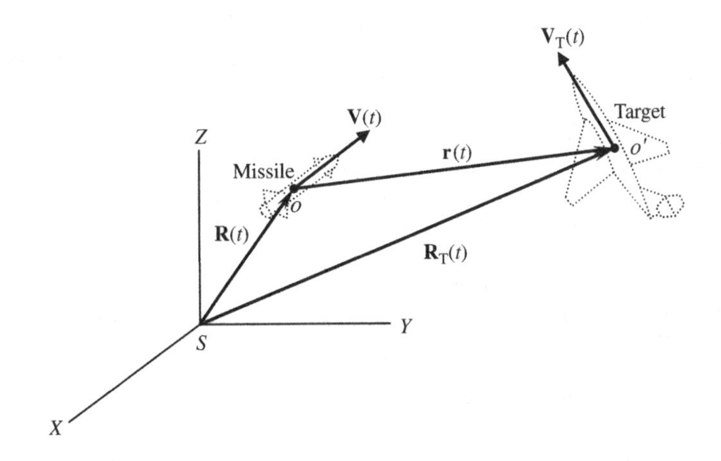

Figure 1.2 Missile guidance for interception of an aerial target

acceleration control input, $\mathbf{u}(t)$, to be applied to the missile:

$$\begin{aligned}\mathbf{u}(t) &= \frac{d\mathbf{V}}{dt} \\ &= \mathbf{K}(t)\left[\mathbf{V}_T(t) - \mathbf{V}(t)\right] + \frac{d\mathbf{K}}{dt}\mathbf{r} \\ &= \mathbf{K}\mathbf{v}(t) + \dot{\mathbf{K}}\mathbf{r}(t).\end{aligned} \tag{1.14}$$

For a successful interception of the target, the relative separation, $\mathbf{r}$, must vanish at some time, $T = t_f - t$, without any regard to the relative velocity, $\mathbf{v}$, prevailing at the time.

A likely choice of state vector for the interception problem is

$$\mathbf{x}(t) = [\mathbf{r}(t), \mathbf{v}(t)]^T, \tag{1.15}$$

which yields the following linear feedback control law of the tracking system:

$$\mathbf{u}(t) = [\dot{\mathbf{K}}(t), \mathbf{K}(t)]\,\mathbf{x}(t). \tag{1.16}$$

The main advantage of the control law given by equation (1.14) is the linear relationship it provides between the required input, $\mathbf{u}$, and the measured outputs, $(\mathbf{r}, \mathbf{v})$, even though the actual plant may have a nonlinear character. Thus the PNG control law is quite simple to implement, and nearly all practical air-to-air missiles are guided by PNG control laws. As the missile is usually rocket powered, its thrust during the engagement is nearly constant. In such a case, PNG largely involves a rotation of the missile's velocity vector through linear feedback relationships between the required normal acceleration components (that are generated by moving aerodynamic fins and/or by thrust vectoring) and the measured relative coordinates and velocity components of the target (obtained by a radar or an infrared sensor mounted on the missile). The proportional navigation gain matrix, $\mathbf{K}(t)$, must be chosen such that the interception $[\mathbf{r}(t_f) \to \mathbf{0}]$ for the largest likely initial relative distance, $\|\, \mathbf{r}(0) \,\|$, takes place within the allowable maximum acceleration, $\|\, \mathbf{u} \,\| \leq u_m$ as well as the maximum time of operation, t_f, of the engines powering the missile. We shall have occasion to discuss the PNG law a little later.

A tracking system with a time-varying reference state, $\mathbf{x}_d(t)$, can be termed successful only if it can maintain the plant's state, $\mathbf{x}(t)$, within a specified percentage error, $\|\, \mathbf{x}_d(t) - \mathbf{x}(t) \,\| \leq \delta$, of the desired reference state at all times. The achieved error tolerance (or corridor about the reference state), δ, thus gives a measure of the control system's performance. The control system's performance is additionally judged by the time taken by the plant's state to reach the desired error tolerance about the reference state, as well as the magnitude of the control inputs required in the process. The behavior of the closed-loop system is divided into the response at large times, $t \to \infty$, called the *steady-state response*, and that at small values of time when large deviations (called *overshoots*) from the desired state could occur. A successful control system is one in which the maximum overshoot is small, and the time taken to reach within a small percentage of the desired state is also reasonably small.

Example 1.2 Consider a third-order tracking system with the state vector $\mathbf{x} = (x_1, x_2, x_3)^T$. A plot of the nominal trajectory, $\mathbf{x}_d(t)$ is shown in Figure 1.3. The tracking error corridor is defined by the Euclidean norm of the off-nominal state deviation as follows:

$$\|\, \mathbf{x}_d(t) - \mathbf{x}(t) \,\| = \sqrt{(x_1 - x_{1d})^2 + (x_2 - x_{2d})^2 + (x_3 - x_{3d})^2} \leq \delta, \tag{1.17}$$

where δ is the allowable error tolerance. The actual trajectory is depicted in Figure 1.3 and the maximum overshoot from the nominal is also indicated.

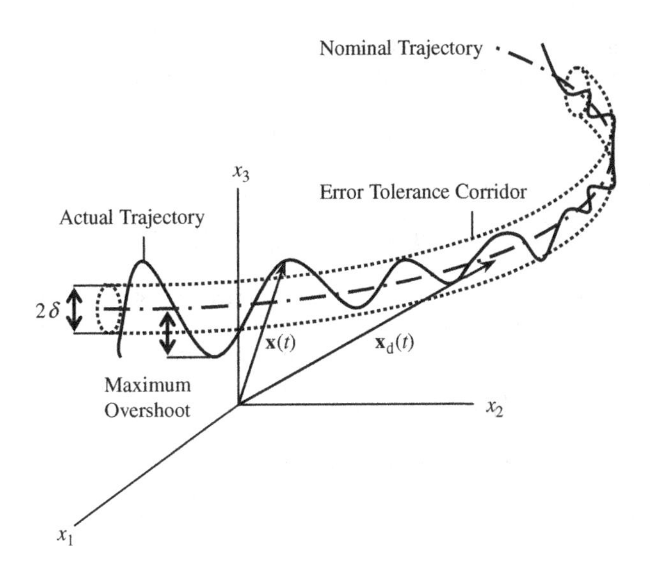

Figure 1.3 Nominal trajectory and tracking error corridor with tolerance, δ

1.2.1 Linear Tracking Systems

With the definition for a successful control system given above, one can usually approximate a nonlinear tracking system by linear differential equations resulting from the assumption of small deviations from the reference state. A first-order Taylor series expansion of the control system's governing (nonlinear) differential equations about the reference state thus yields a linear tracking system (Appendix A), and the given reference solution, $\mathbf{x}_d(t)$, is regarded as the nominal state of the resulting linear system. A great simplification occurs by making such an assumption, because we can apply the principle of *linear superposition* to a linearized system, in order to yield the total output due to a linear combination of several input vectors. Linear superposition also enables us to utilize operational calculus (such as Laplace and Fourier transforms) and linear algebraic methods for design and analysis of control systems. Appendix A briefly presents the linear systems theory, which can be found in detail in any textbook on linear systems, such as Kailath (1980).

Let the control system without disturbance variables be described by the state equation

$$\dot{\boldsymbol{\xi}} = \mathbf{f}[\boldsymbol{\xi}(t), \boldsymbol{\eta}(t), t], \tag{1.18}$$

where $\boldsymbol{\xi}$ is the state vector, and $\boldsymbol{\eta}$, the input vector. The nonlinear vector functional, $\mathbf{f}(.)$, is assumed to possess partial derivatives with respect to state and input variables in the neighborhood of the reference, nominal trajectory, $\boldsymbol{\xi}_0(t)$, which is a solution to equation (1.18) and thus satisfies

$$\dot{\boldsymbol{\xi}}_0(t) = \mathbf{f}[\boldsymbol{\xi}_0(t), \boldsymbol{\eta}_0(t), t], \quad t_i \le t \le t_f, \tag{1.19}$$

where $\boldsymbol{\eta}_0(t)$ is the known input (called the *nominal input*) applied to the system in the interval ($t_i \le t \le t_f$).

In order to maintain the system's state close to a given reference trajectory, the tracking system must possess a special property, namely *stability* about the nominal reference trajectory. While stability can be defined in various ways, for our purposes we will consider *stability in the sense of Lyapunov* (Appendix B), which essentially implies that a small control perturbation from the nominal input results in only a small deviation from the nominal trajectory.

In a tracking system, the system is driven close to the nominal trajectory by the application of the *control input*, $\mathbf{u}(t)$, defined as the difference between the actual and the nominal input vectors:

$$\mathbf{u}(t) = \boldsymbol{\eta}(t) - \boldsymbol{\eta}_0(t), \quad t_i \leq t \leq t_f, \tag{1.20}$$

such that the state deviation, $\mathbf{x}(t)$, from the nominal trajectory, given by

$$\mathbf{x}(t) = \boldsymbol{\xi}(t) - \boldsymbol{\xi}_0(t), \quad t_i \leq t \leq t_f, \tag{1.21}$$

remains small. The assumption of a small control input causing a small state perturbation (which results from the stability about the reference trajectory) is crucial to a successful control system design, and leads to the following Taylor series expansion around the nominal trajectory:

$$\begin{aligned}\dot{\mathbf{x}}(t) &= \tfrac{\partial \mathbf{f}}{\partial \boldsymbol{\xi}}[\boldsymbol{\xi}_0(t), \boldsymbol{\eta}_0(t), t]\mathbf{x}(t) \\ &\quad + \tfrac{\partial \mathbf{f}}{\partial \boldsymbol{\eta}}[\boldsymbol{\xi}_0(t), \boldsymbol{\eta}_0(t), t]\mathbf{u}(t) + \mathcal{O}(2), \quad t_i \leq t \leq t_f,\end{aligned} \tag{1.22}$$

where $\mathcal{O}(2)$ denotes the second- and higher-order terms involving control and state deviations that are neglected due to the small perturbation (stability) assumption.

The Jacobian matrices (Appendix A) of $\mathbf{f}$ with respect to $\boldsymbol{\xi}$ and $\boldsymbol{\eta}$ in the neighborhood of the nominal trajectory are denoted by

$$\mathbf{A}(t) = \frac{\partial \mathbf{f}}{\partial \boldsymbol{\xi}}[\boldsymbol{\xi}_0(t), \boldsymbol{\eta}_0(t), t], \quad \mathbf{B}(t) = \frac{\partial \mathbf{f}}{\partial \boldsymbol{\eta}}[\boldsymbol{\xi}_0(t), \boldsymbol{\eta}_0(t), t]. \tag{1.23}$$

Retaining only the linear terms in equation (1.22), we have the following state-space description of the system as a set of first-order, linear, ordinary differential equations:

$$\dot{\mathbf{x}}(t) = \mathbf{A}(t)\mathbf{x}(t) + \mathbf{B}(t)\mathbf{u}(t), \quad t_i \leq t \leq t_f, \tag{1.24}$$

with the initial condition $\mathbf{x}_i = \mathbf{x}(t_i)$. Often, the system's governing equations are linearized *before* expressing them as a set of first-order, nonlinear, state equations (equation (1.19)) leading to the same result as equation (1.24). For the time being, we are ignoring the disturbance inputs to the system, which can be easily included through an additional term on the right-hand side.

When the applied control input is zero ($\mathbf{u}(t) = \mathbf{0}$), equation (1.24) becomes the following *homogeneous* equation:

$$\dot{\mathbf{x}}(t) = \mathbf{A}(t)\mathbf{x}(t). \tag{1.25}$$

The solution to the homogeneous state equation, with the initial condition $\mathbf{x}_i = \mathbf{x}(t_i)$ is expressed as

$$\mathbf{x}(t) = \boldsymbol{\Phi}(t, t_i)\mathbf{x}(t_i), \quad \text{for all } t. \tag{1.26}$$

where $\boldsymbol{\Phi}(t, t_i)$ is called the *state transition matrix*. The state transition matrix thus has the important property of transforming the state at time t_i to another time t. Other important properties of the state transition matrix are given in Appendix A. Clearly, stability of the system about the nominal trajectory can be stated in terms of the initial response of the homogeneous system perturbed from the nominal state at some time t_i (Appendix B).

The general solution of nonhomogeneous state equation (1.24), subject to initial condition, $\mathbf{x}_i = \mathbf{x}(t_i)$, can be written in terms of $\boldsymbol{\Phi}(t, t_i)$ as follows:

$$\mathbf{x}(t) = \boldsymbol{\Phi}(t, t_i)\mathbf{x}_i + \int_{t_i}^{t} \boldsymbol{\Phi}(t, \tau)\mathbf{B}(\tau)\mathbf{u}(\tau)d\tau, \quad \text{for all } t. \tag{1.27}$$

The derivation of the state transition matrix is a formidable task for a linear system with time-varying coefficient matrices (called a *time-varying linear system*). Only in some special cases can the exact closed-form expressions for $\mathbf{\Phi}(t, t_i)$ be derived. Whenever $\mathbf{\Phi}(t, t_i)$ cannot be obtained in closed form, it is necessary to apply approximate numerical techniques for the solution of the state equation.

The output (or *response*) of the linearized tracking system can be expressed as

$$\mathbf{y}(t) = \mathbf{C}(t)\mathbf{x}(t) + \mathbf{D}(t)\mathbf{u}(t), \tag{1.28}$$

where $\mathbf{C}(t)$ is called the *output coefficient matrix*, and $\mathbf{D}(t)$ is called the *direct transmission matrix*. If $\mathbf{D}(t) = \mathbf{0}$, there is no direct connection from the input to the output and the system is said to be *strictly proper*.

1.2.2 Linear Time-Invariant Tracking Systems

In many flight control applications, the coefficient matrices of the plant linearized about a reference trajectory (equations (1.24) and (1.28)) are nearly constant with time. This may happen because the time scale of the deviations from the reference trajectory is too small compared to the time scale of reference dynamics. Examples of these include orbital maneuvers and attitude dynamics of spacecraft about a circular orbit, and small deviations of an aircraft's flight path and attitude from a straight and level, equilibrium flight condition. In such cases, the tracking system is approximated as a *linear time-invariant system*, with **A**, **B** treated as constant matrices. The state transition matrix of a linear time-invariant (LTI) system is written as

$$\mathbf{\Phi}(t, t_i) = e^{\mathbf{A}(t-t_i)}, \tag{1.29}$$

where $e^{\mathbf{A}(t-t_i)}$ is called the *matrix exponential* of the square matrix, $\mathbf{A}(t - t_i)$, and can be calculated by either Laplace transforms or linear algebraic numerical methods (Appendix A). By substituting equation (1.29) into equation (1.27) with $t_i = 0$, we can write the following general expression for the state of an LTI system in the presence of an arbitrary, Laplace transformable input, which starts acting at time $t = 0$ when the system's state is $\mathbf{x}(0) = \mathbf{x}_0$:

$$\mathbf{x}(t) = e^{\mathbf{A}t}\mathbf{x}_0 + \int_0^t e^{\mathbf{A}(t-\tau)}\mathbf{B}(\tau)\mathbf{u}(\tau)\mathrm{d}\tau. \tag{1.30}$$

Since a linear feedback control system can often be designed for an LTI plant using either the traditional transfer function approach (see Chapter 2 of Tewari 2002), or the multi-variable state-space approach (Chapters 5 and 6 of Tewari 2002), the resulting LTI tracking system can be considered a basic form of all flight control systems. For an LTI plant with state equation

$$\dot{\mathbf{x}} = \mathbf{A}\mathbf{x} + \mathbf{B}\mathbf{u}, \tag{1.31}$$

and output (or measurement) equation

$$\mathbf{y} = \mathbf{C}\mathbf{x} + \mathbf{D}\mathbf{u}, \tag{1.32}$$

a typical linear feedback law for tracking a reference trajectory, $\mathbf{x}(t) = \mathbf{0}$, with a constant regulator gain, **K**, is

$$\mathbf{u} = -\mathbf{K}\mathbf{x}_o, \tag{1.33}$$

where $\mathbf{x}_o(t)$ is the estimated state deviation computed by a linear observer (or Kalman filter) with the observer state, $\mathbf{z}(t)$, observer gain, **L**, and the following observer (or filter) dynamics:

$$\mathbf{x}_o = \mathbf{L}\mathbf{y} + \mathbf{z}, \tag{1.34}$$

$$\dot{\mathbf{z}} = \mathbf{F}\mathbf{z} + \mathbf{G}\mathbf{y} + \mathbf{H}\mathbf{u}. \tag{1.35}$$

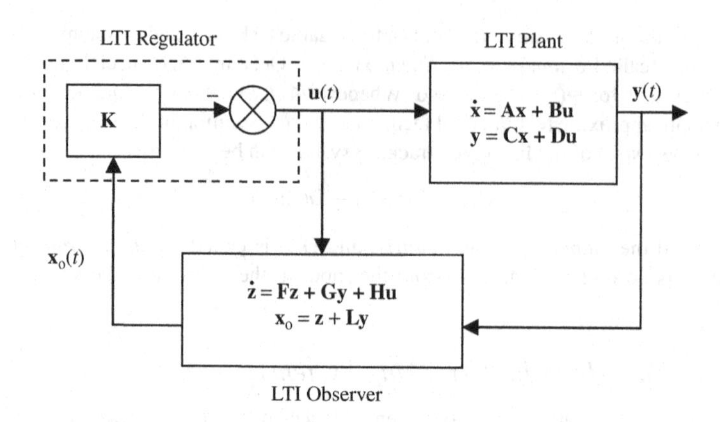

Figure 1.4 Linear time-invariant tracking system

The regulator gain matrix, **K**, the observer gain matrix, **L**, and the constant coefficient matrices of the observer, **F**, **G**, **H**, must satisfy the asymptotic stability criteria for the overall closed-loop system, which translates into separately guaranteeing asymptotic stability of the regulator and observer by the *separation principle* (Kailath 1980). Furthermore, in order to be practical, the design of both the regulator and the observer must be *robust* with respect to modeling uncertainties (process noise) and random external disturbances (measurement noise), both of which are often small in magnitude (thereby not causing a departure from the linear assumption), but occur in a wide spectrum of frequencies. A block diagram of an LTI tracking system with observer is shown in Figure 1.4. The design of such systems for achieving stability and robustness with respect to uncertainties and random disturbances is the topic of several control systems textbooks such as Kwakernaak and Sivan (1972), Maciejowski (1989), Skogestad and Postlethwaite (1996), and Tewari (2002). Examples of the LTI design methods for achieving robustness are *linear quadratic Gaussian* (LQG) and H_∞ techniques that are now standard in modern control design and involve the assumption of a zero-mean *Gaussian white noise*[1] process and measurement noise. A linear observer ensuring the maximum robustness in the state estimation process with respect to a zero-mean Gaussian white noise and measurement noise is called a *Kalman filter*, or *linear quadratic estimator*, and is a part of the LQG controller. Since LQG control, H_∞ control, and the Kalman filter naturally arise out of the optimal control theory, we shall have occasion to consider them in Chapter 2.

1.3 Guidance and Control of Flight Vehicles

Vehicles capable of sustained motion through air or space are termed *flight vehicles*, and are broadly classified as *aircraft*, *spacecraft*, and *rockets*. Aircraft flight is restricted to the atmosphere, and spacecraft flight to exo-atmospheric space, whereas rockets are equally capable of operating both inside and outside the sensible atmosphere. Of these, aircraft have the lowest operating speeds due to the restriction imposed by aerodynamic *drag* (an atmospheric force opposing the motion and proportional to the square of the speed). Furthermore, the requirement of deriving *lift* (a force normal to the flight direction necessary for balancing the weight) and *thrust* (the force along flight direction to counter drag) from the denser regions of the atmosphere restricts aircraft flight to the smallest altitudes of all flight vehicles. In contrast, spacecraft have the highest speeds and altitudes due to the requirement of generating centripetal acceleration entirely

[1] *White noise* is a statistical process with a flat power spectrum, that is, the signal has the same power at any given frequency. A Gaussian process has a normal (bell-shaped) probability density distribution.

by gravity, while the operating speeds and altitudes of rockets (mostly employed for launching the spacecraft) fall somewhere between those of the aircraft and spacecraft.

All flight vehicles require manipulation (i.e., adjustment or *control*) of position, velocity, and attitude (or orientation) for successful and efficient flight. A transport aircraft navigating between points A and B on the Earth's surface must follow a flight path that ensures the smallest possible fuel consumption in the presence of winds. A fighter aircraft has to maneuver in a way such that the normal acceleration is maximized while maintaining the total energy and without exceeding the structural load limits. A spacecraft launch rocket must achieve the necessary orbital velocity while maintaining a particular plane of flight. A missile rocket has to track a maneuvering target such that an intercept is achieved before running out of propellant. An atmospheric entry vehicle must land at a particular point with a specific terminal energy without exceeding the aero-thermal load limits. In all of these cases, precise control of the vehicle's attitude is required at all times since the aerodynamic forces governing an atmospheric trajectory are very sensitive to the body's orientation relative to the flight direction. Furthermore, in some cases, attitude control alone is crucial for the mission's success. For example, a tumbling (or oscillating) satellite is useless as an observation or communications platform, even though it may be in the desired orbit. Similarly, a fighter (or bomber) aircraft requires a stable attitude for accurate weapons delivery.

Flight vehicle control can be achieved by either a pilot, an automatic controller (or *autopilot*), or both acting in tandem. The process of controlling a flight vehicle is called *flying*. A manned vehicle is flown either manually by the pilot, or automatically by the autopilot that is programmed by the pilot to carry out a required task. Unmanned vehicles can be flown either remotely or by onboard autopilots that receive occasional instructions from the ground. It is thus usual to have some form of human intervention in flying, and rarely do we have a fully automated (or *autonomous*) flight vehicle that selects the task to be performed for a given mission and also performs it. Therefore, the job of designing an automatic control system (or autopilot) is quite simple and consists of: (i) generating the required flight tasks for a given mission that are then stored in the autopilot memory as computer programs or specific data points; and (ii) putting in place a mechanism that closely performs the required (or reference) flight tasks at a given time, despite external disturbances and internal imperfections. In (i), the reference tasks are generally a set of positions, velocities, and attitudes to be followed as functions of time, and can be updated (or modified) by giving appropriate signals by a human being (pilot or ground controller). The result is an automatic control system that continually compares the actual position, velocity, and attitude of the vehicle with the corresponding reference values (i) and makes the necessary corrections (ii) in order that the vehicle moves in the desired manner.

Any flight vehicle must have two separate classes of control systems: first, control of position and linear velocity relative to a planet fixed frame, called trajectory control (or more specifically *guidance* that results in the vehicle's *navigation*[2] from one position to another; and second, control of vehicle's orientation (*attitude control*) with respect to a frame of reference. The desired position and velocity – usually derived from the solution of a trajectory optimization problem – could be stored onboard at discrete times, serving as nominal (reference) values against which the actual position and velocity can be compared. A guidance system continually compares the vehicles's actual position and velocity with the nominal ones, and produces linear acceleration commands in order to correct the errors. Most flight vehicles require reorientation (rotation) is realized in practice. Vehicle rotation is performed by applying an angular acceleration external/internal torques. Thus, in such a case, attitude control system becomes subservient (actuator) to the guidance system. In layman terms, the guidance system can be said to "drive the vehicle on an invisible highway in the sky" by using the attitude control system to twist and turn the vehicle. In many flight conditions, a natural stability is inherent in the attitude dynamics so that

[2] The term *navigation* is sometimes applied specifically to the determination of a vehicle's current position. However, in this book, we shall employ the term in its broader context, namely the process of changing the vehicle's position between two given points.

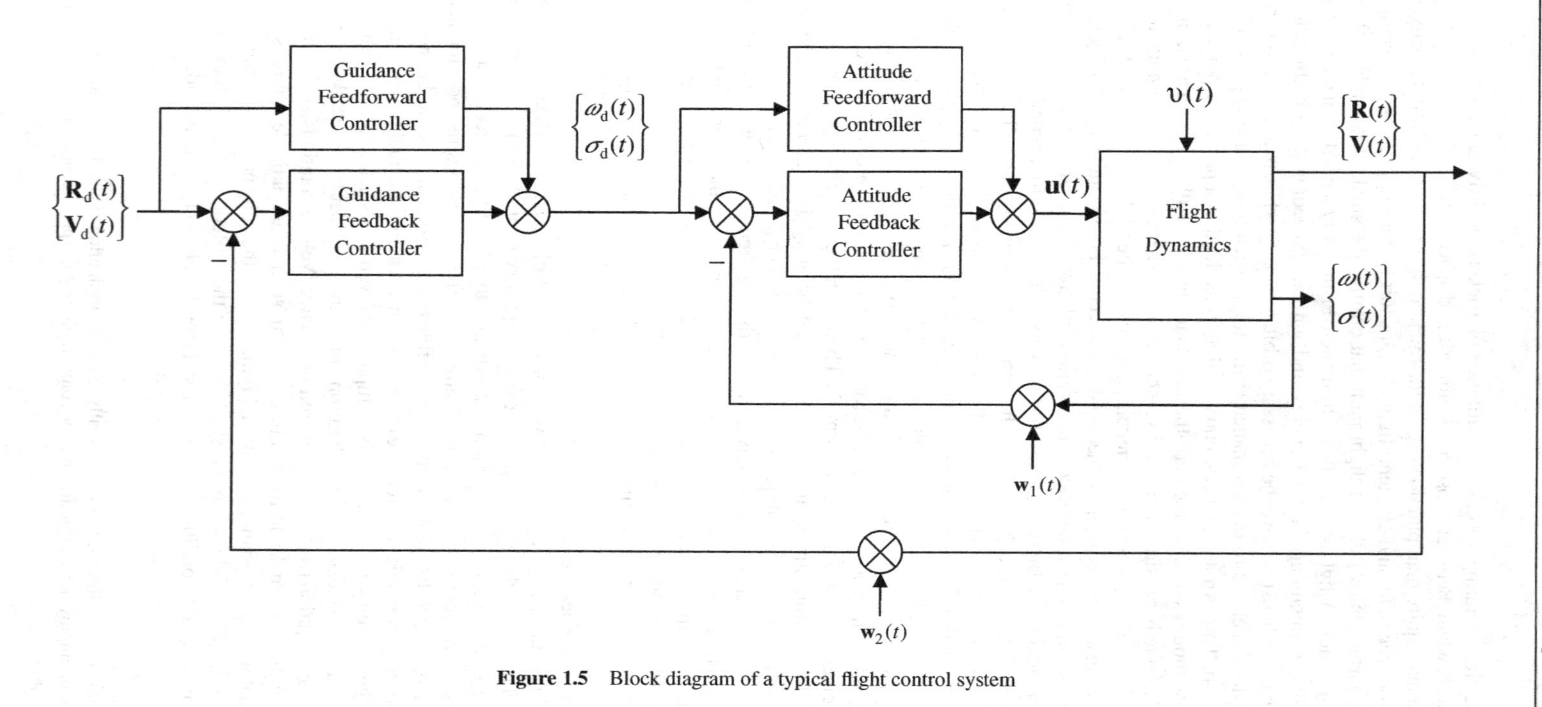

Figure 1.5 Block diagram of a typical flight control system

equilibrium is attained without control effort – that is, the vehicle may, by and large, keep on moving along a desired path on its own. However, even in such cases, it is necessary to augment the stability of the equilibrium point for improved performance. Therefore, we rarely have a flight situation where neither guidance nor attitude control is required.

For most flight vehicles, navigation involves a time scale about an order of magnitude larger than that of attitude dynamics. Hence, navigation along a slowly changing flight path can often be achieved manually (by either a pilot or a ground operator), especially when the time scale of the associated dynamics is larger than a few seconds. When the time scales of control system dynamics are too small to be managed manually, automatic flight control mechanisms become indispensable. Such is the case for all high-performance aircraft, missiles, launch-vehicles, entry vehicles, and some spacecraft. Automatic flight control is also necessary in order to alleviate pilot workload in demanding tasks that require accurate energy and momentum management, and has thus become a common feature of almost all flight vehicles.

A block diagram of a typical flight control system is shown in Figure 1.5. Note the separate feedback loops for guidance and control. The desired position vector, $\mathbf{r}_d(t)$, and the desired velocity vector, $\mathbf{v}_d(t)$ (usually in an inertial reference frame), are fed as a vector input – variously called *setpoint, reference signal*, or *desired output* – to the automatic flight control system. In addition, there are disturbance inputs to the control system in the form of process noise, $\boldsymbol{\nu}(t)$, measurement noise of attitude loop, $\mathbf{w}_1(t)$, and measurement noise of the guidance loop, $\mathbf{w}_2(t)$.

1.4 Special Tracking Laws

While the mathematical derivation of control laws for tracking systems generally requires optimal control theory (Chapter 2) as well as a detailed mathematical model of the plant dynamics, we can discuss certain special tracking laws without resorting to either optimal control, or accurate plant modeling. Such control laws are derived intuitively for guidance and attitude control of flight vehicles, and have been successfully applied in practical flight control situations in the past – such as in early tactical and strategic missiles and spacecraft (Battin 1999). They were especially indispensable at a time when the optimal control theory was not well established. Being extremely simple to implement, such intuitive control methods are still in vogue today.

1.4.1 Proportional Navigation Guidance

We briefly discussed proportional navigation guidance in Example 1.2 for guiding a missile to intercept an aerial target. However, PNG is a more general strategy and can be applied to any situation where the nominal trajectory is well known. Thus we consider $\mathbf{r}(t)$ in equation (1.13) to be the instantaneous separation (or position error) from a given nominal trajectory, which must be brought to zero in time $T - t$. The total time of flight, $t_f = t + T$, is generally specified at the outset. The feedback control law for PNG, equation (1.14) – rewritten below for convenience – implies that the corrective acceleration control, $\mathbf{u}(t)$, continues to be applied to the missile until *both* the position error and the velocity error, $\mathbf{v}(t) = \mathbf{V}_T(t) - \mathbf{V}(t)$, become zero:

$$\mathbf{u}(t) = \mathbf{K}\mathbf{v}(t) + \dot{\mathbf{K}}\mathbf{r}(t). \tag{1.36}$$

Thus PNG provides a strategy for intercepting the nominal trajectory with the necessary nominal velocity, $\mathbf{V}_T(t)$, and the velocity error, $\mathbf{v}(t)$, is referred to as the *velocity to be gained*.

Equation (1.13) indicates that the PNG law essentially guides the missile's velocity vector toward a *projected interception point*, P, which is instantaneously obtained at time t by projecting the target's current velocity, $\mathbf{V}_T(t)$, in a straight line as shown in Figure 1.6. Clearly, P must be a function of time as

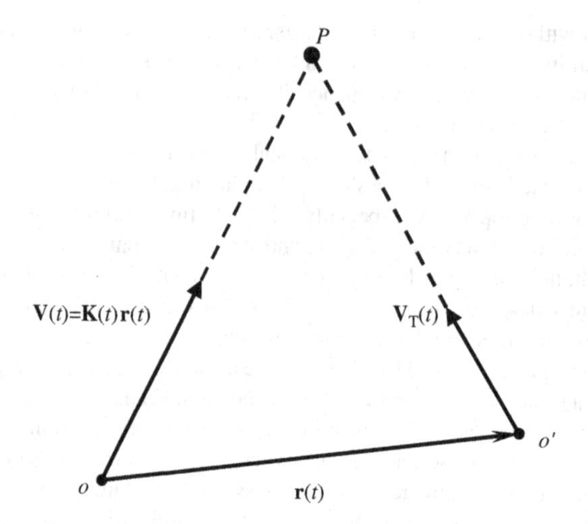

Figure 1.6 Proportional navigation guidance and the projected interception point

neither $\mathbf{r}(t)$ nor $\mathbf{V}_\mathrm{T}(t)$ is constant. Given the velocity triangle shown in Figure 1.6, we have

$$(T-t)\mathbf{K}\mathbf{r} = \mathbf{r} + (T-t)\mathbf{V}_\mathrm{T}, \tag{1.37}$$

where $T = t_\mathrm{f} - t$ is the time of projected interception. At first glance, it would appear that the elements of the PNG gain matrix can be computed from equation (1.15) at each time instant for the current values of the relative separation, $\mathbf{r}$, the target velocity, $\mathbf{V}_\mathrm{T}$, and the desired time before intercept (also called *time to go*), $T - t$. However, the determination of $\mathbf{K}(t)$ from equation (1.15) is not possible as it gives only three scalar equations for the nine unknown elements. Additional conditions should therefore be specified, such as the optimization of certain variables (Chapter 2) for a unique determination of the PNG gains.

Fortunately, it is not necessary to know the time-varying PNG gains explicitly for the implementation of the proportional navigation law. Substituting equation (1.15) into equation (1.36) gives

$$\begin{aligned}\mathbf{u}(t) &= \dot{\mathbf{V}} = \tfrac{\mathrm{d}}{\mathrm{d}t}\{\mathbf{K}\mathbf{r}\} \\ &= \dot{\mathbf{V}}_\mathrm{T} + \tfrac{\mathbf{r}}{(T-t)^2} + \tfrac{\dot{\mathbf{V}}_\mathrm{T}-\dot{\mathbf{V}}}{T-t},\end{aligned} \tag{1.38}$$

or

$$\mathbf{u}(t) = \dot{\mathbf{V}} = \dot{\mathbf{V}}_\mathrm{T} + \frac{\mathbf{r}}{(T-t)(T-t+1)}. \tag{1.39}$$

Thus, given the target's velocity vector and the instantaneous separation vector, $\mathbf{r}(t)$, both of which can be either directly measured, or estimated from other measured variables, one can apply a corrective acceleration input to the interceptor according to equation (1.39). It is to be reiterated that there is no need to solve the equations of motion of the interceptor in real time for application of the PNG tracking law.

Example 1.3 Consider a simple example of tracking a projectile flying relatively slowly so that planetary curvature is negligible. Furthermore, the altitude is small enough for acceleration due to gravity, g, to be constant. Such an approximation is termed the *flat planet assumption*, and implies that the acceleration

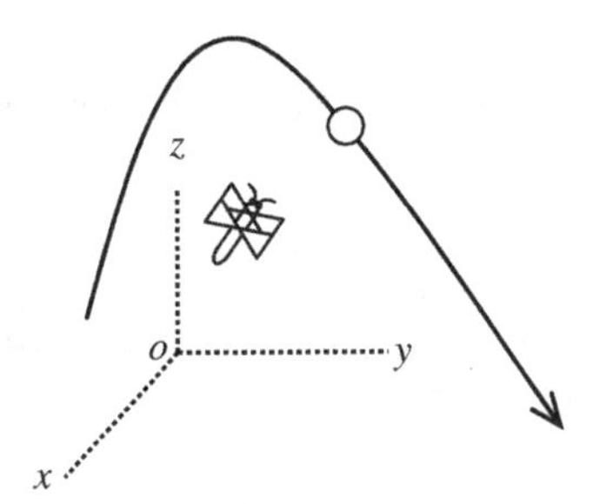

Figure 1.7 Geometry of interception of a ball by a dragonfly

vector due to gravity, $\mathbf{a}_g$, is always vertical. The time of flight is small enough that a frame of reference, $(oxyz)$, fixed to the planetary surface can be considered inertial by neglecting planetary rotation and revolution.

A dragonfly initially at the origin of the frame $(oxyz)$, is trying to alight on a cricket ball that has been hit at an initial velocity, $10\mathbf{i} + 20\mathbf{k}$ m/s, by a batsman at initial coordinates, $(-10, -5, 0.5)$ m (Figure 1.7). Devise a proportional navigation law for the dragonfly such that it makes an in-flight rendezvous with the ball. Neglect atmospheric drag on the ball.

The acceleration of the ball (the target) can be expressed as

$$\ddot{\mathbf{R}}_{\mathrm{T}} = \mathbf{a}_g = -g\mathbf{k}, \tag{1.40}$$

with initial position and velocity vectors, $\mathbf{R}_{\mathrm{T}}(0)$ and $\mathbf{V}_{\mathrm{T}}(0)$, respectively. Since the acceleration is constant, the instantaneous position of the ball is obtained by integration to be

$$\mathbf{R}_{\mathrm{T}}(t) = \mathbf{R}_{\mathrm{T}}(0) + \mathbf{V}_{\mathrm{T}}(0)t - \frac{g}{2}t^2\mathbf{k}. \tag{1.41}$$

The total time of flight, t_{f}, is easily calculated as follows:

$$z_{\mathrm{T}}(t_{\mathrm{f}}) = 0 = z_{\mathrm{T}}(0) + \dot{z}_{\mathrm{T}}(0)t_{\mathrm{f}} - \frac{g}{2}t_{\mathrm{f}}^2, \tag{1.42}$$

or

$$t_{\mathrm{f}} = \frac{\dot{z}_{\mathrm{T}}(0)}{g} + \sqrt{\frac{\dot{z}_{\mathrm{T}}^2(0)}{g^2} + \frac{2z_{\mathrm{T}}(0)}{g}} = 4.1023 \text{ s}.$$

Thus, it is necessary to have interception at a time $T < t_{\mathrm{f}}$. Let us select $T = 30/g$, which yields the following initial velocity for the dragonfly, by virtue of equation (1.15) and the dragonfly's initial position, $\mathbf{R}(0) = \mathbf{0}$:

$$\mathbf{V}(0) = \mathbf{V}_{\mathrm{T}}(0) + \frac{1}{T}\mathbf{R}_{\mathrm{T}}(0) = \begin{pmatrix} 6.7300 \\ -1.6350 \\ 20.1635 \end{pmatrix} \text{ m/s}.$$

The dragonfly must be traveling with this much velocity initially in order to successfully intercept the ball by the PNG approach.

For the simulation of the flight of the dragonfly by a fourth-order Runge–Kutta method, we wrote a MATLAB® program called *run_dragonfly.m* that has the following statements:

```
global g; g=9.81;% m/s^2
global T; T=30/g;% projected intercept time (s)
global RT0; RT0=[-10 -5 0.5]'; % target's initial position (m)
global VT0; VT0=[10 0 20]'; % target's initial velocity (m/s)
V0=VT0+RT0/T % interceptor's initial velocity (m/s)

%Runge-Kutta integration of interceptor's equations of motion:
[t,x]=ode45(@dragonfly,[0 3],[0 0 0 V0']);

% Calculation of instantaneous position error:
RT=RT0*ones(size(t'))+VT0*t'+.5*[0 0 -g]'*(t.^2)';
error=sqrt((x(:,1)-RT(1,:)').^2+(x(:,2)-RT(2,:)').^2+(x(:,3)-RT(3,:)').^2);
```

The equations of motion of the dragonfly are specified in the following file named *dragonfly.m* (which is called by *run_dragonfly.m*):

```
function xdot=dragonfly(t,x)
global g;
global T;
global RT0;
global VT0;
rT=RT0+VT0*t+0.5*[0 0 -g]'*t^2;
r=rT-x(1:3,1);
xdot(1:3,1)=x(4:6,1);
eps=1e-8;
% Avoiding singularity at t=T:
if abs(T-t)>eps
xdot(4:6,1)=[0 0 -g]'+r/((T-t)*(T-t+1));
else
    xdot(4:6,1)=[0 0 -g]';
end
```

In order to avoid the singularity in equation (1.39) at precisely $t = T$, we specify a zero corrective acceleration at that time, so that the dragonfly is freely falling in the small interval, $T - \epsilon < t < T + \epsilon$. The results of the simulation are plotted in Figures 1.8–1.10. Note that due to the non-planar nature of its flight path, the dragonfly is never able to actually sit on the ball, with the smallest position error, $\| \mathbf{r} \|$ (called *miss distance*) being 0.0783 m at $t = 2.295$ s. The miss distance is a characteristic of the PNG approach, which almost never results in a zero error intercept. A relatively small miss distance (such as in the present example) is generally considered to be a successful interception. When applied to a missile, the PNG law is often combined with another approach called *terminal guidance* that takes over near the point of interception in order to further reduce the position error. The maximum speed reached by the dragonfly is its initial speed of 21.3198 m/s, which appears to be within its physical capability of about 80 km/hr.

1.4.2 Cross-Product Steering

It was seen above in the dragonfly example that non-coplanar target and interceptor trajectories caused a near miss instead of a successful rendezvous. A suitable navigational strategy for a rendezvous would then appear to be the one that simultaneously achieves a zero miss distance and a zero relative speed. Intuition suggests that a simple approach for doing so is to turn the velocity vector, $\mathbf{V}(t)$, such that the

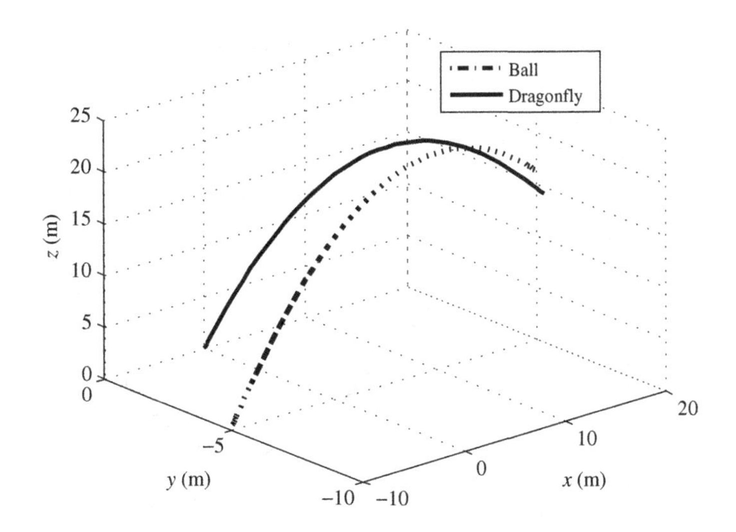

Figure 1.8 Three-dimensional plot of the flight paths taken by the ball and the dragonfly for PNG guidance

instantaneous velocity error (or velocity to be gained), $\mathbf{v}(t) = \mathbf{V}_T(t) - \mathbf{V}(t)$, becomes aligned with the acceleration error,

$$\mathbf{a} = \frac{d\mathbf{v}}{dt} = \frac{d\mathbf{V}_T}{dt} - \frac{d\mathbf{V}}{dt} = \dot{\mathbf{V}}_T - \dot{\mathbf{V}}. \tag{1.43}$$

This implies that the cross-product of velocity and acceleration errors must vanish:

$$\mathbf{v} \times \mathbf{a} = \mathbf{0}. \tag{1.44}$$

Such a navigational law is termed *cross-product steering*.

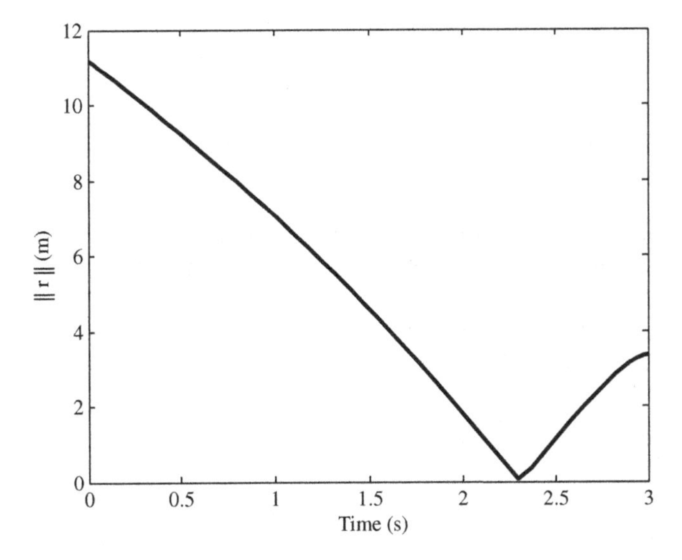

Figure 1.9 Separation error vs. time between the ball and the dragonfly with PNG guidance

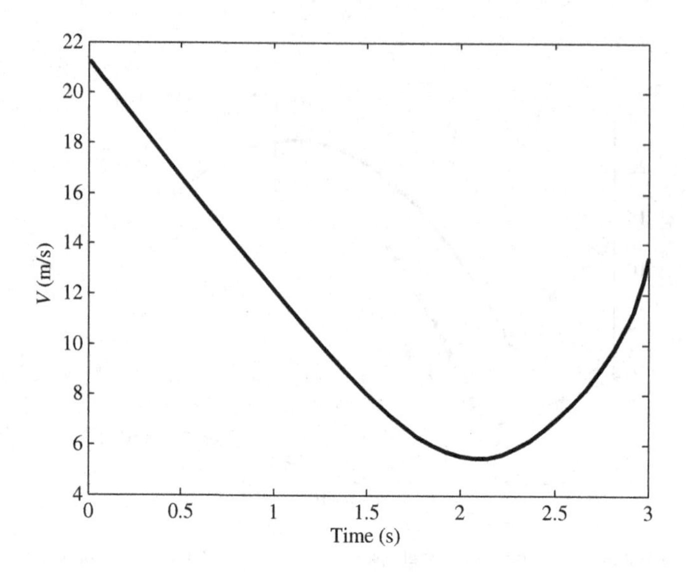

Figure 1.10 Time history of the required speed of the dragonfly for interception by proportional navigation

Since position error is an integral of the velocity error, it is quite possible to make both vanish simultaneously by choosing a constant, c, such that

$$\mathbf{a} = c\mathbf{v}. \tag{1.45}$$

Thus, the required initial velocity of the interceptor is obtained as follows by specifying $\mathbf{r}(T) = \mathbf{0}$ and $\mathbf{v}(T) = \mathbf{0}$:

$$\mathbf{v}(T) - \mathbf{v}(0) = c\,[\mathbf{r}(T) - \mathbf{r}(0)] \tag{1.46}$$

or

$$\mathbf{V}(0) = \mathbf{V}_\mathrm{T}(0) - c\mathbf{R}_\mathrm{T}(0). \tag{1.47}$$

The value of c must be chosen such that the rendezvous takes place with the desired accuracy within the specified time of flight.

Example 1.4 Let us repeat the ball and dragonfly example (Example 1.3) with cross-product steering instead of PNG navigation. To this end, the statements of the file *dragonfly.m* are modified as follows:

```
function xdot=dragonfly_cps(t,x)
global g;
global RT0;
global VT0;
global c;
vT=VT0+[0 0 -g]'*t;
v=vT-x(4:6,1);
rT=RT0+VT0*t+0.5*[0 0 -g]'*t^2;
r=rT-x(1:3,1);
xdot(1:3,1)=x(4:6,1);
% Cross-product steering (to be applied until
```

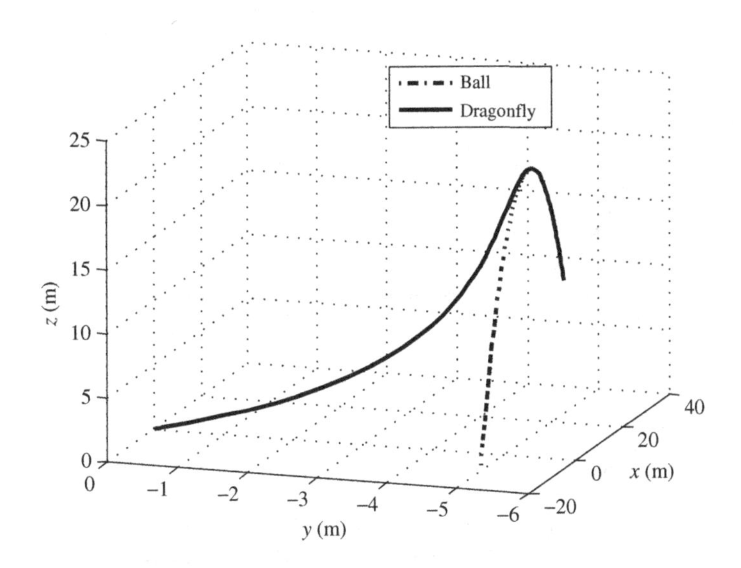

Figure 1.11 Three-dimensional plot of the flight paths taken by the ball and the dragonfly with cross-product steering

```
% a desired positional accuracy is achieved):
if abs(norm(r))>0.0009
xdot(4:6,1)=[0 0 -g]'-c*v;
else
xdot(4:6,1)=[0 0 -g]';
end
```

After some trial and error, $c = -3\ \text{s}^{-1}$ is selected, for which the dragonfly's initial velocity is

$$\mathbf{V}(0) = \mathbf{V}_{\mathrm{T}}(0) - c\mathbf{R}_{\mathrm{T}}(0) = \begin{pmatrix} -20 \\ -15 \\ 21.5 \end{pmatrix} \text{ m/s.}$$

The results of the Runge–Kutta simulation are plotted in Figures 1.11–1.13. The desired rendezvous is achieved in 2.029 s with a zero miss distance (8.35×10^{-6} m) and minimum velocity error 0.0027 m/s. The dragonfly now achieves the same plane of flight as the ball (Figure 1.11) and can alight on the ball at the minimum position error point, after which the velocity error and the positional error remain virtually zero (Figure 1.12). However, the required initial speed of the dragonfly is now increased to 32.97 m/s (Figure 1.13), which appears to be impractical. Therefore, while cross-product steering offers both position and velocity matching (compared with only position matching of PNG), it requires a much larger control input than the PNG navigation law for the same initial errors. A compromise can be achieved by using PNG in the initial phase of the flight, followed by cross-product steering in the terminal phase.

1.4.3 Proportional-Integral-Derivative Control

A large variety of tracking problems for mechanical systems (including flight dynamic systems) involve the simultaneous control of a generalized position vector, $\mathbf{q}(t)$, as well as its time derivative

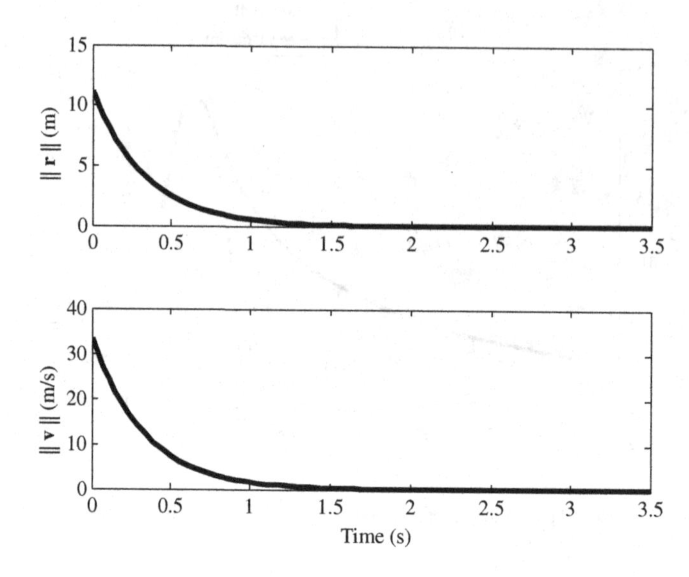

Figure 1.12 Separation and relative speed error vs. time between the ball and the dragonfly with cross-product steering

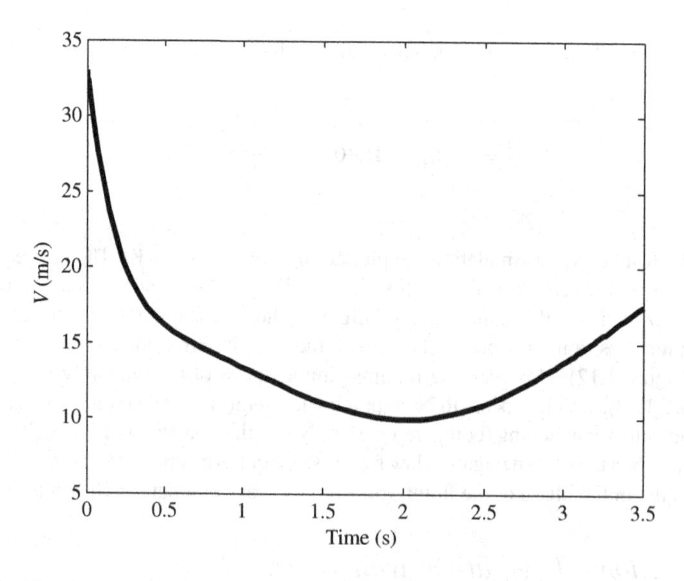

Figure 1.13 Time history of the required speed of the dragonfly for rendezvous by cross-product steering

(generalized velocity), $\dot{\mathbf{q}}(t)$, both of which constitute the nonlinear system's state vector,

$$\xi(t) = \begin{Bmatrix} \mathbf{q}(t) \\ \dot{\mathbf{q}}(t) \end{Bmatrix}. \tag{1.48}$$

For ensuring that the error from a desired (or nominal) generalized trajectory, $\mathbf{q}_d(t)$, defined by

$$\mathbf{e}(t) = \mathbf{q}_d(t) - \mathbf{q}(t), \tag{1.49}$$

is always kept small, a likely feedback control law is the one that applies a corrective control input, $\mathbf{u}(t)$, proportional to the error, the time integral of the error, and its rate as follows:

$$\mathbf{u}(t) = \mathbf{K}_p\mathbf{e}(t) + \mathbf{K}_i \int_0^t \mathbf{e}(\tau)\mathrm{d}\tau + \mathbf{K}_d\dot{\mathbf{e}}(t), \quad t \geq 0, \tag{1.50}$$

where $(\mathbf{K}_p, \mathbf{K}_i, \mathbf{K}_d)$ are (usually constant) gain matrices. In this way, not only can a correction be applied based upon the instantaneous deviation from the trajectory, but also the historically accumulated error as well as its tendency for the future can be corrected through the integral and derivative terms, respectively. Such a control law is termed *proportional-integral-derivative* (PID) control and requires a measurement and feedback of not only the error vector, but also its time integral and time derivative, as shown in the schematic block diagram of Figure 1.14. The process of determination of the gain matrices (often from specific performance objectives) is called *PID tuning*. While the PID control was originally intended for single-variable systems, its application can be extended to multi-variable plants as follows.

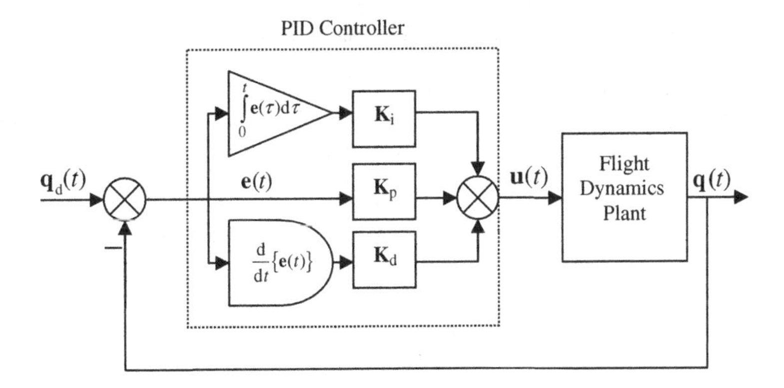

Figure 1.14 Schematic block diagram of PID control system

Since a successful application of PID control would result in a small error vector, the plant's governing equations can be generally linearized about the nominal trajectory resulting in the following governing error equation:

$$\ddot{\mathbf{e}} = \mathbf{A}_1\mathbf{e} + \mathbf{A}_2\dot{\mathbf{e}} + \mathbf{B}\mathbf{u}, \tag{1.51}$$

with the small initial conditions,

$$\mathbf{e}(0) = \mathbf{e}_0, \quad \dot{\mathbf{e}}(0) = \dot{\mathbf{e}}_0. \tag{1.52}$$

Here, $(\mathbf{A}_1, \mathbf{A}_2, \mathbf{B})$ are Jacobian matrices of the linearized plant (Appendix A) and could be varying with time along the nominal trajectory. Substitution of equation (1.50) into equation (1.51) with the error state vector

$$\mathbf{x}(t) = \begin{Bmatrix} \int_0^t \mathbf{e}(\tau)d\tau \\ \mathbf{e}(t) \\ \dot{\mathbf{e}}(t) \end{Bmatrix} \tag{1.53}$$

results in the following dynamic state equation of the closed-loop system:

$$\dot{\mathbf{x}} = \mathbf{A}\mathbf{x}, \tag{1.54}$$

where the closed-loop dynamics matrix is given by

$$\mathbf{A} = \begin{bmatrix} \mathbf{0} & \mathbf{I} & \mathbf{0} \\ \mathbf{0} & \mathbf{0} & \mathbf{I} \\ \mathbf{B}\mathbf{K}_\mathrm{i} & \mathbf{A}_1 + \mathbf{B}\mathbf{K}_\mathrm{p} & \mathbf{A}_2 + \mathbf{B}\mathbf{K}_\mathrm{d} \end{bmatrix}. \tag{1.55}$$

Clearly, for the error to remain small at all times $t \geq 0$ beginning from a small initial state of

$$\mathbf{x}(0) = \begin{Bmatrix} \mathbf{0} \\ \mathbf{e}_0 \\ \dot{\mathbf{e}}_0 \end{Bmatrix}, \tag{1.56}$$

the closed-loop error dynamics (equation (1.54)) must be stable in the sense of Lyapunov (Appendix B). The PID gain matrices must therefore be selected in a way that ensures stability of the closed-loop system. However, such a derivation of the gain matrices can be termed practical only for a time-invariant system (i.e., when $(\mathbf{A}_1, \mathbf{A}_2, \mathbf{B})$ are constant matrices).

Example 1.5 Consider an axisymmetric, rigid spacecraft of principal moments of inertia, $J_{xx} = J_{yy}$ and J_{zz}, equipped with a rotor that can apply a gyroscopic, internal torque input, $\mathbf{u} = (u_x, u_y)^T$, normal to the axis of symmetry. The spacecraft has the following equations of motion:

$$\mathbf{u} = \mathbf{J}\dot{\boldsymbol{\omega}} + \boldsymbol{\omega} \times \mathbf{J}\dot{\boldsymbol{\omega}}, \tag{1.57}$$

where

$$\mathbf{J} = \begin{pmatrix} J_{xx} & 0 & 0 \\ 0 & J_{xx} & 0 \\ 0 & 0 & J_{zz} \end{pmatrix} \tag{1.58}$$

and

$$\boldsymbol{\omega} = \begin{Bmatrix} \omega_x \\ \omega_y \\ \omega_z \end{Bmatrix}. \tag{1.59}$$

When the torque input is zero, the spacecraft has an equilibrium state with pure spin about the axis of symmetry, $\boldsymbol{\omega}^T = (0, 0, n)$. If an initial disturbance with small, off-axis angular velocity components, $\omega_x(0), \omega_y(0)$, is applied at $t = 0$, the spacecraft has the following LTI dynamics linearized about the equilibrium state:

$$\begin{Bmatrix} \dot{\omega}_x \\ \dot{\omega}_y \end{Bmatrix} = \begin{pmatrix} 0 & -k \\ k & 0 \end{pmatrix} \begin{Bmatrix} \omega_x \\ \omega_y \end{Bmatrix} + \frac{1}{J_{xx}} \begin{pmatrix} 1 & 0 \\ 0 & 1 \end{pmatrix} \begin{Bmatrix} u_x \\ u_y \end{Bmatrix}, \tag{1.60}$$

$$\omega_z = n = \text{const.}, \tag{1.61}$$

where

$$k = n\frac{J_{zz} - J_{xx}}{J_{xx}}$$

is real and constant.

It is necessary to devise an active controller that can reduce the off-axis spin error vector, $\boldsymbol{\omega}_{xy} = (\omega_x, \omega_y)^T$, quickly to zero with the help of the feedback control law

$$\mathbf{u} = J_{xx}\left(k_1\boldsymbol{\omega}_{xy} + k_2\dot{\boldsymbol{\omega}}_{xy}\right), \quad t \geq 0, \tag{1.62}$$

where (k_1, k_2) are constant gains. Since the integral term of the PID control law, equation (1.50), is absent here, the resulting feedback law is termed *proportional-derivative* (PD) control. Substituting equation (1.62) into equation (1.60), we have

$$\begin{aligned} J_{xx}\dot{\omega}_x + \omega_y\omega_z(J_{zz} - J_{xx}) &= J_{xx}\left(k_1\omega_x + k_2\dot{\omega}_x\right), \\ J_{xx}\dot{\omega}_y + \omega_x\omega_z(J_{xx} - J_{zz}) &= J_{xx}(k_1\omega_y + k_2\dot{\omega}_y), \\ J_{zz}\dot{\omega}_z &= 0, \end{aligned} \tag{1.63}$$

which implies

$$\frac{\mathrm{d}\omega_{xy}^2}{\mathrm{d}t} = 2k_1\omega_{xy}^2 + k_2\frac{\mathrm{d}\omega_{xy}^2}{\mathrm{d}t}. \tag{1.64}$$

By selecting the state variables of the closed-loop error dynamics as $x_1 = \omega_x$ and $x_2 = \omega_y$, we arrive at the state space form of equation (1.54) with the state dynamics matrix

$$\mathbf{A} = \frac{1}{1 - k_2}\begin{pmatrix} k_1 & -k \\ k & k_1 \end{pmatrix}. \tag{1.65}$$

While the complete solution vector for the off-axis angular velocity components is given by

$$\left\{ \begin{matrix} \omega_x(t) \\ \omega_y(t) \end{matrix} \right\} = e^{\mathbf{A}t}\left\{ \begin{matrix} \omega_x(0) \\ \omega_y(0) \end{matrix} \right\}, \quad t \geq 0, \tag{1.66}$$

the solution for the closed-loop error magnitude, $\omega_{xy} = \sqrt{\omega_x^2 + \omega_y^2}$, is simply obtained as

$$\omega_{xy}(t) = \omega_{xy}(0)e^{\frac{k_1}{1-k_2}t}, \quad t \geq 0. \tag{1.67}$$

For asymptotic stability (Appendix B) we require that

$$\frac{k_1}{1 - k_2} < 0,$$

which can be satisfied by choosing $k_1 < 0$ and $0 < k_2 < 1$. The magnitudes of (k_1, k_2) are limited by actuator torque and sensor sensitivity constraints.

A representative case of $J_{xx} = J_{yy} = 1000$ kg.m^2, $J_{zz} = 3000$ kg.m^2, $n = 0.01$ rad/s, $k_1 = -1$, $k_2 = 0.5$, $\omega_x(0) = -0.001$ rad/s, and $\omega_y(0) = 0.002$ rad/s, computed with the following MATLAB statements:

```
>> k1=-1;k2=.5;n=0.01;k=2*n;
>> A=[k1 -k;k k1]/(1-k2); % CL state dynamics matrix
>> wx0=-0.001;wy0=0.002;
```

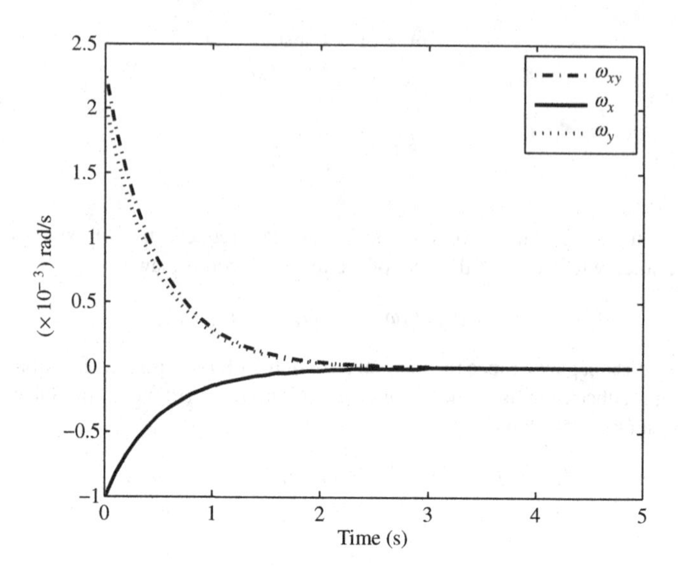

Figure 1.15 Off-axis, closed-loop angular velocity error for an axisymmetric, spin stabilized spacecraft controlled by a rotor that applies a feedback internal torque by the PD scheme

```
>> for i=1:50;
>>    t=(i-1)*0.1;T(:,i)=t;
>>    X(:,i)=sqrt(wx0^2+wy0^2)*exp(k1*t/(1-k2)); %omega_xy magnitude
>>    wxy(:,i)=expm(A*t)*[wx0;wy0]; % Components of omega_xy vector
>> end
```

is plotted in Figure 1.15. Note that the error components decay to almost zero in about 3 s.

1.5 Digital Tracking System

Now we consider how a tracking system is practically implemented. A modern control system drives the actuators through electrical signals as control inputs, while the sensors produce electrical signals as plant outputs. A tracking controller must therefore manipulate the input electrical signals and generate corresponding output electrical signals. While the control system blocks shown in Figure 1.1 represent the input, output, and state variables as continuous functions of time (called *continuous time* or *analog* variables) this is not how an actual control system might work. For example, all modern control systems have at their heart a digital computer, which can process only *discrete time* (or *digital*) signals that are essentially electrical impulses corresponding to the current values (magnitudes and signs) of the time-varying signals. Effectively, this implies that a digital control system can only *sample* time-varying signals at discrete time intervals, called *time steps*. What this means is that, rather than passing prescribed continuous functions of time such as $x(t)$, $y(t)$, one can only send and receive electrical impulses through/from the controller block that correspond to the current values of the variables at a given time instant such as x_k, y_k, where k denotes the kth time instant. Since all modern controllers are digital in character, we require special blocks for converting continuous time signals to digital ones (and vice versa) for communicating with a controller. These essential blocks are called *analog-to-digital* (A/D) and *digital-to-analog* (D/A) *converters* and are shown in Figure 1.16. The A/D and D/A blocks are synchronized by the same clock that sends out an electrical impulse every Δt seconds, which is a fixed duration called the *sampling*

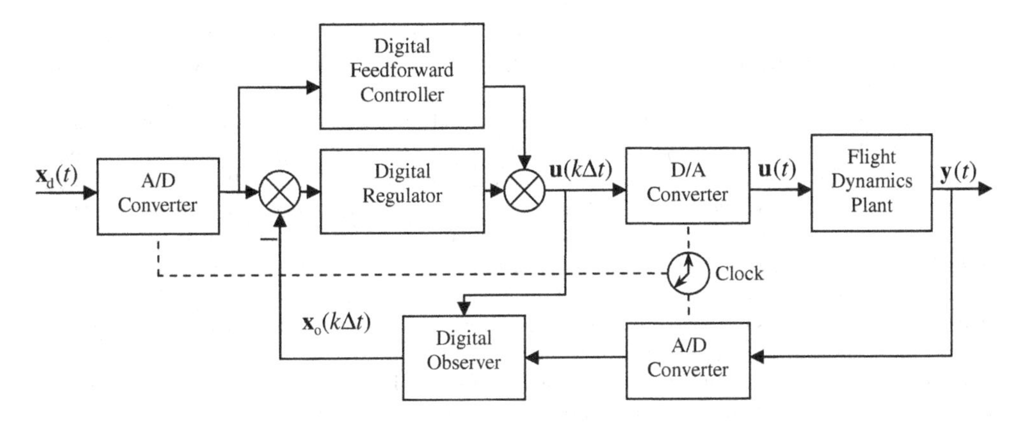

Figure 1.16 Block diagram of a digital flight control system

interval. At the arrival of the clock signal, the A/D block releases its output signal that was *held* at either a constant value (zero-order hold), or an interpolated value of the continuous time input signals received during the sampling interval. In contrast, the D/A block continuously produces its output signal as a weighted average of discrete input signals received (and stored) over the past few sampling intervals.

The digital sampling and holding process involved in A/D conversion mathematically transforms the governing linear differential equations in time into linear algebraic, *difference equations* that describe the evolution of the system's state over a sampling interval. In the frequency domain (Appendix A), the transform method for a discrete time, linear time-invariant system (equivalent to continuous time Laplace transform) is the *z-transform*. For more details about digital control systems, see Chapter 8 of Tewari (2002), or Phillips and Nagle (1995).

A digital control system is unable to respond to signals with frequency greater than the *Nyquist frequency*, defined as half of the sampling frequency and given by $\pi/\Delta t$. Hence, a digital control system has an inherent robustness with respect to high-frequency noise that is always present in any physical system. This primary advantage of a digital control system over an equivalent analog control system is responsible for the replacement of electronic hardware in almost all control applications in favor of digital electronics – ranging from the common digital video recorders to aircraft and spacecraft autopilots.

Since a digital control system is designed and analyzed using techniques equivalent to a continuous time control system, one can easily convert one into the other for a sufficiently small sampling interval, Δt. While many important statistical processes and noise filters are traditionally described as discrete rather than continuous time signals, their continuous time analogs can be easily derived. Therefore, it is unnecessary to discuss both continuous time and discrete time flight control design, and we shall largely restrict ourselves to the former in the remainder of this book.

1.6 Summary

A control system consists of a plant and a controller, along with actuators and sensors that are usually clubbed with the plant. Closed-loop (automatic) control is the only practical alternative in reaching and maintaining a desired state in the presence of disturbances. There are two distinct types of automatic controllers: (a) terminal controllers that take a nonlinear system to a final state in a given time, and (b) tracking controllers which maintain a system close to a nominal, reference trajectory. A tracking controller generally consists of an observer, a regulator, and a feedforward controller. While most plants are governed by nonlinear state equations, they can be linearized about a nominal, reference trajectory, which is a particular solution of the plant's state equations. Linearized state equations are invaluable in designing practical automatic controllers as they allow us to make use of linear systems theory, which easily lends itself to solution by transition matrix as well as providing systematic stability criteria.

All flight vehicles require control of position, velocity, and attitude for a successful mission. Separate control subsystems are required for guiding the vehicle's center of mass along a specific trajectory, and control of vehicle's attitude by rotating it about the center of mass. Due to their greatly different time scales, the guidance system and the attitude control system are usually designed separately and then combined into an overall flight control system. While optimal control techniques are generally required for designing the guidance and attitude control systems, certain intuitive techniques have been successfully applied for practical guidance and control systems in the past, and continue to be in use today due to their simplicity. Examples of intuitive methods include proportional navigation, cross-product steering, and proportional-integral-derivative control.

Many practical control systems are implemented as discrete time (or digital) rather than continuous time (or analog) systems. Due to a bandwidth naturally limited by the sampling rate, a digital control system has an inbuilt robustness with respect to high-frequency noise signals that is not present in analog systems, which has led to the replacement of analog systems by digital systems in all modern control hardware.

Exercises

(1) A system has the following state equations:

$$\begin{aligned} \dot{x}_1 &= -2x_2 + u_1, \\ \dot{x}_2 &= x_3 + u_2, \\ \dot{x}_3 &= -3x_1. \end{aligned}$$

(a) If the initial condition at $t = 0$ is $x_1(0) = 10$, $x_2(0) = 5$, $x_3(0) = 2$ and $u(t) = (t, 1)^T$, solve the state equations for the first 10 s.

(b) Is the system controllable? (See Appendix A.)

(c) Is the system observable with x_1 and x_2 as outputs? (See Appendix A.)

(2) For a system with the following state-space coefficient matrices:

$$\mathbf{A} = \begin{pmatrix} -2 & -3 & 0 \\ 0 & 0 & 1 \\ 0 & 1 & 0 \end{pmatrix}, \quad \mathbf{B} = \begin{pmatrix} 1 & 0 \\ 0 & 0 \\ 0 & -1 \end{pmatrix}$$

(a) Determine the response to a zero initial condition and a unit impulse function applied at $t = 0$ as the first input.

(b) What are the eigenvalues of the matrix $\mathbf{A}$? (*Ans.* 1, −1, −2.)

(c) Analyze the controllability of the system when only the first input is applied.

(d) Analyze the controllability of the system when only the second input is applied.

(3) Can the following be the state transition matrix of a homogeneous linear system with state $\mathbf{x}(t) = (-1, 0, -2)^T$ at time $t = 1$?

$$\mathbf{\Phi}(t, 0) = \begin{pmatrix} 1 - \sin t & 0 & -te^{-2t} \\ 0 & e^{-t}\cos t & 0 \\ -2t & 0 & 1 \end{pmatrix}$$

Why? (*Ans. No.*)

(4) A homogeneous linear system has the following state transition matrix:

$$\mathbf{\Phi}(t,0) = \begin{pmatrix} \cos t & -\sin t \\ \sin t & \cos t \end{pmatrix}.$$

If the state at time $t = 1$ is $\mathbf{x}(t) = (-1.3818, -0.3012)^T$, what was the state at $t = 0$?

(5) Compute and plot the internal control torque components for the rotor-stabilized spacecraft of Example 1.5. What is the maximum torque magnitude required?

(6) Repeat the simulation of Example 1.3 by adding the following drag deceleration on the ball:

$$\dot{\mathbf{V}}_\mathrm{T} = -0.0016(V_\mathrm{T}\mathbf{V}_\mathrm{T})\ (\mathrm{m/s^2})$$

Is there a change in the miss distance and the maximum speed of the dragonfly?

(7) Repeat the simulation of Example 1.4 with the drag term given in Exercise 4. What (if any) changes are observed?

(8) Consider a controller for the plant given in Exercise 2 based upon proportional-integral-derivative feedback from the first state variable to the second control input:

$$u_2(t) = k_\mathrm{p}x_1(t) + k_\mathrm{i}\int_0^t x_1(\tau)\mathrm{d}\tau + k_\mathrm{d}\dot{x}_1(t).$$

(a) Draw a block diagram of the resulting closed-loop system.
(b) Is it possible to select the PID gains, $k_\mathrm{p}, k_\mathrm{i}, k_\mathrm{d}$, such that closed-loop system has all eigenvalues in the left-half s-plane?

(9) Replace the PID controller of Exercise 8 by a full-state feedback to the second control input given by

$$u_2(t) = -k_1x_1(t) - k_2x_2(t) - k_3x_1(t).$$

(a) Draw a block diagram of the resulting closed-loop system.
(b) Select the regulator gains, k_1, k_2, k_3, such that closed-loop system has all eigenvalues at $s = -1$. (*Ans.* $k_1 = -1/3,\ k_2 = -2,\ k_3 = -1$.)
(c) Determine the resulting control system's response to a zero initial condition and a unit impulse function applied at $t = 0$ as the first input, $u_1(t) = \delta(t)$.

(10) Rather than using the full-state feedback of Exercise 9, it is decided to base the controller on the feedback of the first state variable, $x_1(t)$. In order to do so, a full-order observer (Appendix A) must be separately designed for estimating the state vector from the measurement of $x_1(t)$ as well as the applied input, $u_2(t)$, which is then supplied to the regulator designed in Exercise 9. The state equation of the observer is given by

$$\dot{\mathbf{x}}_\mathrm{o} = (\mathbf{A} - \mathbf{LC})\mathbf{x}_\mathrm{o} + \mathbf{B}_2u_2 + \mathbf{L}x_1,$$

where

$$\mathbf{C} = (1,\ 0,\ 0), \quad \mathbf{B}_2 = \begin{pmatrix} 0 \\ 0 \\ -1 \end{pmatrix}, \quad \mathbf{L} = \begin{pmatrix} \ell_1 \\ \ell_2 \\ \ell_3 \end{pmatrix}.$$

(a) Select the observer gains, ℓ_1, ℓ_2, ℓ_3, such that the observer dynamics matrix, $(\mathbf{A} - \mathbf{LC})$, has all eigenvalues at $s = -5$. (*Ans.* $\ell_1 = 13,\ \ell_2 = -76/3,\ \ell_3 = -140/3$.)

(b) Close the loop with $u_2(t) = -(k_1, k_2, k_3)\mathbf{x}_o(t)$ and draw the block diagram of the overall closed-loop system with the plant, observer, and regulator.

(c) Determine the resulting control system's response to zero initial condition and a unit impulse function applied at $t = 0$ as the first input. What difference (if any) is seen compared with the response in Exercise 9?

(11) Repeat Exercise 10 after replacing the full-order observer with a reduced-order observer (Appendix A).

References

Battin, R.H. (1999) *An Introduction to the Mathematics and Methods of Astrodynamics*. American Institute of Aeronautics and Astronautics, Reston, VA.

Kailath, T. (1980) *Linear Systems*. Prentice Hall, Englewood Cliffs, NJ.

Kwakernaak, H. and Sivan, R. (1972) *Linear Optimal Control Systems*. John Wiley & Sons, Inc., New York.

Maciejowski, J.M. (1989) *Multivariable Feedback Design*. Addison-Wesley, Reading, MA.

Phillips, C.L. and Nagle, H.T. (1995) *Digital Control Systems Analysis and Design*. Prentice Hall, Englewood Cliffs, NJ.

Skogestad, S. and Postlethwaite, I. (1996) *Multivariable Feedback Design: Analysis and Control*. John Wiley & Sons, Ltd, Chichester.

Tewari, A. (2002) *Modern Control Design with MATLAB and Simulink*. John Wiley & Sons, Ltd, Chichester.

2

Optimal Control Techniques

2.1 Introduction

A general control system consists of two basic elements: *terminal control* and *tracking control*. Terminal control refers to the task of guiding a system from a given initial state to a desired final state in a prescribed time, often in the presence of constraints. Guidance of rockets and spacecraft, as well as long-range aircraft navigation, requires terminal control. On the other hand, the objective of tracking control is to maintain the plant's state in a close neighborhood of a nominal (or reference) state. Examples of tracking control include aircraft autopilots and flight management systems, orbital control of spacecraft, and attitude control of all aerospace vehicles.

Referring to the plant state equation

$$\dot{\mathbf{x}} = \mathbf{f}[\mathbf{x}(t), \mathbf{u}(t), t], \tag{2.1}$$

a terminal control system would aim to change the plant's state, $\mathbf{x}(t)$, from an initial state, $\mathbf{x}_i = \mathbf{x}(t_i)$, at a given *initial time*, t_i, to a *terminal* (or final) state, $\mathbf{x}_f = \mathbf{x}(t_f)$, at a specified *terminal time*, t_f, by applying a control input, $\mathbf{u}(t)$, in the fixed *control interval*, $(t_i \le t \le t_f)$. Often, the terminal state is not explicitly given, but satisfies a scalar *terminal equality constraint* given by

$$s[\mathbf{x}(t_f), t_f] = 0. \tag{2.2}$$

Furthermore, equality or inequality constraints on the control input and state variables (called *interior point constraints*) may be present at the interior of the control interval, given by

$$\boldsymbol{\alpha}[\mathbf{x}(t), \mathbf{u}(t)] = \mathbf{0}. \tag{2.3}$$

The terminal controller design thus requires an iterative solution of the nonlinear plant state equation (2.1) subject to the initial conditions, terminal and equality constraints, as well as some additional requirements (such as minimization or maximization of a given function of state and control variables). Such a solution procedure is generally referred to as a *two-point boundary value problem*, and produces a reference (or nominal) trajectory, $\mathbf{x}_d(t)$, and the corresponding reference control history, $\mathbf{u}_d(t)$.

The design of a tracking controller is carried out such that the plant's state follows the nominal trajectory, $\mathbf{x}_d(t)$, which satisfies equation (2.1),

$$\dot{\mathbf{x}}_d = \mathbf{f}[\mathbf{x}_d(t), \mathbf{u}_d(t), t], \tag{2.4}$$

---

where $\mathbf{u}_d(t)$ is the *nominal control input*. If the state vector is available for measurement and feedback, a state feedback control law called the *regulator*, of the form

$$\Delta\mathbf{u} = \mathbf{u} - \mathbf{u}_d = \mathbf{h}(\mathbf{e}(t), t), \tag{2.5}$$

is derived by ensuring that the tracking error,

$$\mathbf{e}(t) = \mathbf{x}(t) - \mathbf{x}_d(t), \tag{2.6}$$

is either driven to zero, or in any case kept small, in the presence of modeling errors and external disturbances. However, it is rarely possible to measure the entire state vector; instead only a few output variables, $\mathbf{y}(t)$, can be sensed by available sensors. The plant's output is mathematically given by the *output equation*,

$$\mathbf{y} = \mathbf{g}[\mathbf{x}(t), \mathbf{u}(t), t]. \tag{2.7}$$

In order to be successful, a tracking controller must have a separate subsystem called the *observer* that generates an *estimate* of the plant's state vector, $\mathbf{x}_o(t)$, and supplies it to the regulator according to the following *observer state equation*:

$$\dot{\mathbf{x}}_o = \mathbf{f}_o[\mathbf{x}_o(t), \mathbf{y}(t), \mathbf{u}(t), t], \tag{2.8}$$

which is designed to minimize the estimation error,

$$\mathbf{e}_o(t) = \mathbf{x}_o(t) - \mathbf{x}(t). \tag{2.9}$$

Thus, any flight control task can be practically handled by first deriving a reference trajectory and the corresponding nominal input history, $\mathbf{u}_d(t)$, in the open loop by the solution of a nonlinear, two-point boundary value problem, and then designing a feedback controller for tracking the reference trajectory despite disturbances. The resulting control system therefore has both *feedforward* and *feedback* paths described by

$$\mathbf{u} = \mathbf{u}_d + \mathbf{h}[\mathbf{e}(t) + \mathbf{e}_o(t), t] \tag{2.10}$$

and schematically shown in Figure 2.1. Note the presence of *process noise*, $\boldsymbol{\nu}(t)$, as a disturbance input, which arises due to errors in the plant model. In attempting to measure and feed back the outputs, an additional disturbance called the *measurement noise*, $\mathbf{w}(t)$, is invariably introduced. Both process and measurement noise inputs are assumed to be statistically known, and hence excluded from the nominal plant model used to derive the terminal controller. However, the tracking controller usually incorporates statistical models for *filtering* out the noise signals through an appropriately designed observer.

Owing to the fact that a successful tracking control keeps the plant's state rather close to the reference trajectory, the assumption of small deviations from the nominal can usually be employed, leading to a linearized plant and great simplification in the controller design. This is the most common approach in deriving flight controllers and takes full advantage of the linear systems theory (Appendix A). Typically, one writes the linearized state equation of the plant as

$$\dot{\mathbf{e}} = \mathbf{A}\mathbf{e} + \mathbf{B}\Delta\mathbf{u} + \mathbf{F}\boldsymbol{\nu}, \tag{2.11}$$

where $\boldsymbol{\nu}(t)$ is the process noise vector,

$$\Delta\mathbf{u} = \mathbf{u} - \mathbf{u}_d, \tag{2.12}$$

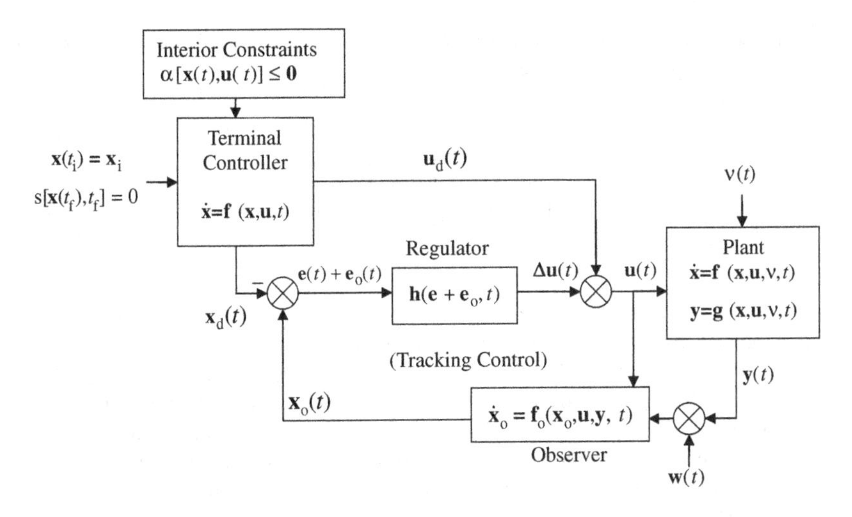

Figure 2.1 A general control system

the corrective tracking control, and $\mathbf{A}(t)$, $\mathbf{B}(t)$, and $\mathbf{F}(t)$ are the *Jacobian matrices* (Appendix A) derived from a first-order Taylor series expansion of equation (2.1) about the reference trajectory, $\mathbf{x}_d(t)$. Taking advantage of the linear systems theory, we can easily derive linear regulator laws of the

$$\Delta\mathbf{u} = -\mathbf{K}\mathbf{e} \tag{2.13}$$

kind (in the case of state feedback) for driving the error vector to zero in the steady state, provided the plant satisfies certain conditions such as *controllability* (Appendix A). A suitable modification of equation (2.13) for the case of observer-based output feedback tracking control is easily obtained through a linear observer, provided the plant is *observable* with the given outputs.

Optimal control theory (Athans and Falb 2007; Bryson and Ho 1975) offers a powerful method of deriving both terminal and tracking controllers, and provides a systematic framework for modern control design methods. *Optimal control* refers to the time history of the control inputs that take the system from an initial state to a desired final state while minimizing (or maximizing) an objective function. The evolution of the system from the initial to final states under the application of optimal control inputs is called an *optimal trajectory*. Finding an optimal trajectory between two system states is the primary task of terminal control that is generally carried out in open loop. A small variation in initial condition – or the presence of small external disturbances – causes a perturbation of the system from its optimal trajectory, tracking which requires a modification of the optimal control for a neighboring optimal trajectory, generally in a closed loop. Thus open-loop optimal control can yield a terminal control input, superimposed on which is the tracking control input calculated by an optimal feedback control law. Therefore, optimization of control inputs – either in an open loop or a closed loop – is integral to modern control design.

2.2 Multi-variable Optimization

We begin with a few preliminaries necessary for studying multi-variable optimization problems. For a scalar function, $L(\mathbf{x}, \mathbf{u})$, called the *objective function*, to have a *minimum* with respect to a set of m

mutually independent, unconstrained scalar variables, $\mathbf{u} = (u_1, u_2, \ldots, u_m)^T$, called *control variables*, the following *necessary condition* must be satisfied:

$$\left(\frac{\partial L}{\partial \mathbf{u}}\right)^T = \begin{Bmatrix} \frac{\partial L}{\partial u_1} \\ \frac{\partial L}{\partial u_2} \\ \vdots \\ \frac{\partial L}{\partial u_m} \end{Bmatrix} = \mathbf{0}. \tag{2.14}$$

A point $\mathbf{u}^*$ that satisfies the necessary condition for a minimum is called an *extremal point* (or *stationary point*) of L. Note that an extremal point is not guaranteed to be a minimum point, as it might also be a maximum point, or neither. In order to ensure that an extremal point is also a minimum point, we require a separate condition to be satisfied by the objective function L at the extremal point. Such a condition would require an increase of L while moving away from the extremal point in any direction, and is therefore related to the sign of the second-order partial derivatives of $L(\mathbf{x}, \mathbf{u})$ with respect to $\mathbf{u}$.

If the square matrix

$$\frac{\partial^2 L}{\partial \mathbf{u}^2} = \begin{pmatrix} \frac{\partial^2 L}{\partial {u_1}^2} & \frac{\partial^2 L}{\partial u_1 \partial u_2} & \cdots & \frac{\partial^2 L}{\partial u_1 \partial u_m} \\ \frac{\partial^2 L}{\partial u_2 \partial u_1} & \frac{\partial^2 L}{\partial {u_2}^2} & \cdots & \frac{\partial^2 L}{\partial u_2 \partial u_m} \\ \vdots & \vdots & \vdots & \vdots \\ \frac{\partial^2 L}{\partial u_m \partial u_1} & \frac{\partial^2 L}{\partial u_m \partial u_2} & \cdots & \frac{\partial^2 L}{\partial u_m \partial u_m} \end{pmatrix}, \tag{2.15}$$

called the *Hessian*, is *positive definite* (i.e., has all real, positive eigenvalues) at an extremal point, then the point is guaranteed to be a minimum of $L(\mathbf{x}, \mathbf{u})$ with respect to $\mathbf{u}$. In other words, the positive definiteness of the Hessian at an extremal point is a *sufficient condition* for the existence of a minimum of L with respect to $\mathbf{u}$, and is expressed as

$$\frac{\partial^2 L}{\partial \mathbf{u}^2} > 0, \tag{2.16}$$

Similarly, if the extremal point satisfies

$$\frac{\partial^2 L}{\partial \mathbf{u}^2} < 0, \tag{2.17}$$

then it is a maximum point $L(\mathbf{x}, \mathbf{u})$ with respect to $\mathbf{u}$. But there can be extremal points that are neither a minimum nor a maximum point. If the Hessian has both positive and negative eigenvalues at an extremal point, then the point concerned is said to be a *saddle point*. Furthermore, if any of the eigenvalues of the Hessian vanishes at an extremal point then we have a *singular point*.

A point $\hat{\mathbf{u}}$ that satisfies the necessary and sufficient conditions for a minimum is called an *optimal point* (or minimum point) of L. It is possible that we may only be able to find an extremal point, but not a minimum point for a given function, $L(\mathbf{x}, \mathbf{u})$.

If $L(\mathbf{x}, \hat{\mathbf{u}}) \leq L(\mathbf{x}, \mathbf{u})$ for all $\mathbf{u} \neq \hat{\mathbf{u}}$, then the optimal point $\hat{\mathbf{u}}$ is said to be a *global* (or *absolute*) *minimum* of $L(\mathbf{x}, \mathbf{u})$ with respect to $\mathbf{u}$. If a minimum point is not a global minimum then it is said to be a *local minimum*.

Example 2.1 Determine the extremal and minimum points of

$$L = x_1^2 - x_2^2 + 3x_3 + u_1^2 - 2\sin u_2$$

with respect to $\mathbf{u} = (u_1, u_2)^T$.

We begin with the extremal point, $\mathbf{u}^*$, for which we have

$$\left(\frac{\partial L}{\partial \mathbf{u}^*}\right)^T = \begin{Bmatrix} 2u_1^* \\ -2\cos u_2^* \end{Bmatrix} = \begin{Bmatrix} 0 \\ 0 \end{Bmatrix},$$

which is identically satisfied by $u_1^* = 0$ and $u_2^* = (2n-1)\pi/2$, $n = 1, 2, 3, \ldots$. For the optimal point, we have the Hessian

$$\frac{\partial^2 L}{\partial \hat{\mathbf{u}}^2} = \begin{pmatrix} 2 & 0 \\ 0 & 2\sin \hat{u}_2 \end{pmatrix},$$

whose eigenvalues are $(2, 2\sin \hat{u}_2)$. Thus, the Hessian is positive definite for any value $\hat{u}_1$ and $\sin \hat{u}_2 > 0$, implying that only the extremal points $[0, (2n-1)\pi/2]$, $n = 1, 3, 5, \ldots$, are the minimum points of the given function with respect to $\mathbf{u}$ having the same minimum value

$$\hat{L} = x_1^2 - x_2^2 + 3x_3 + u_1^2 - 2.$$

The remaining extremal points, $[0, (2n-1)\pi/2]$, $n = 2, 4, 6, \ldots$, produce the Hessian eigenvalues $(2, -2)$, and are thus saddle points of $L(\mathbf{x}, \mathbf{u})$ with respect to $\mathbf{u}$.

The necessary and sufficient conditions, equation (2.14) and inequality (2.16) respectively, for the existence of a minimum point can be easily derived by taking a second-order Taylor series expansion of $L(\mathbf{x}, \mathbf{u})$ with respect to $\mathbf{u}$ about an extremal point (Stengel 1994). If the function $L(\mathbf{x}, \mathbf{u})$ is either discontinuous, or non-smooth with respect to $\mathbf{u}$, it may have a minimum at an extremal point without satisfying the sufficient condition for optimality. Thus, the sufficient condition is *not* a necessary condition. Furthermore, if one is looking for a minimum in a *bounded region* (called the *feasible space* or *admissible space*) of the set of all possible m-dimensional real vectors, $\mathbf{u}$, it is possible that a minimum may occur at a *boundary point*, which satisfies neither the necessary nor the sufficient condition for optimality. This fact is depicted graphically in Figure 2.2 for $L = u_1^2 + u_2$ and a feasible space given by $-1 \le u_1 \le 1$ and $u_2 \ge 0.5$, for which the boundary point, $u_1 = 0$, $u_2 = 0.5$, is the minimum point that satisfies neither equation (2.14) nor inequality (2.16). Therefore, it is necessary to qualify equation (2.14) and inequality (2.16) as being the necessary and sufficient conditions, respectively, for a minimum with respect to either an unbounded $\mathbf{u}$, or at an *interior point* (i.e., not a boundary point) of the feasible space for $\mathbf{u}$.

2.3 Constrained Minimization

As indicated earlier, there could be bounds placed upon the control variables in the form of *inequality constraints* while minimizing an objective function.[1] It is also possible that there exist additional variables, $\mathbf{x}$, upon which the objective function also depends, and which are related to the control variables through equality constraints. The result of the optimization would, in either case, be qualitatively influenced by the presence of the constraints, and therefore must carefully account for them.

[1] Equality constraints upon control variables of the form $\mathbf{g}(\mathbf{u}) = \mathbf{0}$ are considered trivial as they can either be solved uniquely for a control vector, $\mathbf{u}$, thereby rendering the optimization meaningless, or merely reduce the number of control variables to be determined by optimization.

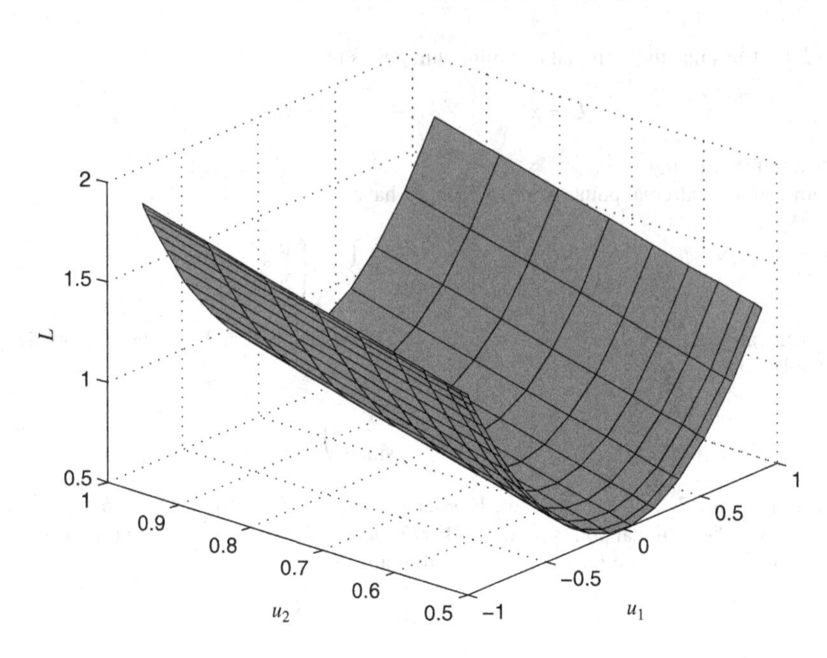

Figure 2.2 Feasible space bounded by $-1 \leq u_1 \leq 1$ and $u_2 \geq 0.5$ for minimization of $L = u_1^2 + u_2$ with respect to $\mathbf{u} = (u_1, u_2)^T$

2.3.1 *Equality Constraints*

In the minimization of $L(\mathbf{x}, \mathbf{u})$ with respect to $\mathbf{u}$, if there are equality constraints representing relationships between the control variables, $\mathbf{u}$, and the other variables, $\mathbf{x}$, of the form

$$\mathbf{f}(\mathbf{x}, \mathbf{u}) = \mathbf{0}, \tag{2.18}$$

the minimum points (if they exist) would be affected by the fact that they must also satisfy the constraints. The equality constraints must be such that it is possible to solve them for $\mathbf{x}$ corresponding to an extremal point, $\mathbf{u}$, thereby yielding the extremal value of L. Therefore, it is necessary to take the equality constraints into account by adjoining them with $L(\mathbf{x}, \mathbf{u})$ into the following *augmented objective function*:

$$\mathcal{J}(\mathbf{x}, \mathbf{u}) = L(\mathbf{x}, \mathbf{u}) + \boldsymbol{\lambda}^T \mathbf{f}(\mathbf{x}, \mathbf{u}), \tag{2.19}$$

where $\boldsymbol{\lambda}$ is a set of parameters of the same size as $\mathbf{x}$ – called *Lagrange multipliers* – that must be solved for so that $\mathcal{J}$ is minimized. By minimizing $\mathcal{J}$, we are simultaneously ensuring that L has a minimum with respect to $\mathbf{u}$ and that the equality constraints of equation (2.18) are also satisfied. The necessary conditions for optimality are thus

$$\frac{\partial \mathcal{J}}{\partial \mathbf{x}} = \mathbf{0}, \quad \frac{\partial \mathcal{J}}{\partial \mathbf{u}} = \mathbf{0}. \tag{2.20}$$

The values of the Lagrange multipliers corresponding to an extremal point, $\boldsymbol{\lambda}^*$, for satisfying the necessary conditions are easily derived from the first part of equation (2.20):

$$\boldsymbol{\lambda}^* = -\left[\left(\frac{\partial \mathbf{f}}{\partial \mathbf{x}}\right)^{-1}\right]^T \left(\frac{\partial L}{\partial \mathbf{x}}\right)^T, \tag{2.21}$$

which requires a non-singular Jacobian, $\partial \mathbf{f}/\partial \mathbf{x}$, at an extremal point (that is also necessary for determining $\mathbf{x}^*$ (and thus L^*) from $\mathbf{u}^*$). Once the extremal Lagrange multipliers are determined, the second part of equation (2.20) yields

$$\frac{\partial L}{\partial \mathbf{u}} + (\boldsymbol{\lambda}^*)^T \frac{\partial \mathbf{f}}{\partial \mathbf{u}} = \mathbf{0}, \tag{2.22}$$

a solution of which yields the extremal point, $\mathbf{u}^*$. Generally, equations (2.18), (2.21), and (2.22) require an iterative solution for the extremal values, $(\mathbf{x}^*, \boldsymbol{\lambda}^*, \mathbf{u}^*)$.

Example 2.2 For the function minimization of Example 2.1, determine the minimum points, subject to the following equality constraints:

$$\begin{aligned} x_1 - 2x_2 + 5x_3 &= 2u_1 - \cos u_2, \\ -2x_1 + x_3 &= 2\sin u_2, \\ -x_1 + 3x_2 + 2x_3 &= 0. \end{aligned}$$

The constraint equations expressed in the form of equation (2.18) are

$$\mathbf{f} = \begin{Bmatrix} f_1 \\ f_2 \\ f_3 \end{Bmatrix} = \begin{Bmatrix} x_1 - 2x_2 + 5x_3 - 2u_1 + \cos u_2 \\ -2x_1 + x_3 - 2\sin u_2 \\ -x_1 + 3x_2 + 2x_3 \end{Bmatrix}.$$

Now we write

$$\left(\frac{\partial L}{\partial \mathbf{x}}\right)^T = \begin{Bmatrix} 2x_1 \\ -2x_2 \\ 3 \end{Bmatrix},$$

$$\frac{\partial \mathbf{f}}{\partial \mathbf{x}} = \begin{pmatrix} \frac{\partial f_1}{\partial x_1} & \frac{\partial f_1}{\partial x_2} & \frac{\partial f_1}{\partial x_3} \\ \frac{\partial f_2}{\partial x_1} & \frac{\partial f_2}{\partial x_2} & \frac{\partial f_2}{\partial x_3} \\ \frac{\partial f_3}{\partial x_1} & \frac{\partial f_3}{\partial x_2} & \frac{\partial f_3}{\partial x_3} \end{pmatrix} = \begin{pmatrix} 1 & -2 & 5 \\ -2 & 0 & 1 \\ -1 & 3 & 2 \end{pmatrix}.$$

Since the determinant $|\,\partial \mathbf{f}/\partial \mathbf{x}\,| = -39 \neq 0$, we can invert the Jacobian matrix $\partial \mathbf{f}/\partial \mathbf{x}$ to derive the extremal Lagrange multipliers as follows:

$$(\boldsymbol{\lambda}^*)^T = -\frac{\partial L}{\partial \mathbf{x}} \left(\frac{\partial \mathbf{f}}{\partial \mathbf{x}}\right)^{-1}$$

or

$$\boldsymbol{\lambda}^* = -\frac{1}{39} \begin{Bmatrix} 6x_1^* + 6x_2^* + 18 \\ -38x_1^* + 14x_2^* + 3 \\ 4x_1^* - 22x_2^* + 12 \end{Bmatrix}.$$

Next, we derive the extremal control variables, $\mathbf{u}^*$, from equation (2.22) as follows:

$$\left(\frac{\partial L}{\partial \mathbf{u}^*}\right)^T = \left\{ \begin{array}{c} 2u_1^* \\ -2\cos u_2^* \end{array} \right\},$$

$$\frac{\partial \mathbf{f}}{\partial \mathbf{u}} = \begin{pmatrix} \frac{\partial f_1}{\partial u_1} & \frac{\partial f_1}{\partial u_2} \\ \frac{\partial f_2}{\partial u_1} & \frac{\partial f_2}{\partial u_2} \\ \frac{\partial f_3}{\partial u_1} & \frac{\partial f_3}{\partial u_2} \end{pmatrix} = \begin{pmatrix} -2 & -\sin u_2 \\ 0 & -2\cos u_2 \\ 0 & 0 \end{pmatrix}.$$

Thus, equation (2.22) yields the following extremal control variables:

$$u_1^* = -\frac{1}{39}(6x_1^* + 6x_2^* + 18),$$

$$u_2^* = \tan^{-1}\left\{ \frac{72 + 38x_1^* - 14x_2^*}{18 + 6x_1^* + 6x_2^*} \right\}.$$

Finally, the extremal values, $\mathbf{x}^*$ and $\mathbf{u}^*$, are related by the equality constraints as follows:

$$\mathbf{x}^* = \left(\frac{\partial \mathbf{f}}{\partial \mathbf{x}}\right)^{-1} \left\{ \begin{array}{c} 2u_1^* - \cos u_2^* \\ 2\sin u_2^* \\ 0 \end{array} \right\}$$

or

$$\left\{ \begin{array}{c} x_1^* \\ x_2^* \\ x_3^* \end{array} \right\} = \frac{1}{39} \begin{pmatrix} 3 & -19 & 2 \\ -3 & -7 & 11 \\ 6 & 1 & 4 \end{pmatrix} \left\{ \begin{array}{c} 2u_1^* - \cos u_2^* \\ 2\sin u_2^* \\ 0 \end{array} \right\}.$$

Clearly, we have to solve for $(\mathbf{x}^*, \boldsymbol{\lambda}^*, \mathbf{u}^*)$ by an iterative scheme. Fortunately, the constraints are linear in $\mathbf{x}$, allowing us to uniquely determine $\mathbf{x}^*$ from $\mathbf{u}^*$ by matrix inversion. If it were not so, a more sophisticated, nonlinear programming algorithm would be necessary (such as the Newton–Raphson method). A simple iterative solution procedure for problems with explicitly solvable constraints is depicted in Figure 2.3, where the convergence criterion is that the Euclidean norm of the variation of the control vector between two successive steps is less than or equal to a specified tolerance, ϵ:

$$\| \mathbf{u}^* - \mathbf{u}_0^* \| \leq \epsilon.$$

The iterative solution depends upon the starting point of the iteration. We demonstrate the solution procedure for this problem, beginning with an initial guess of $u_1^* = 0, u_2^* = \pi/2$ which is a minimum point for the *unconstrained* minimization (Example 2.1). The computational steps with $\epsilon = 10^{-10}$ are programmed in the following MATLAB® statements:

```
>> eps=1e-10;
>> i=0;
>> u10=0; u20=pi/2; % initial guess
>> A=[1 -2 5;-2 0 1;-1 3 2]; % Jacobian, df/dx
>> Ain=inv(A);
```

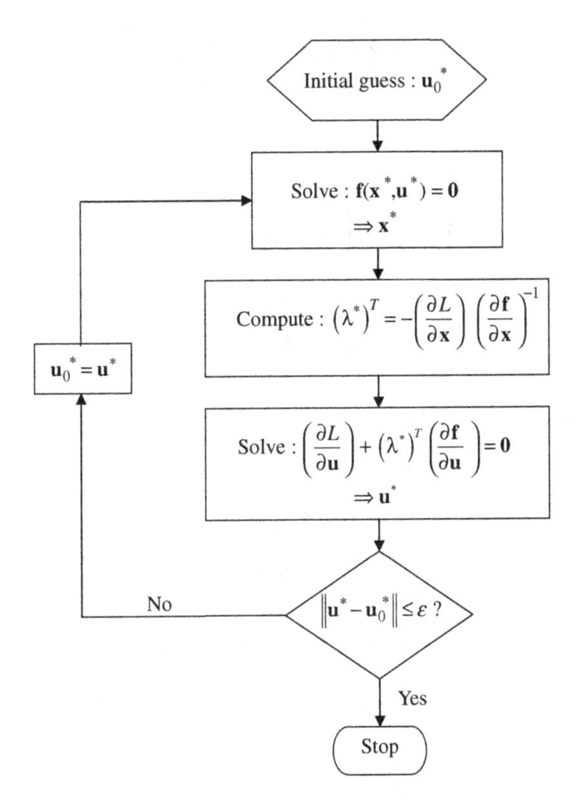

Figure 2.3 A simple iteration procedure for solving the static, constrained minimization problem with explicit constraints, $\mathbf{f}(\mathbf{x}, \mathbf{u}) = \mathbf{0}$

```
>> x=Ain*[2*u10-cos(u20);2*sin(u20);0]; %solving the constraints for x
>> u1=-(6*x(1,1)+6*x(2,1)+18)/39;
>> u2=atan((72+38*x(1,1)-14*x(2,1))/(18+6*x(1,1)+6*x(2,1)));
>> du=[u1-u10;u2-u20];
>> while norm(du)>eps
>>   u10=u1;u20=u2;
>>   x=Ain*[2*u10-cos(u20);2*sin(u20);0];
>>   u1=-(6*x(1,1)+6*x(2,1)+18)/39;
>>   u2=atan((72+38*x(1,1)-14*x(2,1))/(18+6*x(1,1)+6*x(2,1)));
>>   du=[u1-u10;u2-u20]
>>   i=i+1
>> end
```

The resulting solution,

$$\begin{Bmatrix} x_1^* \\ x_2^* \\ x_3^* \end{Bmatrix} = \begin{Bmatrix} -1.00114096910 \\ -0.28584773541 \\ -0.07179888144 \end{Bmatrix},$$

$$\boldsymbol{\lambda}^* = \begin{Bmatrix} -0.26354019931 \\ -0.94978175718 \\ -0.36625862314 \end{Bmatrix},$$

$$\begin{Bmatrix} u_1^* \\ u_2^* \end{Bmatrix} = \begin{Bmatrix} -0.26354019931 \\ 1.30636592883 \end{Bmatrix},$$

is obtained in eight iterations with an error of

$$\| \mathbf{u}^* - \mathbf{u}_0^* \| = 8.97 \times 10^{-11}.$$

2.3.2 *Inequality Constraints*

Consider general inequality constraints of the form

$$\mathbf{f}(\mathbf{x}, \mathbf{u}) \leq \mathbf{0}, \tag{2.23}$$

while minimizing $L(\mathbf{x}, \mathbf{u})$ with respect to $\mathbf{u}$. Extending the concept of Lagrange multipliers, $\boldsymbol{\lambda}$, we adjoin L with the inequality constraint function into the augmented objective function

$$\mathcal{J}(\mathbf{x}, \mathbf{u}) = L(\mathbf{x}, \mathbf{u}) + \boldsymbol{\lambda}^T \mathbf{f}(\mathbf{x}, \mathbf{u}). \tag{2.24}$$

Noting the two possibilities for an optimal point (if it exists), that is, either

$$\mathbf{f}(\hat{\mathbf{x}}, \hat{\mathbf{u}}) < \mathbf{0} \tag{2.25}$$

or

$$\mathbf{f}(\hat{\mathbf{x}}, \hat{\mathbf{u}}) = \mathbf{0}, \tag{2.26}$$

we say that in the first case (inequality (2.25)), the constraint is not effective and thus it can be ignored by putting $\boldsymbol{\lambda} = \mathbf{0}$ in equation (2.24), which is treated as unconstrained minimization discussed above. In the second case (equation (2.26)), we note that the minimum point (if it exists) lies on the constraining boundary, and a small control variation, $\delta\mathbf{u}$, about the minimum point, $(\hat{\mathbf{x}}, \hat{\mathbf{u}})$, would result in either an increase, or no change in L:

$$\delta L = \frac{\partial L}{\partial \mathbf{u}} \delta\mathbf{u} \geq 0, \tag{2.27}$$

where $\delta\mathbf{u}$ must satisfy the constraint inequality (2.23), with $\boldsymbol{\lambda} > \mathbf{0}$:

$$\boldsymbol{\lambda}^T \delta\mathbf{f} = \boldsymbol{\lambda}^T \frac{\partial \mathbf{f}}{\partial \mathbf{u}} \delta\mathbf{u} \leq 0. \tag{2.28}$$

Since $\delta\mathbf{u}$ is arbitrary, there are only the following two possibilities of reconciling inequalities (2.27) and (2.28) (Bryson and Ho 1975), that is, either

$$\frac{\partial L}{\partial \mathbf{u}} = -\boldsymbol{\lambda}^T \frac{\partial \mathbf{f}}{\partial \mathbf{u}} \tag{2.29}$$

or

$$\frac{\partial L}{\partial \mathbf{u}} = \mathbf{0}, \quad \frac{\partial \mathbf{f}}{\partial \mathbf{u}} = \mathbf{0}. \tag{2.30}$$

The three possibilities represented by inequality (2.25) and equations (2.29) and (2.30) can be expressed in the following compact form:

$$\frac{\partial L}{\partial \mathbf{u}} + \boldsymbol{\lambda}^T \frac{\partial \mathbf{f}}{\partial \mathbf{u}} = \mathbf{0}, \tag{2.31}$$

where

$$\boldsymbol{\lambda} \begin{cases} = \mathbf{0}, & \mathbf{f}(\mathbf{x}, \mathbf{u}) < \mathbf{0}, \\ \geq \mathbf{0}, & \mathbf{f}(\mathbf{x}, \mathbf{u}) = \mathbf{0}. \end{cases} \tag{2.32}$$

Equation (2.31) is the necessary condition for optimality and must be solved in an iterative manner to yield an optimal point (if it exists).

Example 2.3 Maximize the steady rate of climb of a jet airplane, $\dot{h} = v \sin\phi$, at standard sea level, $\rho = 1.225\ \mathrm{kg/m^3}$, subject to the following constraints:

$$\sin\phi = \frac{f_\mathrm{T} - D}{mg}, \tag{2.33}$$

$$D = \frac{1}{2}\rho v^2 S (C_{D0} + K C_L^2), \tag{2.34}$$

$$C_L = \frac{2mg\cos\phi}{\rho S v^2}, \tag{2.35}$$

$$C_L \leq C_{L\max}, \tag{2.36}$$

$$v \leq v_{\max}. \tag{2.37}$$

Use the following data:

$$\frac{f_\mathrm{T}}{mg} = 0.8, \quad \frac{mg}{S} = 3000\ \mathrm{N/m^2}, \quad v_{\max} = 255\ \mathrm{m/s},$$

$$C_{D0} = 0.015, \quad K = 0.04, \quad C_{L\max} = 1.2.$$

We express the objective function to be minimized as

$$L(v, \phi) = -v\sin\phi, \tag{2.38}$$

subject to

$$\sin\phi = \frac{(f_\mathrm{T} - D)v}{mg} = \frac{f_\mathrm{T}}{mg} - \frac{1}{2}\rho v^2 \frac{S}{mg} C_{D0} - \frac{2Kmg\cos^2\phi}{\rho S v^2}, \tag{2.39}$$

which incorporates the equality constraint (2.35). Substituting the numerical data, we have

$$\sin\phi = 0.8 - 3.0625 \times 10^{-6} v^2 - \frac{195.9184\cos^2\phi}{v^2}. \tag{2.40}$$

The inequality constraints (2.36) and (2.37), called the *stall limit* and the *structural limit* respectively, are expressed as

$$\cos\phi \leq 0.000245v^2, \quad v \leq 255 \text{ m/s}.$$

Taking $x = \sin\phi$ and $u = v$, we have minimization of

$$L(x, u) = -ux, \tag{2.41}$$

subject to

$$x = 0.8 - 3.0625 \times 10^{-6}u^2 - \frac{195.9184(1 - x^2)}{u^2} \tag{2.42}$$

and

$$\sqrt{1 - x^2} \leq 0.000245u^2, \quad u \leq 255 \text{ m/s}. \tag{2.43}$$

We will first attempt to solve the problem without the two inequality constraints, and then add the inequality constraints graphically. By adjoining the equality constraint (2.42), with the objective function, we have

$$\mathcal{J} = L + \lambda\left[x - 0.8 + 3.0625 \times 10^{-6}u^2 + \frac{195.9184(1 - x^2)}{u^2}\right], \tag{2.44}$$

minimization of which yields

$$\lambda = -\frac{\partial L/\partial x}{\partial f/\partial x} = \frac{u}{2x - \frac{u^2}{195.9184}}$$

and

$$\frac{\partial L}{\partial u} + \lambda\frac{\partial f}{\partial u} = u^4 + 81632.65(x - 1.6)u^2 + 31986677.55x^2 = 0, \tag{2.45}$$

which can be solved for extremal u^2 as follows:

$$u_{1,2}^2 = 40816.3265(1.6 - x) \mp \frac{1}{2}\sqrt{81632.65^2(x - 1.6)^2 - 127946710.204}. \tag{2.46}$$

We plot the two extremal solutions, $v = u_{1,2}$, in the range $0 \leq \sin\phi \leq 1$ representing a positive climb – along with the inequality constraints – in Figure 2.4 and observe that the first inequality constraint in expression (2.43) (stall limit) intersects the low-speed solution, $v = u_1$, at $x = \sin\phi = 0.9875$, while the second (structural limit) intersects with the higher-speed solution, $v = u_2$, at $x = \sin\phi = 0.79959$, corresponding to $\phi = 53.091$ degrees. A plot of the objective function, $L = -v\sin\phi$, for the two extremal solutions as well as the two inequality constraints in Figure 2.5 reveals a minimum value of $L = -203.8957$ m/s at the intersection of $v = u_2$ and $v \leq 255$ m/s corresponding to a climb angle of $\phi = 53.091°$. Therefore, the maximum rate of climb is 203.8957 m/s corresponding to $v = 255$ m/s and $\phi = 53.091°$. The stall limit solution produces a much smaller rate of climb of only 24.8 m/s, and therefore the stall limit is ineffective in the maximum rate of climb solution.

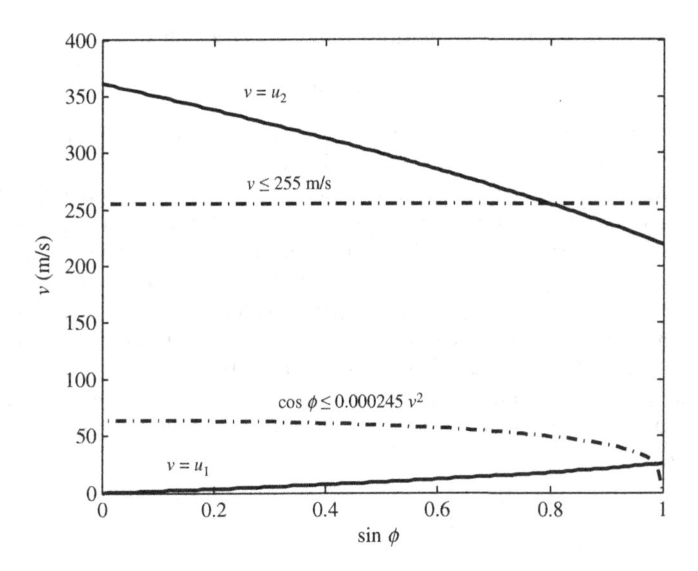

Figure 2.4 Extremal solutions and inequality constraints for the maximum rate of climb of a jet aircraft

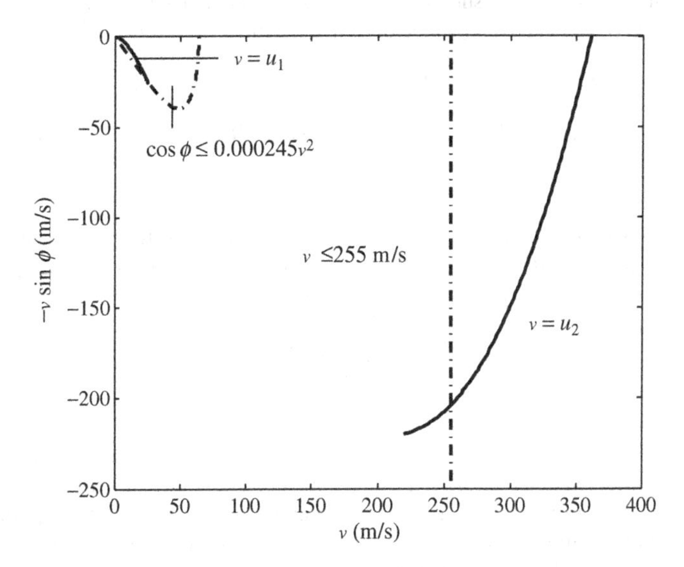

Figure 2.5 Objective function corresponding to the extremal solutions and inequality constraints for the maximum rate of climb of a jet aircraft

2.4 Optimal Control of Dynamic Systems

Thus far we have discussed minimization of cost functions of variables $(\mathbf{x}, \mathbf{u})$ that may be related by static constraints such as equation (2.18). We will now extend the treatment to dynamic systems with $\mathbf{x}(t)$ and $\mathbf{u}(t)$ being the state and control vectors, respectively. Consider a flight dynamic system given by the following continuous time state equation with a known initial condition:

$$\dot{\mathbf{x}} = \mathbf{f}(\mathbf{x}, \mathbf{u}, t), \quad \mathbf{x}(t_0) = \mathbf{x}_0, \tag{2.47}$$

where $\mathbf{u}(t)$ is the control input vector bounded by constraints (to be discussed later) in a given interval, $t_0 \leq t \leq t_f$ (called *admissible control input*), and $\mathbf{f}(.)$ is a continuous functional and has a continuous partial derivative with respect to state, $\partial \mathbf{f}/\partial \mathbf{x}$, in the given interval. The random disturbance inputs, $\boldsymbol{\nu}(t)$, $\mathbf{w}(t)$, are excluded from the optimal control problem, but are taken into account in estimating the state vector from measurements. Let the transient performance objectives be specified in terms of a scalar function of control and state variables, $L[\mathbf{x}(t), \mathbf{u}(t), t]$, called the *Lagrangian*. For an acceptable performance, the system's response, $\mathbf{x}(t)$, to the applied control input, $\mathbf{u}(t)$, should be such that the Lagrangian is minimized with respect to the control input in a control interval, $t_0 \leq t \leq t_f$. Furthermore, the performance at the final (or terminal) time, t_f, is prescribed by another scalar function, $\varphi[\mathbf{x}(t_f), t_f]$, called the *terminal cost*, that also must be minimized. Hence, both transient and terminal performance objectives are combined into the following scalar *objective function*[2] to be minimized with respect to the control input, $\mathbf{u}(t)$:

$$J = \varphi[\mathbf{x}(t_f), t_f] + \int_{t_0}^{t_f} L[\mathbf{x}(t), \mathbf{u}(t), t]\mathrm{d}t. \tag{2.48}$$

However, the state variables and the control variables are related through the dynamic state equation. Therefore, the optimization is subject to the *equality constraint* of equation (2.47) that must be satisfied by $\mathbf{x}(t)$, $\mathbf{u}(t)$ at all times. In order to ensure that the equality constraint is satisfied, the constraint equation is adjoined to the Lagrangian in an augmented objective function, $\mathcal{J}$, as follows:

$$\begin{aligned} \mathcal{J} &= J + \boldsymbol{\lambda}^T(t)\mathbf{f}[\mathbf{x}(t), \mathbf{u}(t), t] \\ &= \varphi[\mathbf{x}(t_f), t_f] + \int_{t_0}^{t_f} \left\{ L[\mathbf{x}(t), \mathbf{u}(t), t] + \boldsymbol{\lambda}^T(t) \left(\mathbf{f}[\mathbf{x}(t), \mathbf{u}(t), t] - \dot{\mathbf{x}} \right) \right\} \mathrm{d}t, \end{aligned} \tag{2.49}$$

where $\boldsymbol{\lambda}(t)$ is a vector of *Lagrange multipliers* (or *co-state* vector) of the same size as the order of system. The co-state vector must be determined from the optimization process and is related to the partial derivative of L with respect to $\mathbf{f}$, when $\mathbf{u}$ is held constant:

$$\boldsymbol{\lambda}^T = -\left(\frac{\partial L}{\partial \mathbf{f}}\right)_{\mathbf{u}}. \tag{2.50}$$

The last term in the integrand of equation (2.49) applies a penalty for deviating from the equality constraint of the state equation. Thus, the optimal control problem can be posed as the minimization of $\mathcal{J}$ with respect to $\mathbf{u}(t)$. Since the control input vector, $\mathbf{u}(t)$, must be uniquely determined at each time, t, in the control interval, $t_0 \leq t \leq t_f$, the optimal control problem has infinitely many parameters of optimization, and is thus said to be of infinite dimension. The final time, t_f, in the objective function (2.49) can either be an additional performance parameter specified in advance, or a dependent variable in the optimization process. When t_f is pre-specified, the optimal control problem is said to be of *fixed interval*; otherwise, it is said to be an *open-interval* problem.

In addition to the dynamic state equation, there may be equality and inequality constraints on the state vector, $\mathbf{x}(t)$ and the control input, $\mathbf{u}(t)$, that must be satisfied during minimization of the objective

[2] An optimal control problem having the objective function of the form given by equation (2.48) is said to be of *Bolza type*. If $L[\mathbf{x}(t), \mathbf{u}(t), t] = 0$, the associated optimal control problem is called *Mayer type*, whereas if $\varphi[\mathbf{x}(t_f), t_f] = 0$ we have a *Lagrange type* problem.

function, $\mathcal{J}$. Inclusion of such constraints generally causes an increase of complexity in the mathematical formulation and solution of the optimal control problem.

2.4.1 Optimality Conditions

To derive the conditions for the existence of a local minimum of $\mathcal{J}$ with respect to $\mathbf{u}(t)$ (called *optimality conditions*), it is useful to expand the objective function in a Taylor series about the optimal value. Let $\hat{\mathcal{J}}$ denote the minimum value of the objective function corresponding to the optimal control input, $\hat{\mathbf{u}}(t)$, for which the solution to the state equation (called the *optimal trajectory*) is $\hat{\mathbf{x}}(t)$. Then a small *variation* in the objective function corresponding to a non-optimal control input, $\mathbf{u}$ (in the vicinity of the optimal control input), is given by a Taylor series expansion about the optimal point (truncated to first-order terms) as follows:

$$\delta\mathcal{J} = \mathcal{J} - \hat{\mathcal{J}} = \frac{\partial \mathcal{J}}{\partial \mathbf{x}}\delta\mathbf{x} + \frac{\partial \mathcal{J}}{\partial \mathbf{u}}\delta\mathbf{u}, \tag{2.51}$$

where

$$\delta\mathbf{u} = \mathbf{u} - \hat{\mathbf{u}}, \quad \delta\mathbf{x} = \mathbf{x} - \hat{\mathbf{x}}$$

are small deviations (variations) in control and state from their respective optimal values, and $\partial\mathcal{J}/\partial\mathbf{x}$, $\partial\mathcal{J}/\partial\mathbf{u}$ are row vectors of sizes $1 \times n$ and $1 \times m$, respectively, evaluated at $\mathbf{u} = \hat{\mathbf{u}}$.

The plant's dynamic state equation (2.47) can also be expanded in a Taylor series about the optimal trajectory, $\hat{\mathbf{x}}(t)$, for which $\mathbf{u}(t) = \hat{\mathbf{u}}(t)$, resulting in the linear approximation

$$\delta\dot{\mathbf{x}} = \frac{\partial \mathbf{f}}{\partial \mathbf{x}}\delta\mathbf{x} + \frac{\partial \mathbf{f}}{\partial \mathbf{u}}\delta\mathbf{u}, \quad \delta\mathbf{x}(t_0) = \mathbf{0}. \tag{2.52}$$

Since a successful optimal control should always maintain the trajectory in the vicinity of the optimal trajectory, only first-order state and control variations need be considered in the derivation of the optimal control input. Consequently, the plant's *Jacobian* matrices,

$$\mathbf{A} = \frac{\partial \mathbf{f}}{\partial \mathbf{x}}, \quad \mathbf{B} = \frac{\partial \mathbf{f}}{\partial \mathbf{u}},$$

of sizes $n \times n$ and $n \times m$, respectively, and evaluated at $\mathbf{u} = \hat{\mathbf{u}}$, are important in the optimal control problem. Therefore, optimal control for a nonlinear plant is based upon its linearization about the optimal trajectory, $\hat{\mathbf{x}}(t)$. However, the optimal trajectory is seldom known *a priori*, but must be solved for iteratively.

It is important to note that the dynamic state equation (2.47) relates the trajectory, $\delta\mathbf{x}(t)$, to control history, $\delta\mathbf{u}(t)$; hence, on the optimal trajectory the state variation, $\delta\mathbf{x}$, is caused entirely due to control variation, $\delta\mathbf{u}$. Thus, we must have

$$\frac{\partial \mathcal{J}}{\partial \mathbf{x}} = \mathbf{0} \tag{2.53}$$

on the optimal trajectory, for which $\mathbf{u} = \hat{\mathbf{u}}$. Furthermore, for $\mathcal{J}$ to be minimum for $\mathbf{u} = \hat{\mathbf{u}}$, we must also have

$$\frac{\partial \mathcal{J}}{\partial \mathbf{u}} = \mathbf{0}. \tag{2.54}$$

Therefore, equations (2.53) and (2.54) are the *necessary conditions* for optimality that produce $\delta J = 0$ from equation (2.51). Thus, $\mathcal{J}$ is said to be *stationary* at a point satisfying the necessary conditions for optimality. A solution $\mathbf{x}^*(t)$ of the state equation for which $\delta\mathcal{J} = 0$ is called the *extremal trajectory*.

Note that stationarity of the objective function is a necessary but not sufficient condition for optimality. Therefore, an extremal trajectory need not be the optimal trajectory. Furthermore, when *inequality constraints* are placed on the control input, it is quite possible that the extremal trajectory does not meet those constraints; thus, the region bound by the inequality constraints (the *feasible region*) might not include a minimum point of $\mathcal{J}$. In such a case, the optimal trajectory lies on the boundary of the feasible region, but does not satisfy the stationarity condition (i.e., it is not an extremal trajectory).

In order to derive the *sufficient condition* for optimality on an extremal trajectory, we need to consider the following Taylor series expansion of $\mathcal{J}$ about the optimal point truncated to the second-order terms, while holding $\delta\mathbf{x} = \mathbf{0}$:

$$\mathcal{J} - \hat{\mathcal{J}} = \frac{\partial \mathcal{J}}{\partial \mathbf{u}}\delta\mathbf{u} + \frac{1}{2}(\delta\mathbf{u})^T \frac{\partial^2 \mathcal{J}}{\partial \mathbf{u}^2}\delta\mathbf{u}. \tag{2.55}$$

Since the partial derivatives are evaluated at the optimal point for which $\partial\mathcal{J}/\partial\mathbf{u}$ must necessarily vanish, and $\mathcal{J} > \hat{\mathcal{J}}$ if $\hat{\mathcal{J}}$ is the minimum value of $\mathcal{J}$, we have the following sufficient condition for minimization of J with respect to $\mathbf{u}$:

$$(\delta\mathbf{u})^T \frac{\partial^2 \mathcal{J}}{\partial \mathbf{u}^2}\delta\mathbf{u} > 0, \tag{2.56}$$

or the square matrix $\partial^2\mathcal{J}/\partial\mathbf{u}^2$ (Hessian) must be positive definite, because $\delta\mathbf{u}$ is an arbitrary variation. Note that equation (2.56) is the sufficient condition for optimality along an extremal trajectory, and is also referred to as the *Legendre–Clebsch condition* (or the *convexity condition*). As pointed out above, in a problem with inequality constraints, the optimal trajectory may not be extremal, but could instead lie on the boundary of the feasible region. In such a case, none of the optimality conditions given by equations (2.53), (2.54), and (2.56) will be satisfied at the minimum point.

2.5 The Hamiltonian and the Minimum Principle

While the optimal, dynamic control problem without inequality constraints involves the minimization of $\mathcal{J}$ with respect to $\mathbf{u}$, it is much more useful to translate the necessary and sufficient conditions into those satisfied by the Lagrangian and the Lagrange multipliers. For this purpose, we combine the transient performance (Lagrangian) and the equality constraint into an augmented scalar function called the *Hamiltonian*, defined by

$$H[\mathbf{x}(t), \mathbf{u}(t), t] = L[\mathbf{x}(t), \mathbf{u}(t), t] + \boldsymbol{\lambda}^T(t)\mathbf{f}[\mathbf{x}(t), \mathbf{u}(t), t]. \tag{2.57}$$

Depending upon whether t_f is fixed and also whether any further constraints are specified for the control input and the state, the optimality conditions based on Hamiltonian are variously posed. In any case, the necessary stationarity condition becomes

$$\frac{\partial H}{\partial \mathbf{u}} = \mathbf{0}, \tag{2.58}$$

while the Legendre–Clebsch sufficiency condition can be expressed as

$$H_{\mathbf{uu}} = \frac{\partial^2 H}{\partial \mathbf{u}^2} > \mathbf{0}, \tag{2.59}$$

which implies that if the new Hessian, $H_{\mathbf{uu}}$, is positive definite, then the extremal trajectory satisfying equation (2.58) is optimal.

There are special cases for which the Legendre–Clebsch sufficient condition is not satisfied with strict inequality. An example is the Hamiltonian with linear dependence on some elements of the control vector, $\mathbf{u}$. In such cases, the determinant of the Hessian, $H_{\mathbf{uu}}$, vanishes at some point along the extremal trajectory, and the associated optimal control problem is said to be *singular*, its sufficient condition reducing to the *weakened Legendre–Clebsch* condition,

$$H_{\mathbf{uu}} \geq \mathbf{0}. \tag{2.60}$$

An optimal trajectory, $\hat{\mathbf{x}}(t)$, the corresponding optimal co-state trajectory, $\hat{\boldsymbol{\lambda}}(t)$, and the corresponding optimal control history, $\hat{\mathbf{u}}(t)$, satisfying both the necessary and sufficient conditions of optimality given by equations (2.58) and (2.59), minimize the Hamiltonian locally. This fact can be expressed by the inequality

$$H[\hat{\mathbf{x}}(t), \mathbf{u}(t), \hat{\boldsymbol{\lambda}}, t] \geq H[\hat{\mathbf{x}}(t), \hat{\mathbf{u}}(t), \hat{\boldsymbol{\lambda}}, t], \tag{2.61}$$

where $\mathbf{u}(t)$ is a non-optimal, admissible control history. Inequality (2.61) is both a necessary and sufficient condition for optimality and is called Pontryagin's *minimum principle* (Pontryagin *et al.* 1962); it states that of all control histories, $\mathbf{u}(t)$, corresponding to the neighborhood of the optimal trajectory, $[\hat{\mathbf{x}}(t), \hat{\boldsymbol{\lambda}}(t)]$, the one that minimizes the Hamiltonian is the optimal history, $\hat{\mathbf{u}}(t)$. A stationary and convex Hamiltonian satisfies the minimum principle. A Hamiltonian that is either non-stationary or non-convex would require additional constraints to be imposed upon the control input, in order to satisfy inequality (2.61). However, since the minimum principle is universal in its applicability, it can be used to test the optimality of a given point, even in those cases where the necessary conditions for optimality and/or the sufficient Legendre–Clebsch condition may fail to give any pertinent information.

2.5.1 Hamilton–Jacobi–Bellman Equation

For an optimal trajectory, $\hat{\mathbf{x}}(t)$, minimizing the performance index,

$$J = \varphi[\mathbf{x}(t_f), t_f] + \int_{t_0}^{t_f} L[\mathbf{x}(t), \mathbf{u}(t), t]\mathrm{d}t, \tag{2.62}$$

subject to

$$\dot{\mathbf{x}} = \mathbf{f}(\mathbf{x}, \mathbf{u}, t), \quad \mathbf{x}(t_0) = \mathbf{x}_0, \tag{2.63}$$

and corresponding to the optimal control history, $\hat{\mathbf{u}}(t)$, one can define the following *optimal return function* for $t_0 \leq t \leq t_f$:

$$\hat{V}[\hat{\mathbf{x}}(t), t] = \varphi[\hat{\mathbf{x}}(t_f), t_f] + \int_{t}^{t_f} L[\hat{\mathbf{x}}(\tau), \hat{\mathbf{u}}(\tau), \tau]\mathrm{d}\tau \tag{2.64}$$

or

$$\hat{V}[\hat{\mathbf{x}}(t), t] = \varphi[\hat{\mathbf{x}}(t_f), t_f] - \int_{t_f}^{t} L[\hat{\mathbf{x}}(\tau), \hat{\mathbf{u}}(\tau), \tau]\mathrm{d}\tau. \tag{2.65}$$

Differentiating $\hat{V}$ with respect to time, we have

$$\frac{\mathrm{d}\hat{V}}{\mathrm{d}t} = -L[\hat{\mathbf{x}}(t), \hat{\mathbf{u}}(t), t], \tag{2.66}$$

and also

$$\frac{\mathrm{d}\hat{V}}{\mathrm{d}t} = \frac{\partial \hat{V}}{\partial t} + \frac{\partial \hat{V}}{\partial \mathbf{x}}\dot{\mathbf{x}} = \frac{\partial \hat{V}}{\partial t} + \frac{\partial \hat{V}}{\partial \mathbf{x}}\mathbf{f}. \tag{2.67}$$

By equating equations (2.66) and (2.67), we have the following partial differential equation, called the *Hamilton–Jacobi–Bellman* (HJB) equation, which must be satisfied by the optimal return function:

$$-\frac{\partial \hat{V}}{\partial t} = L[\hat{\mathbf{x}}(t), \hat{\mathbf{u}}(t), t] + \frac{\partial \hat{V}}{\partial \mathbf{x}}\mathbf{f}. \tag{2.68}$$

The boundary condition for the optimal return function is

$$\hat{V}[\hat{\mathbf{x}}(t_f), t_f] = \varphi[\hat{\mathbf{x}}(t_f), t_f]. \tag{2.69}$$

A solution for the HJB equation in optimal trajectory space, $\hat{\mathbf{x}}$, and time, t, for a set of initial states, $\mathbf{x}_0$, results in a field of optimal solutions. A finite set of small initial perturbations about a given optimal path thus produces a set of *neighboring optimal solutions* by the solution of the HJB equation. However, a solution of the HJB equation is a formidable task requiring simultaneous integration in space and time for neighboring extremal trajectories with finite-difference (or finite volume) methods. A more practical use of the HJB equation is to provide a sufficient condition for testing the optimality of a given return function.

On comparing equations (2.68) and (2.57), we note that

$$-\frac{\partial \hat{V}}{\partial t} = H[\hat{\mathbf{x}}(t), \hat{\boldsymbol{\lambda}}(t), \hat{\mathbf{u}}(t), t], \tag{2.70}$$

where

$$\hat{\boldsymbol{\lambda}}^T = \frac{\partial \hat{V}}{\partial \mathbf{x}}. \tag{2.71}$$

With these identities, the HJB equation can be used to provide optimality conditions, but its main application lies in deriving nonlinear feedback control laws of the form

$$\mathbf{u}(t) = \hat{\mathbf{u}}\,[t, \mathbf{x}(t)]\,, \tag{2.72}$$

from the arguments of the optimal return function, $\hat{V}[\hat{\mathbf{x}}(t), t]$.

For a system described by

$$\dot{\mathbf{x}} = \mathbf{f}(\mathbf{x}, t), \quad \mathbf{x}(t_0) = \mathbf{x}_0, \tag{2.73}$$

whose equilibrium point is $\mathbf{x}_e = \mathbf{0}$, if the optimal return function, $\hat{V}[\hat{\mathbf{x}}(t), t]$, satisfies Lyapunov's theorem for global asymptotic stability in the sense of Lyapunov (Appendix B) for all optimal trajectories, $\hat{\mathbf{x}}(t)$, then the given system is globally asymptotically stable about the origin, $\mathbf{x} = \mathbf{0}$. In this regard, the optimal return function is regarded as a Lyapunov function. Hence, the task of a designer is to find a return function, $V(\mathbf{x}, t)$, that is continuously differentiable with respect to the time and the state variables the system, and satisfies the following sufficient conditions for global asymptotic stability about the equilibrium point at the origin:

$$V(\mathbf{0}, t) = 0, \quad V(\mathbf{x}, t) > 0; \quad \frac{dV}{dt}(\mathbf{x}, t) < 0, \quad \text{for all } \mathbf{x} \neq \mathbf{0}; \tag{2.74}$$

$$\|\, \mathbf{x} \,\| \rightarrow \infty \text{ implies } V(\mathbf{x}, t) \rightarrow \infty. \tag{2.75}$$

2.5.2 Linear Time-Varying System with Quadratic Performance Index

Consider a linear time-varying system given by

$$\dot{\mathbf{x}} = \mathbf{A}(t)\mathbf{x}(t) + \mathbf{B}(t)\mathbf{u}(t), \quad \mathbf{x}(t_0) = \mathbf{x}_0, \tag{2.76}$$

for which the following quadratic objective function is to be minimized:

$$\begin{aligned} J &= \mathbf{x}^T(t_f)\mathbf{Q}_f\mathbf{x}(t_f) \\ &+ \int_{t_0}^{t_f} \left[\mathbf{x}^T(t)\mathbf{Q}(t)\mathbf{x}(t) + 2\mathbf{x}^T(t)\mathbf{S}(t)\mathbf{u}(t) + \mathbf{u}^T(t)\mathbf{R}(t)\mathbf{u}(t)\right] \mathrm{d}t, \end{aligned} \tag{2.77}$$

where $\mathbf{Q}_f$, $\mathbf{Q}(t)$, and $\mathbf{R}(t)$ are symmetric (but $\mathbf{S}(t)$ could be asymmetric) cost coefficient matrices. Since $J \geq 0$, we also require that $\mathbf{Q}_f$, $\mathbf{Q}(t)$, and $\mathbf{R}(t)$ be at least positive semi-definite. Furthermore, because

$$L[\mathbf{x}(t), \mathbf{u}(t), t] = \mathbf{x}^T(t)\mathbf{Q}(t)\mathbf{x}(t) + 2\mathbf{x}^T(t)\mathbf{S}(t)\mathbf{u}(t) + \mathbf{u}^T(t)\mathbf{R}(t)\mathbf{u}(t), \tag{2.78}$$

we propose a quadratic value function with a positive definite, symmetric matrix, $\mathbf{P}(t)$,

$$V[\mathbf{x}(t), t] = \mathbf{x}^T(t)\mathbf{P}(t)\mathbf{x}(t), \tag{2.79}$$

whose optimal value must satisfy the HJB equation with the terminal boundary condition,

$$\hat{V}[\hat{\mathbf{x}}(t_f), t_f] = \mathbf{x}^T(t_f)\mathbf{Q}_f\mathbf{x}(t_f). \tag{2.80}$$

Thus, we have

$$\boldsymbol{\lambda}^T = \frac{\partial V}{\partial \mathbf{x}} = 2\mathbf{x}^T(t)\mathbf{P}(t), \tag{2.81}$$

and the following Hamiltonian:

$$\begin{aligned} H &= \mathbf{x}^T(t)\mathbf{Q}(t)\mathbf{x}(t) + 2\mathbf{x}^T(t)\mathbf{S}(t)\mathbf{u}(t) + \mathbf{u}^T(t)\mathbf{R}(t)\mathbf{u}(t) \\ &\quad + 2\mathbf{x}^T(t)\mathbf{P}(t)[\mathbf{A}(t)\mathbf{x}(t) + \mathbf{B}(t)\mathbf{u}(t)]. \end{aligned} \tag{2.82}$$

To derive the optimal control, we differentiate H with respect to $\mathbf{u}$ and equate the result to zero ($H_{\mathbf{u}} = \mathbf{0}$):

$$\hat{\mathbf{x}}^T(t)\mathbf{S}(t) + \hat{\mathbf{u}}^T(t)\mathbf{R}(t) + \hat{\mathbf{x}}^T(t)\hat{\mathbf{P}}(t)\mathbf{B}(t) = \mathbf{0}, \tag{2.83}$$

or

$$\hat{\mathbf{u}}(t) = -\mathbf{R}^{-1}(t)\left[\mathbf{B}^T(t)\hat{\mathbf{P}}(t) + \mathbf{S}^T(t)\right]\hat{\mathbf{x}}(t). \tag{2.84}$$

Equation (2.84) is perhaps the most important equation in modern control theory, and is called the *linear optimal feedback control law*. Since the matrix $\mathbf{P}(t)$ is chosen to be positive definite, we have $V[\mathbf{x}(t), t] > 0$ for all $\mathbf{x}(t)$, and by substituting the linear optimal feedback control law into

$$\hat{H} = -\frac{\partial \hat{V}}{\partial t} = -\hat{\mathbf{x}}^T\dot{\hat{\mathbf{P}}}\hat{\mathbf{x}}, \tag{2.85}$$

we have the Legendre–Clebsch condition, $H_{\mathbf{uu}} > \mathbf{0}$, implying a negative definite matrix, $\dot{\hat{\mathbf{P}}}(t)$ (i.e., $\dot{V}[\mathbf{x}(t), t] < 0$ for all $\mathbf{x}(t)$). Therefore, the optimal return function given by equation (2.79) with a positive definite $\hat{\mathbf{P}}(t)$ is globally asymptotically stable by Lyapunov's theorem. We also note that the Legendre–Clebsch condition implies that $\mathbf{R}(t)$ is positive definite.

The substitution of equation (2.81) into equation (2.79) results in an optimal Hamiltonian and the following HJB equation (see equation (2.68)):

$$\begin{aligned}\hat{\mathbf{x}}^T\dot{\hat{\mathbf{P}}}\hat{\mathbf{x}} = -\hat{\mathbf{x}}^T[(\mathbf{A}-\mathbf{B}\mathbf{R}^{-1}\mathbf{S}^T)^T\hat{\mathbf{P}}+\hat{\mathbf{P}}(\mathbf{A}-\mathbf{B}\mathbf{R}^{-1}\mathbf{S}^T)\\ -\hat{\mathbf{P}}\mathbf{B}\mathbf{R}^{-1}\mathbf{B}^T\hat{\mathbf{P}}+\mathbf{Q}-\mathbf{S}\mathbf{R}^{-1}\mathbf{S}^T]\hat{\mathbf{x}},\end{aligned} \tag{2.86}$$

which yields the following *matrix Riccati equation* (MRE) to be satisfied by the optimal matrix, $\hat{\mathbf{P}}$:

$$\begin{aligned}-\dot{\hat{\mathbf{P}}} = \mathbf{Q}+(\mathbf{A}-\mathbf{B}\mathbf{R}^{-1}\mathbf{S}^T)^T\hat{\mathbf{P}}+\hat{\mathbf{P}}(\mathbf{A}-\mathbf{B}\mathbf{R}^{-1}\mathbf{S}^T)\\ -\hat{\mathbf{P}}\mathbf{B}\mathbf{R}^{-1}\mathbf{B}^T\hat{\mathbf{P}}-\mathbf{S}\mathbf{R}^{-1}\mathbf{S}^T,\end{aligned} \tag{2.87}$$

that must be solved subject to the boundary condition,

$$\hat{\mathbf{P}}(t_f) = \mathbf{Q}_f. \tag{2.88}$$

The MRE is fundamental to linear optimal control and must be integrated backward in time by an appropriate numerical scheme. Being nonlinear in nature, the MRE solution procedure is termed *nonlinear* (or *dynamic*) *programming*. The existence of a unique, positive definite solution, $\hat{\mathbf{P}}(t)$, is guaranteed if $\mathbf{Q}_f$, $\mathbf{Q}(t)$, $\mathbf{R}(t)$ are positive definite, although less restrictive conditions on the cost coefficients are possible, as we shall see later.

2.6 Optimal Control with End-Point State Equality Constraints

Consider the case where equality constraints (apart from equation (2.47)) are imposed on the state variables at the initial time, t_0, and the final time, t_f. While some state variables are specified by an initial condition

$$x_i(t_0) = x_{0i}, \quad i = 1, \ldots, p \ (p \le n), \tag{2.89}$$

another set of state variables are implicitly specified through the terminal constraint

$$\mathbf{g}[\mathbf{x}(t_f), t_f] = \mathbf{0}, \tag{2.90}$$

where $\mathbf{g}[.]$ is a vector of dimension $(q \times 1)$.[3] As in the case of the equality constraint of equation (2.47) being adjoined to the Hamiltonian equation (2.57) through Lagrange multipliers, $\boldsymbol{\lambda}(t)$, we select an additional set of Lagrange multipliers, $\boldsymbol{\mu}$, to adjoin the terminal equality constraint to the terminal penalty, $\varphi[\mathbf{x}(t_f), t_f]$, as follows:

$$\Phi[\mathbf{x}(t_f), t_f] = \varphi[\mathbf{x}(t_f), t_f] + \boldsymbol{\mu}^T\mathbf{g}[\mathbf{x}(t_f), t_f]. \tag{2.91}$$

It is to be noted that $\boldsymbol{\mu}$ is a constant vector of size $q \times 1$, and the new objective function to be minimized is obtained as follows by replacing φ by the adjointed terminal penalty, Φ, in equation (2.48):

$$\mathcal{J} = \Phi[\mathbf{x}(t_f), t_f] + \int_{t_0}^{t_f} \left\{L[\mathbf{x}(t), \mathbf{u}(t), t] + \boldsymbol{\lambda}^T(t)\left(\mathbf{f}[\mathbf{x}(t), \mathbf{u}(t), t] - \dot{\mathbf{x}}\right)\right\} dt. \tag{2.92}$$

[3] Generally, $q \le n$. However, for problems with $L =$ const., we must have $q \le n-1$ since the constant Lagrangian provides an additional equality constraint to be satisfied by $\mathbf{x}(t)$ at all times, including $t = t_f$.

Taking the differential of equation (2.92), we have

$$\mathrm{d}\mathcal{J} = \left[\left(\frac{\partial \Phi}{\partial t} + L\right)\mathrm{d}t + \frac{\partial \Phi}{\partial \mathbf{x}}\mathrm{d}\mathbf{x}\right]_{t=t_f} + \int_{t_0}^{t_f}\left(\frac{\partial H}{\partial \mathbf{x}}\delta\mathbf{x} + \frac{\partial H}{\partial \mathbf{u}}\delta\mathbf{u} - \boldsymbol{\lambda}^T\delta\dot{\mathbf{x}}\right)\mathrm{d}t, \tag{2.93}$$

where the Hamiltonian, H, is given by equation (2.57) and $\delta\mathbf{x}$, $\delta\mathbf{u}$ are the state and control variations from the optimal trajectory. Writing

$$\mathrm{d}\mathbf{x} = \delta\mathbf{x} + \dot{\mathbf{x}}\mathrm{d}t \tag{2.94}$$

and integrating the last term in the integrand of equation (2.93) by parts, we have

$$\mathrm{d}\mathcal{J} = \left(\frac{\partial \Phi}{\partial t} + L + \boldsymbol{\lambda}^T\dot{\mathbf{x}}\right)_{t=t_f}\mathrm{d}t_f + \left[\left(\frac{\partial \Phi}{\partial \mathbf{x}} - \boldsymbol{\lambda}^T\right)\mathrm{d}\mathbf{x}\right]_{t=t_f} + \boldsymbol{\lambda}^T(t_0)\delta\mathbf{x}(t_0) + \int_{t_0}^{t_f}\left[\left(\frac{\partial H}{\partial \mathbf{x}} + \dot{\boldsymbol{\lambda}}^T\right)\delta\mathbf{x} + \frac{\partial H}{\partial \mathbf{u}}\delta\mathbf{u}\right]\mathrm{d}t. \tag{2.95}$$

Since $\mathcal{J}$ is stationary on the extremal trajectory, we must choose $\boldsymbol{\lambda}(t)$ in order to make $\mathrm{d}\mathcal{J} = 0$, irrespective of the arbitrary variations, $\mathrm{d}t_f$, $\mathrm{d}\mathbf{x}(t_f)$, and $\delta\mathbf{x}$. Therefore, we have the following necessary conditions for optimality, in addition to equations (2.47) and (2.90):

$$\delta x_i(t_0) = 0, \quad i = 1, \ldots, p \ (p \le n), \tag{2.96}$$

$$\lambda_k(t_0) = 0, \quad k = p+1, \ldots, n \ (p < n), \tag{2.97}$$

$$H_u = \frac{\partial H}{\partial \mathbf{u}} = \mathbf{0}, \tag{2.98}$$

$$\dot{\boldsymbol{\lambda}} = -\left(\frac{\partial H}{\partial \mathbf{x}}\right)^T = -H_{\mathbf{x}}{}^T = -\mathbf{A}^T\boldsymbol{\lambda} - \left(\frac{\partial L}{\partial \mathbf{x}}\right)^T, \tag{2.99}$$

$$\boldsymbol{\lambda}^T(t_f) = \left(\frac{\partial \Phi}{\partial \mathbf{x}}\right)_{t=t_f} = \left(\frac{\partial \varphi}{\partial \mathbf{x}} + \boldsymbol{\mu}^T\frac{\partial \mathbf{g}}{\partial \mathbf{x}}\right)_{t=t_f}, \tag{2.100}$$

$$\left(\frac{\mathrm{d}\Phi}{\mathrm{d}t} + L\right)_{t=t_f} = 0, \tag{2.101}$$

where

$$\frac{\mathrm{d}\Phi}{\mathrm{d}t} = \frac{\partial \Phi}{\partial t} + \frac{\partial \Phi}{\partial \mathbf{x}}\dot{\mathbf{x}} = \frac{\partial \Phi}{\partial t} + \frac{\partial \Phi}{\partial \mathbf{x}}\mathbf{f}. \tag{2.102}$$

The differential equations (2.99) for Lagrange multipliers are called *co-state equations* (or *adjoint equations*) as they must be integrated simultaneously with the state equations (2.47) to generate the extremal trajectory. The condition given by equation (2.101) can be expressed in terms of φ, $\mathbf{g}$, and $\boldsymbol{\mu}$ by substituting equations (2.102) and (2.91). The initial boundary conditions given by equations (2.96) and

(2.97) are necessary to satisfy $\boldsymbol{\lambda}^T(t_0)\delta\mathbf{x}(t_0) = 0$ in equation (2.95). We note that for a state variable, x_i, on which initial condition is not specified (i.e., $\delta x_i(t_0) \neq 0$), the corresponding element of the co-state vector, λ_i must vanish at $t = t_0$ so that J remains invariant with $\delta\mathbf{x}(t_0)$ on the extremal trajectory.

2.6.1 Euler–Lagrange Equations

The necessary conditions for optimality derived above are called *Euler–Lagrange equations*. They can be stated as the following theorem:

Theorem 2.1 If an admissible (i.e., finite, piecewise continuous) control, $\mathbf{u}(t)$, of the system

$$\dot{\mathbf{x}} = \mathbf{f}(\mathbf{x}, \mathbf{u}, t)$$

minimizes the Hamiltonian

$$H[\mathbf{x}(t), \mathbf{u}(t), t] = L[\mathbf{x}(t), \mathbf{u}(t), t] + \boldsymbol{\lambda}^T(t)\mathbf{f}[\mathbf{x}(t), \mathbf{u}(t), t],$$

where $\mathbf{f}(.)$ is a continuous functional and has a continuous partial derivative, $\partial\mathbf{f}/\partial\mathbf{x}$, with respect to the state variables, then there exists an absolutely continuous co-state vector, $\boldsymbol{\lambda}(t)$, non-zero in the control interval, $t_0 \leq t \leq t_f$, such that the following conditions are satisfied:

$$\frac{\partial H}{\partial \mathbf{u}} = \mathbf{0},$$

$$\dot{\mathbf{x}} = \frac{\partial H}{\partial \boldsymbol{\lambda}},$$

$$\dot{\boldsymbol{\lambda}} = -\left(\frac{\partial H}{\partial \mathbf{x}}\right)^T,$$

$$\boldsymbol{\lambda}^T(t_f) = \left(\frac{\partial \varphi}{\partial \mathbf{x}} + \boldsymbol{\mu}^T\frac{\partial \mathbf{g}}{\partial \mathbf{x}}\right)_{t=t_f},$$

$$\left(\frac{d\Phi}{dt} + L\right)_{t=t_f} = 0,$$

where Φ, $\mathbf{g}$, $\boldsymbol{\mu}$ are as given above.

Since they are valid on a specific extremal trajectory, $\mathbf{x}(t)$, the Euler–Lagrange equations are *local* conditions for optimality and thus guarantee minimization of the Hamiltonian with respect to only small variations in $\mathbf{u}(t)$. In other words, one can have other extremal trajectories that also satisfy the Euler–Lagrange equations. Because they are the necessary conditions for optimal control, the Euler–Lagrange equations can be solved to produce a specific extremal trajectory and the corresponding optimal control history depending upon the conditions imposed upon the state and co-state variables.

2.6.2 Special Cases

A special class of optimal control problems has some state variables explicitly specified at $t = t_f$, which results in the following terminal boundary conditions:

$$x_j(t_f) = x_{fj}, \quad j = 1, \ldots, q \ (q \leq n). \tag{2.103}$$

Substituting equation (2.103) into equation (2.90), we have

$$\mathbf{g}[\mathbf{x}(t_f), t_f] = \mathbf{G}\mathbf{x}(t_f) - \mathbf{x}_f = \mathbf{0}, \tag{2.104}$$

where $\mathbf{x}_f$ is the $(q \times 1)$ vector of the state variables specified at $t = t_f$, and $\mathbf{G}$ is the following influence coefficient matrix of size $q \times n$:

$$\mathbf{G} = \left[\mathbf{I}_{q\times q} \quad \mathbf{0}_{q\times(n-q)}\right]. \tag{2.105}$$

Only the $(n - q)$ state variables not fixed at $t = t_f$ contribute to the terminal penalty function, φ. Therefore, the necessary condition given by equation (2.100) becomes the following for the case where q state variables are fixed at $t = t_f$:

$$\boldsymbol{\lambda}(t_f) = \left(\frac{\partial \varphi}{\partial \mathbf{x}}\right)^T_{t=t_f} + \mathbf{G}^T \boldsymbol{\mu}, \tag{2.106}$$

where

$$\left(\frac{\partial \varphi}{\partial \mathbf{x}}\right)^T_{t=t_f} = \begin{pmatrix} \mathbf{0}_{q\times 1} \\ \frac{\partial \varphi}{\partial x_{(q+1)}} \\ \vdots \\ \frac{\partial \varphi}{\partial x_n} \end{pmatrix}_{t=t_f}, \tag{2.107}$$

and $\boldsymbol{\mu}$ is a $q \times 1$ Lagrange multiplier vector whose constant value is determined from the other necessary conditions. Thus, equation (2.106) can be written in scalar form as follows:

$$\lambda_j(t_f) = \begin{cases} \mu_j, & j = 1, \ldots, q, \\ \left(\frac{\partial \varphi}{\partial x_j}\right)_{t=t_f}, & j = q+1, \ldots, n. \end{cases}$$

A major simplification occurs for problems with a fixed terminal time, t_f, wherein $dt_f = 0$, hence equation (2.101) is no longer a necessary condition for optimality. In such a case, a solution of equations (2.98)–(2.100) yields the extremal control history, $\mathbf{u}^*(t)$.

Another simplification in the optimal control formulation is for time-invariant systems, that is, systems whose dynamic state equations as well as the Lagrangian do not explicitly depend upon time, t. For a time-invariant system, we can write

$$\dot{\mathbf{x}} = \mathbf{f}(\mathbf{x}, \mathbf{u}), \quad \mathbf{x}(t_0) = \mathbf{x}_0, \tag{2.108}$$

and

$$H[\mathbf{x}(t), \mathbf{u}(t)] = L[\mathbf{x}(t), \mathbf{u}(t)] + \boldsymbol{\lambda}^T(t)\mathbf{f}[\mathbf{x}(t), \mathbf{u}(t)]. \tag{2.109}$$

Differentiating equation (2.109) with respect to time, we have

$$\begin{aligned}\dot{H} &= \frac{\partial L}{\partial t} + \dot{\boldsymbol{\lambda}}^T \mathbf{f} + \boldsymbol{\lambda}^T \left(\frac{\partial \mathbf{f}}{\partial \mathbf{x}} \dot{\mathbf{x}} + \frac{\partial \mathbf{f}}{\partial \mathbf{u}} \dot{\mathbf{u}} \right) \\ &= \left(\frac{\partial L}{\partial \mathbf{x}} + \boldsymbol{\lambda}^T \frac{\partial \mathbf{f}}{\partial \mathbf{x}} \right) \dot{\mathbf{x}} + \left(\frac{\partial L}{\partial \mathbf{u}} + \boldsymbol{\lambda}^T \frac{\partial \mathbf{f}}{\partial \mathbf{u}} \right) \dot{\mathbf{u}} + \dot{\boldsymbol{\lambda}}^T \mathbf{f}\end{aligned} \tag{2.110}$$

or

$$\dot{H} = H_{\mathbf{x}} \dot{\mathbf{x}} + H_{\mathbf{u}} \dot{\mathbf{u}} + \dot{\boldsymbol{\lambda}}^T \mathbf{f}. \tag{2.111}$$

Since all the derivatives are evaluated at the optimal point, for which the necessary conditions (2.98) and (2.99) dictate that

$$H_{\mathbf{u}} = \mathbf{0}, \quad -H_{\mathbf{x}} = \dot{\boldsymbol{\lambda}}^T,$$

substituting which (along with equation (2.108)) into equation (2.111) gives

$$\dot{H} = \dot{\boldsymbol{\lambda}}^T (\mathbf{f} - \dot{\mathbf{x}}) = 0 \tag{2.112}$$

or $H = \text{const}$. Thus, the Hamiltonian remains constant along the optimal trajectory for a time-invariant problem. The most commonly used terminal cost and Lagrangian functions for a time-invariant problem are of the following quadratic forms:

$$\varphi[\mathbf{x}(t_f)] = [\mathbf{x}(t_f) - \mathbf{x}]^T \mathbf{Q}_f [\mathbf{x}(t_f) - \mathbf{x}] \tag{2.113}$$

and

$$L[\mathbf{x}(t), \mathbf{u}(t)] = \left(\mathbf{x}^T, \ \mathbf{u}^T \right) \begin{pmatrix} \mathbf{Q} & \mathbf{S} \\ \mathbf{S}^T & \mathbf{R} \end{pmatrix} \begin{pmatrix} \mathbf{x} \\ \mathbf{u} \end{pmatrix}. \tag{2.114}$$

Here, $\mathbf{Q}_f$, $\mathbf{Q}$, $\mathbf{S}$, and $\mathbf{R}$ are constant matrices known as *cost coefficients* that specify the relative penalties (weighting factors) in minimizing deviations in terminal-state, state, and control variables. Since a quadratic objective function results from a second-order Taylor series expansion about the optimal trajectory, and penalizes large deviations more than small ones, it is a logical choice in any practical optimal control problem.

Many flight control applications can be posed as optimal control problems with end-point constraints, and have closed-form solutions. Optimal roll control of aircraft and missiles (Chapter 3) and single-axis rotation of spacecraft (Chapter 7) are examples of such situations. However, many practical flight control applications involve a set of nonlinear differential equations that cannot be generally solved in closed form. The numerical integration of a set of nonlinear ordinary differential equations with conditions applied at two boundaries of the independent variable is referred to as a *two-point boundary value problem* (2PBVP). Solving a 2PBVP is commonly required in open-loop optimal control formulations with time as the independent variable. We digress a little from our discussion of flight control systems to consider practical methods of solving nonlinear 2PBVPs.

2.7 Numerical Solution of Two-Point Boundary Value Problems

The general 2PBVP occurring in optimal flight control applications without interior point constraints can be expressed as a set of first-order, ordinary differential equations comprising the state and co-state equations, written as

$$\dot{\mathbf{y}} = \mathbf{f}(\mathbf{y}, t), \tag{2.115}$$

where

$$\mathbf{y}(t) = \begin{Bmatrix} \mathbf{x}(t) \\ \boldsymbol{\lambda}(t) \end{Bmatrix}$$

is the vector of dependent variables of dimension $2n \times 1$ (n is the order of the plant) and $\mathbf{f}(.)$ is a mapping vector functional $(2n + 1 \times 1) \rightarrow (2n \times 1)$. The equality boundary conditions applied at the ends of the control interval, $0 \leq t \leq t_f$, are generally expressed as

$$\mathbf{g}_0[\mathbf{y}(0)] = \mathbf{0}, \quad \mathbf{g}_f[\mathbf{y}(t_f)] = \mathbf{0}. \tag{2.116}$$

If the boundary conditions are only specified at $t = 0$, the resulting problem is called an *initial value problem* (IVP) whose integration can be carried out by a variable step time-marching scheme such as the fourth-order, fifth-stage Runge–Kutta (RK-4(5)) method (Atkinson 2001; Fehlberg 1969), based on a Taylor series approximation of equation (2.115). However, a 2PBVP is more difficult to solve than an IVP since boundary conditions must be satisfied at both the ends. Figure 2.6 graphically depicts the distinction between an IVP and a 2PBVP for the case of a second-order system of differential equations,

$$\dot{y} = y_1, \quad \dot{y}_1 = f(t, y, y_1). \tag{2.117}$$

The IVP consists of obtaining the solution $y(t)$, $\dot{y}(t)$ subject to the initial condition, $y(0) = y_{0a}$, $\dot{y}(0) = \dot{y}_{0a}$.

It is not necessary that there should exist a solution to a particular 2PBVP, or that the solution be unique. Even in the scalar case, that is, with the differential equations given by equation (2.117), the existence and uniqueness of the solution to a 2PBVP depend upon the nature of the functional, $f(t, y, y_1)$, at the interior and boundary points. Matters become complicated if $f(t, y, y_1)$ has singularities or discontinuous partial derivatives. The difficulties of computing a solution to a 2PBVP are further compounded by errors in the numerical approximation employed for the purpose. In some cases there may be spurious numerical solutions when none really exist, and in others a particular numerical scheme may be unable to find a solution when one does exist.

To illustrate the existence and uniqueness of 2PBVP solutions, consider the *Sturm–Liouville problem* which has the linear differential equation

$$\ddot{y} + \lambda^2 y = 0,$$

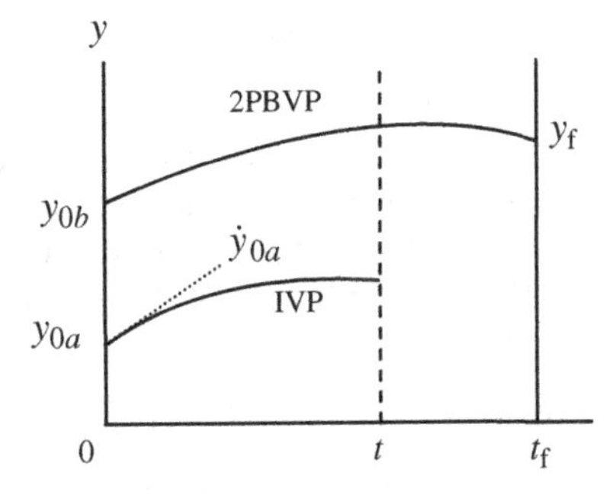

Figure 2.6 Initial value problem (IVP) and a 2PBVP for the scalar case

and *Dirichlet boundary conditions*

$$y(0) = 0, \quad y(\pi) = 0.$$

In addition to the trivial solution, $y(t) = 0$, the Sturm–Liouville problem possesses non-trivial solutions, $y(t) \neq 0$, called *eigenfunctions*, for some values of the parameter, λ, called *eigenvalues*. If $y(t)$ is an eigenfunction for a particular eigenvalue, λ, then $cy(t)$ is also an eigenfunction for the same eigenvalue, where c is a constant. Hence, there are infinitely many solutions to the Sturm–Liouville problem. It can be shown that if a 2PBVP based on a linear system has multiple solutions, then it has infinitely many solutions.

Another example of a non-unique solution to a 2PBVP is the nonlinear differential equation

$$\ddot{y} + |y| = 0,$$

with boundary conditions

$$y(0) = 0, \quad y(a) = b.$$

It can be shown (Shampine *et al.* 2000) that for $a > \pi$ there are two solutions for each negative value of b. Thus, a nonlinear 2PBVP can have finitely many solutions.

Even when a unique solution to a 2PBVP does exist, iterative numerical schemes are necessary for solving the 2PBVP, and are divided into two broad solution methods: shooting methods and collocation methods.

2.7.1 Shooting Method

The idea behind the shooting method is to pose the boundary value problem as if it were an initial value problem (IVP), and then change the initial conditions by trial and error until the boundary conditions are satisfied with a given accuracy. Thus, the shooting method is based upon a repeated integration of the governing differential equations for various initial conditions, using a standard numerical scheme such as the RK-4(5) method (Fehlberg 1969). Simple shooting methods solve the IVP over the entire control interval, $0 \leq t \leq t_f$, and can face convergence problems even if the governing system of equations is stable and well conditioned. In order to remove such difficulties, multiple shooting methods have been developed (Stoer and Bulirsch 2002) that partition the control interval into several grid points, and shooting is carried out in stages.

In order to illustrate a simple shooting method, we consider the 2PBVP based upon the second-order system of equation (2.117) with the following boundary conditions:

$$y(0) = y_0, \quad y(t_f) = y_f.$$

Figure 2.7 graphically depicts a shooting technique where the initial slope, $\dot{y}(0)$, is adjusted and the IVP with initial condition $[y(0), \dot{y}(0)]$ is solved iteratively until the terminal error, $|y(t_f) - y_f|$, is brought within a specified tolerance. Clearly, an algorithm is necessary that modifies the current choice of the initial slope, $\dot{y}^{(k)}(0)$, based upon that at the previous iteration step, $\dot{y}^{(k-1)}(0)$, and the terminal deviation at the previous iteration step, $[y^{(k-1)}(t_f) - y_f]$. A possible choice for correcting the initial slope for the next iteration could be

$$\dot{y}^{(k)}(0) = \dot{y}^{(k-1)}(0) - c\left[y^{(k-1)}(t_f) - y_f\right],$$

where the constant, c, is pre-selected in order to achieve convergence with only a small number of iterations.

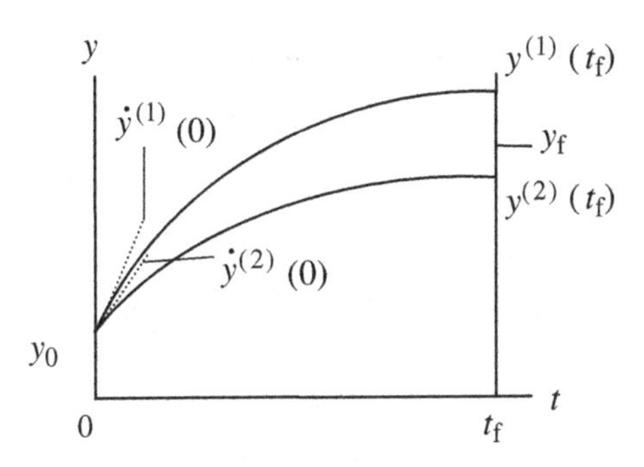

Figure 2.7 A simple shooting method for solving a 2PBVP for the scalar case

Example 2.4 Solve the following 2PBVP using a simple shooting method:

$$\dot{y} = y_1, \quad \dot{y}_1 = -4y^2;$$
$$y(0) = 0, \quad y(1) = 1.$$

We begin by writing the MATLAB code *bvp2ordshoot.m* listed in Table 2.1 that utilizes the fourth-order, fifth-stage Runge–Kutta [RK-4(5)] intrinsic MATLAB IVP solver, *ode45.m*, at each iteration step. The shooting algorithm used here is

$$\dot{y}^{(k)}(0) = \dot{y}^{(k-1)}(0) - 4\left[y^{(k-1)}(t_f) - y_f\right].$$

The terminal error tolerance of $|y(t_f) - y_f| \leq 10^{-6}$ is achieved after seven iterations (tabulated in Table 2.2) and the computed solution is plotted in Figure 2.8.

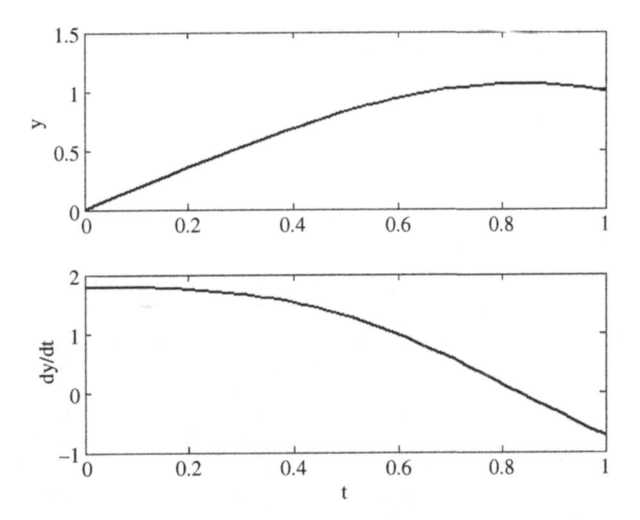

Figure 2.8 Solution to a second-order 2PBVP by a simple shooting method

Table 2.1

```
% Program for solving 2PBVP for a second-order system
% y''=f(t,y,y'),
% with end conditions on dependent variable,
% y(t0)=a; y(tf)=b,
% by a simple shooting method.
% Requires 'ode2ord.m' for specifying the differential equation
% for integration by MATLAB's RK-4(5) IVP code 'ode45.m'.
% (c) 2009 Ashish Tewari
% Solution vector: Y=[y, dy/dt]'; Time vector: T.
function [T,Y]=bvp2ordshoot(t0,tf,a,b)
c=-4; % Constant for shooting algorithm
tol=1e-6; % Terminal error tolerance
yp=(b-a)/(tf-t0) % Initial guess of dy/dt(0)
% Initial value problem solution by 'ode45.m'
[T,Y]=ode45('ode2ord',[t0 tf],[a yp]');
n=size(T,1);
y1=Y(n,1)
y1dot=Y(n,2)
res=abs(y1-b)
i=1;
% Iteration for terminal error follows:
while res>tol
    i=i+1
yp=c*(y1-b)+yp
T=[];Y=[];
[T,Y]=ode45('ode2ord',[t0 tf],[a yp]');
n=size(T,1);
y1=Y(n,1)
y1dot=Y(n,2)
res=abs(y1-b)
end

% Program for specifying the differential equation
% for integration by MATLAB's RK-4(5) IVP code 'ode45.m'.
function dydx=ode2ord(x,y)
dydx=[y(2)
      -4*y(1)^2];
```

Table 2.2 Iterations of a simple shooting method for solving a second-order 2PBVP

k	$\dot{y}^{(k)}(0)$	$y^{(k)}(1)$	$\dot{y}^{(k)}(1)$	$\lvert y^{(k)}(t_f) - y_f \rvert$
1	1.0	0.7210	0.0232	0.2790
2	2.116	1.0655	−1.1188	0.0655
3	1.854	1.0158	−0.8013	0.0158
4	1.7908	1.0013	−0.7278	0.0013
5	1.7855	1.0001	−0.7217	6.318×10^{-5}
6	1.7853	1.000003	−0.72141	2.8556×10^{-6}
7	1.7852	1.000000	−0.7214	1.2868×10^{-7}

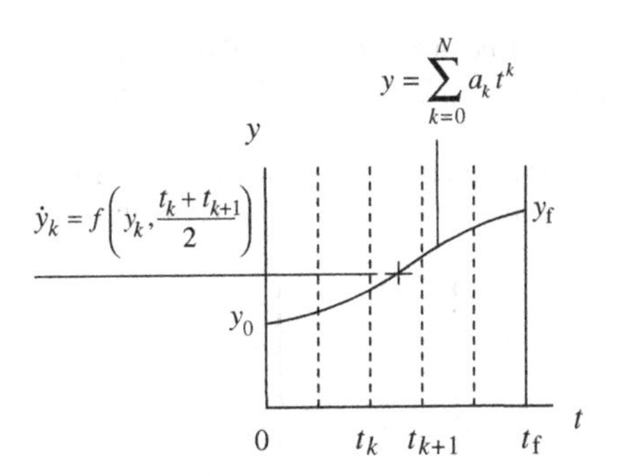

Figure 2.9 A collocation method for solving a 2PBVP for the scalar case

2.7.2 *Collocation Method*

In contrast to the shooting method, the collocation method approximates the solution by a linear combination of a number of piecewise continuous polynomials that are usually defined on a mesh of collocation points (Mattheij and Molenaar 2002; Shampine *et al.* 2003). The approximate solution is then substituted into the system of ordinary differential equations such that the system is exactly satisfied at each collocation point. The number of collocation points plus the number of boundary conditions must equal the number of unknown coefficients in the approximate solution. Figure 2.9 illustrates the discretization and collocation points for a scalar differential equation. The most common choice of approximation is a linear combination of spline functions. In order to achieve a given accuracy, the collocation points must be carefully selected. Let us now consider a simple example of the collocation method applied to solve a 2PBVP.

Example 2.5 Consider the following 2PBVP for a second-order system:

$$\ddot{y} = f(t, y, \dot{y}),$$

$$y(t_0) = y_0, \quad y(t_f) = y_f.$$

We devise a simple collocation technique based upon a polynomial solution,

$$y(t) = a_0 + a_1 t + \cdots + a_n t^n,$$

where the coefficients, $a_0, a_1, \ldots, a_n$, are constants, and apply this approximate solution over the single interval, $t_0 \leq t \leq t_f$, such that the boundary conditions at both ends are met and the governing differential equation is satisfied at both the ends as well as at an interior point, $y(t_1)$, $t_0 \leq t_1 \leq t_f$. Thus we write

$$y(t_0) = a_0 + a_1 t_0 + \cdots + a_n t_0^n = y_0,$$

$$y(t_f) = a_0 + a_1 t_f + \cdots + a_n t_f^n = y_f,$$

$$y(t_1) = a_0 + a_1 t_1 + \cdots + a_n t_1^n = y_1,$$

$$\dot{y}(t) = a_1 + 2a_2 t + \cdots + n a_n t^{n-1},$$

$$\ddot{y}(t) = 2a_2 + 6a_3 t + \cdots + n(n-1) a_n t^{n-2} = f(t, y, \dot{y}).$$

The solution, $[y(t_1), \dot{y}(t_1)]$, at interior point $t = t_1$ (called the collocation point), is obtained by solving the IVP with initial condition $[y(t_0), \dot{y}(t_0)]$, in the interval $t_0 \leq t \leq t_1$, in order to determine the values of the polynomial coefficients.

One can choose a fifth-order polynomial ($n = 5$) which is forced to satisfy the boundary conditions, slopes, and differential equation at both ends of the control interval:

$$\begin{Bmatrix} a_0 \\ a_1 \\ a_2 \\ a_3 \\ a_4 \\ a_5 \end{Bmatrix} = \mathbf{C} \begin{Bmatrix} y_0 \\ \dot{y}_0 \\ \ddot{y}_0 \\ \ddot{y}_f \\ \dot{y}_f \\ y_f \end{Bmatrix},$$

where

$$\mathbf{C} = \begin{pmatrix} 1 & t_0 & t_0^2 & t_0^3 & t_0^4 & t_0^5 \\ 0 & 1 & 2t_0 & 3t_0^2 & 4t_0^3 & 5t_0^4 \\ 0 & 0 & 2 & 6t_0 & 12t_0^2 & 20t_0^3 \\ 0 & 0 & 2 & 6t_f & 12t_f^2 & 20t_f^3 \\ 0 & 1 & 2t_f & 3t_f^2 & 4t_f^3 & 5t_f^4 \\ 1 & t_0 & t_0^2 & t_0^3 & t_0^4 & t_0^5 \end{pmatrix}.$$

By using an IVP solver that generates the solution, $y(t_1)$, at an interior point, $t = t_1$, from the initial condition, $[y_0, \dot{y}_0]$, we can evolve an iterative solution strategy wherein the error in satisfying the differential equation at the single collocation point, $t = t_1$,

$$\epsilon = |\ddot{y}(t_1) - f[t_1, y(t_1), \dot{y}(t_1)]|,$$

is brought to within a specified tolerance by suitably changing the initial condition.

MATLAB contains a more sophisticated, in-built 2PBVP solver called *bvp4c.m* that is based upon a collocation scheme (Shampine *et al.* 2003). It integrates a system of ordinary differential equations of the form

$$\frac{\mathrm{d}\mathbf{y}}{\mathrm{d}x} = \mathbf{f}(x, \mathbf{y}),$$

on the closed interval, $a \leq x \leq b$, subject to general two-point boundary conditions of the form

$$\mathbf{g}_a[(a)] = \mathbf{0}, \quad \mathbf{g}_b[\mathbf{y}(b)] = \mathbf{0},$$

where $\mathbf{g}_a(.)$, $\mathbf{g}_b(.)$ are vector functionals. The solver *bvp4c.m* can also find unknown parameters while iterating for a 2PBVP solution (such as the eigenvalues of a Sturm–Liouville problem). Its solution strategy is based upon dividing the control interval into a number of sub-intervals, and approximating the solution in each sub-interval by a cubic spline collocated at the mid-point.

Example 2.6 Consider the following 2PBVP:

$$\dot{y} = y_1, \quad \dot{y}_1 = t - 2y - \dot{y};$$

$$y(0) = 1, \quad y(1) = 2.$$

Table 2.3

```
% Program for solving 2PBVP for a second-order system, y''=f(t,y,y'),
% with end conditions on dependent variable, y(t0)=a; y(tf)=b,
% by collocation of a quintic polynomial function
% y=c0+c1*t+c2*t^2+c3*t^3+c4*t^4+c5*t^5
% on single interval [t0,tf].
% Requires 'ode2ord.m' for specifying the differential equation
% for integration by MATLAB's RK-4(5) IVP code 'ode45.m'.
% (c) 2009 Ashish Tewari
% Solution vector: c=[c0, c1, c2, c3, c4, c5]';
function c=bvp2ordquintic(t0,tf,a,b)
tol=1e-6;
t1=(tf+t0)/2
y0dot=(b-a)/(tf-t0)
f0=t0-2*a+y0dot;
C1=[1 t0 t0^2 t0^3 t0^4 t0^5; 0 1 2*t0 3*t0^2 4*t0^3 5*t0^4;
    0 0 2 6*t0 12*t0^2 20*t0^3; 0 0 2 6*tf 12*tf^2 20*tf^3;
    0 1 2*tf 3*tf^2 4*tf^3 5*tf^4; 1 tf tf^2 tf^3 tf^4 tf^5];
[T,Y]=ode45('ode2ord',[t0 t1],[a y0dot]');
n=size(T,1); y1=Y(n,1); y1dot=Y(n,2); f1=t1-2*y1+y1dot;
[T,Y]=ode45('ode2ord',[t0 tf],[a y0dot]');
n=size(T,1); yf=Y(n,1); yfdot=Y(n,2); ff=tf-2*b+yfdot;
c=inv(C1)*[a y0dot f0 ff yfdot b]';
c0=c(1,1);c1=c(2,1);c2=c(3,1);c3=c(4,1);c4=c(5,1);c5=c(6,1);
err=f1-2*c2-6*c3*t1-12*c4*t1^2-20*c5*t1^3
while abs(err)>tol
y0dot=y0dot-err
f0=t0-2*a+y0dot;
[T,Y]=ode45('ode2ord',[t0 t1],[a y0dot]');
n=size(T,1); y1=Y(n,1); y1dot=Y(n,2); f1=t1-2*y1+y1dot;
[T,Y]=ode45('ode2ord',[t0 tf],[a y0dot]');
n=size(T,1); yf=Y(n,1); yfdot=Y(n,2); ff=tf-2*b+yfdot;
c=inv(C1)*[a y0dot f0 ff yfdot b]';
c0=c(1,1);c1=c(2,1);c2=c(3,1);c3=c(4,1);c4=c(5,1);c5=c(6,1);
err=f1-2*c2-6*c3*t1-12*c4*t1^2-20*c5*t1^3;
end

% Program for specifying the differential equation
% for integration by MATLAB's RK-4(5) IVP code 'ode45.m'.
function dydx=ode2ord(x,y)
dydx=[y(2)
      x-2*y(1)+y(2)];
```

We shall solve this problem using alternatively the simple collocation scheme presented in Example 2.5 with $t_1 = 0.5$ and a tolerance, $\epsilon \leq 10^{-6}$ (coded in *bvp2ordquintic.m*; see Table 2.3), as well as the MATLAB intrinsic 2PBVP solver *bvp4c.m*, and compare the results. The converged polynomial coefficients of the simple collocation scheme, *bvp2ordquintic.m*, requring 15 iteration steps, are listed in Table 2.4.

In order to utilize MATLAB's intrinsic code *bvp4c.m* for the present problem, a calling program is listed in Table 2.5 which uses five collocation points, imposes the boundary conditions through a listed subroutine, *bc2ord.m*, and solves differential equations specified through *ode2ord.m*, previously listed in Table 2.3. The computed results of the two different collocation codes are plotted in Figure 2.10. Note the excellent agreement between the two methods.

Table 2.4 Converged coefficients of a simple collocation method for solving a second-order 2PBVP

Coefficient	Converged value
a_0	1.0
a_1	1.6538589436
a_2	−0.1730705282
a_3	−0.3844115179
a_4	−0.1789638515
a_5	0.0825869540

Table 2.5

```
% Calling program for solving the 2PBVP of a second-order
% system d^2y/dt^2=f(t,y,dy/dt),
% by collocation method of MATLAB's intrinsic code 'bvp4c.m'.
% Requires 'ode2ord.m' and 'bc2ord.m'.
% (c) 2009 Ashish Tewari
% Collocation solution with 5 points with t0=0, tf=1:
solinit = bvpinit(linspace(0,1,5),[1 0]); % Initial guess
%                                            (y=1, dy/dt=0)
sol = bvp4c(@ode2ord,@bc2ord,solinit);% Solution by bvp4c.m
t = linspace(0,1); % Time vector
x = deval(sol,t); % Solution state vector, [y, dy/dt]'

% Program for specifying boundary conditions for the 2PBVP.
% (To be called by 'bvp4c.m')
function res=bc2ord(ya,yb)
% Boundary conditions follow:
res=[ya(1)-1    % y(t0)=1
     yb(1)-2]; % y(tf)=2
```

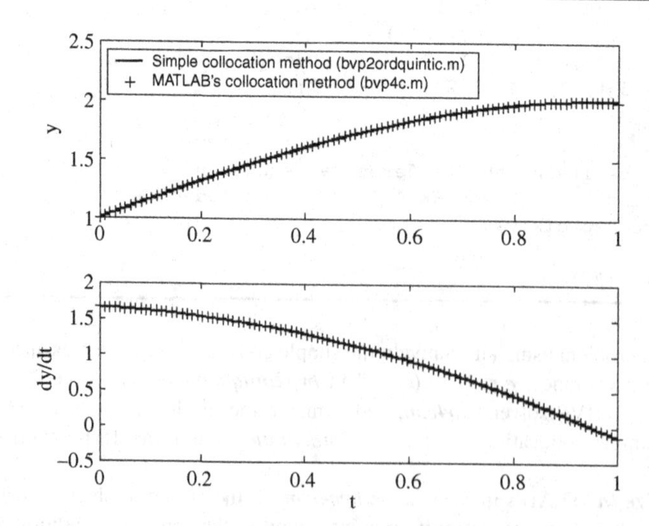

Figure 2.10 Solution to a second-order 2PBVP by a simple collocation method, *bvp2ordquintic.m*, and the MATLAB intrinsic solver, *bvp4c.m*

2.8 Optimal Terminal Control with Interior Time Constraints

We would like to end the treatment of terminal control of dynamic systems by considering equality and inequality constraints on state and control variables in the interior (rather than at the ends) of the control interval, $t_0 \leq t \leq t_f$. As in the case of static optimization problems considered above, such constraints are easily adjoined with the objective function and the dynamic state equation in the Hamiltonian.

Consider a general inequality constraint of the form

$$\mathbf{g}(\mathbf{x}, \mathbf{u}, t) \leq \mathbf{0}, \tag{2.118}$$

while minimizing $\mathcal{J}(\mathbf{x}, \mathbf{u}, t)$ with respect to $\mathbf{u}$. Defining additional Lagrange multipliers, $\boldsymbol{\mu}$, we adjoin the Hamiltonian with the inequality constraint function as follows:

$$\bar{H} = L + \boldsymbol{\lambda}^T \mathbf{f} + \boldsymbol{\mu}^T \mathbf{g} = H + \boldsymbol{\mu}^T \mathbf{g}. \tag{2.119}$$

Note the two possibilities for an optimal point (if it exists): either

$$\mathbf{g}(\hat{\mathbf{x}}, \hat{\mathbf{u}}, t) < \mathbf{0} \tag{2.120}$$

or

$$\mathbf{g}(\hat{\mathbf{x}}, \hat{\mathbf{u}}, t) = \mathbf{0}. \tag{2.121}$$

In the first case (equation (2.120)), the constraint is not effective and thus it can be ignored by putting $\boldsymbol{\mu} = \mathbf{0}$ in equation (2.119), which is exactly as if there were no inequality constraint. In the second case (equation (2.121)), we note that the minimum point (if it exists) lies on the constraining boundary, and a small control variation, $\delta\mathbf{u}$, about the minimum point, $(\hat{\mathbf{x}}, \hat{\mathbf{u}})$, would result in either an increase or no change in H:

$$\delta H = \frac{\partial H}{\partial \mathbf{u}} \delta\mathbf{u} \geq 0, \tag{2.122}$$

where $\delta\mathbf{u}$ must satisfy the constraint, equation (2.118), with $\boldsymbol{\mu} > \mathbf{0}$:

$$\boldsymbol{\mu}^T \delta\mathbf{g} = \boldsymbol{\mu}^T \frac{\partial \mathbf{g}}{\partial \mathbf{u}} \delta\mathbf{u} \leq 0. \tag{2.123}$$

Since $\delta\mathbf{u}$ is arbitrary, there are only the following two possibilities for reconciling inequalities (2.122) and (2.123): either

$$\frac{\partial H}{\partial \mathbf{u}} = -\boldsymbol{\mu}^T \frac{\partial \mathbf{g}}{\partial \mathbf{u}} \tag{2.124}$$

or

$$\frac{\partial H}{\partial \mathbf{u}} = \mathbf{0}, \quad \frac{\partial \mathbf{g}}{\partial \mathbf{u}} = \mathbf{0}. \tag{2.125}$$

The three possibilities represented by inequality (2.120) and equations (2.124) and (2.125) can be expressed in the following compact form:

$$\frac{\partial H}{\partial \mathbf{u}} + \boldsymbol{\mu}^T \frac{\partial \mathbf{g}}{\partial \mathbf{u}} = \mathbf{0}, \tag{2.126}$$

where

$$\boldsymbol{\mu} \begin{cases} = \mathbf{0}, & \mathbf{g}(\mathbf{x}, \mathbf{u}, t) < \mathbf{0}, \\ \geq \mathbf{0}, & \mathbf{g}(\mathbf{x}, \mathbf{u}, t) = \mathbf{0}. \end{cases} \tag{2.127}$$

Equation (2.126) is the necessary condition for optimality and must be solved in an iterative manner to yield an optimal point (if it exists).[4]

The consequence of the optimal control subject to inequality (2.118) on Euler–Lagrange equations (the necessary conditions for optimality) is the latter's modification to the following:

$$\dot{\boldsymbol{\lambda}} = \begin{cases} -\left(\frac{\partial H}{\partial \mathbf{x}}\right)^T, & \mathbf{g}(\mathbf{x}, \mathbf{u}, t) < \mathbf{0}, \\ -\left(\frac{\partial H}{\partial \mathbf{x}}\right)^T - \left(\frac{\partial \mathbf{g}}{\partial \mathbf{x}}\right)^T \boldsymbol{\mu}, & \mathbf{g}(\mathbf{x}, \mathbf{u}, t) = \mathbf{0}; \end{cases} \tag{2.128}$$

$$\mathbf{0} = \begin{cases} \left(\frac{\partial H}{\partial \mathbf{u}}\right)^T, & \mathbf{g}(\mathbf{x}, \mathbf{u}, t) < \mathbf{0}, \\ \left(\frac{\partial H}{\partial \mathbf{u}}\right)^T + \left(\frac{\partial \mathbf{g}}{\partial \mathbf{u}}\right)^T \boldsymbol{\mu}, & \mathbf{g}(\mathbf{x}, \mathbf{u}, t) = \mathbf{0}. \end{cases} \tag{2.129}$$

2.8.1 Optimal Singular Control

An interesting application of the inequality constraints is the solution of the *singular optimal control* problem, so named because the optimal control input cannot be derived from the minimization of the Hamiltonian inside the feasible space. Examples of singular control problems are those that have a non-convex Hamiltonian with respect to control, that is,

$$\frac{\partial^2 H}{\partial \mathbf{u}^2} = \mathbf{0}. \tag{2.130}$$

The optimal points of a singular control problem with a non-convex Hamiltonian with respect to control can only be found on the boundary of the feasible region bounded by inequality constraints. An optimal control problem with a linear plant and a Lagrangian that is either independent of or linear in control is easily seen to have a non-convex Hamiltonian with respect to control, and is thus singular. Since the Legendre–Clebsch condition (equation (2.59)) fails to establish optimality for a singular control problem, one has to look to Pontryagin's minimum principle for an answer.

Consider a *control affine* system defined by

$$\dot{\mathbf{x}}(t) = \mathbf{f}[\mathbf{x}(t), t] + \mathbf{B}(t)\mathbf{u}(t), \tag{2.131}$$

which is bounded in control as follows:

$$u_{\min} \leq \| \mathbf{u}(t) \| \leq u_{\max}, \tag{2.132}$$

and has a Lagrangian that is linear in $\mathbf{u}$,

$$L[\mathbf{x}(t), \mathbf{u}(t), t] = \ell_1[\mathbf{x}(t), t] + \ell_2(t)\mathbf{u}(t). \tag{2.133}$$

Clearly, we have a non-convex Hamiltonian,

$$H = \ell_1 + \boldsymbol{\lambda}^T\mathbf{f} + \left(\ell_2 + \boldsymbol{\lambda}^T\mathbf{B}\right)\mathbf{u}, \tag{2.134}$$

[4] The treatment of an equality constraint, $\mathbf{g}(\mathbf{x}, \mathbf{u}, t) = \mathbf{0}$, follows quite easily from equation (2.126).

with respect to $\mathbf{u}$. If the trajectory and control pair, $\hat{\mathbf{x}}(t)$, $\hat{\mathbf{u}}(t)$, satisfies Pontryagin's minimum principle,

$$H\left[\hat{\mathbf{x}}(t), \mathbf{u}(t), \hat{\boldsymbol{\lambda}}, t\right] \geq H\left[\hat{\mathbf{x}}(t), \hat{\mathbf{u}}(t), \hat{\boldsymbol{\lambda}}, t\right], \tag{2.135}$$

then they constitute the optimal solution. However, since the first part,

$$\ell_1\left[\hat{\mathbf{x}}(t), t\right] + \hat{\boldsymbol{\lambda}}^T \mathbf{f}\left[\hat{\mathbf{x}}(t), t\right] \geq \ell_1\left[\hat{\mathbf{x}}(t), t\right] + \hat{\boldsymbol{\lambda}}^T \mathbf{f}\left[\hat{\mathbf{x}}(t), t\right], \tag{2.136}$$

is trivially satisfied, we are left with the optimality condition

$$\left[\ell_2(t) + \hat{\boldsymbol{\lambda}}^T \mathbf{B}(t)\right] \mathbf{u}(t) \geq \left[\ell_2(t) + \hat{\boldsymbol{\lambda}}^T \mathbf{B}(t)\right] \hat{\mathbf{u}}(t). \tag{2.137}$$

Equation (2.137) is an important condition for optimality and is called the *switching condition*, because it requires switching between the minimum and maximum control bounds given by

$$\| \hat{\mathbf{u}}(t) \| = \begin{cases} u_{\min}, & \ell_2(t) + \hat{\boldsymbol{\lambda}}^T \mathbf{B}(t) > 0, \\ u_{\max}, & \ell_2(t) + \hat{\boldsymbol{\lambda}}^T \mathbf{B}(t) < 0. \end{cases} \tag{2.138}$$

The function $\left[\ell_2(t) + \hat{\boldsymbol{\lambda}}^T \mathbf{B}(t)\right]$ is known as the *switching function* since its sign determines on which control boundary the optimal point would lie at a given time. Interestingly, the optimal control always has either the minimum or the maximum magnitude and is thus referred to as the *bang-bang control*. Time and fuel minimization problems of linear flight control fall into this category.

2.9 Tracking Control

So far, we have deliberated upon the open-loop, optimal control theory wherein an objective function, $\mathcal{J}$, of state and control variables is minimized subject to the flight dynamic state equation (2.47), end-point and interior point constraints, resulting in an optimal control history, $\hat{\mathbf{u}}(t)$, and a corresponding extremal trajectory, $\hat{\mathbf{x}}(t)$, subject to an initial condition, $\mathbf{x}(t_0) = \mathbf{x}_0$. Now, if there is either a variation in the initial condition, $[t_0, \mathbf{x}_0]$, a change in the system's parameters governing equation (2.47), or application of disturbance inputs, there must be an attendant change in the optimal control history, $\hat{\mathbf{u}}(t)$, which minimizes the same objective function subject to the given set of constraints, and must result in a new extremal trajectory different from the original one. Since any point on an extremal trajectory, $[t, \hat{\mathbf{x}}(t)]$, can be regarded as an initial condition for the remainder of the trajectory, the optimal control depends upon the extremal trajectory according to

$$\hat{\mathbf{u}}(t) = \hat{\mathbf{u}}\left[t, \hat{\mathbf{x}}(t)\right]. \tag{2.139}$$

The functional dependence given by equation (2.139) of the optimal control history on a point $[t, \hat{\mathbf{x}}(t)]$ on the extremal trajectory is referred to as the *optimal feedback control law*. Clearly, the optimal control must change if there is any variation in the corresponding trajectory due to a change in the initial condition, a change in the constraints, or presence of unmodeled dynamics in the form of internal and external disturbances.

While an optimal control history can be painstakingly computed for every extremal trajectory, a simpler method would be to consider only small deviations from a given extremal trajectory that can be regarded as a nominal trajectory, resulting in a locally linearized feedback control law. The net control input is assumed to be merely the sum of the original, open-loop optimal control and the linear feedback control perturbation. Such an approximation based upon linearization in the neighborhood of a known optimal solution is called the *neighboring extremal* approach and forms the basis of linear optimal control theory with a quadratic cost functional. The linear quadratic feedback control law is easily derived using the necessary optimality conditions given by the Euler–Lagrange equations. Solution of the time-varying linear quadratic control problem requires nonlinear, dynamic programming methods (Bellman 1957).

2.9.1 Neighboring Extremal Method and Linear Quadratic Control

Consider a flight dynamic plant with state vector, $\boldsymbol{\xi}(t)$, and control input vector, $\boldsymbol{\eta}(t)$ governed by the state equation

$$\dot{\boldsymbol{\xi}} = \mathbf{f}(\boldsymbol{\xi}, \boldsymbol{\eta}, t). \tag{2.140}$$

Let an extremal trajectory, $\boldsymbol{\xi}_{\mathrm{d}}(t)$, and the corresponding extremal control history, $\boldsymbol{\eta}_{\mathrm{d}}(t)$, be available from the necessary conditions for the solution of the optimal control problem minimizing an objective function,

$$J(\boldsymbol{\xi}, \boldsymbol{\eta}) = \varphi[\boldsymbol{\xi}(t_{\mathrm{f}}), t_{\mathrm{f}}] + \int_{t_0}^{t_{\mathrm{f}}} L[\boldsymbol{\xi}(t), \boldsymbol{\eta}(t), t]\mathrm{d}t, \tag{2.141}$$

subject to certain specific constraints. The extremal control and trajectory are the nominal functions satisfying equation (2.140),

$$\dot{\boldsymbol{\xi}}_{\mathrm{d}} = \mathbf{f}(\boldsymbol{\xi}_{\mathrm{d}}, \boldsymbol{\eta}_{\mathrm{d}}, t), \tag{2.142}$$

about which small control and state deviations,

$$\mathbf{u}(t) = \boldsymbol{\eta}(t) - \boldsymbol{\eta}_d(t), \quad t_0 \leq t \leq t_{\mathrm{f}}, \tag{2.143}$$

$$\mathbf{x}(t) = \boldsymbol{\xi}(t) - \boldsymbol{\xi}_d(t), \quad t_0 \leq t \leq t_{\mathrm{f}}, \tag{2.144}$$

are to be minimized. Employing a first-order Taylor series expansion about the nominal trajectory (Appendix A), we have the following linear state equation governing small off-nominal deviations:

$$\dot{\mathbf{x}}(t) = \mathbf{A}(t)\mathbf{x}(t) + \mathbf{B}(t)\mathbf{u}(t), \quad \mathbf{x}(t_0) = \mathbf{x}_0, \tag{2.145}$$

where

$$\mathbf{A}(t) = \left.\frac{\partial \mathbf{f}}{\partial \boldsymbol{\xi}}\right|_{\boldsymbol{\xi}_{\mathrm{d}}, \boldsymbol{\eta}_{\mathrm{d}}}; \quad \mathbf{B}(t) = \left.\frac{\partial \mathbf{f}}{\partial \boldsymbol{\eta}}\right|_{\boldsymbol{\xi}_{\mathrm{d}}, \boldsymbol{\eta}_{\mathrm{d}}} \tag{2.146}$$

are the Jacobian matrices of the expansion. Similarly, the objective function can be expanded about the extremal (nominal) solution, $(\boldsymbol{\xi}_{\mathrm{d}}, \boldsymbol{\eta}_{\mathrm{d}})$, up to the second-order terms

$$J(\boldsymbol{\xi}_{\mathrm{d}} + \mathbf{x}, \boldsymbol{\eta}_{\mathrm{d}} + \mathbf{u}) \simeq J(\boldsymbol{\xi}_{\mathrm{d}}, \boldsymbol{\eta}_{\mathrm{d}}) + \Delta^2 J(\mathbf{x}, \mathbf{u}), \tag{2.147}$$

noting that the first variation of J about the extremal trajectory is identically zero,

$$\Delta J(\mathbf{x}, \mathbf{u}) = \left.\frac{\partial J}{\partial \boldsymbol{\xi}}\right|_{\boldsymbol{\xi}_{\mathrm{d}}, \boldsymbol{\eta}_{\mathrm{d}}} \mathbf{x} + \left.\frac{\partial J}{\partial \boldsymbol{\eta}}\right|_{\boldsymbol{\xi}_{\mathrm{d}}, \boldsymbol{\eta}_{\mathrm{d}}} \mathbf{u} = \mathbf{0}. \tag{2.148}$$

The second variation of J about the extremal trajectory is given by

$$\Delta^2 J(\mathbf{x}, \mathbf{u}) = \frac{1}{2}\mathbf{x}^T(t_{\mathrm{f}})\mathbf{Q}_{\mathrm{f}}\mathbf{x}(t_{\mathrm{f}}) + \frac{1}{2}\int_{t_0}^{t_{\mathrm{f}}} \{\mathbf{x}^T(t), \mathbf{u}^T(t)\} \begin{bmatrix} \mathbf{Q}(t) & \mathbf{S}(t) \\ \mathbf{S}^T(t) & \mathbf{R}(t) \end{bmatrix} \begin{Bmatrix} \mathbf{x}(t) \\ \mathbf{u}(t) \end{Bmatrix}, \tag{2.149}$$

which is a quadratic form with the following cost coefficient matrices:

$$\mathbf{Q}_\mathrm{f} = \left.\frac{\partial^2 \varphi}{\partial \boldsymbol{\xi}^2}\right|_{\boldsymbol{\xi}_\mathrm{d}(t_\mathrm{f})} \quad ; \quad \mathbf{Q}(t) = \left.\frac{\partial^2 L}{\partial \boldsymbol{\xi}^2}\right|_{\boldsymbol{\xi}_\mathrm{d},\boldsymbol{\eta}_\mathrm{d}} \tag{2.150}$$

and

$$\mathbf{S}(t) = \left.\frac{\partial^2 L}{\partial \boldsymbol{\xi} \partial \boldsymbol{\eta}}\right|_{\boldsymbol{\xi}_\mathrm{d},\boldsymbol{\eta}_\mathrm{d}} \quad ; \quad \mathbf{R}(t) = \left.\frac{\partial^2 L}{\partial \boldsymbol{\eta}^2}\right|_{\boldsymbol{\xi}_\mathrm{d},\boldsymbol{\eta}_\mathrm{d}} . \tag{2.151}$$

Thus, the second variation of the objective function, $\Delta^2 J$, about the extremal trajectory is a quadratic cost function that must be minimized subject to a linearized dynamic equation for a neighboring extremal trajectory. This forms the basis of the linear quadratic optimal control problem for neighboring extremal trajectories.

The Hamiltonian corresponding to the quadratic cost function, $\Delta^2 J$, subject to linear dynamic constraint of equation (2.145) is the following:

$$\begin{aligned} H = &\frac{1}{2}\mathbf{x}^T(t)\mathbf{Q}(t)\mathbf{x}(t) + \mathbf{x}^T(t)\mathbf{S}(t)\mathbf{u}(t) + \frac{1}{2}\mathbf{u}^T(t)\mathbf{R}(t)\mathbf{u}(t) \\ &+ \boldsymbol{\lambda}^T(t)[\mathbf{A}(t)\mathbf{x}(t) + \mathbf{B}(t)\mathbf{u}(t)]. \end{aligned} \tag{2.152}$$

The necessary conditions for optimality with a fixed terminal time, t_f, are then given by Euler–Lagrange equations (Theorem 2.1):

$$\dot{\boldsymbol{\lambda}} = -\left(\frac{\partial H}{\partial \mathbf{x}}\right)^T = -\mathbf{Q}(t)\mathbf{x}(t) - \mathbf{S}(t)\mathbf{u}(t) - \mathbf{A}^T(t)\boldsymbol{\lambda}(t), \tag{2.153}$$

$$\boldsymbol{\lambda}(t_\mathrm{f}) = \left(\frac{\partial \varphi}{\partial \mathbf{x}}\right)^T_{t=t_\mathrm{f}} = \mathbf{Q}_\mathrm{f}\mathbf{x}(t_\mathrm{f}), \tag{2.154}$$

$$\frac{\partial H}{\partial \mathbf{u}} = \mathbf{0} = \mathbf{S}^T(t)\mathbf{x}(t) + \mathbf{R}(t)\mathbf{u}(t) + \mathbf{B}^T(t)\boldsymbol{\lambda}(t), \tag{2.155}$$

the last of which is solved for the optimal control as follows:

$$\mathbf{u}(t) = -\mathbf{R}^{-1}(t)\left[\mathbf{S}^T(t)\mathbf{x}(t) + \mathbf{B}^T(t)\boldsymbol{\lambda}(t)\right]. \tag{2.156}$$

Substitution of equation (2.156) into equations (2.145) and (2.153) results in the following set of linear state and co-state equations:

$$\dot{\mathbf{x}} = \left[\mathbf{A}(t) - \mathbf{B}(t)\mathbf{R}^{-1}(t)\mathbf{S}^T(t)\right]\mathbf{x}(t) - \mathbf{B}(t)\mathbf{R}^{-1}(t)\mathbf{B}^T(t)\boldsymbol{\lambda}(t), \tag{2.157}$$

$$\dot{\boldsymbol{\lambda}} = -\left[\mathbf{A}^T(t) - \mathbf{S}(t)\mathbf{R}^{-1}(t)\mathbf{B}^T(t)\right]\boldsymbol{\lambda}(t) + \left[\mathbf{S}(t)\mathbf{R}^{-1}(t)\mathbf{S}^T(t) - \mathbf{Q}(t)\right]\mathbf{x}(t), \tag{2.158}$$

which must be solved subject to the boundary conditions,

$$\mathbf{x}(t_0) = \mathbf{x}_0, \quad \boldsymbol{\lambda}(t_\mathrm{f}) = \mathbf{Q}_\mathrm{f}\mathbf{x}(t_\mathrm{f}). \tag{2.159}$$

The linear 2PBVP given by equations (2.157)–(2.159) must be integrated in time by an iterative, numerical method such that the boundary conditions are satisfied. However, since the state and co-state vectors of the linear system are related by equation (2.154) at the final time, we look for a solution to equations (2.157)–(2.159) that satisfies a *transition matrix* ensuring linear independence of solutions, $\mathbf{x}(t)$ and $\boldsymbol{\lambda}(t)$. Thus, we write the combined state and co-state equations as

$$\begin{Bmatrix} \dot{\mathbf{x}} \\ \dot{\boldsymbol{\lambda}} \end{Bmatrix} = \begin{bmatrix} \mathbf{A} - \mathbf{B}\mathbf{R}^{-1}\mathbf{S}^T & -\mathbf{B}\mathbf{R}^{-1}\mathbf{B}^T \\ \mathbf{S}\mathbf{R}^{-1}\mathbf{S}^T - \mathbf{Q} & -\mathbf{A}^T + \mathbf{S}\mathbf{R}^{-1}\mathbf{B}^T \end{bmatrix} \begin{Bmatrix} \mathbf{x} \\ \boldsymbol{\lambda} \end{Bmatrix}, \tag{2.160}$$

which must satisfy the boundary conditions of equation (2.159). We would like to begin the solution by integrating backward in time from $t = t_f$, for which we write

$$\begin{Bmatrix} \mathbf{x}(t) \\ \boldsymbol{\lambda}(t) \end{Bmatrix} = \begin{bmatrix} \mathbf{X}(t) & \mathbf{0} \\ \mathbf{0} & \boldsymbol{\Lambda}(t) \end{bmatrix} \begin{Bmatrix} \mathbf{x}(t_f) \\ \boldsymbol{\lambda}(t_f) \end{Bmatrix}, \tag{2.161}$$

where $\mathbf{X}(t)$ and $\boldsymbol{\Lambda}(t)$ are transition matrices corresponding to the backward evolution in time of $\mathbf{x}(t)$ and $\boldsymbol{\lambda}(t)$, respectively. Clearly, we must have

$$\boldsymbol{\lambda}(t) = \boldsymbol{\Lambda}(t)\boldsymbol{\lambda}(t_f) = \boldsymbol{\Lambda}(t)\mathbf{Q}_f\mathbf{x}(t_f), \tag{2.162}$$

$$\mathbf{x}(t) = \mathbf{X}(t)\mathbf{x}(t_f), \tag{2.163}$$

and

$$\mathbf{x}(t_0) = \mathbf{X}(t_0)\mathbf{x}(t). \tag{2.164}$$

Inverting equation (2.163) and substituting it into equation (2.162), we have

$$\boldsymbol{\lambda}(t) = \boldsymbol{\Lambda}(t)\mathbf{Q}_f\mathbf{X}^{-1}(t)\mathbf{x}(t). \tag{2.165}$$

Since both the transition matrices must satisfy

$$\mathbf{X}(t_f) = \mathbf{I}, \quad \boldsymbol{\Lambda}(t_f) = \mathbf{I}, \tag{2.166}$$

equation (2.165) represents the *adjoint relationship* between the solutions, $\mathbf{x}(t)$ and $\boldsymbol{\lambda}(t)$, written as

$$\boldsymbol{\lambda}(t) = \mathbf{P}(t)\mathbf{x}(t), \tag{2.167}$$

where $\mathbf{P}(t) = \boldsymbol{\Lambda}(t)\mathbf{Q}_f\mathbf{X}^{-1}(t)$. Substituting equation (2.167) into equations (2.156), we have the linear optimal feedback control law:

$$\mathbf{u}(t) = -\mathbf{R}^{-1}(t)[\mathbf{B}^T(t)\mathbf{P}(t) + \mathbf{S}^T(t)]\mathbf{x}(t). \tag{2.168}$$

Note that equation (2.168) is the same as equation (2.88) which was derived using the HJB equation for a linear time-varying system. Taking the time derivative of equation (2.167) and substituting into equations (2.157), (2.158), and (2.168), we have the following differential equation – called the *matrix Riccati equation* (MRE) – to be satisfied by the matrix, $\mathbf{P}(t)$:

$$\begin{aligned} -\dot{\mathbf{P}} &= \mathbf{Q} + (\mathbf{A} - \mathbf{B}\mathbf{R}^{-1}\mathbf{S}^T)^T\mathbf{P} + \mathbf{P}(\mathbf{A} - \mathbf{B}\mathbf{R}^{-1}\mathbf{S}^T) \\ &\quad -\mathbf{P}\mathbf{B}\mathbf{R}^{-1}\mathbf{B}^T\mathbf{P} - \mathbf{S}\mathbf{R}^{-1}\mathbf{S}^T, \end{aligned} \tag{2.169}$$

which is the same as equation (2.87) derived by the HJB equation, and must be solved subject to the boundary condition,

$$\mathbf{P}(t_f) = \mathbf{Q}_f. \tag{2.170}$$

Figure 2.11 is a block diagram of the linear quadratic regulator based upon state feedback for tracking a nominal, optimal trajectory, $\boldsymbol{\xi}_d(t)$, and is guaranteed to result in a neighboring optimal trajectory,

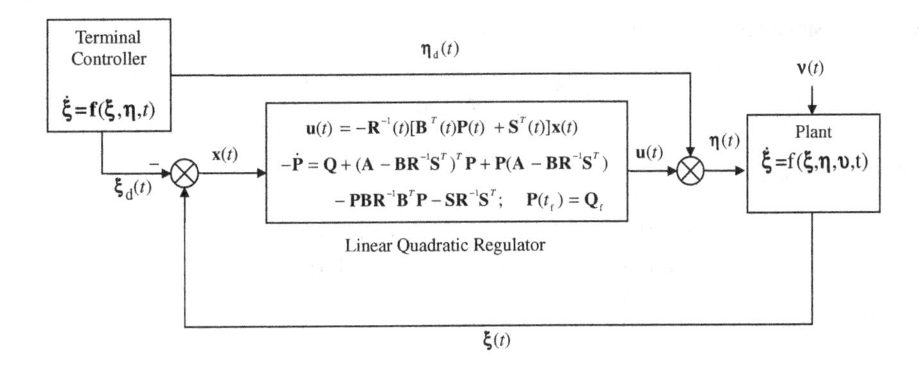

Figure 2.11 A linear quadratic regulator for tracking a nominal, optimal trajectory

$\boldsymbol{\xi}_d(t) + \mathbf{x}(t)$. Since the MRE is derived both from the necessary conditions (Euler–Lagrange equations) and the sufficient condition (HJB equation) of optimality, it must reflect both necessary and sufficient conditions for the existence of an optimal control law for linear systems with a quadratic performance index. As discussed earlier, a sufficient condition for optimality is the existence of a positive definite solution to the MRE, $\mathbf{P}(t)$, for all times in the control interval, $t_0 \leq t \leq t_f$.

The MRE is at the heart of modern control theory (Abou-Kandil *et al.* 2003), and can be regarded as a simplified form of the nonlinear 2PBVP in time represented by the Euler–Lagrange equations, or the nonlinear partial differential equations in space and time derived through the HJB formulation. However, being nonlinear in character, the solution procedure to a general MRE is only slightly simpler, often requiring iterative numerical methods similar to the nonlinear 2PBVP. Although simple iterative schemes based upon repeated linear, algebraic solutions are usually applied when the coefficient matrices are either slowly or periodically varying, the convergence to a positive definite (or even positive semi-definite) solution is not always guaranteed. Given the complexity of an MRE solution, it is much easier to directly solve the linear state and co-state equations (2.157)–(2.159), by an iterative 2PBVP procedure such as the shooting or collocation method. The linearity of the adjoint system of equations (2.160) ensures the existence of a transition matrix, $\mathbf{\Phi}(t, t_0)$, such that

$$\begin{Bmatrix} \mathbf{x}(t) \\ \boldsymbol{\lambda}(t) \end{Bmatrix} = \mathbf{\Phi}(t, t_0) \begin{Bmatrix} \mathbf{x}(t_0) \\ \boldsymbol{\lambda}(t_0) \end{Bmatrix}, \tag{2.171}$$

with the boundary conditions

$$\mathbf{x}(t_0) = \mathbf{x}_0, \quad \boldsymbol{\lambda}(t_f) = \mathbf{Q}_f \mathbf{x}(t_f). \tag{2.172}$$

The transition matrix has the usual properties (Chapter 1) of

$$\mathbf{\Phi}(t_0, t_0) = \mathbf{I}, \quad \mathbf{\Phi}(t_0, t) = \mathbf{\Phi}^{-1}(t, t_0), \tag{2.173}$$

and

$$\dot{\mathbf{\Phi}}(t, t_0) = \begin{bmatrix} \mathbf{A} - \mathbf{B}\mathbf{R}^{-1}\mathbf{S}^T & -\mathbf{B}\mathbf{R}^{-1}\mathbf{B}^T \\ \mathbf{S}\mathbf{R}^{-1}\mathbf{S}^T - \mathbf{Q} & -\mathbf{A}^T + \mathbf{S}\mathbf{R}^{-1}\mathbf{B}^T \end{bmatrix} \mathbf{\Phi}(t, t_0), \tag{2.174}$$

as well as the special property of being a *symplectic matrix*,[5] that is,

$$\boldsymbol{\Phi}^T(t, t_0)\begin{pmatrix} \mathbf{0} & \mathbf{I} \\ -\mathbf{I} & \mathbf{0} \end{pmatrix}\boldsymbol{\Phi}(t, t_0) = \begin{pmatrix} \mathbf{0} & \mathbf{I} \\ -\mathbf{I} & \mathbf{0} \end{pmatrix}. \tag{2.175}$$

Partitioning $\boldsymbol{\Phi}(t, t_0)$ as

$$\begin{Bmatrix} \mathbf{x}(t) \\ \boldsymbol{\lambda}(t) \end{Bmatrix} = \begin{bmatrix} \boldsymbol{\Phi}_{xx}(t, t_0) & \boldsymbol{\Phi}_{x\lambda}(t, t_0) \\ \boldsymbol{\Phi}_{\lambda x}(t, t_0) & \boldsymbol{\Phi}_{\lambda\lambda}(t, t_0) \end{bmatrix} \begin{Bmatrix} \mathbf{x}(t_0) \\ \boldsymbol{\lambda}(t_0) \end{Bmatrix}, \tag{2.176}$$

the symplectic nature of the transition matrix implies that

$$\boldsymbol{\Phi}^{-1}(t, t_0) = \begin{bmatrix} \boldsymbol{\Phi}^T_{\lambda\lambda}(t, t_0) & -\boldsymbol{\Phi}^T_{x\lambda}(t, t_0) \\ -\boldsymbol{\Phi}^T_{\lambda x}(t, t_0) & \boldsymbol{\Phi}^T_{xx}(t, t_0) \end{bmatrix}, \tag{2.177}$$

which is very useful in carrying out the matrix operations required for the solution of the 2PBVP.

In order to prove that $\boldsymbol{\Phi}(t, t_0)$ is symplectic, we note that $\boldsymbol{\Phi}(t_0, t_0) = \mathbf{I}$ satisfies $\boldsymbol{\Phi}^T\mathbf{J}\boldsymbol{\Phi} = \mathbf{J}$. Therefore, if it can be shown that

$$\frac{\mathrm{d}}{\mathrm{d}t}(\boldsymbol{\Phi}^T\mathbf{J}\boldsymbol{\Phi}) = \mathbf{0}, \tag{2.178}$$

then the identity $\boldsymbol{\Phi}^T\mathbf{J}\boldsymbol{\Phi} = \mathbf{J}$ is satisfied for all times. Making use of equation (2.174), which we rewrite as

$$\dot{\boldsymbol{\Phi}} = \begin{pmatrix} \mathbf{E} & \mathbf{F} \\ \mathbf{G} & -\mathbf{E}^T \end{pmatrix}\boldsymbol{\Phi}, \tag{2.179}$$

where

$$\begin{aligned} \mathbf{E} &= \mathbf{A} - \mathbf{B}\mathbf{R}^{-1}\mathbf{S}^T, \\ \mathbf{F} &= -\mathbf{B}\mathbf{R}^{-1}\mathbf{B}^T, \\ \mathbf{G} &= \mathbf{S}\mathbf{R}^{-1}\mathbf{S}^T - \mathbf{Q}, \end{aligned} \tag{2.180}$$

we have

$$\begin{aligned} \frac{\mathrm{d}}{\mathrm{d}t}(\boldsymbol{\Phi}^T\mathbf{J}\boldsymbol{\Phi}) &= \dot{\boldsymbol{\Phi}}^T\mathbf{J}\boldsymbol{\Phi} + \boldsymbol{\Phi}^T\mathbf{J}\dot{\boldsymbol{\Phi}} \\ &= \boldsymbol{\Phi}^T\begin{pmatrix} \mathbf{G} - \mathbf{G}^T & \mathbf{E}^T - \mathbf{E}^T \\ \mathbf{E} - \mathbf{E} & \mathbf{F} - \mathbf{F}^T \end{pmatrix}\boldsymbol{\Phi} = \mathbf{0}, \end{aligned} \tag{2.181}$$

because $\mathbf{F}, \mathbf{G}$ are symmetric matrices. Thus, we have proved that $\boldsymbol{\Phi}(t, t_0)$ is indeed symplectic.

---

[5] A symplectic matrix, $\mathbf{A}$, satisfies $\mathbf{A}^T\mathbf{J}\mathbf{A} = \mathbf{J}$, where

$$\mathbf{J} = \begin{pmatrix} \mathbf{0} & \mathbf{I} \\ -\mathbf{I} & \mathbf{0} \end{pmatrix}$$

is analogous to the *imaginary* number, $\sqrt{-1}$, in complex algebra since $\mathbf{J}^2 = -\mathbf{I}$.

2.10 Stochastic Processes

When process and measurement noise are present in a control system, the closed-loop performance is adversely affected unless the noise signals can be adequately filtered out of the system. The control system must, therefore, have an inbuilt *robustness* in order to perform well in the presence of the unwanted disturbances. However, there is bound to be a deterioration in the control system's performance if the controller is designed by optimal control techniques that do not account for the disturbance signals. Such a deterioration in performance is twofold, being caused by (a) the uncertainties in the plant model (process noise) that prevent the controller from performing as desired in controlling the plant, and (b) errors in measuring the output signals (measurement noise) that get amplified by a feedback loop, thereby applying erroneous input signals to the plant. If the effects of the process and measurement noise can be minimized while designing the controller, the resulting control system will be robust with respect to the expected noise inputs.

This brings us to the question: what are the *expected* disturbances and how can they be modeled? Unfortunately, the disturbance signals are usually indeterminate, that is, they cannot be precisely predicted in advance by a mathematical model. If they were predictable, they would certainly be a part of the plant model and, therefore, we would have neither process nor measurement noise in a control system. Actually, our inability to model physical phenomena that give rise to the process and measurement noise prevents us from precisely controlling a plant as we would like. However, we are not completely in the dark and can generally form rough statistical estimates of a disturbance signal through observation. It is like trying to predict the outcome of a coin toss by having carefully observed a large number of coin tosses in the past. Such an estimate, though imprecise, gives us an idea of what sort of disturbance signal can be expected in future. The mathematically unpredictable signals can be classified as *deterministic* but imprecisely measurable, non-deterministic (or *stochastic*), and completely *random* (having no description whatever), as discussed in Chapter 1 of Tewari (2002) or in Papoulis (1991). However, here we shall interchangeably apply the terms *stochastic* and *random* to describe all phenomena that do not have a precise mathematical model. Such processes are modeled by statistical methods, and an entire branch of mathematics – called *probability theory* (Papoulis 1991) – has evolved to try to predict the outcome of future stochastic events by past observation. In probability theory, one assigns a ratio, $0 \leq p(x) \leq 1$, called the *probability* (or the chance) to the occurrence of a specific outcome identified by a unique number, x, of a future event. The probability is determined from an observation of a total of N events in the past. If the number x turns up in, say, n events, then we have

$$p(x) = \frac{n}{N}. \tag{2.182}$$

The larger the number of observations, N (called the *sample size*), the more accurately the probability of a specific outcome of the event occurring in the future can be determined. A child having observed a total of 100 coin tosses sees that *heads* comes up 47 times. Therefore, he would say that the probability of *heads* occurring in the 101st toss is 0.47. Since a coin toss has only two possible outcomes (either *heads* or *tails*), the sum of the respective probabilities of *heads* and *tails*, $p(x_1)$ and $p(x_2)$ respectively, must be unity. This implies that there is a full probability (or certainty) that the next coin toss will come up with either *heads* or *tails*. Generalizing for an event with M possible outcomes, we write

$$\sum_{i=1}^{M} p(x_i) = \frac{1}{N}\sum_{i=1}^{M} n_i = 1. \tag{2.183}$$

If x is a continuous variable that can take on any value between x_1 and x_2, we define a *probability density function*, $p(x)$, as

$$\int_{x_1}^{x_2} p(x)\mathrm{d}x = 1. \tag{2.184}$$

Equation (2.184) can be generalized by expanding the lower and upper limits of integration to $-\infty$ and ∞, respectively, and noting that $p(x) = 0$ whenever x lies outside its feasible range, $x_1 \leq x \leq x_2$:

$$\int_{-\infty}^{\infty} p(x)\mathrm{d}x = 1. \tag{2.185}$$

Examples of continuous random variables are volume of fuel in a tank at a given time and the direction and intensity of the wind at any given place and time. The *expected value* of a continuous, random variable, x, with a probability density function, $p(x)$, is given by

$$E(x) = \int_{-\infty}^{\infty} x p(x)\mathrm{d}x = \bar{x}, \tag{2.186}$$

where $\bar{x}$, called the *mean value* of x, is the same as its expected value. It is clear from the definition of expected value that

$$E(x - \bar{x}) = \int_{-\infty}^{\infty} x p(x)\mathrm{d}x - \bar{x}\int_{-\infty}^{\infty} p(x)\mathrm{d}x = \bar{x} - \bar{x} = 0. \tag{2.187}$$

One can define the *variance* of a continuous random variable from its mean value as follows:

$$\sigma^2 = E[(x - \bar{x})^2] = \int_{-\infty}^{\infty} (x - \bar{x})^2 p(x)\mathrm{d}x. \tag{2.188}$$

The square root of variance, σ, is called the *standard deviation* of x from its mean value, and has the same units as x.

When dealing with groups of random variables, such as a pair (x, y), we write the *joint probability* of both x and y occurring simultaneously as $p(x, y)$. If x and y are unrelated outcomes of separate, random events (e.g., rolling a pair of dice) then we have

$$p(x, y) = p(x)p(y), \tag{2.189}$$

which is less than (or equal to) both $p(x)$ and $p(y)$. If x and y are continuous random variables (unrelated or not), then we have

$$\int_{-\infty}^{\infty}\int_{-\infty}^{\infty} p(x, y)\mathrm{d}x\mathrm{d}y = 1. \tag{2.190}$$

We can also define a *conditional probability* of x *given* y,

$$p(x|y) = \frac{p(x, y)}{p(y)}, \tag{2.191}$$

from which we note that both conditional and unconditional probabilities of x given y would be the same if x and y are unrelated variables. We also have the following interesting relationship between the conditional probabilities of x given y and y given x, called *Bayes's rule*:

$$p(y|x) = \frac{p(x|y)p(y)}{p(x)}. \tag{2.192}$$

When considering a sum of random variables, x and y,

$$z = x + y, \tag{2.193}$$

we have the following relationships between the respective probabilities:

$$p(z|x) = p(z - x) = p(y) \tag{2.194}$$

and

$$p(z) = \int_{-\infty}^{\infty} p(z|x)p(x)\mathrm{d}x = \int_{-\infty}^{\infty} p(z - x)p(x)\mathrm{d}x. \tag{2.195}$$

The mean and variance of the sum are given by

$$E(z) = \bar{z} = E(x + y) = E(x) + E(y) = \bar{x} + \bar{y} \tag{2.196}$$

and

$$\sigma_z^2 = E[(z - \bar{z})^2] = E[(x - \bar{x})^2] + E[(y - \bar{y})^2] = \sigma_x^2 + \sigma_y^2. \tag{2.197}$$

These results can be extended to a sum of any number of random variables, or even to their linearly weighted sums. There is an important law in statistics called the *central limit theorem* which states that a sum of a large number of random variables (having various arbitrary, unconditional probability density functions) approaches a specific probability density function called the *normal* (or *Gaussian*) *distribution*, given by

$$p(x) = \frac{1}{\sqrt{2\pi}\sigma} e^{-\frac{(x-\bar{x})^2}{2\sigma^2}}. \tag{2.198}$$

Here $\bar{x}$ and σ denote the mean and standard deviation of the normal (Gaussian) distribution, respectively. By using the central limit theorem, a controls engineer can routinely represent a random disturbance by a sum of a large number of random signals that can be regarded as outputs of various, unrelated stochastic processes. By having a simple probability density function, equation (2.198), for the resulting Gaussian noise signal, its effect on the control system can be easily analyzed. A plot of the Gaussian probability density function for $\bar{x} = 0.5$ and $\sigma = 0.6$ is shown in Figure 2.12, displaying the famous bell-shaped

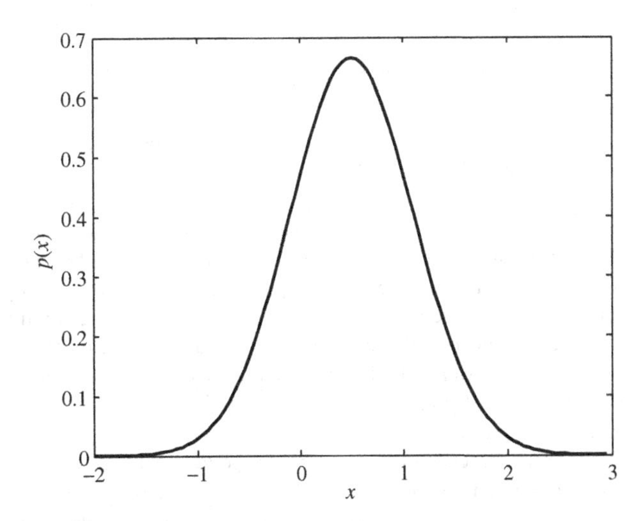

Figure 2.12 Gaussian probability density function for $\bar{x} = 0.5$ and $\sigma = 0.6$

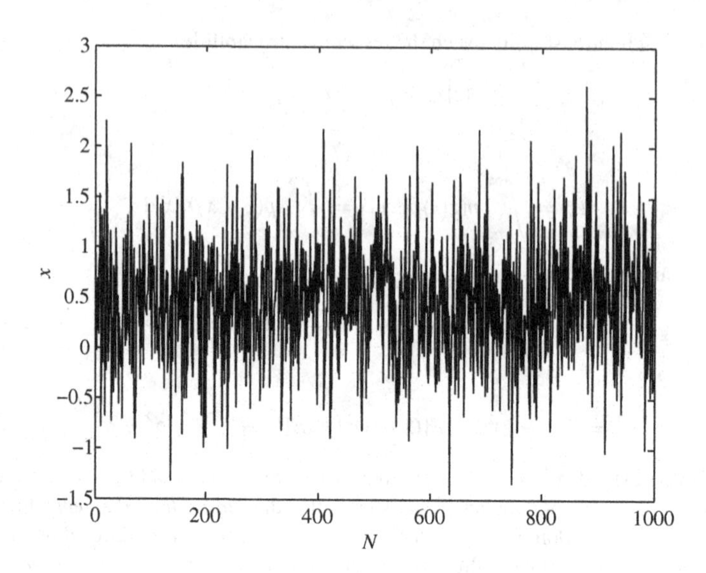

Figure 2.13 Random sequence of 1000 points with a Gaussian distribution of $\bar{x} = 0.5$ and $\sigma = 0.6$

curve. An example of the kind of signal such a distribution would produce is obtained by MATLAB's inbuilt normally distributed random number generator called *randn.m* invoked for 1000 data points and plotted in Figure 2.13 by the following statements:

```
>> r=0.5+0.6*randn(1000,1); %Normal distribution(mean=0.5, s.d.=0.6)
>> plot(r);
```

For a Gaussian process, the probability that x lies between the values $\bar{x} - \xi$ and $\bar{x} + \xi$, (i.e., in an *error band* of $\pm\xi$ about the mean) is given by the following integral:

$$\int_{\bar{x}-\xi}^{\bar{x}+\xi} p(x)\mathrm{d}x = \frac{1}{\sqrt{2\pi}\sigma}\int_{-\xi}^{\xi} e^{-\frac{\eta^2}{2\sigma^2}}\mathrm{d}\eta = \frac{2}{\sqrt{\pi}}\int_0^{\frac{\xi}{\sigma\sqrt{2}}} e^{-u^2}\mathrm{d}u = \mathrm{erf}\left(\frac{\xi}{\sigma\sqrt{2}}\right), \tag{2.199}$$

which is called the *error function*. Values of the error function commonly used in statistical analysis are $\mathrm{erf}(1/\sqrt{2}) = 0.68268949$, $\mathrm{erf}(\sqrt{2}) = 0.95449974$, and $\mathrm{erf}(3/\sqrt{2}) = 0.99730020$ corresponding to $\xi = \sigma$, 2σ, and 3σ, respectively.

Thus far we have concerned ourselves with taking the statistical measures (mean, variance, etc.) of variables that can assume random values at a given time instant. Such snapshot measures are termed *ensemble properties*. For example, an ensemble average (mean) of ages of all the employees in a business firm is found by sampling the respective ages of the various employees at a given time. The variation of the average age with time is neglected. However, if one were to take the ensemble average annually, and then take an average of the data over several years with time, the result will better reflect the average age of an employee in the particular firm at *any* given time. Therefore, time-varying signals essentially require the study of statistical properties over a period of time.

There are special random processes having ensemble properties that are constant with time. Such processes are said to be *stationary*. There are also special stationary random processes – called *ergodic processes* – whose properties sampled over time give rise to exactly the same probability density functions as are obtained by taking the ensemble average. For example, if the age of an employee in a particular business firm were an ergodic phenomenon, it would make no difference to the result whether we take the time average of age of a particular employee over several years, or if we take an ensemble average of ages of all employees that happen to be serving at the present time. One would be hard put to find even one practical example of a stationary (let alone ergodic) stochastic process. Therefore, stationarity and ergodicity are much more useful as mathematical tools than as characteristics of physical phenomena. However, we regularly commit the error of replacing the ensemble average with a time average for processes that are known to be non-ergodic, just because taking an ensemble average would be very difficult (if not impossible). An example of this fact is taking the time-averaged temperature of a particular city from historical data and using it to predict its present temperature by constructing a probability density function. To take the ensemble average at a given time, one would be required to sample the current temperatures of a reasonably large number of cities that have latitude, elevation, distance from sea, population, etc., identical to the chosen city (which is clearly impossible).

There is an interesting property called *correlation* of two (or more) time-varying, random signals defined as the expected value of the product of the two signals, $x(t)$ and $y(t)$, evaluated at different times,

$$\psi_{xy}(t, \tau) = E[x(t)y(\tau)], \quad t \neq \tau. \tag{2.200}$$

If $E[x(t)y(\tau)] = 0$ for all arbitrary times, (t, τ), then $x(t)$ and $y(t)$ are said to be *uncorrelated*. The degree of correlation of a signal with itself at different time instants is defined by the *autocorrelation function*,

$$\psi_{xx}(t, \tau) = E[x(t)x(\tau)], \quad t \neq \tau. \tag{2.201}$$

A random signal, $x(t)$, which is totally uncorrelated with itself at different values of time, that is, where $\psi_{xx}(t, \tau) = 0, \ t \neq \tau$, is called *white noise* and has a special place in control theory, as we shall see later. For a white noise, we have

$$p[x(t)] = p[x(t)|x(\tau)], \quad t \neq \tau, \tag{2.202}$$

that is, the conditional probability of the signal at the present time given its value at some other time is no different from the unconditional probability of the signal at the present time. In other words, knowing the value of the signal at some time does not help us in any way to predict (or determine) its value at any other time. Therefore, in order to find the joint probability of a white noise process evolving from $t = t_0$ with a given probability, $p[x(t_0)]$, to the present time, t, one must know the individual, unconditional probabilities at all times in the interval,

$$\begin{aligned} &p[x(t), x(t-T), x(t-2T), \ldots, x(t_0+T), x(t_0)] \\ &\quad = p[x(t)]p[x(t-T)]p[x(t-2T)]\ldots p[x(t_0+T)]p[x(t_0)], \end{aligned} \tag{2.203}$$

where T is a *sampling interval*, which is a formidable task. For this reason, a white noise process is said to be completely unpredictable.

In contrast to white noise, a signal that is correlated with itself at different values of time, that is, $\psi_{xx}(t, \tau) \neq 0, \ t \neq \tau$, is called *colored noise*. An example of colored noise is a *Markov process*, $x(t)$, beginning at $t = t_0$ with probability, $p[x(t_0)]$, and defined by the following property:

$$\begin{aligned} &p[x(t), x(t-T), x(t-2T), \ldots, x(t_0+T), x(t_0)] \\ &\quad = p[x(t)|x(t-T)]p[x(t-T)|x(t-2T)]\ldots p[x(t_0+T)|x(t_0)]p[x(t_0)]. \end{aligned} \tag{2.204}$$

Thus the joint probability of a Markov process evolving from $t = t_0$ with a given probability, $p[x(t_0)]$, to the present time, t, only depends upon the product of conditional probabilities, $p[x(\tau)|x(\tau - T)]$, $t_0 + T \leq \tau \leq t$, of evolving over one sampling interval (called the *transitional probability*). Clearly, the value of a Markov signal at a given time, $x(t)$, can be predicted from the transitional probabilities,

$p[x(\tau)|x(\tau - T)]$, $t_0 + T \leq \tau \leq t$, and the initial probability, $p[x(t_0)]$. The simplest way of generating a Markov process is by passing a white noise through a linear, time-varying system (called a linear *filter*). The state equation of a scalar Markov process can thus be written as

$$\dot{x}(t) = a(t)x(t) + b(t)w(t), \tag{2.205}$$

where $w(t)$ is a white noise process and $a(t)$, $b(t)$ are filter coefficients.

A Markov process that has a Gaussian probability distribution is termed a *Gauss–Markov process*. It can be generated by passing a Gaussian white noise through a linear filter (equation (2.205)) with a Gaussian initial state, $x(t_0)$.

By subtracting the (ensemble) mean values of the signals in the correlation and autocorrelation functions, one can define *cross-covariance* and *autocovariance* functions as

$$\phi_{xy}(t, \tau) = E[\{x(t) - \bar{x}(t)\}\{y(\tau) - \bar{y}(\tau)\}], \quad t \neq \tau, \tag{2.206}$$

and

$$\phi_{xx}(t, \tau) = E[\{x(t) - \bar{x}(t)\}\{x(\tau) - \bar{x}(\tau)\}], \quad t \neq \tau, \tag{2.207}$$

respectively. Clearly, the variance of a signal is simply obtained by putting $t = \tau$:

$$\phi_{xx}(t, t) = E[\{x(t) - \bar{x}(t)\}^2] = \sigma_x^2. \tag{2.208}$$

Similarly, we can define the *covariance* of two signals by putting $t = \tau$ in the cross-covariance function:

$$\phi_{xy}(t, t) = E[\{x(t) - \bar{x}(t)\}\{y(t) - \bar{y}(t)\}] = \sigma_{xy}. \tag{2.209}$$

For stationary processes, the actual time, t, is immaterial and the results depend only upon the *time-shift*, α, where $\tau = t + \alpha$:

$$\begin{aligned}
\psi_{xy}(\alpha) &= E[x(t)y(t + \alpha)], \\
\psi_{xx}(\alpha) &= E[x(t)x(t + \alpha)], \\
\phi_{xy}(\alpha) &= E[\{x(t) - \bar{x}(t)\}\{y(t + \alpha) - \bar{y}(t + \alpha)\}], \\
\phi_{xx}(\alpha) &= E[\{x(t) - \bar{x}(t)\}\{x(t + \alpha) - \bar{x}(t + \alpha)\}].
\end{aligned} \tag{2.210}$$

The concepts of probability can be extended to vectors of random variables. However, there is a qualitative difference between a scalar random variable and a vector random variable in that the joint probability, $p(\mathbf{x})$, of a vector taking a specific value, say $\mathbf{x} = \mathbf{x}_0$, is much smaller than the unconditional probability of any of its elements, x_i, individually assuming a given value, say x_0, irrespective of the values taken by the remaining elements. Consider the example of drawing 10 identical coins numbered 1 to 10 from a pocket. The probability of drawing, say, the coin numbered 6 at the first attempt is 1/10, while the joint probability of blindly drawing a particular sequence of coins, say 2, 4, 9, 5, 10, 7, 3, 6, 1, 8, is $1/10^{10}$ (one in 10 billion). A vector of continuous, random variables, $\mathbf{x} = (x_1, x_2, \ldots, x_n)^T$, is said to have a *joint probability density* function, $p(\mathbf{x})$, which satisfies the identity

$$\int_{-\infty}^{\infty} \cdots \int_{-\infty}^{\infty} \int_{-\infty}^{\infty} p(\mathbf{x}) \mathrm{d}x_1 \mathrm{d}x_2 \ldots \mathrm{d}x_n = 1 \tag{2.211}$$

and has a mean value given by

$$E(\mathbf{x}) = \bar{\mathbf{x}} = \int_{-\infty}^{\infty} \cdots \int_{-\infty}^{\infty} \int_{-\infty}^{\infty} \mathbf{x} p(\mathbf{x}) \mathrm{d}x_1 \mathrm{d}x_2 \ldots \mathrm{d}x_n. \tag{2.212}$$

An example of a continuous, random vector is a collection of position coordinates, $\mathbf{r} = (x, y, z)^T$ and velocity components, $\mathbf{v} = (u, v, w)^T$, of the center of mass at any given time of an aircraft flying through atmospheric turbulence:

$$\mathbf{x} = (x, y, z, u, v, w)^T. \tag{2.213}$$

The state vector of any physical, stochastic process, $\mathbf{x}(t)$, can be regarded as a random vector whose elements continuously vary with time. It's mean at a given time is given by

$$\bar{\mathbf{x}}(t) = E[\mathbf{x}(t)], \tag{2.214}$$

while the statistical correlation among the elements of a stochastic vector signal, $\mathbf{x}(t)$, is measured by the *correlation matrix* defined by

$$\mathbf{R}_\mathbf{x}(t, \tau) = E[\mathbf{x}(t)\mathbf{x}^T(\tau)]. \tag{2.215}$$

A diagonal correlation matrix indicates that the state variables of a stochastic process are uncorrelated with one another, although they may be autocorrelated with themselves at different times. Putting $t = \tau$ in the correlation matrix produces the *covariance matrix*

$$\mathbf{R}_\mathbf{x}(t, t) = E[\mathbf{x}(t)\mathbf{x}^T(t)], \tag{2.216}$$

which, by definition, is symmetric.

2.10.1 Stationary Random Processes

If we regard all stationary processes as ergodic (which is a big leap, as discussed above), we can easily find the mean and correlation matrix of a stationary process by taking the time averages (rather than ensemble averages) as follows:

$$\bar{\mathbf{x}} = \lim_{T\to\infty} \frac{1}{T} \int_{-\frac{T}{2}}^{\frac{T}{2}} \mathbf{x}(t)\mathrm{d}t, \tag{2.217}$$

$$\mathbf{R}_\mathbf{x}(\alpha) = \lim_{T\to\infty} \frac{1}{T} \int_{-\frac{T}{2}}^{\frac{T}{2}} \mathbf{x}(t)\mathbf{x}^T(t+\alpha)\mathrm{d}t, \tag{2.218}$$

where α is the time-shift. Furthermore, we define a *power spectral density matrix* by taking the *Fourier transform* (Kreyszig 1998) of the correlation matrix, which transforms a time-domain, stationary signal to the frequency domain, as follows:

$$\mathbf{S}_\mathbf{x}(\omega) = \int_{-\infty}^{\infty} \mathbf{R}_\mathbf{x}(\alpha)e^{-i\omega\alpha}\mathrm{d}\alpha. \tag{2.219}$$

The power spectral density (PSD) matrix, $\mathbf{S}_\mathbf{x}(\omega)$, represents the frequency content of the stationary vector random signal, $\mathbf{x}(t)$, that is, the distribution of the power (or intensity) of the various state variables over a range of frequencies, ω. If there are peaks in the magnitude of the elements of the PSD matrix, $||\mathbf{S}_\mathbf{x}(\omega)||$, at some frequencies, they would indicate that the stochastic system can be excited to a large response by applying harmonic inputs at those particular frequencies. The Fourier transform of the random signal itself is given by

$$\mathbf{X}(i\omega) = \int_{-\infty}^{\infty} \mathbf{x}(t)e^{-i\omega t}\mathrm{d}t. \tag{2.220}$$

From equations (2.34)–(2.36), we have the following relationship between $\mathbf{S}_\mathbf{x}(\omega)$ and $\mathbf{X}(i\omega)$:

$$\mathbf{S}_\mathbf{x}(\omega) = \mathbf{X}(i\omega)\mathbf{X}^T(-i\omega). \tag{2.221}$$

An inverse Fourier transform produces the correlation matrix from the PSD matrix as follows:

$$\mathbf{R}_\mathbf{x}(\alpha) = \frac{1}{2\pi}\int_{-\infty}^{\infty} \mathbf{S}_\mathbf{x}(\omega)e^{i\omega\alpha}\mathrm{d}\omega, \tag{2.222}$$

whereas the covariance matrix is simply obtained by putting $\alpha = 0$:

$$\mathbf{R}_\mathbf{x}(0) = \frac{1}{2\pi}\int_{-\infty}^{\infty} \mathbf{S}_\mathbf{x}(\omega)e^{i\omega}\mathrm{d}\omega. \tag{2.223}$$

We saw earlier how the central limit theorem predicts that by superimposing a large number of random signals, one can produce a stationary Gaussian signal. If we further specify that all of the superposed signals are uncorrelated with one another and with themselves in time, and also have zero mean values, then what we have is a mathematical construct called *zero-mean Gaussian white noise* (ZMGWN). While it is impossible to find a ZMGWN signal in practice, the concept itself can be used for a greatly simplified analysis and design of control systems. Let $\mathbf{w}(t)$ be a continuous time, vector ZMGWN signal. Then we have

$$\bar{\mathbf{w}} = \lim_{T\to\infty}\frac{1}{T}\int_{-\frac{T}{2}}^{\frac{T}{2}} \mathbf{w}(t)\mathrm{d}t = \mathbf{0} \tag{2.224}$$

and

$$\mathbf{R}_\mathbf{w}(\alpha) = \lim_{T\to\infty}\frac{1}{T}\int_{-\frac{T}{2}}^{\frac{T}{2}} \mathbf{w}(t)\mathbf{w}^T(t+\alpha)\mathrm{d}t = \mathbf{0}, \quad \alpha \neq 0. \tag{2.225}$$

Furthermore, since

$$\mathbf{W}(i\omega) = \int_{-\infty}^{\infty} \mathbf{w}(t)e^{-i\omega t}\mathrm{d}t = \mathbf{W} = \text{const.}, \tag{2.226}$$

we have

$$\mathbf{S}_\mathbf{w}(\omega) = \mathbf{W}(i\omega)\mathbf{W}^T(-i\omega) = \mathbf{W}\mathbf{W}^T = \mathbf{S}_\mathbf{w} = \text{const.} \tag{2.227}$$

Therefore, the PSD matrix of a ZMGWN is a constant matrix. However, if we take the inverse Fourier transform of a constant $\mathbf{S}_\mathbf{w}$, we end up with a covariance matrix that has all elements tending to infinity:

$$\mathbf{R}_\mathbf{w}(0) = \frac{1}{2\pi}\mathbf{S}_\mathbf{w}\int_{-\infty}^{\infty} e^{i\omega}\mathrm{d}\omega \to \infty. \tag{2.228}$$

We resolve this dilemma by writing

$$\mathbf{R}_\mathbf{w}(\alpha) = \mathbf{S}_\mathbf{w}\delta(\alpha), \tag{2.229}$$

where $\delta(\alpha)$ is the following *Dirac delta* (or unit impulse) function:

$$\delta(\alpha) = \begin{cases} \infty, & \alpha = 0, \\ 0, & \alpha \neq 0, \end{cases} \tag{2.230}$$

with the property that

$$\int_{-\infty}^{\infty} \delta(\theta) \mathrm{d}\theta = 1. \tag{2.231}$$

2.10.2 *Filtering of Random Noise*

When a random, scalar signal is passed through a linear filter, its statistical properties are modified. An example is the coloring of a white noise signal as it passes through a linear time-varying filter. Linear filters are commonly employed in order to reduce the effect of noise in a control system. A linear control system has a finite range of frequencies, called *bandwidth*, in which it can effectively respond to applied inputs. The input applied to a control system is often a noisy reference signal that can be described as a linear superposition of a deterministic signal and a random signal. The simplest description of random noise is the zero-mean Gaussian white noise, which, by definition, is generated by an infinite-bandwidth process. Clearly, it is impossible to block an infinite-bandwidth process by a finite bandwidth filter. However, traditional methods of suppressing noise in a certain frequency range have been successfully employed for single-variable systems. In most practical cases, (non-white) noise has a predominantly high-frequency content (i.e., its power spectral density has more peaks at higher frequencies compared with the control system's bandwidth). In order to filter out the high-frequency noise, a special filter called a *low-pass filter* is generally placed before the control system, which blocks all signal frequencies above a specified *cut-off frequency*, $\omega \geq \omega_0$, thereby reducing the bandwidth of the overall system. The output of a low-pass filter thus contains only lower frequencies, which implies a smoothing of the raw random signal fed to the filter. Sometimes it may be prudent to suppress both high- and low-frequency contents of a noisy signal. This is achieved by passing the signal through a *band-pass filter*, which allows only a specified band of frequencies, $\omega_1 \leq \omega \leq \omega_2$, to pass through. Similarly, a *high-pass filter* blocks all frequencies below a specified cut-off frequency, $\omega \leq \omega_0$, and has an output containing only the higher frequencies. The magnitude of a filtered signal above or below a given frequency can be generally made to decay rapidly with frequency. Such a decay of signal magnitudes with frequency is called *attenuation* or *roll-off*. The output of a linear filter not only has a frequency content different from the input signal, but also certain other characteristics of the filter, such as a phase-shift or a change in magnitude. In other words, the desired reference signal passing through a filter is also distorted by the filter, which is undesirable as it leads to a degradation of control system performance. It is thus inevitable that a filter would produce an output signal based upon its own characteristics that are different from the desired closed-loop characteristics. Usually it is observed that a greater attenuation of noise also leads to a greater distortion of the filtered signal (Tewari 2002). However, a filter can be designed to achieve a desired set of filtering objectives such as cut-off frequencies, desired attenuation (roll-off) of signals above or below the cut-off frequencies, etc., while reducing distortion of the reference signals passing through the filter.

Generally, it is impossible for a multi-variable filter to perfectly block the undesirable vector signals as they simultaneously appear at a wide range of frequencies, without significantly degrading control system performance. Therefore, traditional methods of filtering commonly fail for a multivariate control system, and a qualitatively different approach must be adopted for filter design that strikes an optimum trade-off between the conflicting requirements of noise attenuation and minimum signal distortion (or performance degradation).

2.11 Kalman Filter

A linear, deterministic plant becomes a stochastic system in the presence of process and measurement noise. The simplest model for a linear, stochastic plant is obtained by passing white noise through an appropriate linear system. Consider such a plant with state vector, $\mathbf{x}(t)$, and output vector, $\mathbf{y}(t)$, having the following linear, time-varying state-space representation:

$$\dot{\mathbf{x}}(t) = \mathbf{A}(t)\mathbf{x}(t) + \mathbf{B}(t)\mathbf{u}(t) + \mathbf{F}(t)\boldsymbol{\nu}(t),$$
$$\mathbf{y}(t) = \mathbf{C}(t)\mathbf{x}(t) + \mathbf{D}(t)\mathbf{u}(t) + \mathbf{w}(t), \tag{2.232}$$

where $\boldsymbol{\nu}(t)$ is the process noise vector which arises due to modeling errors such as neglecting nonlinear, high-frequency, or stochastic dynamics, and $\mathbf{w}(t)$ is the measurement noise vector. By assuming $\boldsymbol{\nu}(t)$ and $\mathbf{w}(t)$ to be white noises with a zero mean, we can greatly simplify the model of the stochastic plant. Since the time-varying stochastic system is a non-stationary process, the random noises, $\boldsymbol{\nu}(t)$ and $\mathbf{w}(t)$, are also assumed to be non-stationary white noises for generality. A non-stationary white noise can be simply derived by passing a stationary white noise through an amplifier with a time-varying gain. The correlation matrices of non-stationary white noises, $\boldsymbol{\nu}(t)$ and $\mathbf{w}(t)$, are expressed as follows:

$$\mathbf{R}_\nu(t, \tau) = \mathbf{S}_\nu(t)\delta(t - \tau),$$
$$\mathbf{R}_\mathbf{w}(t, \tau) = \mathbf{S}_\mathbf{w}(t)\delta(t - \tau), \tag{2.233}$$

where $\mathbf{S}_\nu(t)$ and $\mathbf{S}_\mathbf{w}(t)$ are the time-varying power spectral density matrices of $\boldsymbol{\nu}(t)$ and $\mathbf{w}(t)$, respectively. Note the infinite-covariance matrices, $\mathbf{R}_\nu(t, t)$ and $\mathbf{R}_\mathbf{w}(t, t)$, respectively.

A full-state feedback control system cannot be designed for a stochastic plant because its state vector, $\mathbf{x}(t)$, is unknown at any given time. Instead, one has to rely upon an *estimated state vector*, $\mathbf{x}_o(t)$, that is derived from the measurement of the output vector, $\mathbf{y}(\tau)$, over a previous, finite time interval, $t_0 \le \tau \le t$. Thus a subsystem of the controller must be dedicated to form an accurate estimate of the state vector from a finite record of the outputs. Such a subsystem is called an *observer* (or state estimator). The performance of the control system depends upon the accuracy and efficiency with which a state estimate can be supplied by the observer to the control system, despite the presence of process and measurement noise.

In order to take into account the fact that the measured outputs as well as the state variables are random variables, we need an observer that estimates the state vector based upon statistical (rather than deterministic) description of the vector output and plant state. Such an observer is the *Kalman filter*. The Kalman filter is an optimal observer that minimizes the covariance matrix of the *estimation error*,

$$\mathbf{e}_o(t) = \mathbf{x}(t) - \mathbf{x}_o(t). \tag{2.234}$$

In order to understand why it may be useful to minimize the covariance of estimation error, we note that the estimated state, $\mathbf{x}_o(t)$, is based on the measurement of the output, $\mathbf{y}(\tau)$, while knowing the applied input vector, $\mathbf{u}(\tau)$, for a finite time, $t_0 \le \tau \le t$. However, being a non-stationary signal, an accurate average of $\mathbf{x}(t)$ would require measuring the output for an infinite time, that is, gathering infinite number of samples from which the expected value of $\mathbf{x}(t)$ could be derived. Therefore, the best estimate that the Kalman filter can obtain for $\mathbf{x}(t)$ is not the true mean, but a *conditional mean*, $\mathbf{x}_m(t)$, based on only a finite time record of the output, written as

$$\mathbf{x}_m(t) = E[\mathbf{x}(t)|\mathbf{y}(\tau), t_0 \le \tau \le t] \tag{2.235}$$

The deviation of the estimated state vector, $\mathbf{x}_o(t)$, from the conditional mean, $\mathbf{x}_m(t)$, is given by

$$\Delta\mathbf{x}(t) = \mathbf{x}_o(t) - \mathbf{x}_m(t). \tag{2.236}$$

The conditional covariance matrix is defined as the covariance matrix based on a finite record of the output and can be written as follows:

$$\mathbf{R}_e(t, t) = E\left[\mathbf{e}_o(t)\mathbf{e}_o^T(t)\,\middle|\,\mathbf{y}(\tau), t_0 \le \tau \le t\right]$$
$$= E\left[\{\mathbf{x}(t) - \mathbf{x}_o(t)\}\left\{\mathbf{x}^T(t) - \mathbf{x}_o^T(t)\right\}\,\middle|\,\mathbf{y}(\tau), t_0 \le \tau \le t\right], \tag{2.237}$$

which is simplified using equations (2.235) and (2.236) as follows:

$$\begin{aligned}\mathbf{R}_e(t,t) &= E[\mathbf{x}(t)\mathbf{x}^T(t)] - \mathbf{x}_o(t)\mathbf{x}_m^T(t) - \mathbf{x}_o^T(t)\mathbf{x}_m(t) + \mathbf{x}_o(t)\mathbf{x}_o^T(t) \\ &= E[\mathbf{x}(t)\mathbf{x}^T(t)] - \mathbf{x}_m(t)\mathbf{x}_m^T(t) + \Delta\mathbf{x}(t)\Delta\mathbf{x}^T(t). \end{aligned} \tag{2.238}$$

It is evident from equation (2.238) that the best estimate of the state vector, $\Delta\mathbf{x}(t) = \mathbf{0}$, would result in a minimization of the conditional covariance matrix, $\mathbf{R}_e(t, t)$. In other words, minimization of $\mathbf{R}_e(t, t)$ yields the optimal (i.e., the best) observer, which is the Kalman filter.

The state equation of the full-order Kalman filter is that of a time-varying observer, and is expressed as

$$\dot{\mathbf{x}}_o(t) = \mathbf{A}_o(t)\mathbf{x}_o(t) + \mathbf{B}_o(t)\mathbf{u}(t) + \mathbf{L}(t)\mathbf{y}(t), \tag{2.239}$$

where $\mathbf{L}(t)$ is the gain matrix and $\mathbf{A}_o(t)$, $\mathbf{B}_o(t)$ the coefficient matrices of the Kalman filter. Substituting equations (2.239) and (2.232) into equation (2.234), we have the following state equation for the estimation error dynamics:

$$\begin{aligned}\dot{\mathbf{e}}_o(t) = {} & \mathbf{A}_o(t)\mathbf{e}_o(t) + [\mathbf{A}(t) - \mathbf{L}(t)\mathbf{C}(t) - \mathbf{A}_o(t)]\mathbf{x}(t) \\ & + [\mathbf{B}(t) - \mathbf{L}(t)\mathbf{D}(t) - \mathbf{B}_o(t)]\mathbf{u}(t) + \mathbf{F}(t)\boldsymbol{\nu}(t) - \mathbf{L}(t)\mathbf{w}(t). \end{aligned} \tag{2.240}$$

In order that the estimation error dynamics be independent of the state and control variables, we require

$$\begin{aligned}\mathbf{A}_o(t) &= \mathbf{A}(t) - \mathbf{L}(t)\mathbf{C}(t), \\ \mathbf{B}_o(t) &= \mathbf{B}(t) - \mathbf{L}(t)\mathbf{D}(t), \end{aligned} \tag{2.241}$$

which yields

$$\dot{\mathbf{e}}_o(t) = [\mathbf{A}(t) - \mathbf{L}(t)\mathbf{C}(t)]\mathbf{e}_o(t) + \mathbf{F}(t)\boldsymbol{\nu}(t) - \mathbf{L}(t)\mathbf{w}(t). \tag{2.242}$$

Given that $\boldsymbol{\nu}(t)$ and $\mathbf{w}(t)$ are white, non-stationary signals with zero mean values, their linear combination is also a white and non-stationary random signal with zero mean:

$$\mathbf{z}(t) = \mathbf{F}(t)\boldsymbol{\nu}(t) - \mathbf{L}(t)\mathbf{w}(t). \tag{2.243}$$

This drives the estimation error dynamics of the Kalman filter:

$$\dot{\mathbf{e}}_o(t) = [\mathbf{A}(t) - \mathbf{L}(t)\mathbf{C}(t)]\mathbf{e}_o(t) + \mathbf{z}(t). \tag{2.244}$$

A solution to the error state equation for a given initial error, $\mathbf{e}_o(t_0)$, and a corresponding initial error covariance, $\mathbf{R}_e(t_0, t_0) = E\left[\mathbf{e}_o(t_0)\mathbf{e}_o^T(t_0)\right] = \mathbf{R}_{e0}$, can be expressed as follows:

$$\mathbf{e}_o(t) = \mathbf{\Phi}(t, t_0)\mathbf{e}_o(t_0) + \int_{t_0}^{t} \mathbf{\Phi}(t, \lambda)\mathbf{z}(\lambda)\mathrm{d}\lambda, \tag{2.245}$$

where $\mathbf{\Phi}(t, t_0)$ is the error transition matrix corresponding to the homogeneous error state equation,

$$\dot{\mathbf{e}}_o(t) = [\mathbf{A}(t) - \mathbf{L}(t)\mathbf{C}(t)]\mathbf{e}_o(t) = \mathbf{A}_o(t)\mathbf{e}_o(t). \tag{2.246}$$

The conditional error covariance is then derived (dropping the conditional notation for simplicity) by noting that $\mathbf{z}(t)$ is a zero mean, white noise, i.e., $E[\mathbf{z}(t)] = \mathbf{0}$ and

$$\mathbf{R}_z(t, \tau) = \mathbf{S}_z(t)\delta(t - \tau), \tag{2.247}$$

where $\mathbf{S}_z(t)$ is the time-varying power spectral density matrix of $\mathbf{z}(t)$, resulting in the following (for details, see Chapter 7 of Tewari (2002)):

$$\begin{aligned}\mathbf{R}_e(t,t) &= E[\mathbf{e}_o(t)\mathbf{e}_o^T(t)] \\ &= \mathbf{\Phi}(t,t_0)\mathbf{R}_e(t_0,t_0)\mathbf{\Phi}^T(t,t_0) + \int_{t_0}^{t} \mathbf{\Phi}(t,\lambda)\mathbf{S}_z(\lambda)\mathbf{\Phi}^T(t,\lambda)\mathrm{d}\lambda.\end{aligned} \tag{2.248}$$

Differentiation of equation (2.248) with respect to time, along with the fact that the error transition matrix satisfies its own homogeneous state equation,

$$\frac{\mathrm{d}}{\mathrm{d}t}\{\mathbf{\Phi}(t,t_0)\} = \mathbf{A}_o(t)\mathbf{\Phi}(t,t_0), \tag{2.249}$$

yields the following matrix Riccati equation (MRE) for the optimal error covariance matrix:

$$\frac{\mathrm{d}}{\mathrm{d}t}\{\mathbf{R}_e(t,t)\} = \mathbf{A}_o(t)\mathbf{R}_e(t,t) + \mathbf{R}_e(t,t)\mathbf{A}_o^T(t) + \mathbf{S}_z(t). \tag{2.250}$$

The MRE for the Kalman filter must be integrated forward in time, subject to the initial condition,

$$\mathbf{R}_e(t_0,t_0) = \mathbf{R}_{e0}. \tag{2.251}$$

When the process and measurement noise are cross-correlated, we have

$$\mathbf{R}_{\nu w}(t,\tau) = E[\boldsymbol{\nu}(t)\mathbf{w}^T(\tau)] = \mathbf{S}_{\nu w}(t)\delta(t-\tau), \tag{2.252}$$

where $\mathbf{S}_{\nu w}(t)$ is the *cross-spectral density matrix* of $\boldsymbol{\nu}(t)$ and $\mathbf{w}(t)$. Substituting equation (2.243) into equation (2.247) and using the correlation matrix definitions, equations (2.233) and (2.252), we have the following expression for the measurement noise power spectral density:

$$\mathbf{S}_w = \mathbf{F}\mathbf{S}_\nu\mathbf{F}^T - 2\mathbf{F}\mathbf{S}_{\nu w}\mathbf{L}^T + \mathbf{L}\mathbf{S}_w\mathbf{L}^T. \tag{2.253}$$

Substitution of equation (2.253) into equation (2.250), along with the first part of equation (2.241), yields the following optimal Kalman filter gain matrix:

$$\mathbf{L} = \left[\mathbf{R}_e\mathbf{C}^T + \mathbf{F}\mathbf{S}_{\nu w}\right]\mathbf{S}_w^{-1}, \tag{2.254}$$

where the optimal error covariance matrix is the positive definite solution of the following MRE:

$$\begin{aligned}\frac{\mathrm{d}}{\mathrm{d}t}\{\mathbf{R}_e(t,t)\} &= \mathbf{A}_G(t)\mathbf{R}_e(t,t) + \mathbf{R}_e(t,t)\mathbf{A}_G^T(t) \\ &\quad -\mathbf{R}_e(t,t)\mathbf{C}^T(t)\mathbf{S}_w^{-1}(t)\mathbf{C}(t)\mathbf{R}_e(t,t) + \mathbf{F}(t)\mathbf{S}_G(t)\mathbf{F}^T(t),\end{aligned} \tag{2.255}$$

with

$$\begin{aligned}\mathbf{A}_G(t) &= \mathbf{A}(t) - \mathbf{F}(t)\mathbf{S}_{\nu w}(t)\mathbf{S}_w^{-1}(t)\mathbf{C}(t), \\ \mathbf{S}_G(t) &= \mathbf{S}_\nu(t) - \mathbf{S}_{\nu w}(t)\mathbf{S}_w^{-1}(t)\mathbf{S}_{\nu w}^T(t).\end{aligned} \tag{2.256}$$

It is interesting to note the following one-to-one equivalence between the Kalman filter (equations (2.254) and (2.255)) and the linear quadratic regulator (equations (2.168) and (2.169)), in that both the LQR gain and the Kalman filter gain are based upon the solution to a matrix Riccati equation:

$$
\begin{aligned}
\mathbf{A}^T(t) &\Leftrightarrow \mathbf{A}(t),\\
\mathbf{C}^T(t) &\Leftrightarrow \mathbf{B}(t),\\
\mathbf{S_w}(t) &\Leftrightarrow \mathbf{R}(t),\\
\mathbf{F}(t)\mathbf{S}_\nu(t)\mathbf{F}^T(t) &\Leftrightarrow \mathbf{Q}(t),\\
\mathbf{F}(t)\mathbf{S}_{\nu\mathbf{w}}(t) &\Leftrightarrow \mathbf{S}(t).
\end{aligned}
$$

Hence, the Kalman filter is called the *dual* of the LQR regulator.

A major simplification in Kalman filter design occurs for process and measurement noise uncorrelated with each other, that is, $\mathbf{S}_{\nu\mathbf{w}}(t) = \mathbf{0}$ (a common situation), for which we have

$$\mathbf{L} = \mathbf{R}_e\mathbf{C}^T\mathbf{S}_\mathbf{w}^{-1}, \tag{2.257}$$

where

$$
\begin{aligned}
\frac{\mathrm{d}}{\mathrm{d}t}\{\mathbf{R}_e(t,t)\} &= \mathbf{A}(t)\mathbf{R}_e(t,t) + \mathbf{R}_e(t,t)\mathbf{A}^T(t)\\
&\quad -\mathbf{R}_e(t,t)\mathbf{C}^T(t)\mathbf{S}_\mathbf{w}^{-1}(t)\mathbf{C}(t)\mathbf{R}_e(t,t) + \mathbf{F}(t)\mathbf{S}_\nu(t)\mathbf{F}^T(t).
\end{aligned}
\tag{2.258}
$$

A Kalman filter updates its estimation error by the linear state equation (2.246). However, in cases where the error dynamics is essentially nonlinear, a much more accurate state estimate can be obtained using the nonlinear plant state equation (2.47), solved for the nominal case, rather than the linearized version, equation (2.232), while the matrix Riccati equation is still based upon the linear Kalman filter, equation (2.246). Such an implementation of Kalman filter is called an *extended Kalman filter* (EKF). Many long-range (or long-duration) navigational problems of spacecraft utilize the EKF for improved accuracy, since the underlying gravity perturbations are essentially nonlinear and well understood (Chapter 6).

2.12 Robust Linear Time-Invariant Control

Most tracking control problems require dissipation of all errors to zero when the time becomes large compared with the time scale of plant dynamics. Since the control interval is quite large, it is not important to account for the relatively much faster variation of plant parameters, which can – in many cases – be averaged out over a long period, and essentially approximated by linear time-invariant (LTI) systems where the plant coefficient matrices, $\mathbf{A}$, $\mathbf{B}$, $\mathbf{C}$, $\mathbf{D}$, are constants. Such a problem with the objective of a zero, *steady-state* (i.e., as $t \to \infty$) error is referred to as *infinite-horizon control*, because the control interval can be taken to be infinite. In such cases, both LQR and Kalman filter designs are greatly simplified by having an infinite control interval, for which the corresponding solutions, $\mathbf{P}(t)$, $\mathbf{R}_e(t,t)$, approach steady-state values given by $\mathbf{P}(\infty)$, $\mathbf{R}_e(\infty,\infty)$, expressed simply as the constants, $\mathbf{P}$, $\mathbf{R}_e$. The governing equation for a time-invariant LQR problem is derived easily by putting $\dot{\mathbf{P}} = \mathbf{0}$ in the matrix Riccati equation (2.169), resulting in the following *algebraic Riccati equation* (ARE):

$$
\begin{aligned}
\mathbf{0} &= \left(\mathbf{A} - \mathbf{B}\mathbf{R}^{-1}\mathbf{S}^T\right)^T\mathbf{P} + \mathbf{P}\left(\mathbf{A} - \mathbf{B}\mathbf{R}^{-1}\mathbf{S}^T\right)\\
&\quad -\mathbf{P}\mathbf{B}\mathbf{R}^{-1}\mathbf{B}^T\mathbf{P} + \mathbf{Q} - \mathbf{S}\mathbf{R}^{-1}\mathbf{S}^T.
\end{aligned}
\tag{2.259}
$$

The optimal feedback control law is obtained from the algebraic Riccati solution,

$$\mathbf{u}(t) = -\mathbf{R}^{-1}\left(\mathbf{B}^T\mathbf{P} + \mathbf{S}^T\right)\mathbf{x}(t), \tag{2.260}$$

where the cost coefficient matrices, $\mathbf{Q}, \mathbf{R}, \mathbf{S}$, are constants. For asymptotic stability of the regulated system, all the eigenvalues of the closed-loop dynamics matrix,

$$\mathbf{A} - \mathbf{B}\mathbf{R}^{-1}\left(\mathbf{B}^T\mathbf{P} + \mathbf{S}^T\right),$$

must be in the left-half s-plane (Appendix A), which requires that the ARE solution, $\mathbf{P}$ must be a symmetric, positive semi-definite matrix (Tewari 2002), that is, a matrix with all eigenvalues greater than or equal to zero. There may not always be a positive semi-definite solution; on the other hand, there could be several such solutions of which it cannot be determined which is to be regarded as the best. However, it can be proved (Glad and Ljung 2002) that if the following sufficient conditions are satisfied, there exists a unique, symmetric, positive semi-definite solution to the ARE:

- The control cost coefficient matrix, $\mathbf{R}$, is symmetric and positive definite, the matrix $\mathbf{Q} - \mathbf{S}\mathbf{R}^{-1}\mathbf{S}^T$ is symmetric and positive semi-definite, and the pair $(\mathbf{A} - \mathbf{B}\mathbf{R}^{-1}\mathbf{S}^T, \mathbf{Q} - \mathbf{S}\mathbf{R}^{-1}\mathbf{S}^T)$ is detectable (Appendix A), if not observable.
- The pair $(\mathbf{A}, \mathbf{B})$ is either controllable, or at least stabilizable (Appendix A).

In the stochastic sense, a constant error covariance matrix, $\mathbf{R}_e$, implies a stationary white noise process. If the estimation error of a linear system is stationary, the system must be driven by stationary processes. Therefore, an LTI Kalman filter essentially involves the assumption of stationary ZMGWN models for both process noise, $\boldsymbol{\nu}(t)$, and measurement noise, $\mathbf{w}(t)$. Thus, the error covariance matrix must now satisfy the following ARE:

$$\mathbf{0} = \mathbf{A}_G\mathbf{R}_e + \mathbf{R}_e\mathbf{A}_G^T - \mathbf{R}_e\mathbf{C}^T\mathbf{S}_w^{-1}\mathbf{C}\mathbf{R}_e + \mathbf{F}\left(\mathbf{S}_\nu - \mathbf{S}_{\nu w}\mathbf{S}_w^{-1}\mathbf{S}_{\nu w}^T\right)\mathbf{F}^T, \tag{2.261}$$

where $\mathbf{S}_w, \mathbf{S}_\nu, \mathbf{S}_{\nu w}$ are constant spectral density matrices and

$$\mathbf{A}_G = \mathbf{A} - \mathbf{F}\mathbf{S}_{\nu w}\mathbf{S}_w^{-1}\mathbf{C}. \tag{2.262}$$

The constant Kalman filter gain matrix is

$$\mathbf{L} = \left(\mathbf{R}_e\mathbf{C}^T + \mathbf{F}\mathbf{S}_{\nu w}\right)\mathbf{S}_w^{-1}. \tag{2.263}$$

Clearly, the algebraic Riccati equation for the Kalman filter must also have a symmetric, positive semi-definite solution, $\mathbf{R}_e$, for an asymptotically stable Kalman filter dynamics. Furthermore, by satisfying sufficient conditions that are dual to those stated above for the LQR problem, a unique, positive semi-definite solution to the ARE can be found.

The ARE is thus at the heart of both LQR and Kalman filter design for LTI systems. Being a nonlinear algebraic equation, it must be solved numerically, for example by iteration of the following *Lyapunov equation* for a symmetric matrix, $\mathbf{X}$:

$$\mathbf{A}\mathbf{X} + \mathbf{X}\mathbf{A}^T + \mathbf{Q} = \mathbf{0}. \tag{2.264}$$

There are several efficient algorithms for iteratively solving the ARE (Arnold and Laub 1984), such as the one programmed in the MATLAB function, *are.m*.

2.12.1 LQG/LTR Method

The procedure by which a linear quadratic regulator (LQR) and a Kalman filter are designed separately for a linear time-invariant plant, and then put together to form a feedback compensator – as shown in Figure 2.14 – is referred to as the *linear quadratic Gaussian* (LQG) method. The resulting feedback compensator is called an LQG compensator. Figure 2.14 depicts a general case where the output vector, $\mathbf{y}(t)$,

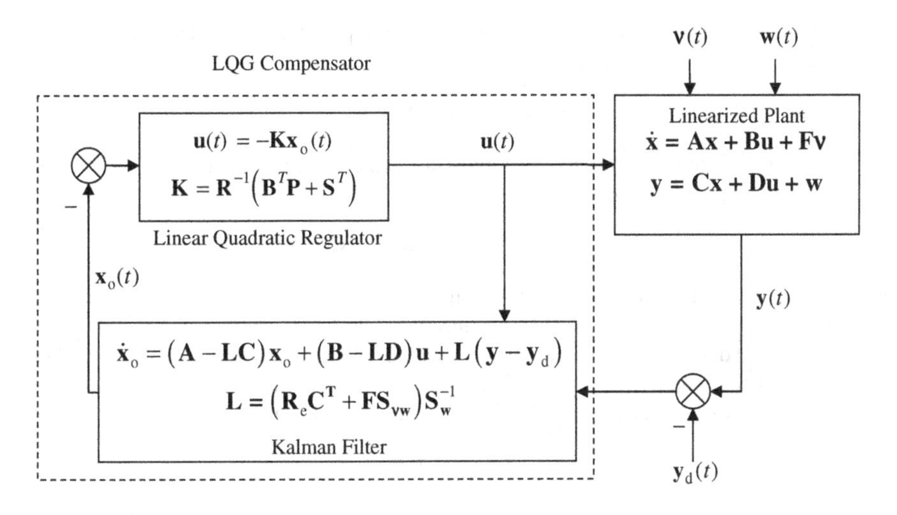

Figure 2.14 A linear quadratic Gaussian compensator for tracking a reference output

is to match a desired output (also called a reference, or commanded output), $\mathbf{y}_d(t)$, in the steady state. Such a reference output is usually commanded by the terminal controller (not shown in the figure). Clearly, the measured signal given to Kalman filter is $[\mathbf{y}(t) - \mathbf{y}_d(t)]$, based upon which (as well as the known input vector, $\mathbf{u}(t)$) it supplies the estimated state for feedback to the LQR regulator. Since the design of the LQG compensator – specified by the gain matrices, $(\mathbf{K}, \mathbf{L})$ – depends upon the chosen LQR cost parameters, $\mathbf{Q}$, $\mathbf{R}$, $\mathbf{S}$, and the selected Gaussian white-noise spectral densities, $\mathbf{S}_w$, $\mathbf{S}_\nu$, $\mathbf{S}_{\nu w}$, it is possible to design infinitely many compensators for a given plant. Usually, there are certain performance and robustness requirements specified for the closed-loop system that indirectly restrict the choice of the cost parameters to a given range. Being based upon optimal control, an LQG compensator has excellent performance features for a given set of cost parameters, but its robustness is subject to the extent the performance is degraded by state estimation through the Kalman filter. If the filter gains are too small, the estimation error does not tend to zero fast enough for the feedback to be accurate. On the other hand, if the Kalman filter has very large gains, there is an amplification of process and measurement noise by feedback, thereby reducing the overall robustness of the control system. Clearly, a balance must be struck in selecting the Kalman filter design parameters, $\mathbf{S}_w$, $\mathbf{S}_\nu$, $\mathbf{S}_{\nu w}$, such that good robustness is obtained without unduly sacrificing performance.

In order to study the robustness of LQG compensated system, refer to the block diagram of the control system transformed to the Laplace domain in a negative feedback configuration, as shown in Figure 2.15.

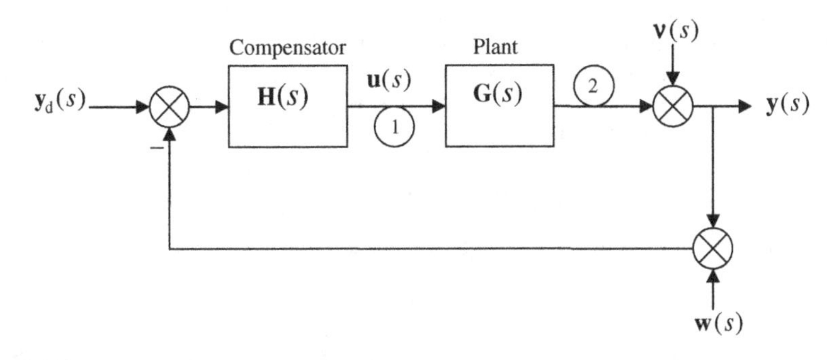

Figure 2.15 Transfer matrix representation of LQG compensated system

For simplicity, consider a *strictly proper* plant (i.e., $\mathbf{D} = \mathbf{0}$ (Appendix A)) of order n represented by the transfer matrix,

$$\mathbf{G}(s) = \mathbf{C}(s\mathbf{I} - \mathbf{A})^{-1}\mathbf{B},$$

of dimension $\ell \times m$, where ℓ is the number of outputs and m the number of inputs. An LQG compensator of dimension $m \times \ell$ has transfer matrix

$$\mathbf{H}(s) = -\mathbf{K}(s\mathbf{I} - \mathbf{A} + \mathbf{BK} + \mathbf{LC})^{-1}\mathbf{L}.$$

The process noise is represented by a ZMGWN disturbance, $\boldsymbol{\nu}(s)$, appearing at the plant's output, while the ZMGWN measurement noise, $\mathbf{w}(s)$, affects the feedback loop as shown. The overall system's transfer matrix, $\mathbf{T}(s)$, from the desired output to the output, is called the *transmission matrix*. On the other hand, the effect of the process noise on the output is given by the transfer matrix, $\mathbf{S}(s)$, called the *sensitivity matrix*. Both $\mathbf{T}(s)$ and $\mathbf{S}(s)$ are derived (with reference to Figure 2.15) as follows:

$$\mathbf{y} = \boldsymbol{\nu} + \mathbf{Gu} = \boldsymbol{\nu} + \mathbf{G}\left[\mathbf{H}(\mathbf{y}_d - \mathbf{y} - \mathbf{w})\right] \tag{2.265}$$

or

$$(\mathbf{I} + \mathbf{GH})\mathbf{y} = \boldsymbol{\nu} + \mathbf{GH}(\mathbf{y}_d - \mathbf{w}), \tag{2.266}$$

thereby implying

$$\mathbf{y} = (\mathbf{I} + \mathbf{GH})^{-1}\boldsymbol{\nu} + (\mathbf{I} + \mathbf{GH})^{-1}\mathbf{GH}(\mathbf{y}_d - \mathbf{w}) \tag{2.267}$$

or

$$\mathbf{y}(s) = \mathbf{S}(s)\boldsymbol{\nu}(s) + \mathbf{T}(s)\left[\mathbf{y}_d(s) - \mathbf{w}(s)\right], \tag{2.268}$$

where

$$\begin{aligned} \mathbf{S}(s) &= [\mathbf{I} + \mathbf{G}(s)\mathbf{H}(s)]^{-1} \\ \mathbf{T}(s) &= [\mathbf{I} + \mathbf{G}(s)\mathbf{H}(s)]^{-1}\mathbf{G}(s)\mathbf{H}(s). \end{aligned} \tag{2.269}$$

Because it can be easily shown that $\mathbf{T}(s) = \mathbf{I} - \mathbf{S}(s)$, the transmission matrix is also called *complementary sensitivity*.

The sensitivity and transmission matrices give us criteria for defining robustness. For a good robustness with respect to the process noise, the sensitivity matrix, $\mathbf{S}(s)$, should have a small "magnitude", while a good robustness with respect to the measurement noise requires that the transmission matrix, $\mathbf{T}(s)$, must have a small "magnitude". However, it is unclear what we mean by the magnitude of a matrix. One can assign a scalar measure, such as the Euclidean norm (Chapter 1) for vectors, to matrices. Since we are now dealing with square arrays of complex numbers, $\mathbf{M}(s)$, where $s = i\omega$, suitable norms for such a matrix, $\mathbf{M}(i\omega)$, at a given frequency, ω, are based upon its *singular values* (also called *principal gains*), defined as the positive square roots of the eigenvalues of the real matrix

$$\mathbf{M}^T(-i\omega)\mathbf{M}(i\omega).$$

The *Hilbert* (or spectral) norm is the largest singular value at a given frequency, ω, denoted by

$$\bar{\sigma}\left\{\mathbf{M}(i\omega)\right\},$$

or simply as $\bar{\sigma}(\mathbf{M})$. The magnitude of a frequency-dependent complex matrix is thus represented by its largest singular value. Clearly, for robustness with respect to the process noise, we require $\bar{\sigma}(\mathbf{S})$ to be as small as possible, whereas robustness with respect to the measurement noise can be achieved by minimizing $\bar{\sigma}(\mathbf{T})$. However, there is a serious conflict between simultaneously minimizing both $\bar{\sigma}(\mathbf{S})$ and $\bar{\sigma}(\mathbf{T})$, because the minimization of one results in the maximization of the other (and vice versa). A design compromise is obtained by choosing different ranges of frequencies for the minimization of $\bar{\sigma}(\mathbf{S})$ and $\bar{\sigma}(\mathbf{T})$.

The process noise, $\boldsymbol{\nu}(s)$, is generally experienced due to modeling errors, which – while being present at all frequencies – usually have their maximum effect on a physical model at the lowest frequencies. An example is the neglect of higher-order dynamics while deriving the plant model, such as structural flexibility of a flight vehicle. Of all the neglected modes, that with the smallest natural frequency typically has the largest contribution to the transfer matrix. As the natural frequency increases, the modes concerned have successively smaller magnitude contributions. Thus it makes sense to minimize $\bar{\sigma}(\mathbf{S})$ in the low-frequency range. On the other hand, measurement noise has its largest contribution at high frequencies, and thus $\bar{\sigma}(\mathbf{T})$ should be minimized at high frequencies. However, we should also remember that $\mathbf{T}(s)$ is the overall system's transfer matrix. A minimization of $\bar{\sigma}(\mathbf{T})$ at higher frequencies has the additional effect of slowing down the closed-loop system's performance in tracking a reference signal, which implies a larger tracking error at a given time. Therefore, in the interests of maintaining a reasonable performance, the minimization of $\bar{\sigma}(\mathbf{T})$ must take place outside the desired *bandwidth* (Appendix A) of the control system.

As discussed above, the gains of the regulator and the Kalman filter must be selected in such a way that there is as little loss as possible of both the performance and robustness by combining the two in a single compensator. Referring to Figure 2.15, we note that the transfer matrix, $\mathbf{H}(s)\mathbf{G}(s)$, denotes the transfer of the input, $\mathbf{u}(s)$, back to itself if the loop is broken at the point marked "1", and is thus called the *return ratio at input*. If all the states are available for measurement, there is no need for a Kalman filter and the ideal return ratio at input is given by

$$\mathbf{H}(s)\mathbf{G}(s) = -\mathbf{K}\,(s\mathbf{I} - \mathbf{A})^{-1}\,\mathbf{B}.$$

On the other hand, if the feedback loop is broken at the point marked "2" (Figure 2.15), the transfer matrix, $\mathbf{G}(s)\mathbf{H}(s)$, represents the transfer of the output, $\mathbf{y}(s)$, back to itself and is called the *return ratio at output*. If there is no regulator in the system, then the ideal return ratio at output is given by

$$\mathbf{G}(s)\mathbf{H}(s) = -\mathbf{C}\,(s\mathbf{I} - \mathbf{A})^{-1}\,\mathbf{L}.$$

If one can recover the ideal return ratio at either the plant's input or the output by suitably designing the LQG compensator, the best possible combination of the LQR regulator and the Kalman filter is achieved and the design process is called *loop-transfer recovery* (LTR). For simplicity in the following discussion, we assume there is no cross-correlation between process and measurement noise, that is, $\mathbf{S}_{\nu\mathbf{w}} = \mathbf{0}$. It can be proved (Maciejowski 1989) that by selecting

$$\mathbf{F} = \mathbf{B}, \quad \mathbf{S}_{\nu} = \rho\mathbf{S}_{\mathbf{w}},$$

and making the positive scalar parameter ρ arbitrarily large, the LQG return ratio at input,

$$\mathbf{H}(s)\mathbf{G}(s) = -\mathbf{K}\,(s\mathbf{I} - \mathbf{A} + \mathbf{BK} + \mathbf{LC})^{-1}\,\mathbf{LC}\,(s\mathbf{I} - \mathbf{A})^{-1}\,\mathbf{B},$$

can be made to approach the ideal return ratio at input. The following procedure for LTR at the input is thus commonly applied:

- Select an LQR regulator by a suitable choice of the weighting matrices $\mathbf{Q}$, $\mathbf{R}$, $\mathbf{S}$ such that good performance and robustness with state feedback are obtained.
- Select $\mathbf{F} = \mathbf{B}$, $\mathbf{S}_{\nu} = \rho\mathbf{S}_{\mathbf{w}}$ and increase ρ until the desired state feedback properties are recovered in the closed-loop system.

A variation of this approach can be applied for LTR at the plant's output, beginning with the design of Kalman filter, and then iterating for the LQR gain until the ideal return ratio at output is recovered:

- Select a Kalman filter by a suitable choice of the weighting matrices, $\mathbf{S}_\nu$, $\mathbf{S}_\mathbf{w}$, $\mathbf{S}_{\nu\mathbf{w}}$, such that a satisfactory return ratio at output obtained.
- Select $\mathbf{Q} = \rho\mathbf{R}$ and increase ρ until the ideal return ratio at output is recovered in the closed-loop system.

Further details of LQG/LTR methods can be found in Maciejowski (1989) and Glad and Ljung (2002). The Control System Toolbox of MATLAB® provides the functions *lqr.m* and *lqe.m* for carrying out the LQR and Kalman filter designs, respectively. Alternatively, the Robust Control Toolbox has specialized functions *lqg.m* for "hands-off" LQG compensator design, and *ltru.m* and *ltry.m* for LTR at plant input and output, respectively. However, for a beginner we recommend the use of a basic ARE solver, such as *are.m*, in order that the relevant design steps are clearly understood.

Example 2.7 Consider a remotely piloted aircraft with the following longitudinal dynamics plant:

$$\mathbf{A} = \begin{pmatrix} -0.025 & -40 & -20 & -32 \\ 0.0001 & -2 & 1 & -0.0007 \\ 0.01 & 12 & -2.5 & 0.0009 \\ 0 & 0 & 1 & 0 \end{pmatrix}, \quad \mathbf{B} = \begin{pmatrix} 3.25 & -0.8 \\ -0.2 & -0.005 \\ -32 & 22 \\ 0 & 0 \end{pmatrix},$$

$$\mathbf{C} = \begin{pmatrix} 0 & 1 & 0 & 0 \\ 0 & 0 & 0 & 1 \end{pmatrix}, \quad \mathbf{D} = \begin{pmatrix} 0 & 0 \\ 0 & 0 \end{pmatrix}.$$

Here, both of the outputs as well as the two control inputs are in radians. It is desired to design an LQG compensator such that the closed-loop response to initial condition, $\mathbf{x}(0) = (0, 0.1, 0, 0)^T$, settles down in about 10 s, without requiring control input magnitudes greater than 5°.

We begin by noting that the plant is unstable, with eigenvalues computed by MATLAB function *damp.m*:

```
>> A=[-0.025 -40  -20  -32; 0.0001  -2  1  -0.0007;
      0.01  12  -2.5  0.0009; 0   0   1   0];
>> damp(A)

  Eigenvalue          Damping        Freq. (rad/s)
   8.56e-001      -1.00e+000          8.56e-001
   5.78e-001      -1.00e+000          5.78e-001
  -2.39e-001       1.00e+000          2.39e-001
  -5.72e+000       1.00e+000          5.72e+000
```

Since the plant is controllable, an LQR regulator can be designed with suitable cost parameters, such as $\mathbf{Q} = 0.01\mathbf{I}$, $\mathbf{R} = \mathbf{I}$, $\mathbf{S} = \mathbf{0}$, as follows:

```
>> B=[3.25  -0.8; -0.2  -0.005; -32  22; 0  0];
>> rank(ctrb(A,B)) %Rank of controllability test matrix (App.-A)
ans =      4

>> Q=0.01*eye(4);R=eye(2); %LQR cost parameters
>> P=are(A,B*inv(R)*B',Q) % Algebraic Riccati solution matrix
```

```
P =
     0.0016245   -0.0052635   -0.0024045   -0.0040841
    -0.0052635     0.058418     0.020835      0.02621
    -0.0024045     0.020835    0.0089661     0.012974
    -0.0040841      0.02621     0.012974     0.034684

>> k=inv(R)*B'*P %Regulator gain matrix
k =
      0.083277     -0.69551      -0.2989      -0.4337
     -0.054172      0.46229      0.19907      0.28857

>> damp(A-B*k) %Eigenvalues of the regulated system

          Eigenvalue                Damping     Freq. (rad/s)
  -6.60e-001                      1.00e+000       6.60e-001
  -5.15e+000                      1.00e+000       5.15e+000
  -6.55e+000 + 5.65e+000i         7.57e-001       8.65e+000
  -6.55e+000 - 5.65e+000i         7.57e-001       8.65e+000

>> C=[0 1 0 0;0 0 0 1]; D=zeros(2,2);
>> sys=ss(A-B*k,B,C,D); %LQR regulated system
>> [y,t,x]=initial(sys,[0 0.1 0 0]'); u=-k*x'; %initial response
```

The regulated system is thus stable and has adequate damping of the complex poles. The regulated initial response (Figure 2.16) shows a settling time of about 8 s with control inputs limited to $\pm 4^\circ$. Now, we would like to recover the state feedback regulated performance with a Kalman filter based upon the measurement of the two outputs. Since the plant is observable with the given outputs (verify), one can easily design such a Kalman filter. We select $\mathbf{F} = \mathbf{B}$, $\mathbf{S_w} = \mathbf{I}$ and put $\mathbf{S}_\nu = \rho \mathbf{S_w}$. Singular values of

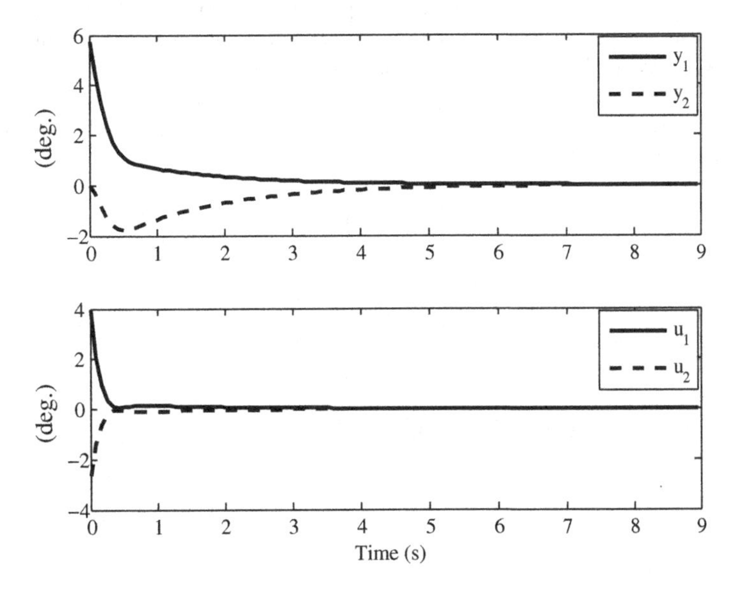

Figure 2.16 Initial response of LQR regulated system

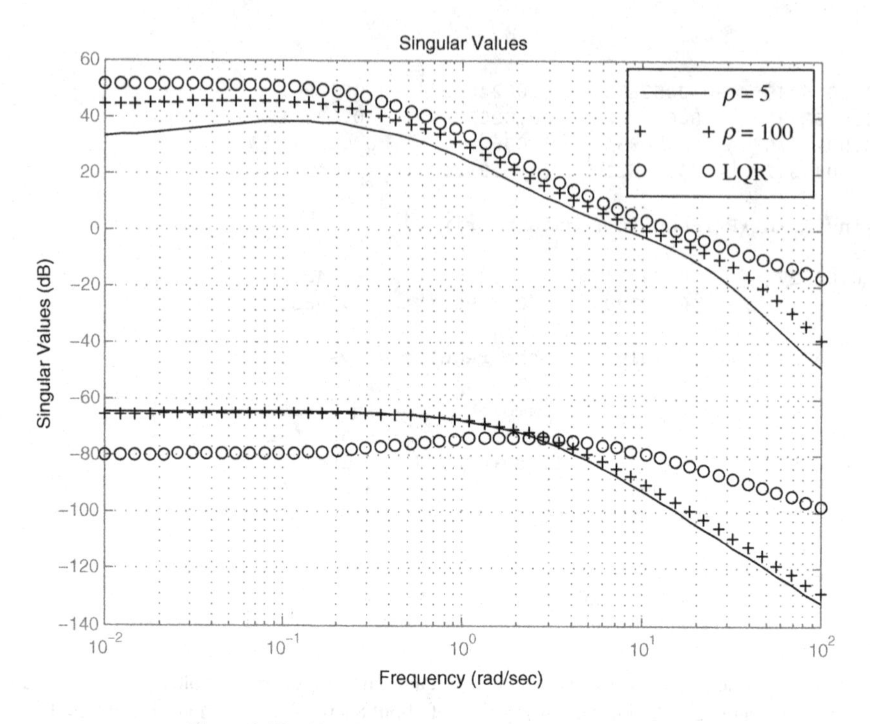

Figure 2.17 Singular values of return ratio at plant's input

the return ratio at the plant's input for $\rho = 5$ and $\rho = 100$ are compared with those of the ideal (LQR) return ratio in Figure 2.17. While a better loop-transfer recovery is obtained for $\rho = 100$ (as expected), this Kalman filter design is unacceptable because it requires control magnitudes twice the specifications (Figure 2.18). Due to a larger observer gain with $\rho = 100$, there is a faster response with a settling time of about 5 s, but an increased susceptibility to measurement noise as the largest singular value shows a smaller decline with frequency (called *roll-off*) at high frequencies. In contrast, $\rho = 5$ fully recovers the LQR performance and also provides a larger roll-off at high frequency. The steps for Kalman filter and LQG design are coded as follows:

```
>> rho=5; Sw=eye(2); Snu=rho*Sw; %Noise spectral densities
>> Re=are(A',C'*inv(Sw)*C,B*Snu*B') %Error covariance matrix
Re =
       27106      -143.88      -1075.1      -215.13
     -143.88       6.0467       50.701       6.3886
     -1075.1       50.701       747.16       49.285
     -215.13       6.3886       49.285       7.5997

>> L=Re*C'*inv(Sw) %Kalman filter gain matrix
L =
     -143.88      -215.13
      6.0467       6.3886
      50.701       49.285
      6.3886       7.5997

>> sys=ss(A,B,C,D); %Plant
```

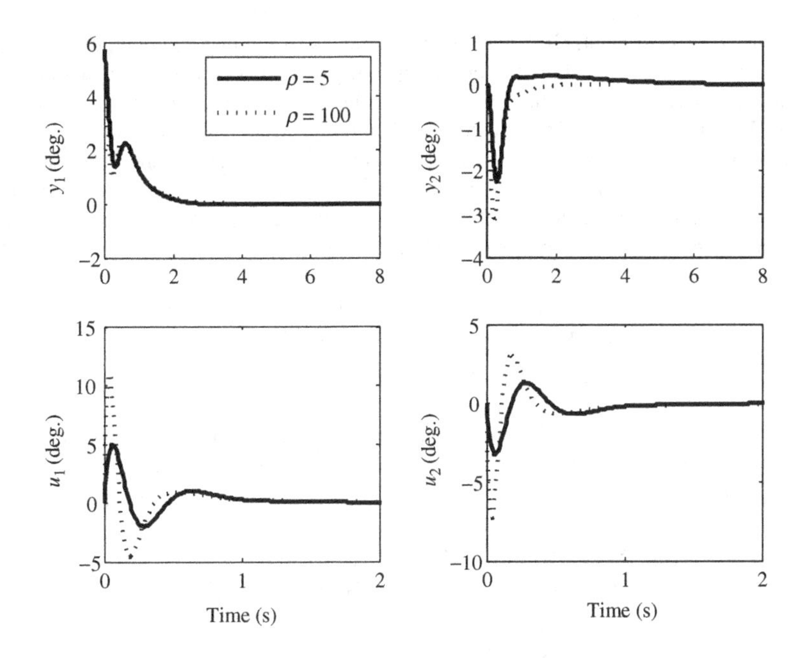

Figure 2.18 Initial response of LQG compensated systems

```
>> sysk=ss(A,B,-k,D); %LQR return ratio
>> Ac=A-B*k-L*C;
>> Bc=L;
>> Cc=-k;
>> Dc=D;
>> sysc=ss(Ac,Bc,Cc,Dc); %LQG compensator
>> sysKG=series(sysc,sys); %LQG return ratio
>> w=logspace(-2,2); sigma(sysKG,w) %singular values of return ratio
>> ACL=[A -B*k;L*C Ac];
>> BCL=[B*k;B*k];
>> sysCL=ss(ACL,BCL,[C zeros(2,4)],zeros(2,4)); %Closed-loop system
>> [y,t,x]=initial(sysCL,[0 0.1 0 0 zeros(1,4)]'); %initial response
>> u=-k*x(:,5:8)'; %closed-loop control inputs
```

2.12.2 H_2/H_∞ Design Methods

Rather than indirectly designing a feedback compensator via a regulator and a Kalman filter, an alternative design approach is to extend the frequency domain design methodology commonly used for single-variable (SISO) systems, to the multi-variable control system. The regulator and Kalman filter then result naturally from such a design, in which the singular value spectra replace the Bodé gain plot, and parameters analogous to the gain and phase margins of an SISO system (cf. Chapter 2 of Tewari (2002)) are addressed for a robust, multi-variable design directly in the frequency domain. In deriving such a compensator, one minimizes a combined, frequency-weighted measure of sensitivity, complementary sensitivity, and transfer matrix from disturbance to plant output, integrated over a range of frequencies. There are two such frequency domain, optimal control methods that only differ by the matrix norm sought to be minimized, namely the H_2 and H_∞ synthesis.

Consider a strictly proper plant, $\mathbf{G}(s)$, with control inputs, $\mathbf{u}(s)$, and measured outputs, $\mathbf{y}(s)$. All other inputs to the control system, namely the reference output, $\mathbf{y}_d(s)$, process noise, $\boldsymbol{\nu}(s)$, and measurement noise, $\mathbf{w}(s)$, are clubbed together in a single vector, $\mathbf{d}(s)$, called *external disturbances*. An output feedback compensator, $\mathbf{H}(s)$, is to be designed for simultaneously minimizing an integrated measure of sensitivity, $\mathbf{S}(s)$, and complementary sensitivity, $\mathbf{T}(s)$, of the output with respect to the disturbance, and transfer matrix, $\mathbf{G_u}(s) = -\mathbf{H}(s)\mathbf{S}(s)$, from disturbance to plant input. However, since these objectives are contradictory (as mentioned above), different ranges of frequency are specified for the minimization of each objective. This is practically implemented through weighting the concerned matrix by *frequency weights* as follows:

$$\begin{aligned}\mathbf{W_S}(s)\mathbf{S}(s) &= \mathbf{W_S}(s)\left[\mathbf{I}+\mathbf{G}(s)\mathbf{H}(s)\right]^{-1},\\ \mathbf{W_T}(s)\mathbf{T}(s) &= \mathbf{W_T}(s)\left[\mathbf{I}+\mathbf{G}(s)\mathbf{H}(s)\right]^{-1}\mathbf{G}(s)\mathbf{H}(s),\\ \mathbf{W_u}(s)\mathbf{G_u}(s) &= -\mathbf{W_u}(s)\mathbf{H}(s)\mathbf{S}(s).\end{aligned} \tag{2.270}$$

Here, $\mathbf{W_S}(s)$ is a strictly proper, square transfer matrix, and $\mathbf{W_T}(s)$, $\mathbf{W_u}(s)$ are square transfer matrices. The plant is thus augmented by the frequency weighting matrices with the additional outputs, $\mathbf{z}_1(s)$, $\mathbf{z}_2(s)$, $\mathbf{z}_3(s)$, called *error signals*:

$$\begin{aligned}\mathbf{y}(s) &= \mathbf{G}(s)\mathbf{u}(s)+\mathbf{d}(s),\\ \mathbf{z}_1(s) &= \mathbf{W_S}(s)\mathbf{y}(s) = \mathbf{W_S}(s)\left[\mathbf{G}(s)\mathbf{u}(s)+\mathbf{d}(s)\right],\\ \mathbf{z}_2(s) &= \mathbf{W_T}(s)\mathbf{G}(s)\mathbf{u}(s),\\ \mathbf{z}_3(s) &= \mathbf{W_u}(s)\mathbf{u}(s).\end{aligned} \tag{2.271}$$

The output feedback control law is given by

$$\mathbf{u}(s) = -\mathbf{H}(s)\mathbf{y}(s). \tag{2.272}$$

The overall control system with the frequency weights and the augmented plant is depicted in Figure 2.19. Note that the compensator design is entirely dependent on the chosen frequency weights (rather than on the LQR cost parameters and noise spectral densities of the LQG case). Furthermore, the controller design is based upon output feedback, which requires the following inherent observer dynamics that is part of the augmented plant:

$$\dot{\mathbf{x}}_o = (\mathbf{A}-\mathbf{FC})\mathbf{x}_o + \mathbf{Bu} + \mathbf{Fy}. \tag{2.273}$$

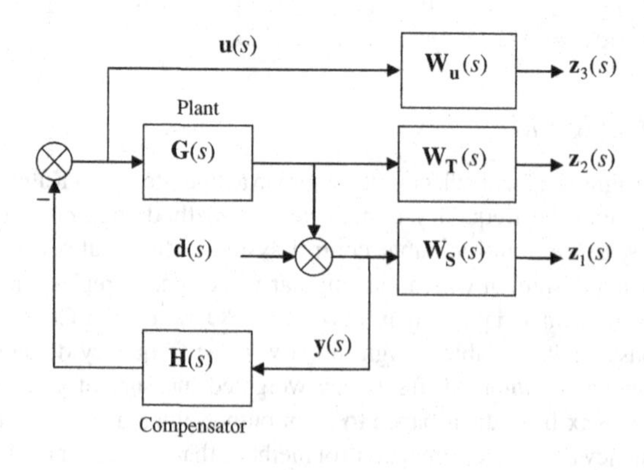

Figure 2.19 Augmented plant and closed-loop system for H_2/H_∞ design

Since the coefficients $\mathbf{A}, \mathbf{F}, \mathbf{C}$ depend upon the chosen frequency weights, a stable observer requires a judicious choice of the frequency weights.

The overall closed-loop transfer matrix, $\mathbf{G}_c(s)$, from the disturbance, $\mathbf{d}$, to the error vector, $\mathbf{z} = (\mathbf{z}_1, \mathbf{z}_2, \mathbf{z}_3)^T$, is thus given by

$$\mathbf{z}(s) = \mathbf{G}_c(s)\mathbf{d}(s) = \begin{bmatrix} \mathbf{W_S}(s)\mathbf{S}(s) \\ -\mathbf{W_T}(s)\mathbf{T}(s) \\ -\mathbf{W_u}(s)\mathbf{G_u}(s) \end{bmatrix} \mathbf{d}(s). \tag{2.274}$$

Operator Norms of Transfer Matrix

We now define the following frequency integral of a proper and asymptotically stable transfer matrix, $\mathbf{Q}(i\omega)$, as its H_2 *norm*:

$$\| \mathbf{Q} \|_2 = \sqrt{\frac{1}{2\pi} \int_{-\infty}^{\infty} \text{tr} \left[\mathbf{Q}(i\omega)\mathbf{Q}^T(-i\omega) \right] d\omega}, \tag{2.275}$$

where tr(.) refers to the *trace* of a square matrix (i.e., the sum of its diagonal elements). Another related operator norm for $\mathbf{Q}(i\omega)$, is its H_∞ *norm*, defined by

$$\| \mathbf{Q} \|_\infty = \sup_\omega \left[\bar{\sigma} \left\{ \mathbf{Q}(i\omega) \right\} \right], \tag{2.276}$$

where $\sup_\omega(.)$ is the supremum (maximum value) with respect to the frequency. Both the operator norms, $\| . \|_2$ and $\| . \|_\infty$, provide a single, positive real number measuring the largest possible magnitude of a frequency-dependent matrix. Therefore, one can employ either of them for minimizing the combined sensitivity reflected by the closed-loop transfer matrix, $\mathbf{G}_c(s)$.

Design Procedure

The H_2 norm of $\mathbf{G}_c(s)$ is

$$\| \mathbf{G}_c \|_2 = \| \mathbf{W_S S} \|_2 + \| \mathbf{W_T T} \|_2 + \| \mathbf{W_u G_u} \|_2 . \tag{2.277}$$

Thus, minimization of $\| \mathbf{G}_c \|_2$ guarantees a simultaneous minimization of the H_2 norm of the weighted sensitivity, complementary sensitivity, and $\mathbf{G_u}(s)$. We also note that the power spectral density of the error vector, $\mathbf{z}(t)$, when the disturbance, $\mathbf{d}(t)$, is a white noise of unit intensity (i.e., has the identity matrix as its power spectral density) is given by equation (2.221) as

$$\mathbf{S_z}(\omega) = \mathbf{Z}(i\omega)\mathbf{Z}^T(-i\omega) = \mathbf{G}_c(i\omega)\mathbf{G}_c^T(-i\omega). \tag{2.278}$$

Thus, the minimization of $\| \mathbf{G}_c \|_2$ directly results in a minimization of the error power spectral density, which is an objective quite similar to that of the LQG compensator. Therefore, one can proceed to derive the compensator that minimizes the H_2 norm of the error in much the same manner as the LQG case.

The augmented plant, equation (2.271), can be represented in an LTI state-space form as follows:

$$\begin{aligned} \dot{\mathbf{x}} &= \mathbf{Ax} + \mathbf{Bu} + \mathbf{Fd}, \\ \mathbf{y} &= \mathbf{Cx} + \mathbf{d}, \\ \mathbf{z} &= \mathbf{Mx} + \mathbf{Nu}. \end{aligned} \tag{2.279}$$

The optimal H_2 synthesis then consists of deriving an LQR regulator with $\mathbf{Q} = \mathbf{M}^T\mathbf{M}$, $\mathbf{R} = \mathbf{I}$ and gain $\mathbf{K} = \mathbf{B}^T\mathbf{P}$ such that

$$\mathbf{H}(s) = -\mathbf{K}\left(s\mathbf{I} - \mathbf{A} + \mathbf{B}\mathbf{B}^T\mathbf{P} + \mathbf{F}\mathbf{C}\right)^{-1}\mathbf{F}, \tag{2.280}$$

where $\mathbf{P}$ is a symmetric, positive semi-definite solution to the Riccati equation

$$\mathbf{0} = \mathbf{A}^T\mathbf{P} + \mathbf{P}\mathbf{A} - \mathbf{P}\mathbf{B}\mathbf{B}^T\mathbf{P} + \mathbf{M}^T\mathbf{M}. \tag{2.281}$$

For simplicity of discussion, we assume

$$\mathbf{N}^T\mathbf{M} = \mathbf{0}, \quad \mathbf{N}^T\mathbf{N} = \mathbf{I}. \tag{2.282}$$

A good way to ensure a stable observer dynamics $(\mathbf{A} - \mathbf{F}\mathbf{C})$ is by replacing $\mathbf{F}$ with a Kalman filter gain, $\mathbf{L}$, which requires a particular structure for the frequency weights. Evidently, a major part of H_∞ design involves selection of suitable frequency weights.

An optimization procedure can be adopted to minimize the H_∞ norm of $\mathbf{G}_c(s)$ based upon the augmented plant, equation (2.279). However, there being no direct relationship between the H_∞ norm of the closed-loop transfer matrix and the objective function of an LQG design, it is difficult to know in advance what is the minimum value of $||\mathbf{G}_c||_\infty$ that would lead to an acceptable design. A practical method (Glover and Doyle 1988) is to choose a positive number, γ, and aim to derive a stabilizing compensator that achieves

$$||\mathbf{G}_c||_\infty = \sup_\omega \left[\bar{\sigma}\left\{\mathbf{G}_c(i\omega)\right\}\right] \leq \gamma. \tag{2.283}$$

By decreasing γ until the compensator fails to stabilize the system, one has found the limit on the minimum value of $||\mathbf{G}_c||_\infty$. Note that if γ is increased to a large value, the design approaches that of the optimal H_2 (or LQG) compensator. Thus, one can begin iterating for γ from a value corresponding to a baseline H_2 (or LQG) design.

It is possible to demonstrate that an optimal controller is derived by satisfying equation (2.283). Define the following function for the plant of equation (2.279) subject to condition (2.282):

$$f(t) = \mathbf{x}(t)^T\mathbf{P}\mathbf{x}(t) + \int_0^t \left[\mathbf{z}(\tau)^T\mathbf{z}(\tau) - \gamma^2\mathbf{d}(\tau)^T\mathbf{d}(\tau)\right] \mathrm{d}\tau, \tag{2.284}$$

where $\mathbf{P}$ is a symmetric, positive semi-definite matrix and $\gamma > 0$. Assuming a zero initial condition, $\mathbf{x}(0) = \mathbf{0}$, we have $f(0) = 0$. We now take the time derivative of $f(t)$ as follows:

$$\begin{aligned}
\dot{f} &= \dot{\mathbf{x}}^T\mathbf{P}\mathbf{x} + \mathbf{x}^T\mathbf{P}\dot{\mathbf{x}} + \mathbf{z}^T\mathbf{z} - \gamma^2\mathbf{d}^T\mathbf{d} \\
&= \mathbf{x}^T\mathbf{A}^T\mathbf{P}\mathbf{x} + \mathbf{u}^T\mathbf{B}^T\mathbf{P}\mathbf{x} + \mathbf{d}^T\mathbf{F}^T\mathbf{P}\mathbf{x} + \mathbf{x}^T\mathbf{P}\mathbf{A}\mathbf{x} + \mathbf{x}^T\mathbf{P}\mathbf{B}\mathbf{u} + \mathbf{x}^T\mathbf{P}\mathbf{F}\mathbf{d} \\
&\quad + \mathbf{x}^T\mathbf{M}^T\mathbf{M}\mathbf{x} + \mathbf{u}^T\mathbf{u} - \gamma^2\mathbf{d}^T\mathbf{d} \\
&= \mathbf{x}^T\left[\mathbf{A}^T\mathbf{P} + \mathbf{P}\mathbf{A} + \mathbf{M}^T\mathbf{M} + \mathbf{P}\left(\frac{1}{\gamma^2}\mathbf{F}\mathbf{F}^T - \mathbf{B}\mathbf{B}^T\right)\mathbf{P}\right]\mathbf{x} \\
&\quad + (\mathbf{u} + \mathbf{B}^T\mathbf{P}\mathbf{x})^T(\mathbf{u} + \mathbf{B}^T\mathbf{P}\mathbf{x}) - \gamma^2\left(\mathbf{d} - \frac{1}{\gamma^2}\mathbf{F}^T\mathbf{P}\mathbf{x}\right)^T\left(\mathbf{d} - \frac{1}{\gamma^2}\mathbf{F}^T\mathbf{P}\mathbf{x}\right).
\end{aligned} \tag{2.285}$$

If we choose $\mathbf{P}$ as a symmetric, positive semi-definite solution of the Riccati equation

$$\mathbf{0} = \mathbf{A}^T\mathbf{P} + \mathbf{P}\mathbf{A} + \mathbf{P}\left(\frac{1}{\gamma^2}\mathbf{F}\mathbf{F}^T - \mathbf{B}\mathbf{B}^T\right)\mathbf{P} + \mathbf{M}^T\mathbf{M}, \tag{2.286}$$

and select a feedback control law such that

$$\mathbf{u}(t) = -\mathbf{B}^T\mathbf{P}\mathbf{x}(t), \tag{2.287}$$

then we have $\dot{f}(t) \leq 0$, thereby implying $f(t) \leq 0$ for all t. Since the first term on the right-hand side of equation (2.284) is always non-negative (due to $\mathbf{P}$ being positive semi-definite), we must have the integral term always less than or equal to zero, which implies

$$\mathbf{z}(t)^T\mathbf{z}(t) \leq \gamma^2\mathbf{d}(t)^T\mathbf{d}(t), \tag{2.288}$$

for all t and for any disturbance signal, $\mathbf{d}(t)$. Note that minimization of γ implies minimization of the H_2 norm (same as Euclidean norm) of the error vector for any given disturbance signal. Hence, the optimality of a compensator satisfying equation (2.283) is guaranteed, provided there exists a positive semi-definite solution of the Riccati equation (2.286).

There is a serious drawback with the H_∞ approach in that it does not automatically produce a stabilizing compensator. We have just now demonstrated the existence of a non-positive *Lyapunov function*, $f(t)$, which shows that a compensator designed by the H_∞ method does not meet the sufficient conditions for stability in the sense of Lyapunov (Appendix B). Therefore, stability of the H_∞ compensator must be separately ensured by requiring that the dynamics matrix, $\mathbf{A} - \mathbf{B}\mathbf{B}^T\mathbf{P}$, must have all the eigenvalues in the left-half s-plane (Appendix A).

The following optimization procedure by Glover and Doyle (1988) is generally implemented for the solution of the H_∞ design problem:

(a) Select a set of frequency weights, $\mathbf{W_S}(s)$, $\mathbf{W_T}(s)$, $\mathbf{W_u}(s)$, for augmenting the plant, $\mathbf{G}(s)$. These are typically diagonal matrices of proper transfer functions. Also, ensure that the chosen frequency weights yield a stable observer dynamics ($\mathbf{A} - \mathbf{F}\mathbf{C}$).
(b) Select a value for γ (usually 1).
(c) Solve equation (2.286) for a symmetric, positive semi-definite matrix, $\mathbf{P}$. If no such solution exists, go back to (b) and increase γ. If a symmetric, positive semi-definite $\mathbf{P}$ exists which yields a stable dynamics matrix ($\mathbf{A} - \mathbf{B}\mathbf{B}^T\mathbf{P}$), try to find a better solution by going back to (b) and decreasing γ.
(d) If the closed-loop properties are satisfactory, stop. Otherwise, go back to (a) and modify the frequency weights.

The Robust Control Toolbox function of MATLAB, *hinfopt.m*, automates the H_∞ design procedure of Glover and Doyle (1988), even for those cases where equation (2.282) does not hold. However, it replaces γ by its reciprocal, and aims to maximize the reciprocal for a stabilizing controller. The Robust Control Toolbox contains another function, *h2LQG.m*, for automated H_2 synthesis. Both *hinfopt.m* and *h2LQG.m* first require the model to be augmented by adding frequency weights through a dedicated function, *augtf.m*.

Example 2.8 Redesign the compensator for the remotely piloted aircraft of Example 2.7 by the optimal H_∞ method and the following revised output equation:

$$\mathbf{C} = \begin{pmatrix} 1 & 0 & 0 & 0 \\ 0 & 1 & 0 & 0 \end{pmatrix}, \quad \mathbf{D} = \begin{pmatrix} 0 & 0 \\ 0 & 0 \end{pmatrix}.$$

The first output, $y_1(t)$, now has units of speed (ft/s), while the second output, $y_2(t)$, is in radians. The closed-loop performance objectives are now revised such that $y_1(t)$ and $y_2(t)$ must settle down in about 20 s, without either input exceeding $\pm 5^\circ$ when an initial condition of $\mathbf{x}(0) = (0, 0.1, 0, 0)^T$ is applied.

We note that due to the unstable plant, the choice of the frequency weights is crucial. After some trials, we select the following frequency weights:

$$\mathbf{W_S}(s) = \begin{pmatrix} \frac{s+100}{10s+1} & 0 \\ 0 & \frac{s+10}{10s+1} \end{pmatrix}, \quad \mathbf{W_T}(s) = (s+1)\begin{pmatrix} 1 & 0 \\ 0 & 1 \end{pmatrix},$$

$$\mathbf{W_u}(s) = \begin{pmatrix} \frac{1}{s+1} & 0 \\ 0 & \frac{4.5}{s+1} \end{pmatrix}.$$

We begin by deriving the augmented plant through *augtf.m* as follows:

```
>> a = [-0.025 -40  -20  -32; 0.0001  -2  1  -0.0007;
         0.01  12  -2.5  0.0009; 0   0   1   0];
>> b = [3.25  -0.8; -0.2  -0.005; -32  22; 0  0];
>> c = [1 0 0 0;0 1 0 0];
>> d=[0  0; 0 0];
>> w1=[1 100;10 1;1 10;10 1]; %Frequency weight W_S
>> w2=[0 1;1 1;0 4.5;1 1]; %Frequency weight W_u
>> w3=[1 1;0 1;1 1;0 1]; %Frequency weight W_T
%% Augmented plant follows:
>> [A,B1,B2,C1,C2,D11,D12,D21,D22]=augtf(a,b,c,d,w1,w2,w3)
A =

   -0.0250  -40.0000  -20.0000  -32.0000         0         0         0         0
    0.0001   -2.0000    1.0000   -0.0007         0         0         0         0
    0.0100   12.0000   -2.5000    0.0009         0         0         0         0
         0         0    1.0000         0         0         0         0         0
   -1.0000         0         0         0   -0.1000         0         0         0
         0   -1.0000         0         0         0   -0.1000         0         0
         0         0         0         0         0         0   -1.0000         0
         0         0         0         0         0         0         0   -1.0000

B1 =

     0     0
     0     0
     0     0
     0     0
     1     0
     0     1
     0     0
     0     0

B2 =

    3.2500   -0.8000
   -0.2000   -0.0050
  -32.0000   22.0000
         0         0
         0         0
         0         0
    1.0000         0
         0    1.0000
```

```
C1 =
   -0.1000         0         0         0    9.9900         0         0         0
         0   -0.1000         0         0         0    0.9900         0         0
         0         0         0         0         0         0    1.0000         0
         0         0         0         0         0         0         0    4.5000
    0.9750  -40.0000  -20.0000  -32.0000         0         0         0         0
    0.0001   -1.0000    1.0000   -0.0007         0         0         0         0

C2 =
    -1     0     0     0     0     0     0     0
     0    -1     0     0     0     0     0     0

D11 =
    0.1000         0
         0    0.1000
         0         0
         0         0
         0         0
         0         0

D12 =
         0         0
         0         0
         0         0
         0         0
    3.2500   -0.8000
   -0.2000   -0.0050

D21 =
     1     0
     0     1

D22 =
     0     0
     0     0

%% Automated H_infinity iterations follow:
>> [gamopt,acp,bcp,ccp,dcp,acl,bcl,ccl,dcl]=hinfopt(A,B1,B2,C1,C2,D11,D12,D21,D22)
<< H-Infinity Optimal Control Synthesis >>

  No      Gamma      D11<=1    P-Exist    P>=0    S-Exist    S>=0   lam(PS)<1     C.L.
 ------------------------------------------------------------------------------------
   1 1.0000e+000     OK         FAIL      FAIL      OK       FAIL       OK         UNST
   2 5.0000e-001     OK         FAIL      FAIL      OK       FAIL      FAIL        UNST
   3 2.5000e-001     OK         FAIL       OK       OK        OK       FAIL        UNST
   4 1.2500e-001     OK         FAIL       OK       OK        OK        OK         STAB
   5 6.2500e-002     OK          OK        OK       OK        OK        OK         STAB
   6 9.3750e-002     OK          OK        OK       OK        OK        OK         STAB
   7 1.0938e-001     OK         FAIL       OK       OK        OK        OK         STAB
   8 1.0156e-001     OK         FAIL       OK       OK        OK        OK         STAB
   9 9.7656e-002     OK          OK        OK       OK        OK        OK         STAB
  10 9.9609e-002     OK          OK        OK       OK        OK        OK         STAB
  11 1.0059e-001     OK          OK        OK       OK        OK        OK         STAB

      Iteration no. 11 is your best answer under the tolerance:   0.0100 .
gamopt =
    0.1006
```

```
acp =
   -0.5518   -0.4460    0.0323   -2.9563   -0.3065    3.5477   -9.2524  -34.0226
   -0.4102   -1.9614    1.0573   -6.9924   -0.9144   17.4240   -8.8076  -55.5082
   -0.2041    0.9167   -0.9071    3.6392    0.5202   -9.9459    5.3254   32.7666
   -1.5257   -4.8025    3.3243  -61.7372   -6.8206   70.0469 -238.2181 -844.8057
   -0.0784   -0.2665    0.1836   -3.0554   -0.4398    3.8673  -10.6609  -39.0291
    0.4133    3.1583   -2.1712   18.0132    2.4924  -50.0146   20.2450  142.9396
   -0.6034   -0.8240    0.5705  -31.4728   -3.4754   10.3659 -204.7381 -642.2426
   -0.0143   -0.0260    0.0201   -0.6273   -0.2617    0.3690   -3.4897  -20.9749

bcp =
    0.0105    0.5093
   -0.0025   -0.4237
    0.0126   -0.7350
   -0.0441   -0.0308
    1.8164   -0.0031
   -0.0187   -0.0548
    0.1064   -0.0633
   -0.2404   -0.0000

ccp =
   -0.1129   -1.1499    0.8523   -6.5519   -0.8749   17.8805   -4.9434  -44.4021
   -0.1014   -1.0532    0.7474   -2.2752   -0.4073   16.7901   23.2167   43.6303

dcp =
  1.0e-014 *
   -0.1166         0
   -0.4756         0

>> syscl=ss(acl,bcl,ccl,dcl); %Closed-loop system
>> [y,t,x]=initial(syscl,[0 .1 0 0 zeros(1,12)]',0:0.1:20);% Initial response
%% Closed-loop control inputs:
>> sysc=ss(acp,bcp,ccp,dcp);
>> dy=[];
>> for i=1:size(y,1)
>> dy(i,:)=-y(i,5:6)/gamopt;
>> end
>> u=lsim(sysc,dy,t);
```

The smallest value of γ achieved in the iterations for a stabilizing controller is thus $\gamma = 1/0.1006 = 9.9404$. The closed-loop initial response and control inputs are plotted in Figures 2.20 and 2.21, respectively. The design specifications are seen to be satisfied.

2.13 Summary

Optimal control theory can be applied to the design of both terminal and tracking control systems. In other words, a nominal trajectory to be followed and the tracking method both arise naturally from optimal control. Necessary and sufficient conditions for the minimization of an objective function of state and control variables, subject to the dynamic constraint of the plant's state equation, are readily derived. A terminal control system changes the plant's state from an initial state to a terminal state by applying a control input in either a free or fixed control interval. The terminal state may not be explicitly known, but should satisfy a terminal equality constraint. Furthermore, interior point constraints – apart from plant's state equation – on the control input and state variables may also be imposed. Terminal controller design

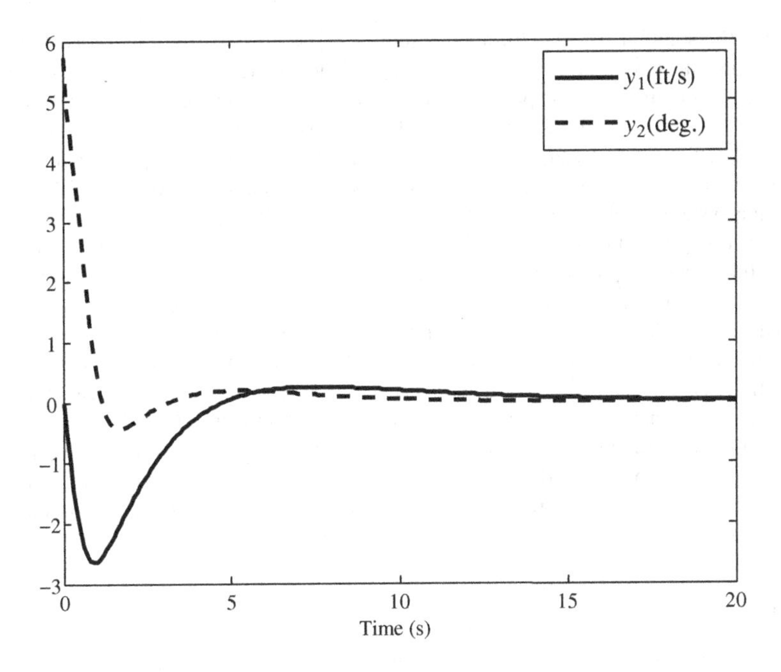

Figure 2.20 Initial response of H_∞ compensated system

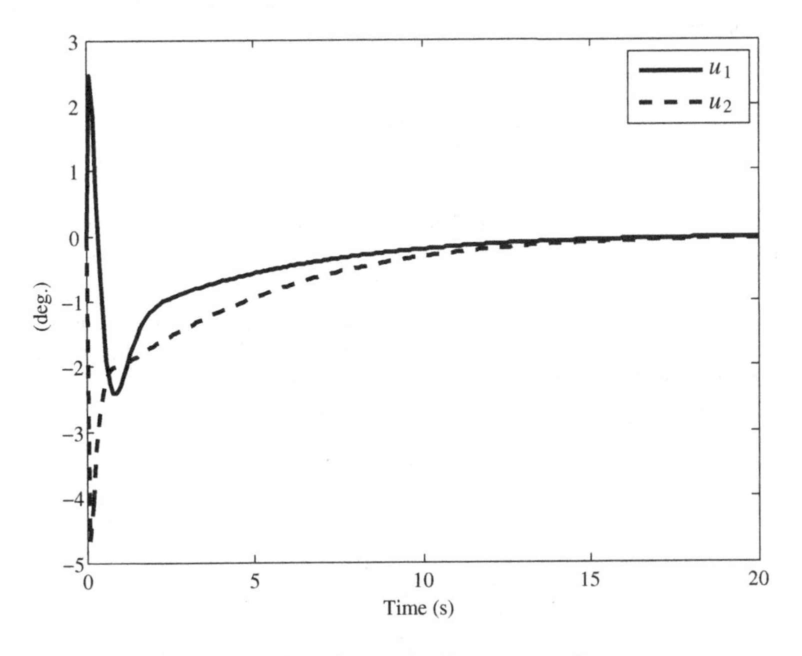

Figure 2.21 Control inputs for H_∞ compensated system

requires an iterative solution of the nonlinear plant state equation subject to the initial conditions, terminal and equality constraints, as well as minimization of the objective function. Such a solution procedure is called a two-point boundary value problem and produces a nominal trajectory and the corresponding nominal control history. The 2PBVP can be solved by either shooting or collocation methods.

An optimal tracking system can be derived from the linearization of the plant about the nominal trajectory and minimization of a quadratic objective function with a linear feedback of the state vector. Such a tracking system is called a linear quadratic regulator (LQR) and can be derived from either the necessary (Euler–Lagrange) or sufficient (Hamilton–Jacobi–Bellman) conditions of optimality. When stochastic disturbances are present in the control system, an additional optimal subsystem – called a Kalman filter – is required to supply an accurate estimate of the plant's state to the regulator. The LQR regulator and Kalman filter are dual to each other, both requiring a positive semi-definite solution to the matrix Riccati equation, and can be designed separately. In the linear time-invariant control system, the Riccati equation becomes algebraic, the disturbances are stationary zero-mean Gaussian white noise, and the resulting combination of LQR and Kalman filter is called the linear quadratic Gaussian (LQG) compensator, a robust design of which can be iteratively derived for loop-transfer recovery at either in the input or the output of the plant. Alternatively, direct frequency domain design methods can be applied to the multi-variable plant to minimize either the H_2 or the H_∞ norm of a frequency-weighted, mixed sensitivity transfer matrix.

Exercises

(1) Determine the extremal and minimum points of

$$L = 3x_1^3 + x_2 + x_3^2 + \tan u_1 - 2u_2^3$$

with respect to $\mathbf{u} = (u_1, u_2)^T$.

(2) For the function minimization of Exercise 1, determine the minimum points, subject to the following equality constraints:

$$\begin{aligned} x_1 - x_3 &= -\sin u_1, \\ 2x_2 + x_3 &= 4\cos u_1, \\ -5x_1 + 3x_2 &= \frac{1}{2}u_2. \end{aligned}$$

(3) For the airplane of Example 2.3 what is the maximum angle of climb, ϕ, at standard sea level, subject to the given constraints? Also, determine the airspeed corresponding to the maximum angle of climb, v.

(4) For the airplane of Example 2.3, instead of maximizing the rate of climb, it is required to minimize the total fuel consumed for a steady climb beginning at sea level ($h_0 = 0$) to a final altitude, $h_f = 5$ km. The additional constraints for the variation of atmospheric density, $\rho(h)$, thrust, $f_T(h)$, airplane weight, $W = mg$, and altitude are the following:

$$\rho = 1.225\left(1 - \frac{0.0065}{288.15}h\right)^{4.258644} \quad (\text{kg/m}^3), \tag{2.289}$$

$$f_T \leq 98000\rho \text{ (N)}, \tag{2.290}$$

$$\dot{W} = -\frac{0.8}{3600}f_T \text{ (N/s)}, \tag{2.291}$$

$$\dot{h} = v \sin \phi. \tag{2.292}$$

Clearly, equations (2.291) and (2.292) constitute dynamic constraints for the minimization of the following objective function:

$$\mathcal{J} = W_{\text{fuel}} = -\int_0^{h_f} \frac{\dot{W}}{\dot{h}} dh. \tag{2.293}$$

(a) Formulate the dynamic optimization problem with $x_1 = v$ and $x_2 = \sin \phi$ as state variables and the thrust, $u = f_T$, as the control input. Instead of the time, t, use the altitude, h, as the independent variable.
(b) Write MATLAB code based upon the boundary-value problem solver, *bvp4c.m*, for the solution of the optimal control problem.
(c) Determine the extremal trajectory, $v(h)$ =const., $\phi(h)$, and the optimal control history, $f_T(h)$, for the initial weight, $W(0) = 150$ kN.

(5) For the landing of a lunar spacecraft on the Moon's surface ($R_0 = 1737.1$ km) from a lunar orbit of initial radius $r_0 = 1850$ km, the following constraints must be satisfied:

$$\dot{r} = v \sin \phi, \tag{2.294}$$

$$\dot{v} = u - \frac{\mu}{r^2} \sin \phi, \tag{2.295}$$

$$v\dot{\phi} = \left(\frac{v^2}{r} - \frac{\mu}{r^2} \right) \cos \phi, \tag{2.296}$$

where $\mu = 4902.8$ km^3/s^2.
(a) Devise a terminal optimal control problem with (r, v, ϕ) as state variables and thrust acceleration, u, as the control input.
(b) Write MATLAB code based upon the boundary-value problem solver, *bvp4c.m*, for the solution of the optimal control problem.
(c) Determine the extremal trajectory and the optimal control history for the boundary conditions, $r(0) = r_0$, $v(0) = 2.128$ km/s, $\phi(0) = -1.4°$, $v(t_f) = 0.005$ km/s, and adjust t_f until a final radius of about $r(t_f) = R_0 + 0.2$ km is obtained. What is the terminal flight-path angle, $\phi(t_f)$? (*Ans.* $t_f = 1345$ *s*, $\phi(t_f) = -90°$. *See Figure 2.22.*)

(6) Consider the minimization of

$$\mathcal{J} = q \,\|\, \mathbf{x}(t_f) \,\|^2 + \int_0^{t_f} \|\, \mathbf{u}(t) \,\|^2 \, dt,$$

subject to

$$\dot{\mathbf{x}}(t) = \mathbf{B}(t)\mathbf{u}(t)$$

and

$$\|\, \mathbf{u}(t) \,\| \leq 1,$$

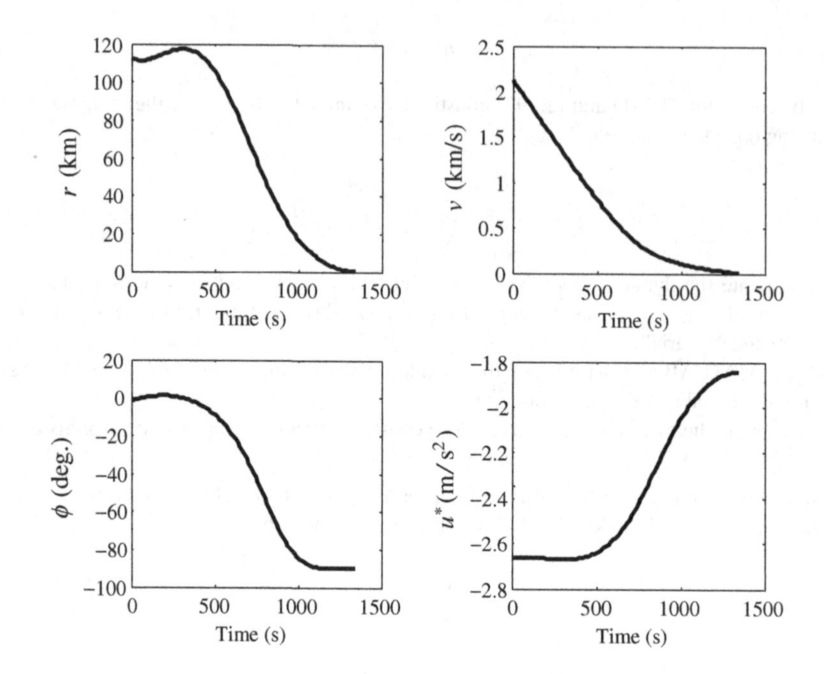

Figure 2.22 Optimal trajectory for lunar lander of Exercise 5

where $q > 0$. Solve for the singular optimal control $\mathbf{u}^*(t)$ for the initial condition, $\mathbf{x}(0) = \mathbf{x}_0$. What is the switching function of this problem?

(7) For a second-order plant given by

$$\ddot{x} = u$$

and

$$-1 \leq u(t) \leq 1,$$

solve the singular, time-optimal control problem (i.e., $L = 1$), subject to boundary conditions $x(0) = x_0$, $x(t_f) = x_f$. What is the switching function of this problem?

(8) Design a linear quadratic regulator for the lunar lander of Exercise 5 using the derived optimal trajectory and control history as nominal values for linearizing the plant. Examine the stability and controllability of the resulting linearized plant. The LQR design is to be based upon fixed gains derived from infinite-horizon control of the plant dynamics at $t = 600$ s. (Such an application of the infinite-horizon solution for controlling a time-varying plant is termed the *frozen gain method*).

(9) Let $x(t)$ and $y(t)$ be Gaussian processes with mean values $\bar{x}, \bar{y}$ and standard deviations σ_x, σ_y, respectively. Show that the sum $z(t) = x(t) + y(t)$ is also a Gaussian process. What is the probability distribution of $z(t)$?

(10) Consider a scalar Markov process described by the state equation

$$\dot{x}(t) = a[w(t) - x(t)], \tag{2.297}$$

where a is a positive, real constant and $w(t)$ is a stationary zero-mean white noise with variance σ_w^2. What is the autocorrelation function of $x(t)$, $\psi_{xx}(t, t+T)$, for values separated by a sampling interval, T? What happens to $\psi_{xx}(t, t+T)$ in the limit $a \to \infty$?

(11) Suppose the white noise input, $w(t)$, in Exercise 10 is a stationary zero-mean Gaussian process with variance σ_w^2. If the initial state of the resulting Markov process has a Gaussian probability distribution, $p[x(t_0)]$, determine the mean and variance of the Markov process.

(12) Design a compensator for a flexible bomber aircraft with the following state-space representation by the LQG/LTR method:

$$\mathbf{A} = \begin{pmatrix} 0.4158 & 1.025 & -0.00267 & -0.0001106 & -0.08021 & 0 \\ -5.5 & -0.8302 & -0.06549 & -0.0039 & -5.115 & 0.809 \\ 0 & 0 & 0 & 1 & 0 & 0 \\ -1040 & -78.35 & -34.83 & -0.6214 & -865.6 & -631 \\ 0 & 0 & 0 & 0 & -75 & 0 \\ 0 & 0 & 0 & 0 & 0 & -100 \end{pmatrix},$$

$$\mathbf{B} = \begin{pmatrix} 0 & 0 \\ 0 & 0 \\ 0 & 0 \\ 0 & 0 \\ 75 & 0 \\ 0 & 100 \end{pmatrix}, \quad \mathbf{D} = \begin{pmatrix} 0 & 0 \\ 0 & 0 \end{pmatrix},$$

$$\mathbf{C} = \begin{pmatrix} -1491 & -146.43 & -40.2 & -0.9412 & -1285 & -564.66 \\ 0 & 1 & 0 & 0 & 0 & 0 \end{pmatrix}.$$

The inputs are the desired elevator deflection, $u_1(t)$ (rad), and the desired canard deflection, $u_2(t)$ (rad), while the outputs are the normal acceleration, $y_1(t)$ (m/s^2), and the pitch rate, $y_2(t)$ (rad/s). The design should result in a normal acceleration within ± 2 m/s^2, pitch rate within ± 0.03 rad/s, and should settle in about 5 s, while requiring elevator and canard deflections within $\pm 10°$, for an initial condition of $\mathbf{x}(0) = (0,\ 0.1\ \text{rad/s},\ 0,\ 0,\ 0,\ 0)^T$. The design must have an adequate margin of stability with respect to process noise occurring in the control system bandwidth, and should reject high-frequency measurement noise with a roll-off of 20 dB/decade.

(13) Redesign the compensator of Exercise 12 by the H_∞ method. What possible frequency weights can satisfy the design requirements?

References

Abou-Kandil, H., Freiling, G., Ionescu, V., and Jank, G. (2003) *Matrix Riccati Equations in Control and Systems Theory*. Birkhäuser, Basel.

Arnold, W.F. and Laub, A.J. (1984) Generalized eigenproblem algorithms and software for algebraic Riccati equations. *Proceedings of the IEEE* **72**, 1746–1754.

Athans, M. and Falb, P.L. (2007) *Optimal Control*. Dover, New York.

Atkinson, K.E. (2001) *An Introduction to Numerical Analysis*. John Wiley & Sons, Inc., New York.

Bellman, R. (1957) *Dynamic Programming*. Princeton University Press, Princeton, NJ.
Bryson, A.E. Jr and Ho, Y.C. (1975) *Applied Optimal Control*. Hemisphere, Washington, DC.
Fehlberg, E. (1969) *Low Order Classical Runge-Kutta Formulas with Stepsize Control and Their Application to Some Heat Transfer Problems*. NASA Technical Report R-315.
Glad, T. and Ljung, L. (2002) *Control Theory*. Taylor and Francis, New York.
Glover, K. and Doyle, J.C. (1988) State-space formulae for all stabilizing controllers that satisfy an H_∞ norm bound and relations to risk sensitivity. *Systems and Control Letters* **11**, 167–172.
Kreyszig, E. (1998) *Advanced Engineering Mathematics*. John Wiley & Sons, Inc., New York.
Maciejowski, J.M. (1989) *Multivariable Feedback Design*. Addison-Wesley, Wokingham.
Mattheij, R.M.M. and Molenaar, J. (2002) *Ordinary Differential Equations in Theory and Practice*, SIAM Classics in Applied Mathematics **43**. SIAM, Philadelphia.
Papoulis, A. (1991) *Probability, Random Variables, and Stochastic Processes*. McGraw-Hill, New York.
Pontryagin, L.S., Boltyanskii, V., Gamkrelidze, R., and Mischenko, E. (1962) *The Mathematical Theory of Optimal Processes*. Interscience, New York.
Shampine, L.F., Reichelt, M.W., and Kierzenka, J. (2000) *Solving Boundary Value Problems for Ordinary Differential Equations in MATLAB with bvp4c*. http://www.mathworks.com/access/pub/bvp.zip
Shampine, L.F., Gladwell, I., and Thompson, S. (2003) *Solving ODEs with MATLAB*. Cambridge University Press, Cambridge.
Stengel, R.F. (1994) *Optimal Control and Estimation*. Dover, New York.
Stoer, J. and Bulirsch R. (2002) *Introduction to Numerical Analysis*. Springer-Verlag, New York.
Tewari, A. (2002) *Modern Control Design with MATLAB and Simulink*. John Wiley & Sons, Ltd, Chichester.

3

Optimal Navigation and Control of Aircraft

Aircraft flight is confined to the denser regions of the atmosphere for necessary aerodynamic lift and airbreathing propulsion, and thus takes place below 30 km altitude. The fastest aircraft speed is only a few multiples of the speed of sound (flight Mach number below 3.0), which is a fraction of the orbital speed at a given altitude. Therefore, for attitudinal motion as well as for short-range flight, one can safely neglect the flight-path curvature and Coriolis acceleration due to the Earth's rotation. However, in long-range flight, these must be accurately taken into account. We address aircraft flight control tasks by separating them into (a) navigation (or guidance) systems that involve both terminal and tracking control in following optimal trajectories obtained from the solution of two-point boundary value problems (2PBVPs), and (b) autopilots (or set-point regulators) that maintain the vehicle in a prescribed, constant state despite disturbances. In the former category we have the sophisticated *flight management systems* that carefully guide the aircraft through a set of computed *waypoints* (i.e., latitude–longitude pairs), which are sensed from either ground-based radio signals, *global positioning system* data, or an onboard *inertial navigation system* computer. On the other hand, autopilots are tasked to hold a specific constant speed, heading (azimuth), and altitude, or even make an automatic landing by regulating the glide-slope, azimuth, speed, and height above the runway by intercepting and following a ground-based radio signal that constitutes an *instrument landing system*. Even the most advanced autopilot is merely a regulator whose reference state is set by the pilot, and its design is easily carried out by linear control methods.

Here it is necessary to mention that a third (and much more sophisticated) class of aircraft control systems can be envisaged in which an unmanned, fully automated, robotic vehicle charts its own course using navigation by terminal and tracking control, and by multi-functional autopilot with set-points changed automatically when required. Such an aircraft is termed *autonomous* and does not require any human intervention during its mission, apart from initially entering the desired mission in the onboard flight computer. An autonomous flight control system must simultaneously account for changing weather and air traffic en route, while achieving optimal fuel consumption for the mission despite winds and unforeseen diversions. Needless to say, an autonomous flight controller must possess the most adaptive capability that can be imagined of any control system. At present, we do not have any operational example of fully autonomous aircraft, but a similar concept is being explored for forthcoming Martian missions.

We will address each of the aforementioned kinds of aircraft flight control system using optimal control techniques discussed in Chapter 2. While the complete set of translational equations of motion should be employed in determining the optimal flight trajectories for long-range navigation, the translational

---

Advanced Control of Aircraft, Spacecraft and Rockets, Ashish Tewari.

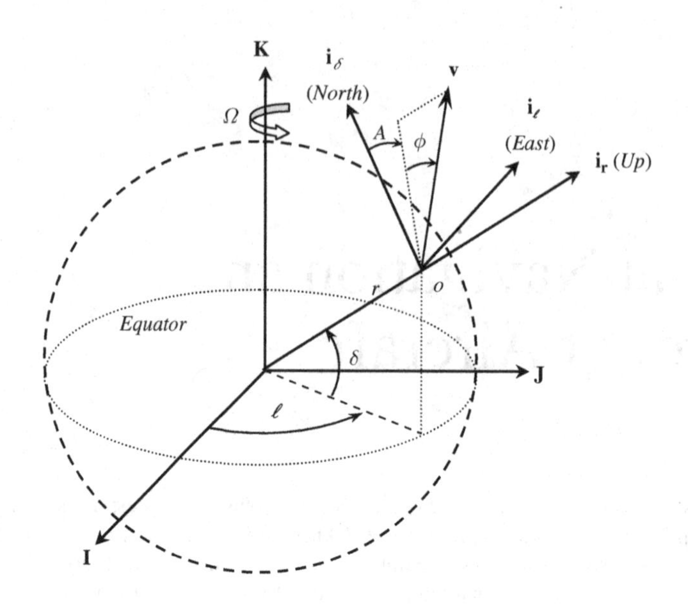

Figure 3.1 Planet fixed frame, $(\mathbf{I}, \mathbf{J}, \mathbf{K})$, and the local horizon frame, $(\mathbf{i_r}, \mathbf{i}_\ell, \mathbf{i}_\delta)$

kinematic behavior is separated from the much faster translational and rotational kinetics when deriving the aircraft plant for autopilot design. Such an approximation enables us to treat the local horizon frame virtually as an inertial reference frame for an aircraft autopilot. On the other hand, the rotational kinetics are unimportant in deriving the plant for a navigation system, these being much faster than the time scale of a long-range flight. Therefore, it is customary to either completely neglect the rotational dynamics by assuming instantaneous changes of attitude, or to treat the aircraft attitude dynamics as an actuator for the navigational control system.

3.1 Aircraft Navigation Plant

Aircraft navigation refers to planning and following a specific flight path that can take the aircraft between two given horizontal locations in the presence of winds and planetary rotation. Since aircraft navigation takes place relative to a rotating, essentially spherical planet, the plant must account for the equations of motion for the translation of the vehicle's center of mass with reference to a planet fixed coordinate frame, $(\mathbf{I}, \mathbf{J}, \mathbf{K})$, attached to the center of a spherical, rotating planet of radius R_0 and angular speed Ω (Figure 3.1). The first two axes, $\mathbf{I}$, $\mathbf{J}$, are in the equatorial plane and the third, $\mathbf{K}$, toward the North Pole. Defining a *local horizon frame*, $(\mathbf{i}_\ell, \mathbf{i}_\delta, \mathbf{i_r})$, with axes along *East longitude*, $\ell(t)$, *latitude*, $\delta(t)$, and *local vertical*, $\mathbf{r}(t)$, respectively (Figure 3.1), the *relative velocity*, $\mathbf{v}(t)$, is expressed as follows:

$$\mathbf{v} = v(\sin\phi \mathbf{i_r} + \cos\phi \sin A \mathbf{i}_\ell + \cos\phi \cos A \mathbf{i}_\delta), \tag{3.1}$$

where $\phi(t)$ is the *flight-path angle*, and $A(t)$ is the *heading angle* (or *velocity azimuth*), as shown in Figure 3.1. The coordinate transformation between the planet-fixed and local horizon frames is given by Euler angle sequence (Tewari 2006) $(\ell)_3, (-\delta)_2$ as follows:

$$\begin{Bmatrix} \mathbf{i_r} \\ \mathbf{i}_\ell \\ \mathbf{i}_\delta \end{Bmatrix} = \mathbf{C}_{\mathrm{LH}} \begin{Bmatrix} \mathbf{I} \\ \mathbf{J} \\ \mathbf{K} \end{Bmatrix}, \tag{3.2}$$

where

$$\mathbf{C}_{\mathrm{LH}} = \mathbf{C}_2(-\delta)\mathbf{C}_3(\ell) = \begin{pmatrix} \cos\delta\cos\ell & \cos\delta\sin\ell & \sin\delta \\ -\sin\ell & \cos\ell & 0 \\ -\sin\delta\cos\ell & -\sin\delta\sin\ell & \cos\delta \end{pmatrix} \quad (3.3)$$

With planetary angular velocity, $\mathbf{\Omega} = \Omega\mathbf{K}$, and the angular velocity of the local horizon frame relative to the planet-fixed frame given by

$$\boldsymbol{\omega}_{\mathrm{LH}} = \dot{\ell}\mathbf{K} - \dot{\delta}\mathbf{i}_\ell, \quad (3.4)$$

we can express the relative and inertial velocities, $\mathbf{v}(t)$, $\mathbf{v}_o(t)$, respectively, as follows:

$$\mathbf{v} = \dot{h}\mathbf{i}_\mathbf{r} + (R_0 + h)(\dot{\ell}\cos\delta\mathbf{i}_\ell + \dot{\delta}\mathbf{i}_\delta), \quad (3.5)$$

$$\mathbf{v}_o = \mathbf{v} + \Omega(R_0 + h)\cos\delta\mathbf{i}_\ell. \quad (3.6)$$

Here, $r(t) = R_0 + h(t)$ is the radial coordinate of the center of mass, with R_0 being the (constant) radius of the spherical planet, and $h(t)$, the altitude.

Comparing equation (3.5) with equation (3.1), we have the following translational kinematics equations:

$$\dot{r} = v\sin\phi, \quad (3.7)$$

$$\dot{\delta} = \frac{v\cos\phi\cos A}{r}, \quad (3.8)$$

$$\dot{\ell} = \frac{v\cos\phi\sin A}{r\cos\delta}. \quad (3.9)$$

Substituting equation (3.6) into Newton's second law of motion, expressed as

$$\mathbf{F} = m\dot{\mathbf{v}}_o + m(\boldsymbol{\omega}\times\mathbf{v}_o), \quad (3.10)$$

with the vehicle's net angular velocity, $\boldsymbol{\omega} = \boldsymbol{\omega}_{\mathrm{LH}} + \mathbf{\Omega}$, we can write the translational kinetics equations as follows:

$$\dot{v} = \frac{F_{vx}}{m} - g\sin\phi - \Omega^2 r\cos\delta(\cos\phi\cos A\sin\delta - \sin\phi\cos\delta), \quad (3.11)$$

$$\begin{aligned} v\dot{A}\cos\phi = {} & \frac{F_{vy}}{m} + \frac{v^2\cos^2\phi\sin A}{r}\tan\delta + \Omega^2 r\sin A\sin\delta\cos\delta \\ & -2\Omega v(\sin\phi\cos A\cos\delta - \cos\phi\sin\delta), \end{aligned} \quad (3.12)$$

$$\begin{aligned} v\dot{\phi} = {} & \frac{F_{vz}}{m} + \left(\frac{v^2}{r} - g\right)\cos\phi + \Omega^2 r\cos\delta(\sin\phi\cos A\sin\delta + \cos\phi\cos\delta) \\ & + 2\Omega v\sin A\cos\delta, \end{aligned} \quad (3.13)$$

where the external forces other than spherical gravity are clubbed into the vector $F_v = (F_{vx}, F_{vy}, F_{vz})^T$.

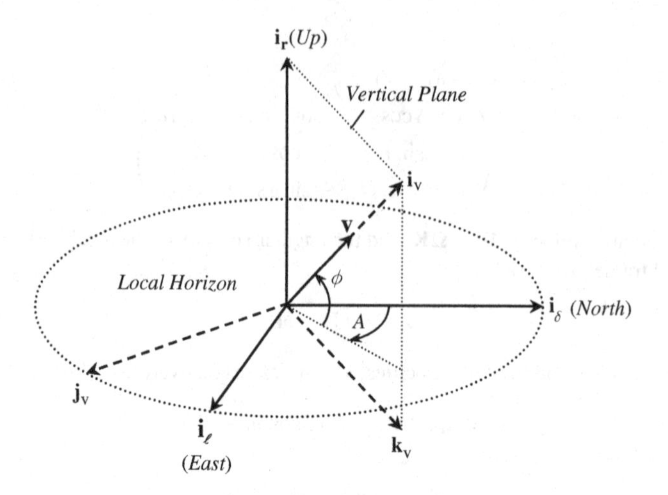

Figure 3.2 The wind axes, $(\mathbf{i}_v, \mathbf{j}_v, \mathbf{k}_v)$, and the local horizon frame, $(\mathbf{i}_r, \mathbf{i}_\ell, \mathbf{i}_\delta)$

Thus, the position of the aircraft's center of mass is described by the spherical coordinates (r, δ, ℓ), while its relative velocity is given by the spherical coordinates, (v, ϕ, A).

In equations (3.11)–(3.13) the aerodynamic and thrust forces are combined into the force vector, F_v, resolved in a moving, right-handed *wind axes* coordinate frame, $(\mathbf{i}_v, \mathbf{j}_v, \mathbf{k}_v)$, with the axis $\mathbf{i}_v$ along the relative velocity vector, and the plane $(\mathbf{i}_v, \mathbf{k}_v)$ being a vertical plane (as shown in Figure 3.2). Note that the orientation of the wind axes is especially chosen such that the unit vector, $\mathbf{j}_v$, denotes the direction of increasing velocity azimuth, A, while $-\mathbf{k}_v$ indicates the direction of increasing flight-path angle, ϕ.

The aerodynamic forces of *lift*, $\mathcal{L}$, *drag*, $\mathcal{D}$, and *sideforce*, f_Y, depend upon the orientation of the aircraft relative to the velocity vector. The lift is the aerodynamic force component that acts normal to the reference wing-plane as well as the flight direction, $\mathbf{i}_v$, while the drag acts opposite to the flight direction. An aircraft generally has a geometric *plane of symmetry*, and is shaped such that the lift always lies in the plane of symmetry, and the smallest aerodynamic drag is achieved when the flight direction relative to the atmosphere, $\mathbf{i}_v$, lies in the plane of symmetry. The sideforce, f_Y, is the sideways component of the net aerodynamic force acting normal to the plane of symmetry. The condition in which the flight direction is confined to the plane of symmetry also leads to a zero sideforce and is called *coordinated flight*. To achieve coordinated flight and hence the maximum aerodynamic efficiency, an aircraft requires precise attitude control – a topic to be discussed later. In the absence of a sideforce, the only way the aircraft can be made to turn horizontally in a coordinated flight is by *banking* the wings, that is, tilting the plane of symmetry such that it makes an angle, σ, (called the *bank angle*) with the local vertical plane, $(\mathbf{i}_v, \mathbf{k}_v)$, as shown in Figure 3.3. In such a case, the lift force, $\mathcal{L}$, acts in the $(\mathbf{j}_v, \mathbf{k}_v)$ plane, thereby creating a centripetal force component, $\mathcal{L} \sin \sigma$, along $\mathbf{j}_v$ (Figure 3.3).

In order to achieve the maximum fuel economy, the direction of thrust, $\mathbf{f}_T$, must lie in the plane of symmetry, and should be nearly aligned with the flight direction, $\mathbf{i}_v$. Thus, we assume that the thrust makes an angle, ϵ (called the *thrust angle*) with $\mathbf{i}_v$, as shown in Figure 3.3. Except for a few special aircraft that are capable of *thrust vectoring* (i.e., changing the thrust angle in flight by thrust deflection), all conventional aircraft have a small but fixed thrust angle due to the engines being rigidly bolted to the airframe. In such a case, the thrust, $\mathbf{f}_T$, approximately acts along the longitudinal axis of the aircraft.

The sum of aerodynamic, $\mathbf{f}_v$, and thrust, $\mathbf{f}_T$, force vectors for coordinated flight resolved in the wind axes can be expressed as follows:

$$\mathbf{F}_v = \mathbf{f}_T + \mathbf{f}_v = (f_T \cos \epsilon - \mathcal{D})\,\mathbf{i}_v + \mathcal{L} \sin \sigma \mathbf{j}_v - (f_T \sin \epsilon + \mathcal{L} \cos \sigma)\,\mathbf{k}_v, \tag{3.14}$$

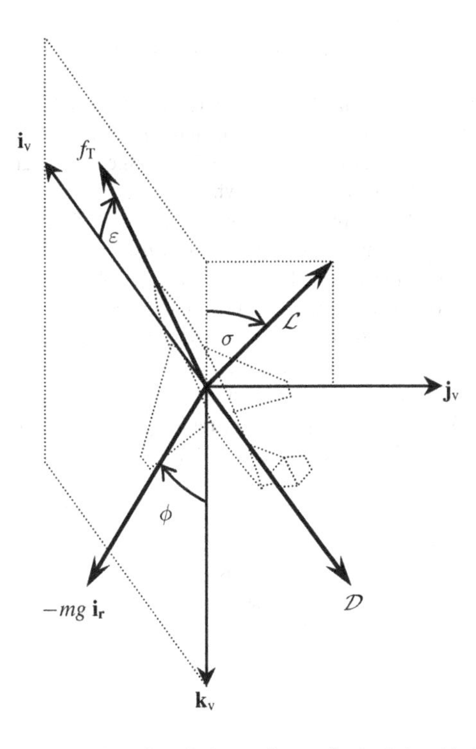

Figure 3.3 The external forces resolved in the wind axes frame, $(\mathbf{i}_v, \mathbf{j}_v, \mathbf{k}_v)$, with flight-path angle, ϕ, and bank angle, σ

with the coordinate transformation between the wind axes and the local horizon frame being given by

$$\begin{Bmatrix} \mathbf{i}_v \\ \mathbf{j}_v \\ \mathbf{k}_v \end{Bmatrix} = \mathbf{C}_W \begin{Bmatrix} \mathbf{i}_r \\ \mathbf{i}_\ell \\ \mathbf{i}_\delta \end{Bmatrix}. \tag{3.15}$$

Here, the rotation matrix is

$$\begin{aligned} \mathbf{C}_W &= \mathbf{C}_2\left(\phi - \tfrac{\pi}{2}\right)\mathbf{C}_1(-A) \\ &= \begin{pmatrix} \sin\phi & \cos\phi\sin A & \cos\phi\cos A \\ 0 & \cos A & -\sin A \\ -\cos\phi & \sin\phi\sin A & \sin\phi\cos A \end{pmatrix}. \end{aligned} \tag{3.16}$$

With the radial location of the vehicle's center of mass given by

$$r = R_0 + h, \tag{3.17}$$

the acceleration due to spherical gravity field is expressed as follows:

$$-g\mathbf{i}_r = -\frac{g_0 R_0^2}{(R_0 + h)^2}\mathbf{i}_r. \tag{3.18}$$

However, since $h \ll R_0$ for an aircraft, the approximation of a constant acceleration due to gravity, $g \simeq g_0$ is valid for all airbreathing aircraft.[1]

An airplane is primarily designed to *cruise*, that is, fly level and steadily at a constant thrust (or power) setting by maintaining a nearly constant airspeed (or Mach number) and altitude, for either the maximum range, or endurance. For subsonic jet airliners, the cruising airspeed is slightly smaller than the speed of sound in order to maximize the lift-to-drag ratio which yields the longest range (Tewari 2006). The aircraft mass slowly decreases with time due to consumption of fuel. The fuel consumption of a modern jet engine per unit thrust is nearly a minimum at the *tropopause*, a standard altitude of 11 km (Tewari 2006). Propeller-driven airplanes have a much lower cruising altitude, with the piston-driven ones flying below 5 km. As the flight progresses at a nearly constant airspeed and the fuel is consumed at an optimal rate, the airplane is allowed to climb gradually in accordance with a reduction of its weight such that thrust and drag are always nearly equal. Consequently, the cruising flight can be approximated to be a *quasi-level* flight, at a nearly constant (*quasi-steady*) airspeed, v, and the equations of motion for the aircraft navigation plant can be derived with the small-angle approximation of $\sin\phi \approx \phi$ and $\cos\phi \approx 1$, and by neglecting terms involving the product of ϕ with Ω:

$$\dot{h} \simeq v\phi, \tag{3.19}$$

$$\dot{\delta} \simeq \frac{v\cos A}{R_0 + h}, \tag{3.20}$$

$$\dot{\ell} = \frac{v\sin A}{(R_0 + h)\cos\delta}, \tag{3.21}$$

$$\dot{v} = \frac{f_T\cos\epsilon - \mathcal{D}}{m} - g_0\phi - \frac{1}{2}\Omega^2(R_0 + h)\cos A\sin 2\delta, \tag{3.22}$$

$$v\dot{\phi} = \frac{\mathcal{L}\cos\sigma + f_T\sin\epsilon}{m} + \frac{v^2}{R_0 + h} - g_0 + \Omega^2(R_0 + h)\cos^2\delta + 2\Omega v\sin A\cos\delta, \tag{3.23}$$

$$v\dot{A} = \frac{\mathcal{L}\sin\sigma}{m} + \frac{v^2\sin A}{r}\tan\delta + \Omega^2 r\sin A\sin\delta\cos\delta + 2\Omega v\sin\delta. \tag{3.24}$$

Note that an additional state equation is necessary for the rate of fuel consumption, which is determined from either a constant *thrust-specific fuel consumption* (TSFC), c_T, of jet powered airplanes,

$$\dot{m} = -\frac{c_T f_T}{g_0}, \tag{3.25}$$

or a constant *power-specific fuel consumption* (PSFC), c_P, of propeller engined airplanes given by

$$\dot{m} = -\frac{c_P f_T v}{g_0}. \tag{3.26}$$

The thrust (and thus the TSFC or PSFC) is generally a function of the airspeed and the atmospheric density (the cruising altitude) as well as the *throttle setting*, μ, which is usually given by the ratio of

---

[1] Exceptions are rocket-powered, experimental aircraft such as the Bell/NASA X-15 which established an altitude record of 80 km in 1962.

engine's rotational speed, n, and the maximum rotational speed, $n_{\max}$,

$$\mu = \frac{n}{n_{\max}}.$$

Thus, we can write the functional equations for the thrust and TSFC as

$$f_T = f_T(h, v, \mu), \quad c_T = c_T(h, v, \mu). \tag{3.27}$$

The power and PSFC can be expressed by similar functional relationships. When a constant throttle setting is maintained at a given speed and altitude, the rate of fuel consumption is constant.

The magnitude of the lift and drag generated by an aircraft depends upon the *angle of attack*, α, defined as the angle made by the flight direction with a reference geometric line in the plane of symmtery, such as the *mean aerodynamic chord* line (Figure 3.4). The variation of lift and drag with the angle of attack in a cruising flight is given by the following linear non-dimensional aerodynamic relationships:

$$C_L = \frac{\mathcal{L}}{\frac{1}{2}\rho v^2 S} = C_{L\alpha}(\alpha - \alpha_{0L}), \tag{3.28}$$

$$C_D = \frac{\mathcal{D}}{\frac{1}{2}\rho v^2 S} = C_{D0} + KC_L^2, \tag{3.29}$$

where the atmospheric density, ρ, is a function of the altitude, h, and can be represented by a standard atmospheric model (Tewari 2006). The lift and drag coefficients, C_L and C_D respectively, are based upon the reference area, S, which is usually the *wing planform area* shown in Figure 3.4. The aerodynamic parameters, $C_{L\alpha}$, α_{0L}, C_{D0} and K, depend upon the flight Mach number and Reynolds number that are essentially constant when cruising. Thus, the lift and drag are modulated by changing the angle

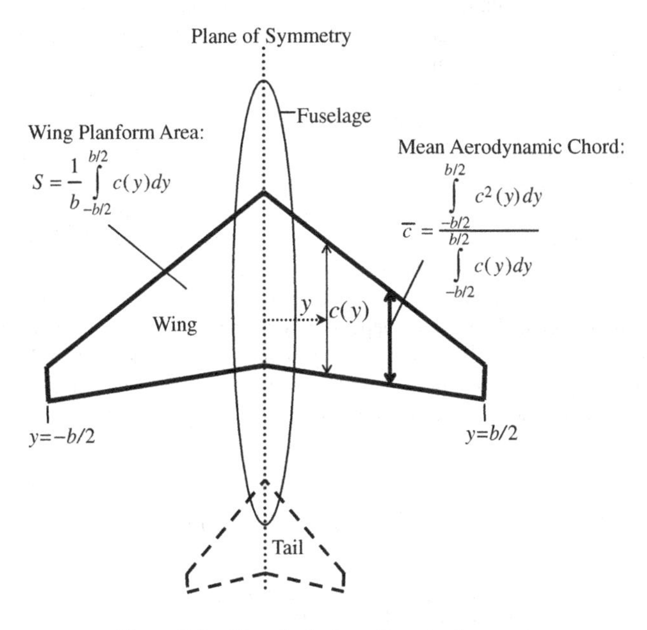

Figure 3.4 Aircraft planform (top view) geometry

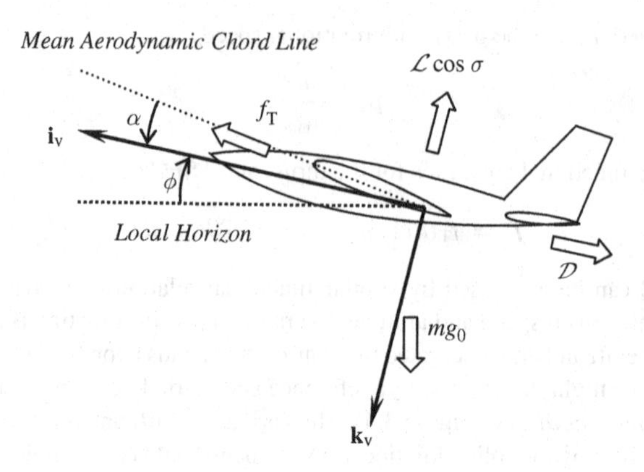

Figure 3.5 External forces in the vertical plane containing the relative velocity vector

of attack with nearly constant airspeed and constant atmospheric density of a cruising flight. For a conventional aircraft without the capability of thrust vectoring, one can readily approximate $\epsilon \simeq \alpha$, which implies that the thrust nearly acts along the longitudinal reference line in the plane of symmetry used for measuring the angle of attack. Thus, for the aircraft navigational plant we shall adopt this convention, which results in the forces in the vertical plane containing the velocity vector, $(\mathbf{i}_v, \mathbf{k}_v)$, as shown in Figure 3.5.

3.1.1 Wind Speed and Direction

The atmosphere is always in motion due to combined effects of planetary rotation and thermodynamic variations in local atmospheric temperature and pressure, leading to strong horizontal air currents, called *winds*. If unaccounted for, winds can cause large errors in the position of airplanes, which can lead to a shortfall in range and fuel starvation. Therefore, careful periodic observations of the wind velocities above selected weather stations along the route, yielding hourly *winds-aloft* data at a series of altitudes, are essential for accurate navigation. Near the cruising altitudes of airplanes, winds of nearly constant speed and direction can be encountered for most of the flight. Examples of such winds are the prevailing winds, such as the easterly *trade winds* in the sub-tropical belt near the equator, the *westerlies* at the mid-latitudes, and the *jet stream*.[2] Many eastward flights are planned to encounter the westerlies or the jet stream, even at the cost of deviating from the *great circle route*.

In a coordinated flight with a constant angle of attack, an airplane does not "feel" the presence of a steady wind, as the aerodynamic lift and drag, which depend only on the velocity relative to the atmosphere, are unaffected. However, a change in the wind strength, or direction, manifests itself as changes in the angle of attack and sideslip, causing a change in the aerodynamic forces. The ground track, on the other hand, is affected even by a steady wind. Let the aiplane's velocity relative to the rotating planet be $\mathbf{v}$, while a horizontal wind of velocity

$$\mathbf{v}_w = v_w(\sin A_w \mathbf{i}_\ell + \cos A_w \mathbf{i}_\delta) \tag{3.30}$$

[2] The jet stream is a very strong west-to-east "river of air" confined to a narrow altitude range near the tropopause and has a width of a few kilometers, but covers almost the entire globe, with core wind speeds reaching 500 km/hr.

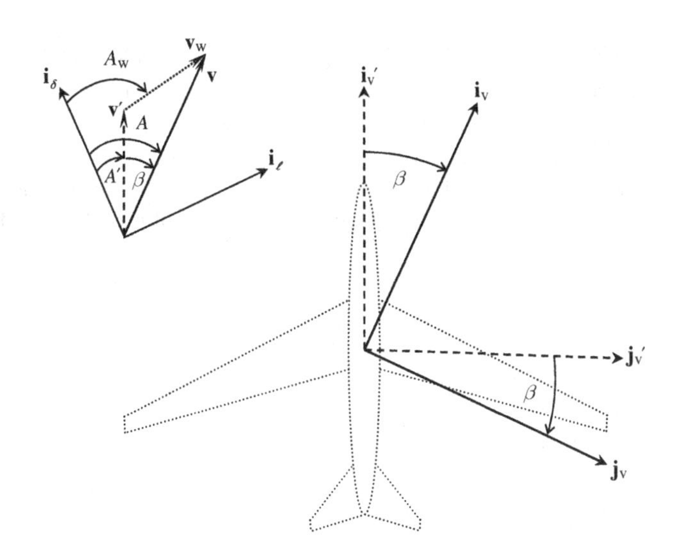

Figure 3.6 The relative velocity vector, $\mathbf{v}'$, in the presence of wind velocity, $\mathbf{v}_w$, and the modified wind axes, $\mathbf{i}'_v, \mathbf{j}'_v, \mathbf{k}_v$

is blowing at a steady speed, v_w, and velocity azimuth, A_w, as shown in Figure 3.6. The airplane's velocity relative to the atmosphere, $\mathbf{v}'$, is given by

$$\begin{aligned}\mathbf{v}' &= \mathbf{v} - \mathbf{v}_w \\ &= v \sin\phi \mathbf{i}_r + (v\cos\phi \sin A - v_w \sin A_w)\mathbf{i}_\ell \\ &\quad + (v\cos\phi\cos A - v_w \cos A_w)\mathbf{i}_\delta,\end{aligned} \tag{3.31}$$

and the *effective sideslip angle*, β, due to the wind is

$$\beta = A - A' \tag{3.32}$$

(Figure 3.6), where A' is the airplane's velocity azimuth relative to the atmosphere. The coordinate transformation between the modified wind axes, $(\mathbf{i}'_v, \mathbf{j}'_v, \mathbf{k}_v)$, and the original (zero-wind) wind axes, $(\mathbf{i}_v, \mathbf{j}_v, \mathbf{k}_v)$, is given according to Figure 3.6 by

$$\begin{Bmatrix} \mathbf{i}_v \\ \mathbf{j}_v \end{Bmatrix} = \begin{pmatrix} \cos\beta & \sin\beta \\ -\sin\beta & \cos\beta \end{pmatrix} \begin{Bmatrix} \mathbf{i}'_v \\ \mathbf{j}'_v \end{Bmatrix}. \tag{3.33}$$

One can express the airplane's relative velocity in the modified wind axes as $\mathbf{v}' = v'\mathbf{i}'_v$ where the airplane's airspeed is given by

$$v' = \sqrt{v^2 + v_w^2 - 2vv_w \cos\phi\cos(A - A_w)}, \tag{3.34}$$

according to the velocity triangle depicted in Figure 3.6.

The governing kinematic equations (3.19)–(3.21) are modified by the wind in so far as the ground speed, v, changes due to the wind velocity, (v_w, A_w), and the velocity azimuth, A, according to equation (3.34). Similarly, the aerodynamic and propulsive forces (Figures 3.3 and 3.5) with the approximation $\epsilon \simeq \alpha$ – resolved in the axes $\mathbf{i}_v, \mathbf{j}_v, \mathbf{k}_v$ in equation (3.14) – are changed due to the change in airspeed (equation (3.34)) and the wind axes, $\mathbf{i}'_v, \mathbf{j}'_v, \mathbf{k}_v$. Let the modified aerodynamic and propulsive forces in a coordinated flight be expressed as follows:

$$
\begin{aligned}
\mathbf{F}_v &= \left(f_T' \cos\alpha' - \mathcal{D}'\right) \mathbf{i}_v' + \mathcal{L}' \sin\sigma \mathbf{j}_v' - \left(f_T' \sin\alpha' + \mathcal{L}' \cos\sigma\right) \mathbf{k}_v \\
&= (f_T \cos\alpha - \mathcal{D} + f_{Xw})\mathbf{i}_v + (\mathcal{L}\sin\sigma + f_{Yw})\mathbf{j}_v - (f_T \sin\alpha + \mathcal{L}\cos\sigma)\mathbf{k}_v,
\end{aligned} \tag{3.35}
$$

where prime denotes the wind modified variables, and f_{Xw}, f_{Yw} are effective force components caused by the wind. Taking the scalar product of equation (3.31) with $\mathbf{i}_v$ and substituting equation (3.15), we have the following relationship for the effective sideslip angle in terms of v, ϕ, A, v_w, and A_w:

$$
\begin{aligned}
v' \cos\beta &= v \sin^2\phi + (v\cos\phi \sin A - v_w \sin A_w)\cos\phi \sin A \\
&\quad + (v\cos\phi\cos A - v_w \cos A_w)\cos\phi\cos A.
\end{aligned} \tag{3.36}
$$

Taking the scalar product of equation (3.35) with each of the unit vectors, $\mathbf{i}_v$, $\mathbf{j}_v$, $\mathbf{k}_v$, we have

$$
\begin{aligned}
f_{Xw} + f_T \cos\alpha - \mathcal{D} &= \left(f_T' \cos\alpha' - \mathcal{D}'\right)\cos\beta + \mathcal{L}\sin\sigma\sin\beta, \\
f_{Yw} + \mathcal{L}\sin\sigma &= \mathcal{L}'\sin\sigma\cos\beta - \left(f_T'\cos\alpha' - \mathcal{D}'\right)\sin\beta, \\
f_T\sin\alpha + \mathcal{L}\cos\sigma &= f_T'\sin\alpha' + \mathcal{L}'\cos\sigma.
\end{aligned} \tag{3.37}
$$

Since α, α' and β are generally small angles, we can approximate equation (3.37) by the following:

$$
\begin{aligned}
f_T - \mathcal{D} &\simeq f_T' - \mathcal{D}', \\
f_{Xw} &\simeq \beta\mathcal{L}'\sin\sigma, \\
f_{Yw} &\simeq -(f_T - \mathcal{D})\beta, \\
\mathcal{L} &\simeq \mathcal{L}'.
\end{aligned} \tag{3.38}
$$

Thus, the change in lift, drag, and thrust magnitudes is negligible, but the wind effectively adds a forward force, $f_{Xw} \simeq \beta\mathcal{L}\sin\sigma$, as well as a sideforce, $f_{Yw} \simeq -(f_T - \mathcal{D})\beta$, in the original (zero-wind) axes, even though the flight is coordinated with reference to the modified wind axes, $\mathbf{i}_v'$, $\mathbf{j}_v'$. The effective forward force and sideforce can appreciably modify the flight path, if not properly compensated for by a navigational system.

3.1.2 Navigational Subsystems

The ability to turn horizontally by making coordinated turns is the crux of aircraft navigation. It is primarily this feature of an airplane that has revolutionized flight over the last century – its absence in balloons and airships left them to a great extent at the mercy of the winds. Airplane navigation primarily consists of charting and following a horizontal course between any two locations on the Earth's surface in the presence of arbitrary winds. The shape and size of the trajectory depend upon whether the time or fuel consumption (or some other criterion) is to be minimized and thus could be markedly different from the way "a crow flies". The determination of such optimal flight paths in the open loop is the objective of the present section.

The primary task of a navigational system is to maintain a steady airspeed at a nearly level altitude by varying the throttle setting, μ, and the angle of attack, α, while maneuvering the aircraft by the bank angle, σ, in such a way as to follow a desired horizontal track, $[\delta(t), \ell(t)]$. The aircraft navigation plant is traditionally separated into two single-input subsystems: (a) the subsystem for maneuvering the aircraft in the horizontal plane, described by the state variables, δ, ℓ, A, m, and the input variable, σ; and (b) the subsystem for modulating thrust and lift by throttle and angle of attack in order to maintain a quasi-level flight at a constant airspeed, v', with the state variables, h, v', ϕ. Once a nominal trajectory is derived for subsystem (a), it is provided as a reference to subsystem (b), which is essentially subservient to subsystem (a). The effects of planetary rotation and curvature are commonly neglected in deriving the plant for subsystem (b), leading to the approximate state equations:

$$\dot{v}' \simeq \frac{f_T - \mathcal{D}}{m} - g_0\phi, \tag{3.39}$$

$$\dot{h} \simeq v'\phi, \tag{3.40}$$

and

$$v'\dot{\phi} \simeq \frac{\mathcal{L}\cos\sigma + f_T\alpha}{m} - g_0. \tag{3.41}$$

By eliminating the flight-path angle, ϕ, from equations (3.39) and (3.40), we have the following approximate *energy equation* (which can also be derived for a general flight path by assuming a flat, non-rotating planet):

$$\dot{h} + \frac{v'}{g_0}\dot{v}' = v'\frac{f_T - \mathcal{D}}{mg_0}. \tag{3.42}$$

Therefore, an automatic controller for subsystem (b) aims simultaneously to achieve

$$\mathcal{L}\cos\sigma + f_T\alpha \simeq mg_0, \quad f_T \simeq \mathcal{D}(\alpha),$$

for a given profile, $h(t)$, $v'(t)$, $m(t)$, $\sigma(t)$.

In summary, we can write the state equations for the two subsystems as follows, and represent them by the block diagrams shown in Figure 3.7:

(a)

$$\dot{\delta} = \frac{v\cos A}{R_0 + h}, \tag{3.43}$$

$$\dot{\ell} = \frac{v\sin A}{(R_0 + h)\cos\delta}, \tag{3.44}$$

$$\dot{A} = \frac{\mathcal{L}\sin\sigma + f_{Y\mathrm{w}}}{mv} + \frac{v}{r}\sin A\tan\delta + \frac{1}{2v}\Omega^2 r\sin A\sin 2\delta + 2\Omega\sin\delta, \tag{3.45}$$

and either

$$\dot{m} = -\frac{c_T f_T}{g_0} \tag{3.46}$$

or

$$\dot{m} = -\frac{c_P f_T v}{g_0}. \tag{3.47}$$

Here, the ground speed, v, is related to the constant airspeed, v', and the constant wind velocity, $(v_\mathrm{w}, A_\mathrm{w})$, according to the velocity triangle shown in Figure 3.6 as follows:

$$v' = \sqrt{v^2 + v_\mathrm{w}^2 - 2vv_\mathrm{w}\cos(A_\mathrm{w} - A)} \tag{3.48}$$

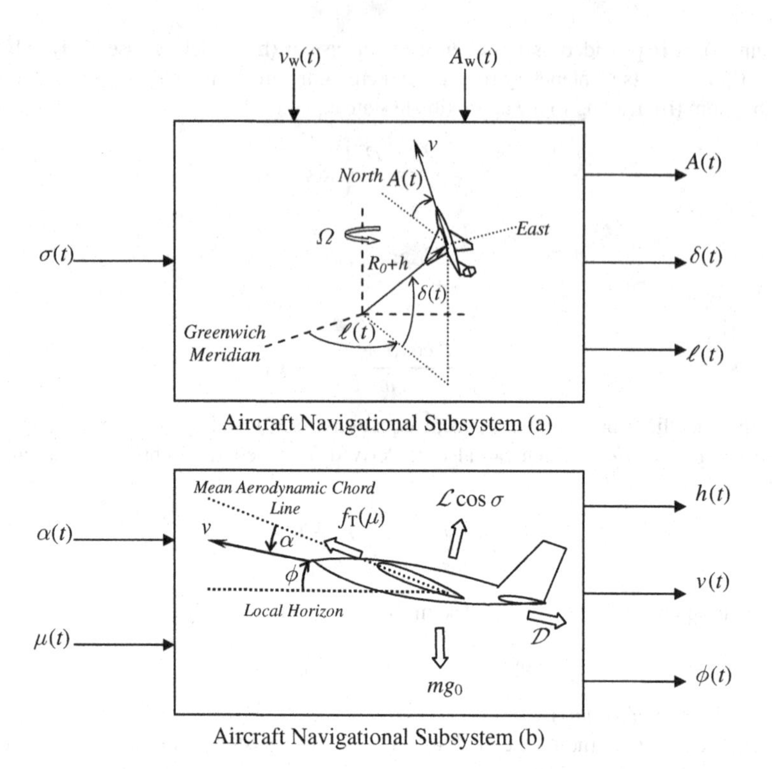

Figure 3.7 Navigational subsystems for an aircraft

or

$$v = \sqrt{(v')^2 - v_w^2 \sin^2(A_w - A)} + v_w \cos(A_w - A). \tag{3.49}$$

(b)

$$\dot{h} = v'\phi, \tag{3.50}$$

$$\dot{v}' = \frac{f_T - \mathcal{D}}{m} - g_0\phi \tag{3.51}$$

$$\dot{\phi} = \frac{f_T\alpha + \mathcal{L}\cos\sigma - mg_0}{mv'}, \tag{3.52}$$

where $\sigma(t)$, $m(t)$ are the reference solutions for subsystem (a).

The design of an automatic navigation system thus consists of determining and following a nominal trajectory for subsystem (a) which is then provided as a reference to be followed by subsystem (b). Consequently, while both terminal and tracking control are necessary for subsystem (a), only a regulator needs to be designed for subsystem (b) for maintaining a constant airspeed and altitude.

3.2 Optimal Aircraft Navigation

Using the state equations for subsystem (a) as the nonlinear plant, one can derive an optimal, open-loop trajectory by the terminal control methodology (Chapter 2), such that it may be followed by a linearized, closed-loop tracking system, as well as by the regulator for subsystem (b). Therefore, we select our state vector as

$$\mathbf{x}(t) = [\delta(t), \ell(t), A(t), m(t)]^T, \tag{3.53}$$

and the normal acceleration,

$$u(t) = \frac{\mathcal{L}}{m} \sin\sigma, \tag{3.54}$$

as the sole control input, and express the state equations (3.43)–(3.47) (with constant values of r and v maintained by subsystem (b)) as follows:

$$\dot{\mathbf{x}} = \mathbf{f}[\mathbf{x}(t), u(t), t], \tag{3.55}$$

$$\mathbf{f} = \begin{Bmatrix} \frac{v}{r}\cos A \\ \frac{v\sin A}{r\cos\delta} \\ \frac{u}{v} + \frac{f_{Y\mathrm{w}}}{mv} + \frac{v}{r}\sin A\tan\delta + 2\Omega\sin\delta + \frac{\Omega^2 r}{2v}\sin A\sin 2\delta \\ -c \end{Bmatrix}. \tag{3.56}$$

Here the constant rate of fuel consumption, c, is given by either

$$c = -\frac{c_\mathrm{T} f_\mathrm{T}}{g_0}, \tag{3.57}$$

or

$$c = -\frac{c_\mathrm{P} f_\mathrm{T} v}{g_0}, \tag{3.58}$$

depending upon whether the aircraft has jet or propeller engines, respectively. Assuming that subsystem (b) does its job of maintaining a quasi-level flight with a constant airspeed, we can approximate

$$f_{Y\mathrm{w}} = -(f_\mathrm{T} - \mathcal{D})\sin\beta \simeq 0. \tag{3.59}$$

It is to be noted that in equation (3.56) the ground speed, v, is not constant, but is related to the constant airspeed, v', constant wind velocity, $(v_\mathrm{w}, A_\mathrm{w})$, and velocity azimuth, A, by equation (3.49), which can either be regarded as the following equality constraint for the optimization problem,

$$v - \sqrt{(v')^2 - v_\mathrm{w}^2\sin^2(A_\mathrm{w} - A)} - v_\mathrm{w}\cos(A_\mathrm{w} - A) = 0, \tag{3.60}$$

or directly substituted into equation (3.56) for v as a function of A.

3.2.1 Optimal Navigation Formulation

We initially pose the aircraft navigation problem as an optimal control problem with end-point constraints (in addition to the state equation (3.53) and equality constraint (3.60)) with a fixed terminal time, t_f, calculated from the fuel mass, $m_f - m_0$, and the optimal rate of fuel consumption, c, which depends upon the pre-selected, optimal airspeed and altitude, as follows:

$$t_f = \frac{m_f - m_0}{c}. \tag{3.61}$$

In order to minimize the control effort at the constant airspeed and altitude, a quadratic Lagrangian cost is selected, Ru^2 (where $R > 0$, the control cost coefficient), resulting in the following Hamiltonian:

$$H = Ru^2 + \boldsymbol{\lambda}^T \mathbf{f}(\mathbf{x}, \mathbf{u}). \tag{3.62}$$

The co-state equations are then derived as follows from the necessary conditions for optimality (Euler–Lagrange equations, Chapter 2):

$$\dot{\boldsymbol{\lambda}} = -\left(\frac{\partial H}{\partial \mathbf{x}}\right)^T = -H_{\mathbf{x}}^T = -\mathbf{A}^T \boldsymbol{\lambda}, \tag{3.63}$$

where

$$\mathbf{A} = \frac{\partial \mathbf{f}}{\partial \mathbf{x}} = \begin{pmatrix} 0 & 0 & a_{13} & 0 \\ a_{21} & 0 & a_{23} & 0 \\ a_{31} & 0 & a_{33} & 0 \\ 0 & 0 & 0 & 0 \end{pmatrix}, \tag{3.64}$$

$$a_{13} = -\frac{v}{r}\sin A + \frac{\cos A}{r}\frac{\partial v}{\partial A}, \tag{3.65}$$

$$a_{21} = \frac{v}{r}\sin A \tan\delta \sec\delta, \tag{3.66}$$

$$a_{23} = \frac{\sec\delta}{r}\left(v\cos A + \frac{\partial v}{\partial A}\sin A\right), \tag{3.67}$$

$$a_{31} = \frac{v}{r}\sin A \sec^2\delta + \Omega^2 \frac{r}{v}\sin A \cos 2\delta + 2\Omega\cos\delta, \tag{3.68}$$

$$\begin{aligned} a_{33} = {} & \frac{\tan\delta}{r}\left(v\cos A + \frac{\partial v}{\partial A}\sin A\right) - \frac{u^*}{v^2}\frac{\partial v}{\partial A} \\ & + \frac{\Omega^2 r}{2}\sin 2\delta\left(\frac{\cos A}{v} - \frac{\sin A}{v^2}\frac{\partial v}{\partial A}\right), \end{aligned} \tag{3.69}$$

with

$$v = \sqrt{(v')^2 - v_w^2 \sin^2(A_w - A)} + v_w \cos(A_w - A) \tag{3.70}$$

and

$$\frac{\partial v}{\partial A} = \left[1 - \frac{v_w \cos(A_w - A)}{\sqrt{(v')^2 - v_w^2 \sin^2(A_w - A)}}\right] v_w \sin(A_w - A). \tag{3.71}$$

Equation (3.63) yields the following scalar co-state equations:

$$\dot{\lambda}_1 = a_{21}\lambda_2 + a_{31}\lambda_3, \tag{3.72}$$

$$\dot{\lambda}_2 = 0, \tag{3.73}$$

$$\dot{\lambda}_3 = a_{13}\lambda_1 + a_{23}\lambda_2 + a_{33}\lambda_3, \tag{3.74}$$

$$\dot{\lambda}_4 = 0. \tag{3.75}$$

The optimal control input history, $u^*(t)$, is also computed from Euler–Lagrange equations as follows:

$$H_u = \frac{\partial H}{\partial u} = 2Ru + \boldsymbol{\lambda}^T \mathbf{B} = 0, \tag{3.76}$$

where

$$\mathbf{B} = \frac{\partial \mathbf{f}}{\partial u} = \begin{pmatrix} 0 \\ 0 \\ \frac{1}{v} \\ 0 \end{pmatrix}, \tag{3.77}$$

resulting in

$$u^*(t) = -\frac{\lambda_3(t)}{2Rv}. \tag{3.78}$$

Therefore, the optimal bank angle history for the aircraft navigational problem is

$$\sigma^* = -\sin^{-1}\left\{\frac{m(t)\lambda_3(t)}{2Rv}\right\}, \tag{3.79}$$

where $m(t)$ and $\lambda_3(t)$ satisfy equations (3.46) (or (3.47)) and (3.74), respectively.

The usual boundary conditions for the aircraft navigational problem are the initial conditions on the state variables,

$$\mathbf{x}(0) = [\delta_0, \ell_0, A_0, m_0]^T, \tag{3.80}$$

the terminal condition on mass (the fourth state variable),

$$m(t_f) = m_f, \tag{3.81}$$

and the final conditions on the co-state variables, $\boldsymbol{\lambda}$, derived from the terminal state cost, $\varphi[\mathbf{x}(t_f), t_f]$, as follows:

$$\lambda_i(t_f) = \left(\frac{\partial \varphi}{\partial x_i}\right)_{t=t_f}, \quad i = 1, 2, 3. \tag{3.82}$$

Apparently, the terminal state cost can be chosen in one of two ways, according to whether the range beginning from an initial location, (δ_0, ℓ_0), is to be maximized, or the final distance from a specific destination, (δ_f, ℓ_f), is to be minimized. If the range is to be maximized, then we have

$$\varphi[\mathbf{x}(t_f), t_f] = -Q\left\{[\delta(t_f) - \delta_0]^2 + [\ell(t_f) - \ell_0]^2\right\}, \tag{3.83}$$

leading to

$$\begin{aligned}\lambda_1(t_f) &= -2Q[\delta(t_f) - \delta_0],\\ \lambda_2(t_f) &= -2Q[\ell(t_f) - \ell_0],\\ \lambda_3(t_f) &= 0.\end{aligned} \tag{3.84}$$

If the distance to a specific destination is to be minimized, then

$$\varphi[\mathbf{x}(t_f), t_f] = Q\left\{[\delta(t_f) - \delta_f]^2 + [\ell(t_f) - \ell_f]^2\right\}, \tag{3.85}$$

resulting in

$$\begin{aligned}\lambda_1(t_f) &= -2Q[\delta_f - \delta(t_f)],\\ \lambda_2(t_f) &= -2Q[\ell_f - \ell(t_f)],\\ \lambda_3(t_f) &= 0.\end{aligned} \tag{3.86}$$

In either case, $Q > 0$ is the terminal state cost parameter. However, since the final solution, $[\delta(t_f), \ell(t_f)]$, is unknown *a priori*, neither equation (3.83) nor equation (3.85) can be implemented in practice. It is much more practical instead to combine the two and select the best trajectory as the one that requires the smallest positive value of t_f to pass through the initial and final locations. This is tantamount to solving the *free terminal time* problem,

$$\varphi[\mathbf{x}(t_f), t_f] = t_f^2, \tag{3.87}$$

with the attendant terminal boundary conditions (Chapter 2)

$$\begin{aligned}\delta(t_f) &= \delta_f,\\ \ell(t_f) &= \ell_f,\\ \lambda_3(t_f) &= 0,\end{aligned} \tag{3.88}$$

and

$$\left(\frac{\partial \varphi}{\partial t} + Ru^2 + \boldsymbol{\lambda}^T \mathbf{f}\right)_{t=t_f} = 0. \tag{3.89}$$

Clearly, the last boundary condition, equation (3.89) – which should ideally give us the required optimal time, t_f – also cannot be implemented due to the final solution, $\mathbf{x}(t_f)$, being unknown *a priori*. Hence, as a practical alternative, we iterate for the smallest possible value of t_f that results in an extremal trajectory passing through the initial and desired final locations. However, the final time must now obey the constraint

$$t_f \leq \frac{m_f - m_0}{c}. \tag{3.90}$$

Other than determining the upper limit on the final time by equation (3.90), the aircraft mass, $m(t)$, has no role to play in designing the optimal trajectory, and can be dropped as a state variable. The bank angle, $\sigma(t)$, can then be computed from the normal acceleration input, $u(t)$, by eliminating the mass from

$$u = \frac{\mathcal{L}}{m} \sin \sigma \tag{3.91}$$

and

$$\mathcal{L} \cos \sigma = mg, \tag{3.92}$$

leading to

$$\sigma = \tan^{-1} \frac{u}{g}. \tag{3.93}$$

3.2.2 *Extremal Solution of the Boundary-Value Problem: Long-Range Flight Example*

The extremal navigational trajectory for level flight is the solution of the state equations,

$$\begin{Bmatrix} \dot{\delta} \\ \dot{\ell} \\ \dot{A} \end{Bmatrix} = \begin{Bmatrix} \frac{v}{r} \cos A \\ \frac{v \sin A}{r \cos \delta} \\ \frac{u}{v} + \frac{v}{r} \sin A \tan \delta + 2\Omega \sin \delta + \frac{\Omega^2 r}{2v} \sin A \sin 2\delta \end{Bmatrix}, \tag{3.94}$$

the co-state equations,

$$\begin{aligned} \dot{\lambda}_1 &= a_{21}\lambda_2 + a_{31}\lambda_3, \\ \dot{\lambda}_2 &= 0, \\ \dot{\lambda}_3 &= a_{13}\lambda_1 + a_{23}\lambda_2 + a_{33}\lambda_3, \end{aligned} \tag{3.95}$$

with coefficients a_{ij} given by equations (3.65)–(3.69), and the ground speed, v, constrained by equation (3.70), subject to the two-point boundary conditions,

$$\begin{aligned} \delta(0) &= \delta_0, \\ \ell(0) &= \ell_0, \\ A(0) &= A_0, \\ \delta(t_f) &= \delta_f, \\ \ell(t_f) &= \ell_f, \\ \lambda_3(t_f) &= 0, \end{aligned} \tag{3.96}$$

and resulting in the optimal normal acceleration control input,

$$u^*(t) = -\frac{\lambda_3(t)}{2Rv}, \tag{3.97}$$

with the associated optimal bank angle,

$$\sigma^*(t) = \tan^{-1}\left\{\frac{u^*(t)}{g}\right\}. \tag{3.98}$$

The two-point boundary value problem concerned can be numerically solved by either a shooting method or a collocation technique (Chapter 2), which we illustrate by the following example of optimal long-range navigation for east to west, transatlantic flight in presence of a steady wind.

Example 3.1 Consider a level flight from London ($\delta = 51.5°$, $\ell = 0$) at a standard altitude of 11 km, with a constant airspeed, $v' = 270$ m/s, and initial velocity azimuth, $A_0 = 287.2°$, to New York's JFK airport ($\delta = 40.4°$, $\ell = -73.5°$). The fuel capacity gives a total endurance of 6.5 hours at the given airspeed and altitude, including reserve fuel for 45 minutes flight to cover diversions due to weather and traffic. Thus, we are looking for optimal trajectories with $t_f \leq 5.75$ hr in the presence of a steady wind, (v_w, A_w).

The program in Table 3.1 was written to solve the 2PBVP for a specified final time, t_f, by the collocation method encoded in MATLAB® in the standard *bvp4.m* routine (Chapter 2). The differential equations and the boundary conditions are supplied by the subroutines, *airnavode.m* and *airnavbc.m* (Table 3.2), respectively. The optimal solution is obtained by iterating for the smallest value of t_f that satisfies the initial and final conditions.

For optimal maneuvering we choose the control cost parameter $R = 0.001$ and begin the solution for the zero-wind case, $v_w = 0$, which yields the corresponding minimum flight time of $t_f = 5.7445$ hr after a few iterations. The resulting plots of latitude, $\delta(t)$, longitude, $\ell(t)$, velocity azimuth, $A(t)$, normal acceleration input, $u^*(t)$, and associated optimal bank angle, $\sigma(t)$ – as well as the ground track, (δ, ℓ) – can be seen in Figures 3.8–3.10. The final destination is seen to be reached with minimal maneuvering (bank angle within $\pm 0.12°$) and with the necessary reserve fuel, thus setting the benchmark for trajectory in the presence of wind. Figure 3.10 shows that the optimal, zero-wind ground track is quite close to the *great circle route*, which is a circle of the same radius as the Earth passing between the initial and final locations. The smaller arc of the great circle gives the shortest distance between any two points along Earth's surface. The equation for the great circle passing between two given points on a sphere, (ℓ_1, δ_1) and (ℓ_2, δ_2), is the following (Wertz 1978):

$$\tan\delta = \begin{cases} \frac{\tan\delta_1 \sin(\ell - \ell_1 + \eta)}{\sin\eta}, & \delta_1 \neq 0, \\ \frac{\tan\delta_2 \sin(\ell - \ell_1 + \eta)}{\sin(\ell_2 - \ell_1)}, & \delta_1 = 0, \end{cases} \tag{3.99}$$

where

$$\cot\eta = \frac{\tan\delta_2}{\tan\delta_1 \sin(\ell_2 - \ell_1)} - \cot(\ell_2 - \ell_1). \tag{3.100}$$

We will later consider great circle routes in detail. The proximity of the optimal trajectory to the great circle path appears to indicate its minimum distance (thus minimum time) between the given locations. However, this has to be confirmed for the optimal trajectories in the presence of a wind.

In order to study the impact of a steady wind on optimal navigational routes, consider a wind speed of $v_w = 20$ m/s in various (constant) directions, A_w . In each case, the minimum time of flight between the given destinations, t_f, is different, and is derived by iterative solutions of the 2PBVP problem, as

Table 3.1

```
% Calling program for solving the 2PBVP for optimal aircraft
% navigation by collocation method using MATLAB's
% intrinsic code 'bvp4c.m'.
% Requires 'airnavode.m' and 'airnavbc.m'.
% (c) 2010 Ashish Tewari
%  dy/dx=f(y,x); a<=x<=b
%  y(x=a), y(x=b): Boundary conditions
%  y(1,1)= Latitude (rad.)
%  y(2,1)= Longitude (rad.)
%  y(3,1)= Velocity azimuth (rad.)
%  y(4:6,1)=lambda (Lagrange multipliers vector)

global del0; del0=51.5*pi/180; %initial latitude
global lam0;lam0=0; %initial longitude
global delf; delf=40.4*pi/180; %final latitude
global lamf;lamf=-73.5*pi/180; %final longitude
global A0;A0=287.2*pi/180; %initial velocity azimuth
global vw;vw=20; %wind speed
global Aw;Aw=45*pi/180; %wind velocity azimuth
global R; R=0.001; %control cost coefficient
global vprime; vprime=270; %airspeed (m/s)
global r; r=6378.14e3+11000; %radius (m)
global Omega; Omega=7.2921e-5; %Earth's rotational rate (rad/s)
tf=6.12*3600; %terminal time

% Collocation points & initial guess follow:
solinit = bvpinit(linspace(0,tf,50),[del0 lam0 A0 0 0 0]);

% 2PBVP solution by collocation method:
sol = bvp4c(@airnavode,@airnavbc,solinit);
x = linspace(0,tf);
y = deval(sol,x);

% Ground speed (m/s) follows:
v=sqrt(vprime^2-vw^2*sin(Aw-y(3,:)).^2)+vw*cos(Aw-y(3,:));

% Normal acceleration input and bank angle profile follow:
u=-y(6,:)./(2*R*v);
sigma=atan(u/9.81);
```

discussed earlier. Table 3.3 shows the value of the minimum flight time for each wind velocity, while Figures 3.11–3.14 compare the velocity azimuth, ground speed, ground track, and optimal bank angle for the various cases. These figures reinforce the conclusion drawn for the zero-wind case, namely that the minimum flight time in the presence of a steady wind also corresponds to a trajectory lying very close to the great circle route. The optimal trajectory followed by the aircraft has a deviation from the great circle route due to the initially chosen velocity azimuth, which is the same ($A_0 = 287.2^\circ$) for all cases.

3.2.3 Great Circle Navigation

The shortest distance between any two points, P_1 and P_2, on the surface of a spherical planet is the smaller arc of a *great circle* passing through the two points (Figure 3.15). Clearly, an aircraft flying a great circle

Table 3.2

```
% Program for specifying governing ODEs for optimal navigation
% expressed as state equations for the 2PBVP
%(to be called by 'bvp4c.m')
% (c) 2010 Ashish Tewari
function dydx=airnavode(x,y)
global vprime;
global vw;
global Aw;
global r;
global R;
global Omega;
v=sqrt(vprime^2-vw^2*sin(Aw-y(3))^2)+vw*cos(Aw-y(3));
u=-0.5*y(6)/(R*v);
dvdA=sin(Aw-y(3))*(1-cos(Aw-y(3))/sqrt(vprime^2-vw^2*sin(Aw-y(3))^2));
dydx=[v*cos(y(3))/r
  v*sin(y(3))/(r*cos(y(1)))
  v*sin(y(3))*tan(y(1))/r+u/v+Omega^2*r*sin(y(3))*sin(2*y(1))/(2*v)+ ...
  2*Omega*sin(y(1))
  v*sin(y(3))*tan(y(1))*sec(y(1))*y(5)/r+(v*sin(y(3))*sec(y(1))^2/r+ ...
  Omega^2*r*sin(y(3))*cos(2*y(1))/v+2*Omega*cos(y(1)))*y(6)
  0
  (dvdA*cos(y(3))-v*sin(y(3)))*y(4)/r + ...
  (v*cos(y(3))+dvdA*sin(y(3)))*y(5)/(r*cos(y(1))) + ...
  (tan(y(1))*(v*cos(y(3))+dvdA*sin(y(3)))/r-u*dvdA/v^2+ ...
  0.5*Omega^2*r*sin(2*y(1))*(cos(y(3))/v-dvdA*sin(y(3))/v^2))*y(6)];

% Program for specifying boundary conditions for 2PBVP of
% optimal aircraft navigation.
% (To be called by 'bvp4c.m')
function res=airnavbc(ya,yb)
global lam0;
global lamf;
global del0;
global delf;
global A0;
res=[ya(1)-del0
    ya(2)-lam0
    ya(3)-A0
    yb(1)-delf
    yb(2)-lamf
    yb(6)];
```

route while maintaining a constant airspeed (i.e., speed relative to the planetary surface in the absence of winds) would travel the shortest distance between any two given points on the rotating planet. If the airspeed is an optimum cruising speed corresponding to the maximum range (Tewari 2006), the fuel required for the flight would be a minimum along the great circle. Conversely, given the fuel capacity and the optimum cruising airspeed, the great circle route would result in the smallest time of flight, which has been shown by the optimal trajectories derived in Example 3.1. However, as also shown in Example 3.1, flying the great circle route requires a continuous adjustment of the trajectory by banking the aircraft in order to compensate for planetary rotation, which would otherwise deflect the flight path westward due to the centripetal and Coriolis acceleration terms in equation (3.45). Although the centripetal and Coriolis

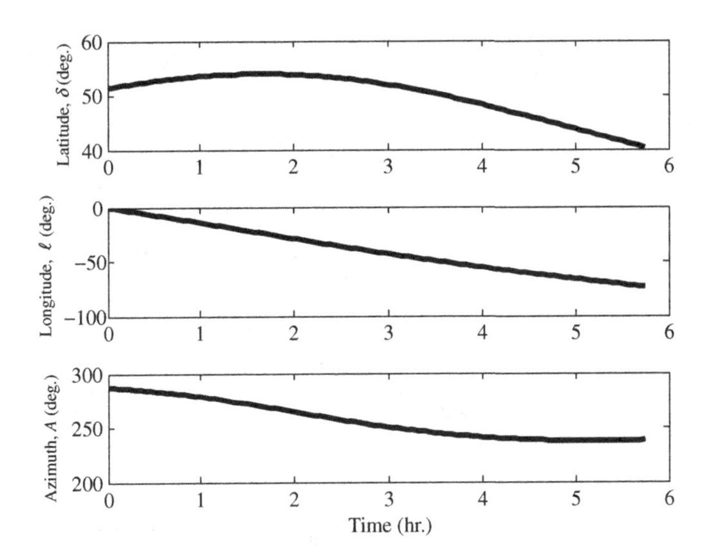

Figure 3.8 Optimal aircraft trajectory (zero-wind case)

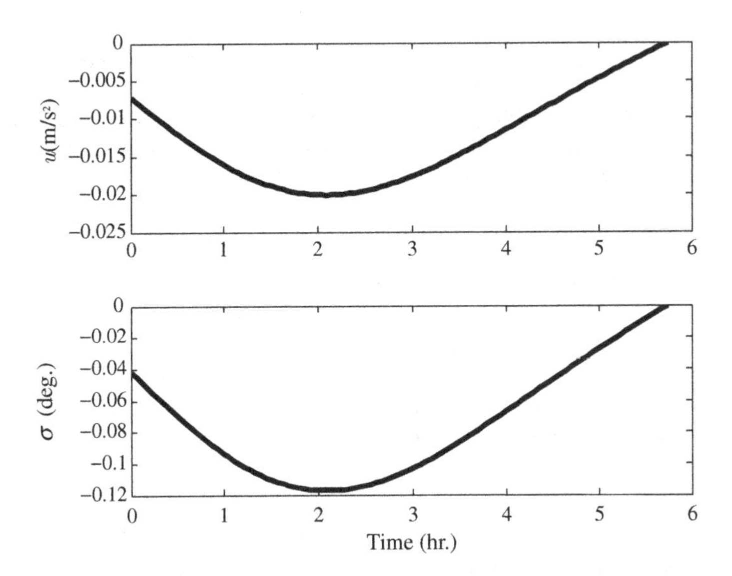

Figure 3.9 Optimal aircraft normal acceleration input and bank angle control history (zero-wind case)

accelerations due to planetary rotation are much smaller for an aircraft than for a spacecraft – owing to the former's smaller flight speed and altitude, as well as the fact that an aircraft maintains a nearly constant airspeed relative to the atmosphere (and thus the rotating planet) – the situation is compounded by the presence of winds. In the presence of even a steady wind, a continuous correction must be made to the flight path in order to fly the great circle route. Thus, a significant navigational error would result if the planetary rotation, Ω, and wind velocity, (v_w, A_w), were not taken into account while planning a flight along the great circle, so much so that an aircraft may either end up hundreds of kilometers away from the desired destination, or face fuel starvation.

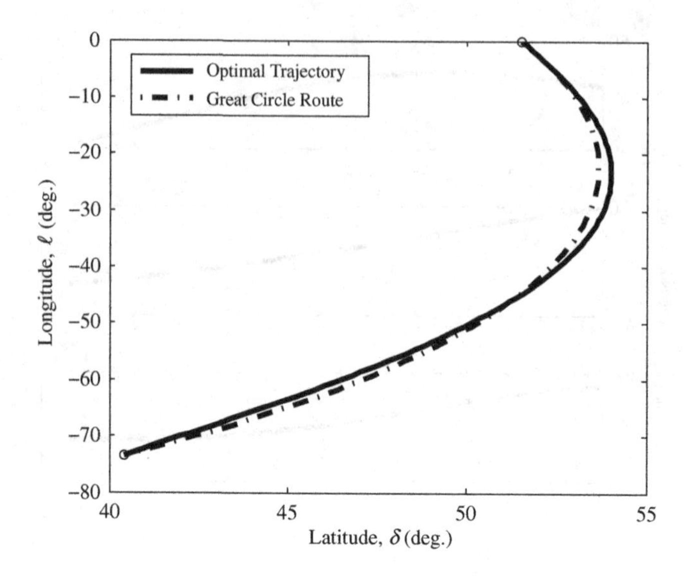

Figure 3.10 Optimal aircraft ground track (zero-wind case)

Consider a great circle of inclination i, with the equator, passing through two points, P_1 and P_2, on a spherical planet (Figure 3.15). The latitude–longitude pairs for the the two points are (δ_1, ℓ_1) and (δ_2, ℓ_2), respectively. The *ascending node* of the great circle is its point of intersection with the equator crossing from south to north, and is given by the unit vector, $\mathbf{n}$, which makes an angle ℓ_0 (called the *longitude of ascending node*) from the zero meridian, $\mathbf{I}$ (the Greenwich meridian for Earth). Thus, we have

$$\mathbf{n} \cdot \mathbf{I} = \cos \ell_0. \tag{3.101}$$

The unit normal to the plane of the great circle is given by

$$\mathbf{p} = \frac{\mathbf{i}_{r1} \times \mathbf{i}_{r2}}{| \mathbf{i}_{r1} \times \mathbf{i}_{r2} |}, \tag{3.102}$$

where $\mathbf{i}_{r1}$ and $\mathbf{i}_{r2}$ are unit vectors from the center passing through the points, P_1 and P_2, respectively. Clearly, $\mathbf{p}$ denotes the direction along the great circle from P_1 to P_2 according to the right-hand rule, and makes angle i with the unit polar axis, $\mathbf{K}$, that is,

$$\mathbf{p} \cdot \mathbf{K} = \cos i. \tag{3.103}$$

The angles (ℓ_0, i) (or the unit vectors $(\mathbf{n}, \mathbf{p})$) uniquely specify a great circle.

Table 3.3 Optimal flight time for various steady-wind conditions

v_w (m/s)	A_w (deg)	t_f (hr)
0	–	5.7445
20	45	6.12
20	90	6.1745
20	235	5.3945
20	270	5.377
20	315	5.549

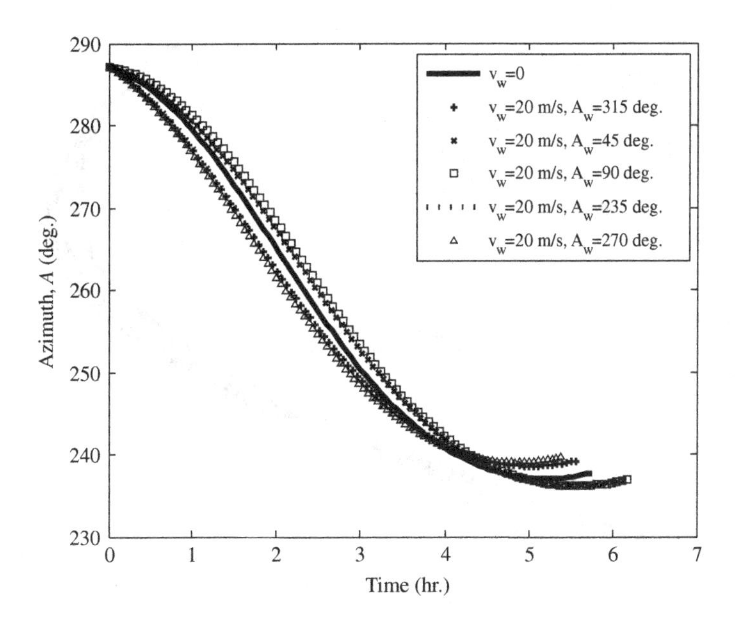

Figure 3.11 Optimal velocity azimuth for various steady-wind conditions

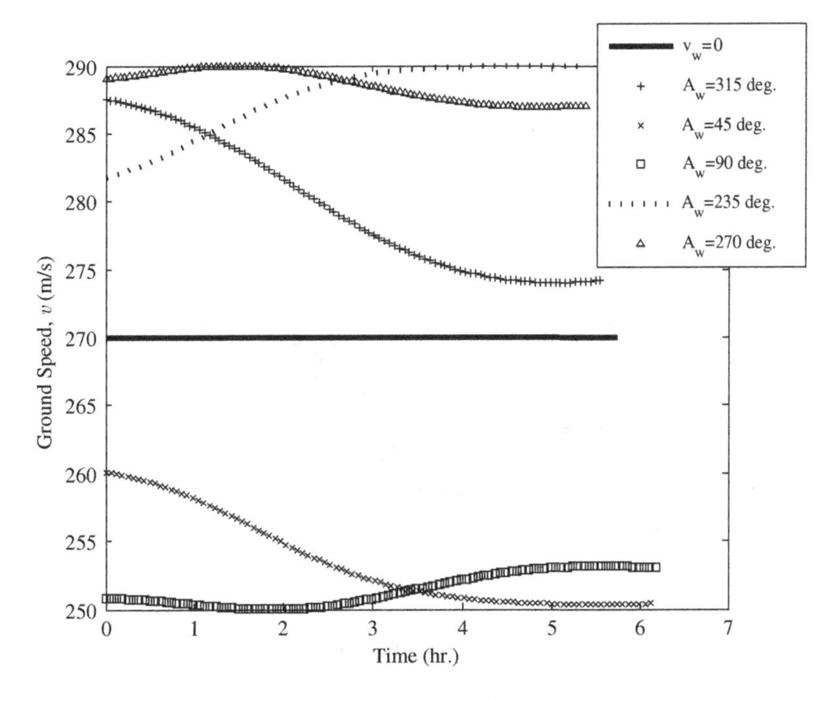

Figure 3.12 Optimal aircraft ground speed for various steady-wind conditions

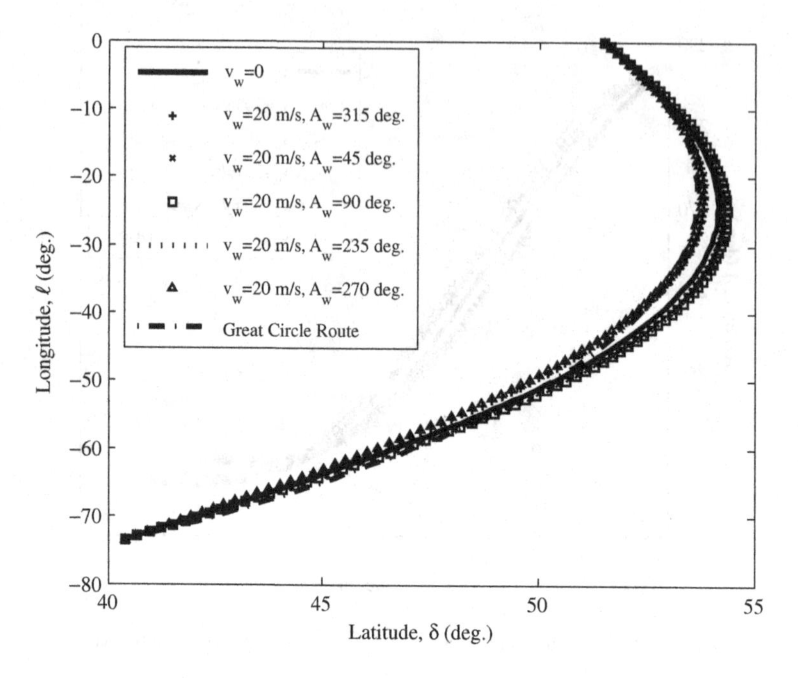

Figure 3.13 Optimal aircraft ground track for various steady-wind conditions

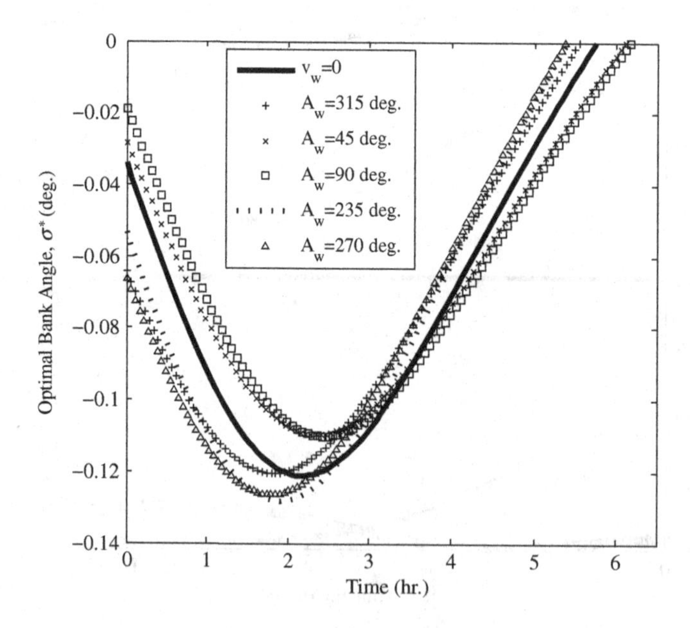

Figure 3.14 Optimal aircraft bank angle for various steady-wind conditions

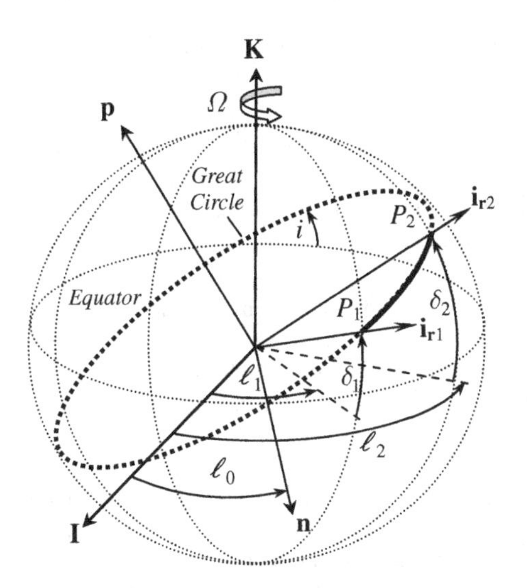

Figure 3.15 Great circle route from P_1 to P_2 along a rotating planet's surface

The direction of flight must be chosen so as to give the shorter of the two paths between the given points. The unit normal vector indicating the direction of flight, **p**, can also be derived from the instantaneous position and velocity vectors, **r**, **v**, as follows:

$$\mathbf{p} = \frac{\mathbf{r} \times \mathbf{v}}{|\mathbf{r} \times \mathbf{v}|}, \tag{3.104}$$

which, substituted into equation (3.103), yields the following spherical trigonometric relationship:

$$\cos i = \cos \delta \sin A. \tag{3.105}$$

Since the inclination, i, is constant for a great circle, we have the following upon differentiation of equation (3.105) with respect to time:

$$0 = -\dot{\delta} \sin \delta \sin A + \dot{A} \cos \delta \cos A \tag{3.106}$$

or

$$\dot{A} = \dot{\delta} \tan A \tan \delta. \tag{3.107}$$

Thus, the flight direction, A, along the great circle must vary with the latitude, δ, according to equation (3.107), which, after substituting the kinematic equation (3.8), becomes

$$\dot{A} = \frac{v}{r} \sin A \tan \delta. \tag{3.108}$$

Comparing equations (3.108) and (3.45), the latter written for the level and coordinated flight case ($f_{Y\text{w}} = 0, u = g \tan \sigma$) as

$$\dot{A} = \frac{v}{r} \sin A \tan \delta + \frac{u}{v} + \frac{1}{2v} \Omega^2 r \sin A \sin 2\delta + 2\Omega \sin \delta, \quad (3.109)$$

it is clear that, for a flight along the great circle, the normal acceleration control input, $u(t)$, (and thus the bank angle, $\sigma(t)$) must be varied such that

$$u = -\frac{1}{2} \Omega^2 r \sin A \sin 2\delta - 2\Omega v \sin \delta. \quad (3.110)$$

Hence, equation (3.110) is the control law for a flight along the great circle, and specifies the required normal acceleration for compensating for the centripetal and Coriolis acceleration terms due to planetary rotation.

Example 3.2 We reconsider Example 3.1 for the level and coordinated flight from London to New York, such that the aircraft follows a great circle route in the presence of a given steady wind velocity, (v_w, A_w). Rather than being an arbitrary number, the initial velocity azimuth, A_0, must now satisfy equation (3.105), which, for the given initial and final locations, (δ_1, ℓ_1) and (δ_2, ℓ_2), is calculated as follows:

$$\mathbf{i}_{r1} = \mathbf{I} \cos \delta_1 \cos \ell_1 + \mathbf{J} \cos \delta_1 \sin \ell_1 + \mathbf{K} \sin \delta_1 = 0.6225\mathbf{I} + 0.7826\mathbf{K},$$

$$\mathbf{i}_{r2} = \mathbf{I} \cos \delta_2 \cos \ell_2 + \mathbf{J} \cos \delta_2 \sin \ell_2 + \mathbf{K} \sin \delta_2 = 0.2163\mathbf{I} - 0.7302\mathbf{J} + 0.6481\mathbf{K},$$

$$\mathbf{p} = \frac{\mathbf{i}_{r1} \times \mathbf{i}_{r2}}{| \mathbf{i}_{r1} \times \mathbf{i}_{r2} |} = 0.7452\mathbf{I} - 0.3054\mathbf{J} - 0.5928\mathbf{K},$$

$$\cos i = \mathbf{p} \cdot \mathbf{K} = -0.5928,$$

$$A_0 = \sin^{-1} \frac{\cos i}{\cos \delta_0} = 287.783^\circ.$$

A program (Table 3.4) was written in order to simulate the great circle trajectory and the necessary bank angle for each wind velocity by a fourth-order Runge–Kutta integration of the equations of motion specified in the subroutine *airnav.m* (Table 3.5). The time of flight for a given wind velocity is the same as the minimum flight time obtained from Example 3.1 and tabulated in Table 3.3. The results of the simulation are plotted in Figures 3.16–3.19. As expected, the ground track for each wind condition is seen (Figure 3.16) to collapse exactly to the great circle route, which also requires different velocity azimuth (Figure 3.17), ground speed (Figure 3.18), and bank angle (Figure 3.19) profiles when compared to Example 3.1. The bank angle is now smaller in each case when compared to the optimal (non-great-circle) values of Figure 3.14 due to the aircraft already beginning with the correct velocity azimuth for great circle navigation.

3.3 Aircraft Attitude Dynamics

The translational kinematic equations of aircraft flight (equations (3.7)–(3.9)) involve a much larger time scale when compared to the translational and rotational kinetics, and are commonly neglected in deriving the aircraft plant for attitude control. Such an approximation is tantamount to treating the local horizon frame, ($\mathbf{i}_r$, $\mathbf{i}_\ell$, $\mathbf{i}_\delta$), as an inertial reference frame for aircraft attitude dynamics. Thus, the orientation of an

Table 3.4

```
% Calling program for simulating the great circle trajectory for
% navigation by fourth-order Runge-Kutta solution.
% Requires 'airnav.m'.
% (c) 2010 Ashish Tewari
%  y(1,1)= Latitude (rad.)
%  y(2,1)= Longitude (rad.)
%  y(3,1)= Velocity azimuth (rad.)

global del0; del0=51.5*pi/180; %initial latitude
global lam0;lam0=0; %initial longitude
global delf; delf=40.4*pi/180; %final latitude
global lamf;lamf=-73.5*pi/180; %final longitude
global vw;vw=20; %wind speed
global Aw;Aw=45*pi/180; %wind velocity azimuth
global vprime; vprime=270; %airspeed (m/s)
global r; r=6378.14e3+11000; %radius (m)
global Omega; Omega=7.2921e-5; %Earth's rotational rate (rad/s)
tf=6.12*3600; %terminal time
dtr=pi/180;
% Computation of the initial velocity azimuth, A_0:
r0=[cos(del0)*cos(lam0) cos(del0)*sin(lam0) sin(del0)]
rf=[cos(delf)*cos(lamf) cos(delf)*sin(lamf) sin(delf)]
n=cross(r0,rf)/norm(cross(r0,rf))
cosi=dot([0 0 1],n)
sinA0=cosi/cos(del0);
A0=asin(sinA0)
v0=sin(A0)*[-sin(lam0) cos(lam0) 0]+ ...
   cos(A0)*[-sin(del0)*cos(lam0) ...
   -sin(del0)*sin(lam0) cos(del0)];
h0=cross(r0,v0)/norm(cross(r0,v0))
if abs(dot(h0,n))<=0.99
    A0=pi-A0
end

% 4th order, 5th stage Runge-Kutta integration:
[x,y]=ode45(@airnav,[0 tf],[del0 lam0 A0]');

% Ground speed (m/s) follows:
v=sqrt(vprime^2-vw^2*sin(Aw-y(:,3)).^2)+vw*cos(Aw-y(:,3));

% Normal acceleration input and bank angle profile follow:
u=-0.5*Omega^2*r*sin(2*y(:,1)).*sin(y(:,3))-2*Omega*v.*sin(y(:,1));
sigma=atan(u/9.81);
```

aircraft's body-fixed frame, $(\mathbf{i}, \mathbf{j}, \mathbf{k})$ – where $(\mathbf{i}, \mathbf{k})$ is the plane of symmetry – relative to an inertial, *north, east, down* (NED) local horizon frame, $(\mathbf{I}, \mathbf{J}, \mathbf{K})$, where

$$\begin{aligned} \mathbf{I} &= \mathbf{i}_\delta, \\ \mathbf{J} &= \mathbf{i}_\ell, \\ \mathbf{K} &= -\mathbf{i}_\mathbf{r}, \end{aligned} \tag{3.111}$$

Table 3.5

```
% Program for specifying governing ODEs of great circle navigation
% expressed as state equations (to be called by 'ode45.m')
% (c) 2010 Ashish Tewari
function dydx=airnav(x,y)
global vprime;
global vw;
global Aw;
global r;
global Omega;
v=sqrt(vprime^2-vw^2*sin(Aw-y(3))^2)+vw*cos(Aw-y(3));
dydx=[v*cos(y(3))/r
      v*sin(y(3))/(r*cos(y(1)))
      v*sin(y(3))*tan(y(1))/r];
```

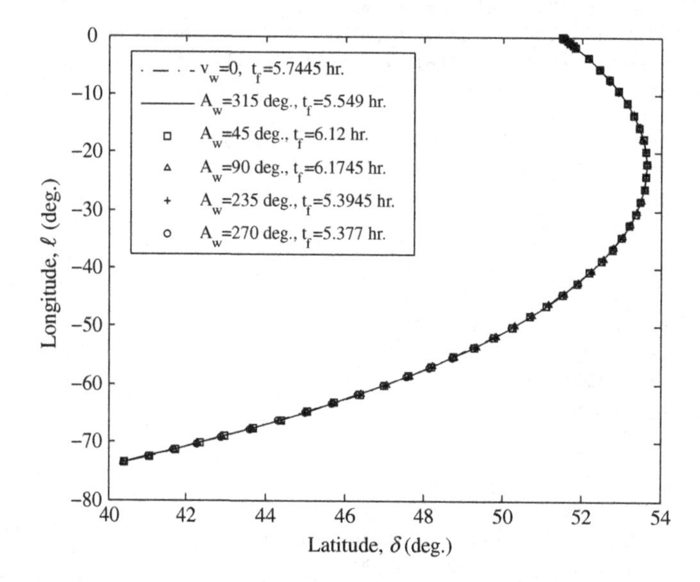

Figure 3.16 Ground track with great circle navigation for various steady-wind conditions

is expressed through the Euler angles, $(\Psi)_3$, $(\Theta)_2$, $(\sigma)_1$ (Figure 3.20), as

$$\begin{Bmatrix} \mathbf{i} \\ \mathbf{j} \\ \mathbf{k} \end{Bmatrix} = \mathbf{C} \begin{Bmatrix} \mathbf{I} \\ \mathbf{J} \\ \mathbf{K} \end{Bmatrix}. \tag{3.112}$$

Here, $\mathbf{C}$ is the rotation matrix

$$\mathbf{C} = \begin{pmatrix} \cos\Theta\cos\Psi & \cos\Theta\sin\Psi & -\sin\Theta \\ \sin\sigma\sin\Theta\cos\Psi - \cos\sigma\sin\Psi & \sin\sigma\sin\Theta\sin\Psi + \cos\sigma\cos\Psi & \sin\sigma\cos\Theta \\ \cos\sigma\sin\Theta\cos\Psi + \sin\sigma\sin\Psi & \cos\sigma\sin\Theta\sin\Psi - \sin\sigma\cos\Psi & \cos\sigma\cos\Theta \end{pmatrix},$$

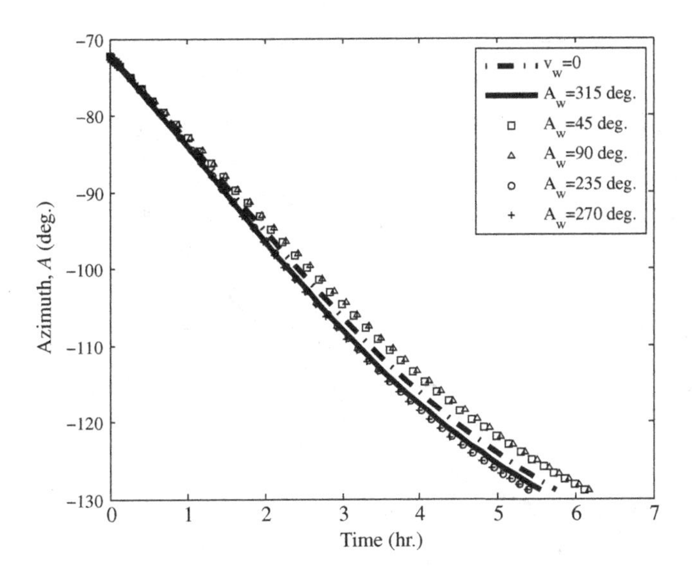

Figure 3.17 Velocity azimuth for great circle navigation for various steady-wind conditions

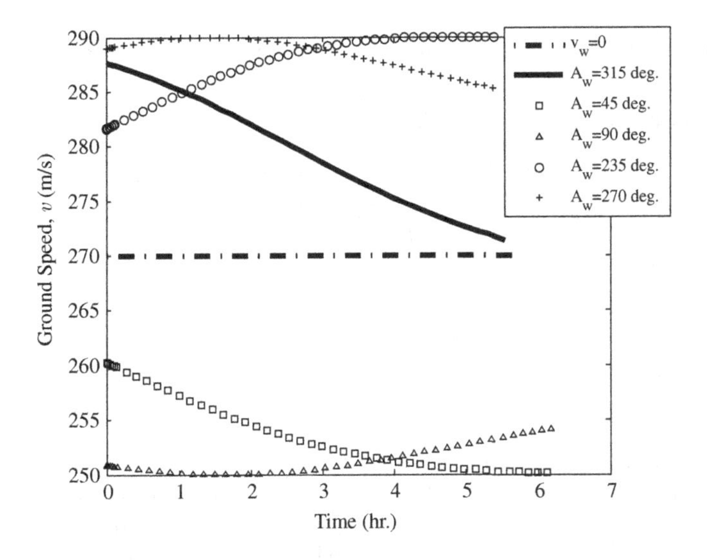

Figure 3.18 Ground speed for great circle navigation for various steady-wind conditions

where σ is the *roll* (or *bank*) angle, Θ the *pitch* angle, and Ψ the *yaw* angle. The angular velocity of the rigid aircraft about its center of mass is given by

$$\omega = P\mathbf{i} + Q\mathbf{j} + R\mathbf{k}, \tag{3.113}$$

where P is the *roll rate*, Q the *pitch rate*, and R the *yaw rate*. Using the Euler angle representation $(\Psi)_3, (\Theta)_2, (\sigma)_1$ as shown in Figure 3.20, we have the rotational kinematics equation

$$\omega(t) = \dot{\sigma}\mathbf{i} + \dot{\Theta}\mathbf{J}' + \dot{\Psi}\mathbf{K}, \tag{3.114}$$

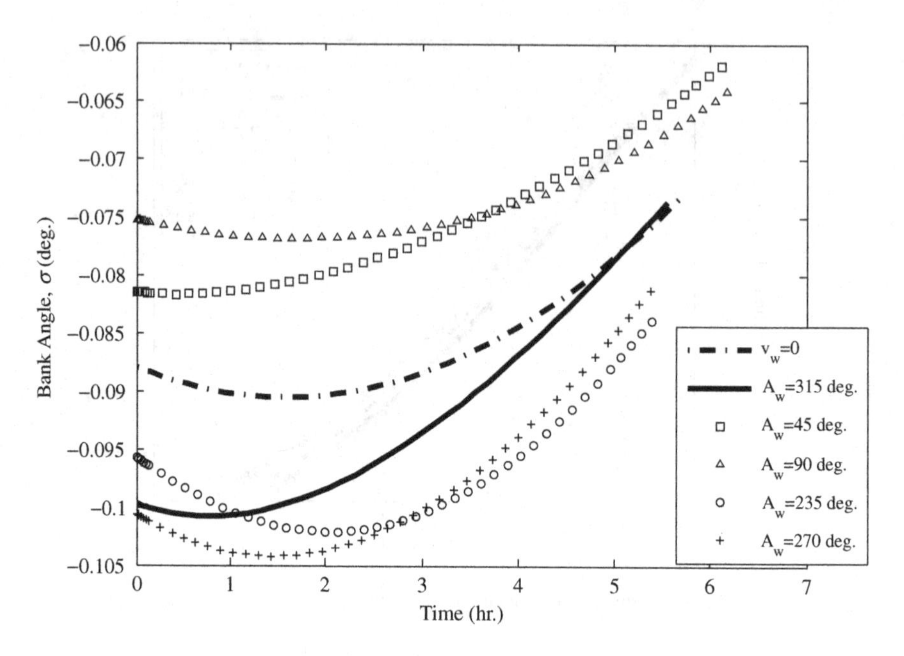

Figure 3.19 Optimal aircraft bank angle for great circle navigation for various steady-wind conditions

which, considering the elementary rotations involved, becomes

$$\omega(t) = \begin{Bmatrix} P \\ Q \\ R \end{Bmatrix} = \begin{Bmatrix} \dot{\sigma} - \dot{\Psi} \sin \Theta \\ \dot{\Theta} \cos \sigma + \dot{\Psi} \sin \sigma \cos \Theta \\ -\dot{\Theta} \sin \sigma + \dot{\Psi} \cos \sigma \cos \Theta \end{Bmatrix} \tag{3.115}$$

or

$$\begin{Bmatrix} \dot{\sigma} \\ \dot{\Theta} \\ \dot{\Psi} \end{Bmatrix} = \frac{1}{\cos \Theta} \begin{pmatrix} \cos \Theta & \sin \sigma \sin \Theta & \cos \sigma \sin \Theta \\ 0 & \cos \sigma \cos \Theta & -\sin \sigma \cos \Theta \\ 0 & \sin \sigma & \cos \sigma \end{pmatrix} \begin{Bmatrix} P \\ Q \\ R \end{Bmatrix}. \tag{3.116}$$

It is evident from equation (3.116) that the Euler angle representation given by the sequence $(\Psi)_3, (\Theta)_2, (\sigma)_1$ is singular for $\Theta = \pm\pi/2$. However, since an aircraft normally does not approach the vertical attitude, this particular Euler angle sequence is commonly applied to aircraft stability and control applications. The exceptional case of vertical flight (such as a high-performance fighter aircraft) can be treated in a manner similar to that of a rocket through an alternative, non-singular, four-parameter attitude representation such as the quaternion (Tewari 2006).

3.3.1 *Translational and Rotational Kinetics*

For an aircraft, the aerodynamic forces and moments vary with a changing attitude of the aircraft (rotation about the center of mass) relative to the velocity vector, which itself is changing in both magnitude and direction due to a translation of the center of mass. Because the time scales of translational and rotational kinetics are generally of the same order of magnitude, it is necessary to study the coupled translational and rotational dynamics for aircraft attitude dynamics. The net external force is the sum of the gravity force, $\mathbf{F}_g(t)$,

$$\mathbf{F}_g = mg\mathbf{K} = mg\left(-\sin\Theta\mathbf{i} + \sin\sigma\cos\Theta\mathbf{j} + \cos\sigma\cos\Theta\mathbf{k}\right), \tag{3.117}$$

and the aerodynamic and propulsive forces, $\mathbf{F}_v(t)$, which are resolved in the body-fixed frame as follows:

$$\mathbf{F}_v = X\mathbf{i} + Y\mathbf{j} + Z\mathbf{k}, \tag{3.118}$$

where X is called the *forward force*, Y the *sideforce*, and Z the *downward force*. Since the NED reference frame (local horizon frame), ($\mathbf{I}$, $\mathbf{J}$, $\mathbf{K}$), is considered instantaneously fixed to the atmosphere, the velocity of the center of mass, $\mathbf{v}(t)$, is taken to be the aircraft's linear velocity relative to the atmosphere (called *relative velocity*), and can be resolved in the body-fixed frame as

$$\mathbf{v} = U\mathbf{i} + V\mathbf{j} + W\mathbf{k}, \tag{3.119}$$

where U is the *forward speed*, V the *sideslip velocity*, and W the *downward* (or *plunge*) velocity. Thus, the translational kinetics equation can be written as

$$\mathbf{F}_v + \mathbf{F}_g = m\dot{\mathbf{v}} + m\left(\boldsymbol{\omega} \times \mathbf{v}\right), \tag{3.120}$$

or, collecting the individual components, we have the following equations of translational dynamics resolved in the body-fixed frame:

$$\begin{aligned} X - mg\sin\Theta &= m(\dot{U} + QW - RV), \\ Y + mg\sin\sigma\cos\Theta &= m(\dot{V} + RU - PW), \\ Z + mg\cos\sigma\cos\Theta &= m(\dot{W} + PV - QU). \end{aligned} \tag{3.121}$$

Equations (3.121) for aircraft attitude dynamics are equivalent to equations (3.11)–(3.13) for translational dynamics, which were resolved in the wind axes for aircraft navigation. The difference between the usage of the two sets of equations is that the mass is assumed constant during the relatively small time scale of attitude dynamics, while the bank angle and the angle of attack – variables of attitude dynamics – are taken to be either constants or external inputs in the navigational plant.

The external torque vector generated by aerodynamic and propulsive means is expressed for an aircraft as

$$\boldsymbol{\tau} = L\mathbf{i} + M\mathbf{j} + N\mathbf{k}, \tag{3.122}$$

where L is the *rolling moment*, M the *pitching moment*, and N the *yawing moment* (Figure 3.20). All aircraft have at least one plane of symmetry, which we shall denote here by the plane, oxz, bound by the body-fixed axes, ($\mathbf{i}$, $\mathbf{k}$), as shown in Figure 3.20. By virtue of the plane of symmetry, we have $J_{xy} = 0 = J_{yz}$ and the rotational kinetics equation can be expressed as

$$\boldsymbol{\tau} = \begin{Bmatrix} L \\ M \\ N \end{Bmatrix} = \mathbf{J}\dot{\boldsymbol{\omega}} + \mathbf{S}(\boldsymbol{\omega})\mathbf{J}\boldsymbol{\omega}, \tag{3.123}$$

where

$$\mathbf{S}(\boldsymbol{\omega}) = \begin{pmatrix} 0 & -R & Q \\ R & 0 & -P \\ -Q & P & 0 \end{pmatrix} \tag{3.124}$$

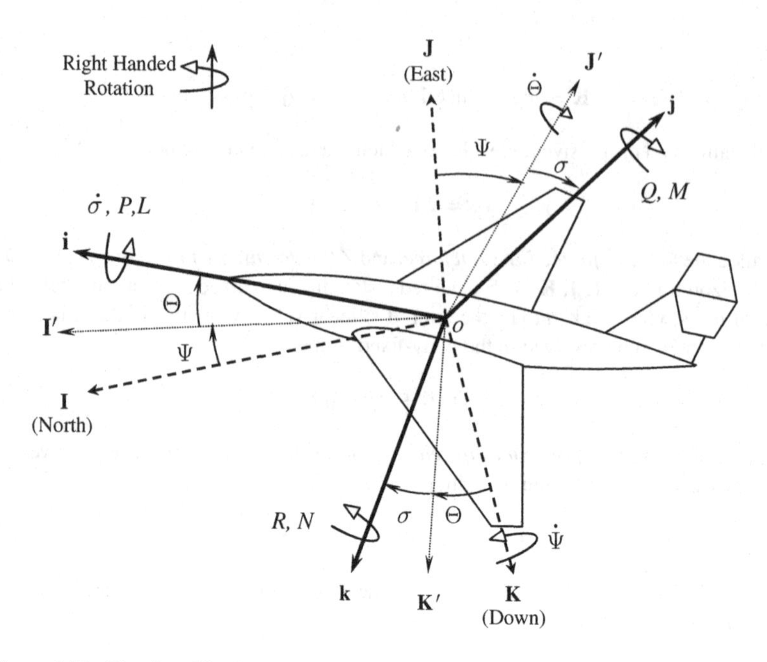

Figure 3.20 The aircraft body axes, (**i**, **j**, **k**), relative to the NED local horizon frame, (**I**, **J**, **K**)

and

$$\mathbf{J} = \begin{pmatrix} J_{xx} & 0 & J_{xz} \\ 0 & J_{yy} & 0 \\ J_{xz} & 0 & J_{zz} \end{pmatrix}. \tag{3.125}$$

Substituting equations (3.124) and (3.125) into equation (3.123) and collecting terms, we have the following scalar equations of rotational kinetics:

$$L = J_{xx}\dot{P} + J_{xz}(\dot{R} + PQ) + \left(J_{zz} - J_{yy}\right) QR, \tag{3.126}$$

$$M = J_{yy}\dot{Q} + J_{xz}(R^2 - P^2) + (J_{xx} - J_{zz})\, PR, \tag{3.127}$$

$$N = J_{zz}\dot{R} + J_{xz}(\dot{P} - QR) + \left(J_{yy} - J_{xx}\right) PQ. \tag{3.128}$$

Equations (3.126)–(3.128) exhibit nonlinear, inertial coupling among the rolling motion (i.e., rotation about **i**), the pitching motion (rotation about **j**), and the yawing motion (rotation about **k**), which must be solved in tandem with the translational kinetics equation (3.13) and the rotational kinematics equation (3.17). Note that we have regarded the aircraft as a rigid body and thus neglected gyroscopic terms arising due to the fixed angular momenta of engines. Such terms are usually negligible, unless the aircraft is a small one but has an engine with a large angular momentum (such as some rotary-engined fighter aircraft of the First World War). In such a case, there would be an additional inertial coupling between pitch and yaw, which is independent of the roll rate, P.

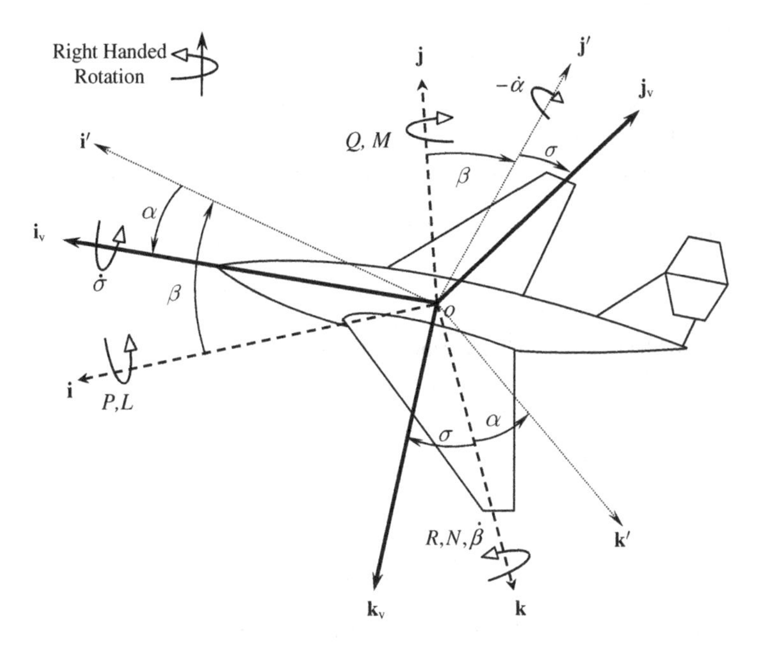

Figure 3.21 The aircraft body axes, $(\mathbf{i}, \mathbf{j}, \mathbf{k})$, relative to the wind axes, $(\mathbf{i}_v, \mathbf{j}_v, \mathbf{k}_v)$

3.3.2 Attitude Relative to the Velocity Vector

For a rigid aircraft, the aerodynamic and propulsive forces, (X, Y, Z), and moments, (L, M, N), are functions of the relative linear velocity vector, that is, the airspeed, v, as well as the relative flow angles – the angle of attack, α, and the sideslip angle, β, arising out of the orientation of the aircraft's body axes relative to the velocity vector (or in other words, the linear velocity vector resolved in the body axes).[3] We shall employ the relative flow angles, α, β, and the geometric bank angle, σ, as Euler angles in order to derive a coordinate transformation between the wind axes, $(\mathbf{i}_v, \mathbf{j}_v, \mathbf{k}_v)$, and the body axes, $(\mathbf{i}, \mathbf{j}, \mathbf{k})$, as shown in Figure 3.21. Recall from Figure 3.3 that the plane $(\mathbf{i}_v, \mathbf{k}_v)$ is the vertical plane, and that $(\mathbf{i}, \mathbf{k})$ is the plane of symmetry (the plane containing the lift vector). Therefore, the bank angle, σ, is merely the angle between the two planes $(\mathbf{i}_v, \mathbf{k}_v)$ and $(\mathbf{i}, \mathbf{k})$ or between the axes $\mathbf{j}_v$ and $\mathbf{j}$ (Figure 3.21). Clearly, the Euler sequence for the necessary coordinate transformation is $(\beta)_3$, $(-\alpha)_2$, $(\sigma)_1$ (Figure 3.21), resulting in

$$\begin{Bmatrix} \mathbf{i}_v \\ \mathbf{j}_v \\ \mathbf{k}_v \end{Bmatrix} = \mathbf{C}_v \begin{Bmatrix} \mathbf{i} \\ \mathbf{j} \\ \mathbf{k} \end{Bmatrix}, \tag{3.129}$$

where $\mathbf{C}_v$ is the following rotation matrix:

[3] The vector (X, Y, Z, L, M, N) also depends upon the relative angular velocity vector, (P, Q, R).

$\mathbf{C}_v =$

$$\begin{pmatrix} \cos\alpha\cos\beta & \cos\alpha\sin\beta & \sin\alpha \\ -\sin\sigma\sin\alpha\cos\beta - \cos\sigma\sin\beta & -\sin\sigma\sin\alpha\sin\beta + \cos\sigma\cos\beta & \sin\sigma\cos\alpha \\ -\cos\sigma\sin\alpha\cos\beta + \sin\sigma\sin\beta & -\cos\sigma\sin\alpha\sin\beta - \sin\sigma\cos\beta & \cos\sigma\cos\alpha \end{pmatrix}.$$

Since the relative velocity vector can be resolved in the body axes as

$$\mathbf{v} = v(\cos\alpha\cos\beta\mathbf{i} + \cos\alpha\sin\beta\mathbf{j} + \sin\alpha\mathbf{k}) \tag{3.130}$$

(Figure 3.21), we have the following relationships among the flow angles, (α, β), and the relative velocity components, (U, V, W) (equation (3.119)):

$$\alpha = \sin^{-1}\frac{W}{\sqrt{U^2+V^2+W^2}}, \tag{3.131}$$

$$\beta = \tan^{-1}\frac{V}{U}. \tag{3.132}$$

In a coordinated flight ($\beta = 0$), with a typically small angle of attack, we have

$$\mathbf{C}_v \simeq \begin{pmatrix} 1 & 0 & \alpha \\ -\alpha\sin\sigma & \cos\sigma & \sin\sigma \\ -\alpha\cos\sigma & -\sin\sigma & \cos\sigma \end{pmatrix} \tag{3.133}$$

or

$$\begin{aligned} \mathbf{i}_v &\simeq \mathbf{i} + \alpha\mathbf{k}, \\ \mathbf{j}_v &\simeq -\alpha\mathbf{i}\sin\sigma + \mathbf{j}\cos\sigma + \mathbf{k}\sin\sigma, \\ \mathbf{k}_v &\simeq -\alpha\mathbf{i}\cos\sigma - \mathbf{j}\sin\sigma + \mathbf{k}\cos\sigma. \end{aligned} \tag{3.134}$$

3.4 Aerodynamic Forces and Moments

Flow of atmospheric gases relative to the body's external surface gives rise to aerodynamic force, $\mathbf{f}_v(t)$, and torque, $\boldsymbol{\tau}(t)$, the fundamental sources of both of which are the surface distributions of static pressure, $p(x, y, z, t)$, acting normal to the surface, and shear stress, $f(x, y, z, t)$, along the relative flow direction:[4]

$$\mathbf{f}_v = \int_A \{-p\mathbf{n} + f(\mathbf{n} \times [-\mathbf{i}_v \times \mathbf{n}])\}\, dA, \tag{3.135}$$

$$\boldsymbol{\tau} = \int_A \mathbf{d} \times \{-p\mathbf{n} + f(\mathbf{n} \times [-\mathbf{i}_v \times \mathbf{n}])\}\, dA, \tag{3.136}$$

[4] Flow far upstream of the body is opposite to the relative velocity vector, $\mathbf{v}$, and is termed *freestream*.

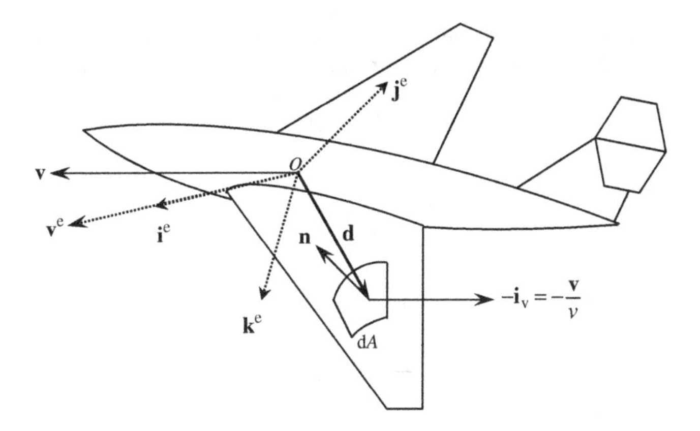

Figure 3.22 The stability axes, $(\mathbf{i}^e, \mathbf{j}^e, \mathbf{k}^e)$, local surface normal, $\mathbf{n}$, and relative flow direction, $-\mathbf{i}_v$

where $\mathbf{d}(x, y, z)$ locates an elemental area, dA, with respect to the center of mass (Figure 3.22), $\mathbf{n}(x, y, z, t)$ denotes the local surface normal (which is changing in time due to the angular velocity, (P, Q, R)), and relative flow direction is given by $-\mathbf{i}_v$, where

$$\mathbf{i}_v = \frac{\mathbf{v}}{v}. \tag{3.137}$$

The static pressure and shear stress distributions in a continuous (continuum) flow are themselves functions of the flow velocity far upstream, (v, α, β) (also called *freestream velocity*), the angular velocity of the rigid body relative to the freestream, (P, Q, R), and the thermodynamic properties of the freestream, namely density, ρ, static temperature, T, dynamic viscosity, $\bar{\mu}$, and specific-heat ratio, γ. The continuous flow properties can be combined into the following non-dimensional parameters governing the flow-field:

(a) *Mach number*, $Ma \doteq v/a_\infty$, where a_∞ is the speed of sound far upstream;
(b) *Reynolds number*, $Re \doteq \rho v \ell_c/\bar{\mu}$ (where ℓ_c is a characteristic length), representing the ratio of inertial and viscous forces;
(c) *Strouhal numbers*, $P\ell_1/v$, $Q\ell_2/v$, $R\ell_3/v$ (where ℓ_1, ℓ_2, ℓ_3 are characteristic lengths along the respective body axes), representing non-dimensional (reduced) frequencies of rotation.

In a rarefied flow, the *mean free-path* of gas molecules, $\bar{\lambda}$, is an additional flow parameter and *Knudsen number*, is the governing non-dimensional parameter given by $Kn \doteq \frac{\bar{\lambda}}{l_c}$, where l_c is a characteristic length. Based upon Knudsen number, the flow regimes are classified as follows:

(i) *free-molecular* flow, $Kn \geq 10$;
(ii) *transition* flow, $0.01 \leq Kn \leq 10$;
(iii) *continuum* (or continuous) flow, $Kn \leq 0.01$.

When the nominal trajectory is a steady maneuver, the aerodynamic forces and moments are balanced by other forces, and thus have constant equilibrium values, $\mathbf{f}_v^e$, $\boldsymbol{\tau}^e$. In such a case, a special body-fixed frame – called *stability axes* – is generally employed, with one axis initially aligned with the nominal flight direction (Figure 3.22). A small perturbation $\Delta\mathbf{v}$, $\Delta\boldsymbol{\omega}$ about the nominal trajectory, $\mathbf{v}^e$, $\boldsymbol{\omega}^e$,

$$\begin{aligned} \mathbf{v} &= \mathbf{v}^e + \Delta\mathbf{v}, \\ \boldsymbol{\omega} &= \boldsymbol{\omega}^e + \Delta\boldsymbol{\omega}, \end{aligned} \tag{3.138}$$

causes small changes in the flow speed, angle of attack, sideslip angle,[5] and the body-rates, (P, Q, R), measured with respect to the stability axes, resulting in the linearized loads by Taylor series expansion:

$$\mathbf{f}_v = \mathbf{f}_v^e + \Delta\mathbf{f}_v, \tag{3.139}$$

$$\boldsymbol{\tau} = \boldsymbol{\tau}^e + \Delta\boldsymbol{\tau}, \tag{3.140}$$

where

$$\Delta\mathbf{f}_v = \sum_{i=1}^{9}\sum_{k=1}^{\infty}\frac{\partial^k \mathbf{f}_v}{\partial \zeta_i^k}\Delta\zeta_i^k, \tag{3.141}$$

$$\Delta\boldsymbol{\tau} = \sum_{i=1}^{9}\sum_{k=1}^{\infty}\frac{\partial^k \boldsymbol{\tau}}{\partial \zeta_i^k}\Delta\zeta_i^k, \tag{3.142}$$

and $\zeta_i(t)$ is the ith element of the *relative motion vector*, $\boldsymbol{\zeta}(t)$:

$$\boldsymbol{\zeta}(t) = \left(v, \dot{v}, \alpha, \dot{\alpha}, \beta, \dot{\beta}, P, Q, R\right)^T. \tag{3.143}$$

Clearly, $\boldsymbol{\zeta}(t)$ represents the motion of the body relative to the atmosphere. Since the perturbation in the motion variables, $\Delta\zeta_i$, from their equilibrium values is small, one can neglect the second- and higher-order terms in equations (3.141) and (3.142), resulting in the following linearized aerodynamic perturbations:

$$\begin{aligned}\Delta\mathbf{f}_v &= \sum_{i=1}^{9}\frac{\partial \mathbf{f}_v}{\partial \zeta_i}\Delta\zeta_i,\\ \Delta\boldsymbol{\tau} &= \sum_{i=1}^{9}\frac{\partial \boldsymbol{\tau}}{\partial \zeta_i}\Delta\zeta_i.\end{aligned} \tag{3.144}$$

Equations (3.144) are very useful in the derivation of atmospheric flight dynamics plant as they can employ the closed-form results of small perturbation aerodynamic theories.

The aerodynamic force and torque vectors can be separated into forces and moments arising due to relative velocity (both linear, $\mathbf{v}$, and angular, $\boldsymbol{\omega}$) of the rigid aircraft with respect to the atmosphere (equation (3.144)), and the control forces and moments generated by the *control surfaces*, namely the *elevator*, the *ailerons*, and the *rudder*, as shown in Figure 3.23. The control surfaces are generally deflected by small angles, thus the aerodynamic control forces and moments obey linear relationships with control surface deflections. Furthermore, the control surfaces are driven by actuators with a dynamic response about ten times faster than that of the vehicle's rigid body dynamics. Therefore, flow unsteadiness due to control surface deflections can often be neglected when considering rigid body aerodynamics. Hence, a quasi-steady approximation of aerodynamic loads linearly varying with control surface deflection is usually employed.[6] The aerodynamic control surface for longitudinal control is the *elevator* with rotation axis parallel to (oy). The elevator can be mounted either forward or behind (oy), usually on the trailing

[5] If there is a freestream rotation (swirl), such as that due to a large propeller, or tip-vortices of a large preceding aircraft, it must also be taken into account in equation (3.138).

[6] When the vehicle's structural flexibility is taken into account, it is necessary to include the unsteady aerodynamic effects of control surface deflections, that is, dependence of aerodynamic forces and moments on deflection rates, $\dot{\delta}_E, \dot{\delta}_A, \dot{\delta}_R$.

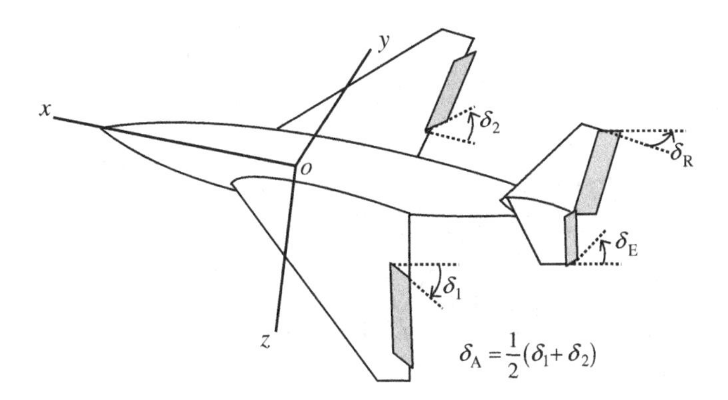

Figure 3.23 The aircraft control surfaces, the elevator, the ailerons, and the rudder

edge of a larger stabilizing surface in the (xy) plane, as shown in Figure 3.21. Elevator deflection, δ_E, creates a forward force, $X_\delta \delta_E$, downforce, $Z_\delta \delta_E$, and a control pitching moment, $M_\delta \delta_E$.

A pair of control surfaces in the (xy) plane, located symmetrically about the axis (ox) and deflected in mutually opposite directions by angles δ_1 and δ_2 (Figure 3.23) are called *ailerons* that are used as roll control devices. The *aileron deflection* is the average of the two separate deflections, $\delta_A = (\delta_1 + \delta_2)/2$, and is designed such that a control rolling moment, $L_A \delta_A$, is produced along with a much smaller, undesirable yawing moment, $N_A \delta_A$. A control surface mounted on a fin in the (xz) plane behind the (oy) axis is called the *rudder*. A rudder deflection, δ_R, creates a sideforce, $Y_R \delta_R$, control yawing moment, $N_R \delta_R$, and a much smaller rolling moment, $L_R \delta_R$. The rudder is primarily used to correct a lateral flight asymmetry.

3.5 Longitudinal Dynamics

A special solution to the aircraft dynamics equations (3.17), (3.13), and (3.126)–(3.128) is identified to be $\sigma = P = R = V = 0$, for which we have

$$\begin{aligned}
\dot{\Theta} &= Q, \\
\dot{\Psi} &= 0, \\
X - mg \sin \Theta &= m(\dot{U} + QW), \\
Y &= 0, \\
Z + mg \cos \Theta &= m(\dot{W} - QU), \\
L &= 0, \\
M &= J_{yy} \dot{Q}, \\
N &= 0,
\end{aligned} \tag{3.145}$$

that is, the motion is confined to the plane of symmetry and can be entirely represented by the variables $\Theta(t)$, $U(t)$, $W(t)$, $Q(t)$, with $\Psi =$ const. Such a motion has a special place in aircraft dynamics and is referred to as *longitudinal motion*. Consequently, the plane of symmetry, oxz, is also called the longitudinal plane (Figure 3.24).

Navigational subsystem (b) for cruising flight of an aircraft essentially involves longitudinal motion wherein a constant airspeed and altitude must be maintained despite disturbances due to atmospheric gusts. The small rolling (banking) motion required for subsystem (a) can be neglected, leading to a flight

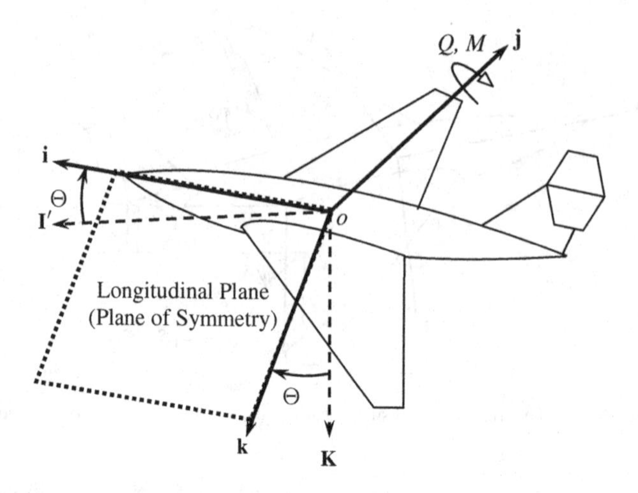

Figure 3.24 Aircraft's plane of symmetry (the longitudinal plane) and the longitudinal motion, ($\sigma = P = R = V = 0$)

confined to the plane of symmetry. Other examples of longitudinal dynamics are climb and descent, as well as take-off and landing.

Longitudinal motion is normally represented by a small displacement from an equilibrium (unaccelerated) flight condition in the longitudinal plane. The flight variables in such an equilibrium are denoted with a subscript e: $U_e, W_e, \theta_e, Q_e, X_e, Z_e, M_e$. A common equilibrium point is an essentially straight-line flight path in the longitudinal plane, for which $Q_e = 0$ and equation (3.145) yields

$$\begin{aligned}\dot{\theta}_e &= 0,\\ X_e - mg\sin\theta_e &= 0,\\ Z_e + mg\cos\theta_e &= 0,\\ M_e &= 0.\end{aligned} \tag{3.146}$$

A convenient reference frame for longitudinal motion is the *stability axes*, $(\mathbf{i}_e, \mathbf{j}_e, \mathbf{k}_e)$, defined such that the axis $\mathbf{i}_e$ is along the equilibrium flight direction, $W_e = 0$. Thus, a longitudinal displacement from equilibrium can be described by the displaced body-fixed frame, $(\mathbf{i}, \mathbf{j}_e, \mathbf{k})$ (Figure 3.25), as well as perturbations in flight variables and aerodynamic forces and moment given by:

$$\begin{aligned}\Theta &= \theta_e + \theta,\\ U &= U_e + u,\\ W &= w,\\ Q &= q,\\ X &= X_e + \bar{X},\\ Z &= Z_e + \bar{Z},\\ M &= \bar{M}.\end{aligned} \tag{3.147}$$

In such a case, the vertical velocity perturbation, w (called *downwash*), is represented as a change in the *angle of attack* (Figure 3.25) given by

$$\alpha = \tan^{-1}\frac{w}{U}. \tag{3.148}$$

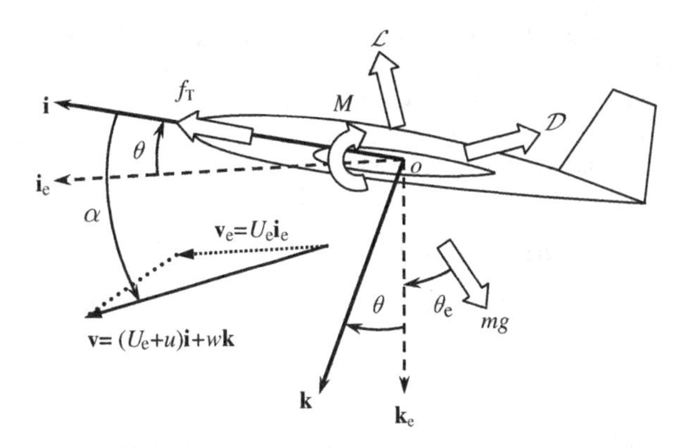

Figure 3.25 The stability axes, $(\mathbf{i}_e, \mathbf{k}_e)$, the displaced body axes $(\mathbf{i}, \mathbf{k})$, and the forces and moment for longitudinal motion

The perturbation quantities, θ, u, w, q, $\bar{X}$, $\bar{Z}$, $\bar{M}$, can be considered small to begin with. Since a successful flight control system would prevent the small perturbations from growing large with time, one need not consider large aerodynamic perturbations. This is fortunate because aerodynamic forces and moments have a rather complex character in the case of large perturbations, which can be very difficult to analyze in closed form. A small perturbation from the equilibrium results in the following approximations:

$$\sin\Theta \simeq \sin\theta_e + \theta\cos\theta_e,$$

$$\cos\Theta \simeq \cos\theta_e - \theta\sin\theta_e,$$

$$u \ll U_e, \quad w \ll U_e, \quad \alpha \simeq \tan^{-1}\frac{w}{U_e} \simeq \frac{w}{U_e},$$

and thus the rotational kinematics equation (3.17) becomes

$$\dot{\theta} = q. \tag{3.149}$$

Since $U = U_e + u \simeq U_e$, we will henceforth drop the subscript on U_e.

Continuing with the small perturbation assumption, we express the longitudinal aerodynamic forces and moments as linear functions of the perturbation variables as follows:

$$\begin{aligned} \bar{X} &= X_u u + X_\alpha \alpha + X_{\dot{\alpha}}\dot{\alpha} + X_q q, \\ \bar{Z} &= Z_u u + Z_\alpha \alpha + Z_{\dot{\alpha}}\dot{\alpha} + Z_q q, \\ \bar{M} &= M_u u + M_\alpha \alpha + M_{\dot{\alpha}}\dot{\alpha} + M_q q. \end{aligned} \tag{3.150}$$

Here, the coefficients

$$X_u = \left(\frac{\partial \bar{X}}{\partial u}\right)_e, \quad \text{etc.}$$

arise due to a Taylor series expansion about the equilibrium flight condition (truncated to first order) and are called *stability derivatives*. Clearly, the stability of the motion depends upon the values of the stability derivatives. Stability derivatives are functions of the aircraft's geometry, size, airspeed, flow regime (i.e., non-dimensional parameters describing the relative flow, such as equilibrium angle of attack, sideslip,

Mach number, and Reynolds number), and atmospheric properties. These functional relationships can either be estimated by semi-empirical methods (Hoak *et al.* 1978), or carefully obtained from wind-tunnel tests. Since a discussion of stability derivatives is beyond the scope of this book, we refer the reader to Etkin and Reid (1995) for the physical phenomena represented by each stability derivative. We briefly note that generally $X_{\dot{\alpha}} \simeq 0$ and $X_q = 0$ in a conventional airplane.

3.5.1 Longitudinal Dynamics Plant

Substituting equations (3.23)–(3.27) into equation (3.22) and neglecting products of small perturbation quantities, we have

$$\begin{aligned} m\dot{u} &= X_u u + X_\alpha \alpha + X_q q - mg\theta \cos\theta_e, \\ mU(\dot{\alpha} - q) &= Z_u u + Z_\alpha \alpha + Z_{\dot{\alpha}}\dot{\alpha} + Z_q q - mg\theta \sin\theta_e, \\ \dot{\theta} &= q, \\ J_{yy}\dot{q} &= M_u u + M_\alpha \alpha + M_q q + M_{\dot{\alpha}}\dot{\alpha}, \end{aligned} \tag{3.151}$$

which can be expressed in a state-space form as

$$\begin{Bmatrix} \dot{u} \\ \dot{\alpha} \\ \dot{\theta} \\ \dot{q} \end{Bmatrix} = \mathbf{A}_{\text{Long}} \begin{Bmatrix} u \\ \alpha \\ \theta \\ q \end{Bmatrix}, \tag{3.152}$$

where

$$\mathbf{A}_{\text{Long}} =$$

$$\begin{bmatrix} \frac{X_u}{m} & \frac{X_\alpha}{m} & -g\cos\theta_e & 0 \\ \frac{Z_u}{mU - Z_{\dot{\alpha}}} & \frac{Z_\alpha}{mU - Z_{\dot{\alpha}}} & -\frac{mg\sin\theta_e}{mU - Z_{\dot{\alpha}}} & \frac{mU + Z_q}{mU - Z_{\dot{\alpha}}} \\ 0 & 0 & 0 & 1 \\ \frac{M_u}{J_{yy}} + \frac{M_{\dot{\alpha}} Z_u}{J_{yy}(mU - Z_{\dot{\alpha}})} & \frac{M_\alpha}{J_{yy}} + \frac{M_{\dot{\alpha}} Z_\alpha}{J_{yy}(mU - Z_{\dot{\alpha}})} & -\frac{M_{\dot{\alpha}}(mg\sin\theta_e)}{J_{yy}(mU - Z_{\dot{\alpha}})} & \frac{M_q}{J_{yy}} + \frac{M_{\dot{\alpha}}(mU + Z_q)}{J_{yy}(mU - Z_{\dot{\alpha}})} \end{bmatrix}$$

is the state dynamics matrix for longitudinal, small-disturbance motion.

The longitudinal modes are determined from the quartic characteristic equation that can be factored into a pair of distinct quadratic modes:

$$|s\mathbf{I} - \mathbf{A}_{\text{Long}}| = \left(s^2 + 2\zeta_p\omega_p s + \omega_p^2\right)\left(s^2 + 2\zeta_s\omega_s s + \omega_s^2\right) = 0, \tag{3.153}$$

where (ω_p, ζ_p) are the natural frequency and damping ratio of the smaller frequency (or *long-period*[7]) mode, and (ω_s, ζ_s) are those of the higher frequency (or *short-period*) mode.

[7] The long-period mode is also called the *phugoid mode* in standard aeronautical terminology.

Example 3.3 Consider an airplane flying straight and level ($\theta_e = 0$) with a constant speed, $U = 55$ m/s. The controls fixed longitudinal parameters of the airplane in this flight condition are as follows:

$$\frac{X_u}{m} = -0.045/\text{s}, \quad \frac{Z_u}{m} = -0.36/\text{s}, \quad \frac{X_\alpha}{m} = 1.96\ \text{m/s}^2,$$

$$\frac{Z_\alpha}{m} = -108\ \text{m/s}^2, \quad \frac{M_\alpha}{J_{yy}} = -8.6/\text{s}^2, \quad \frac{M_{\dot{\alpha}}}{J_{yy}} = -0.9/\text{s}, \quad \frac{M_q}{J_{yy}} = -2/\text{s},$$

while the stability derivatives, Z_q, $Z_{\dot{\alpha}}$, M_u, are negligible in comparison with the other terms.

The longitudinal state dynamics matrix of the airplane in the given equilibrium flight condition is calculated as

$$\mathbf{A}_{\text{Long}} = \begin{pmatrix} -0.045 & 1.96 & -9.81 & 0 \\ -0.0065 & -1.9636 & 0 & 1 \\ 0 & 0 & 0 & 1 \\ 0.0059 & -6.8327 & 0 & -2.9 \end{pmatrix}.$$

The longitudinal modes are computed using MATLAB's Control System Toolbox command *damp* as follows:

```
>> A=[-0.045       1.96    -9.81        0
       -0.0065  -1.9636        0        1
             0        0        0        1
        0.0059  -6.8327        0     -2.9];

>> damp(A)

        Eigenvalue                 Damping       Freq. (rad/s)
  -1.73e-002 + 2.09e-001i         8.26e-002        2.09e-001
  -1.73e-002 - 2.09e-001i         8.26e-002        2.09e-001
  -2.44e+000 + 2.57e+000i         6.88e-001        3.54e+000
  -2.44e+000 - 2.57e+000i         6.88e-001        3.54e+000
```

Thus, $\omega_p = 0.209$ rad/s, $\zeta_p = 0.0826$ for the phugoid mode and $\omega_s = 3.54$ rad/s, $\zeta_s = 0.688$ for the short-period mode of this particular airplane in the given straight and level flight condition. Note that the phugoid mode is lightly damped, while the short-period mode has a near critical damping ratio.

Since the phugoid and the short-period modes generally have widely different frequencies, there is little chance of interaction between the two, and thus each mode can be approximated by a distinct second-order system. The approximate phugoid dynamics consists of a slow oscillation of airspeed, u, and pitch angle, θ, at a nearly constant angle of attack ($\alpha = 0$). The small damping of the phugoid mode is due to a small magnitude of the derivative, X_u, which is predominantly the change in drag (itself a small force) due to variation in forward speed. Thus, the approximate phugoid dynamics can be obtained by neglecting the pitching moment equation and putting $\alpha = 0$ in the axial and normal force equations (Tewari 2006). For a small equilibrium pitch angle, θ_e, the phugoid natural frequency, ω_p, is directly influenced by the derivative, Z_u.

In contrast with the phugoid mode, the approximate short-period dynamics consists of a rapid, well-damped oscillation in α and θ, but a negligible change in airspeed, u. Thus the short-period dynamics is approximated by neglecting the axial force equation and putting $u = 0$ in the remaining equations. The short-period damping is dominated by the derivatives Z_α and M_q.

The control of longitudinal dynamics is exercised through the deflection of the *elevator* and the *engine throttle* that can change the thrust, T, (within the capacity of the engine) by regulating the fuel supply. While the elevator alone can effectively control the changes in pitch angle and the angle of attack, the control of the airspeed additionally requires the engine throttle input. A positive elevator deflection, δ_{E}, is taken to be the one that causes a decrease in the tail lift, thereby applying a positive pitching moment about the center of mass, o. Generally, the engines are mounted very close to the center of mass and hence do not have an appreciable pitching moment contribution when a throttle input is applied, changing the thrust by a small amount. Furthermore, the thrust direction is generally along the flight direction in a well-designed conventional[8] airplane.

Assuming small elevator deflection from equilibrium, the change in total aircraft lift and pitching moment due to the elevator can be expressed by the following positive control derivatives due to the elevator:

$$Z_\delta = \frac{\partial Z}{\partial \delta_{\mathrm{E}}} \tag{3.154}$$

and

$$M_\delta = \frac{\partial M}{\partial \delta_{\mathrm{E}}}. \tag{3.155}$$

The primary longitudinal control derivative is M_δ and is thus called *elevator effectiveness*. Similarly, the primary control derivative for throttle input corresponds to a small change in forward force, given by

$$X_{\mathrm{T}} = \left(\frac{\partial \bar{X}}{\partial \beta_{\mathrm{T}}}\right)_{\mathrm{e}}. \tag{3.156}$$

A well-designed airplane must have negligible normal force and pitching moment thrust contributions, $Z_{\mathrm{T}} = M_{\mathrm{T}} = 0$, otherwise there would be a degraded closed-loop transient response to throttle input.

The actuating subsystems (called *servos*) for elevator and throttle must faithfully and rapidly follow the commands of the control system. The settling time of the actuator must be much smaller than that of the dominant dynamics being controlled. Generally, the elevator actuator must have a settling time quite small compared to that of the short-period mode, while the engine actuator must respond in a fraction of the phugoid time period. In both cases, there should be neither a large overshoot nor a substantial steady-state error. Usually the servo dynamics – being much faster than those of the plant – are either neglected, or modeled as lag effects through first-order transfer functions.

The aircraft plant for longitudinal control is derived by simply adding the linear control contributions to axial force, normal force, and pitching moment equations through the control derivatives, Z_δ, M_δ, X_{T}, etc.

[8] Some special airplanes can have a thrust vectoring engine for better maneuverability, or a shorter runway length requirement.

The longitudinal controls coefficient matrix is expressed as follows:

$$\mathbf{B}_{\text{Long}} = \begin{pmatrix} \frac{X_\delta}{m} & \frac{X_T}{m} \\ \frac{Z_\delta}{mU - Z_{\dot{\alpha}}} & \frac{Z_T}{mU - Z_{\dot{\alpha}}} \\ 0 & 0 \\ \frac{M_\delta}{J_{yy}} + \frac{M_{\dot{\alpha}} Z_\delta}{J_{yy}(mU - Z_{\dot{\alpha}})} & \frac{M_T}{J_{yy}} + \frac{M_{\dot{\alpha}} Z_T}{J_{yy}(mU - Z_{\dot{\alpha}})} \end{pmatrix}. \tag{3.157}$$

3.6 Optimal Multi-variable Longitudinal Control

A typical aircraft flight mainly consists of take-off, climb, level cruise, descent, and landing, each of which have nearly steady, straight-line trajectories in the longitudinal plane. Automatic flight control systems for aircraft are thus largely tasked with controlling the longitudinal dynamics. The longitudinal flight control systems can be generally divided into two categories: autopilots and stability augmentation systems (SAS). Autopilots are control systems that maintain the vehicle close to a set nominal flight path despite atmospheric disturbances and fuel consumption, thereby relieving pilot fatigue in a long-duration flight. An autopilot is also necessary in performing a high-precision task that may be too demanding for a pilot, such as automatic landing in poor visibility conditions. The SAS are essentially tracking systems that take commands from either the pilot or an autopilot, and manipulate the controls in order to achieve the desired task, while compensating for any undesirable behavior in the two longitudinal modes (phugoid and short-period). In a fully automated aircraft, the autopilot serves as the outer loop that provides reference signals to be tracked by the inner SAS loop. In both the loops, feedback of certain output variables is necessary for an adequate closed-loop performance. The sensors usually consist of motion sensors comprising a rate gyro for measuring q, vertical gyro for θ, and accelerometer for normal acceleration at a sensor location, ℓ_z, forward of the center of mass,

$$a_z = U(\dot{\alpha} - q) - \ell_z \dot{q}, \tag{3.158}$$

as well as airflow sensors (called *air data* sensors) that measure static and stagnation pressures and temperatures and relative flow direction. The air data is processed by a *central air data computer* (CADC) in order to generate flow parameters such as airspeed, U, dynamic pressure, $\bar{q}$, Mach number, $\mathcal{M}$, and angle of attack, α.

Since the aircraft plant changes dramatically with aerodynamic parameters governed by airspeed and atmospheric properties, a single control law is seldom adequate for the entire range of speeds and altitudes (called the *flight envelope*) that an aircraft is expected to encounter in its mission. Consequently, the feedback gains for the autopilot and SAS must be varied according to the airspeed, altitude, air temperature, etc., in order to ensure acceptable closed-loop behavior at all times. Such a control system that continuously adapts itself to a changing set of flight variables is rather challenging to design, and is called an *adaptive control system*. While the controllers designed for the longitudinal autopilot and SAS are essentially linear, the self-adaptive controller gains render the entire control system a nonlinear one. A theory of nonlinear adaptive control systems has been developed (Åström and Wittenmark 1995), but there is no uniform procedure in existence that can be generally applied to a given system. A common adaptive control system is the *gain scheduler* where pre-determined functions of controller gains with certain flight parameters are stored onboard a flight control computer. The flight variables concerned, such as the dynamic pressure and Mach number, are computed and fed back from sensed air data via a CADC, and used to determine the controller gains at each time instant. Figure 3.26 shows a block diagram of a general longitudinal flight control system. Note the autopilot switch between manual and

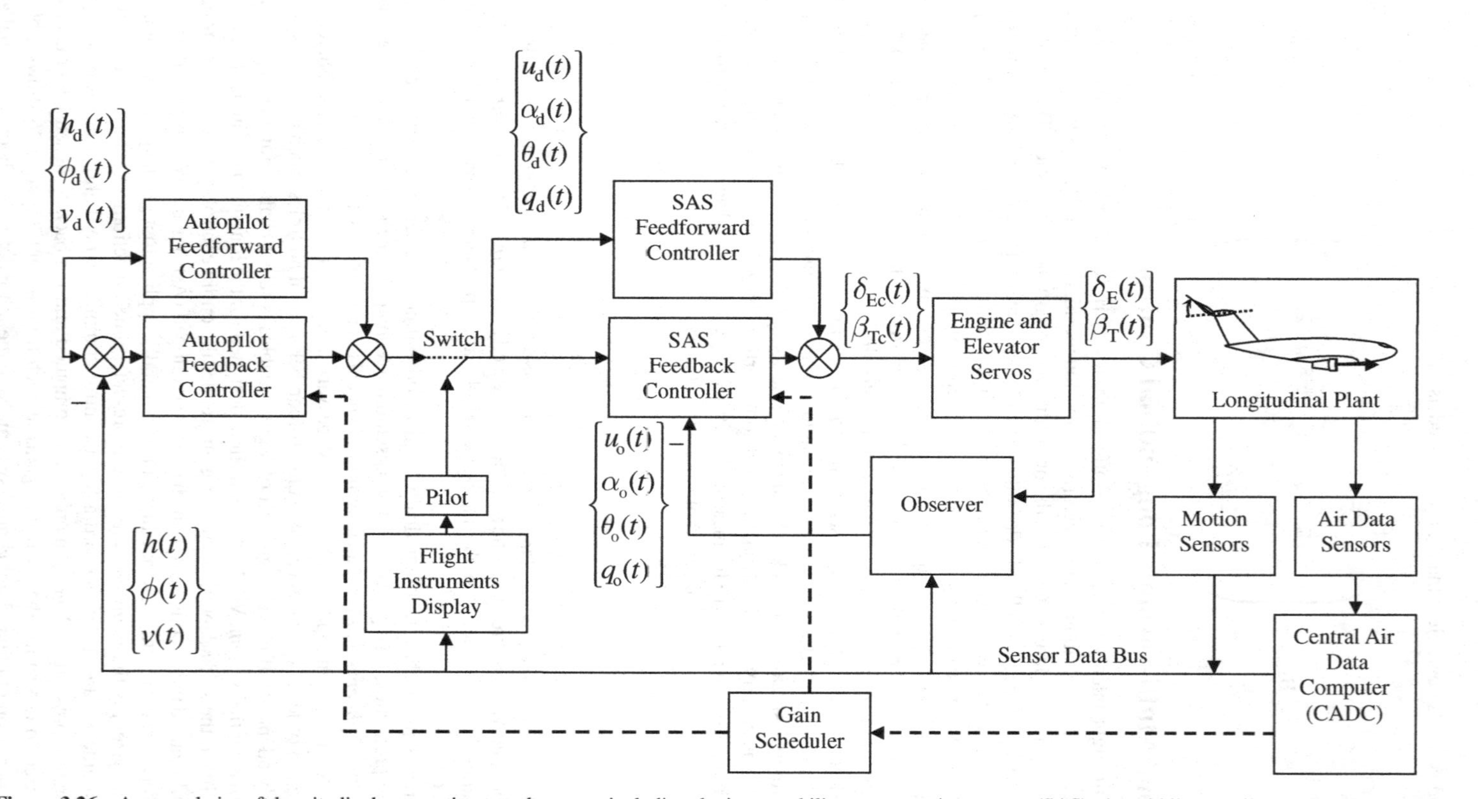

Figure 3.26 A general aircraft longitudinal automatic control system, including the inner stability augmentation system (SAS), the middle autopilot loop, and the outer, nonlinear gain scheduler for the SAS controller gains based upon flight parameters

automatic reference commands for the SAS. If the pilot sets constant reference commands to be flown by the autopilot, we have what is called a *set-point regulation system* where the SAS loop is just a regulator for maintaining zero deviations from the set-point.

3.7 Multi-input Optimal Longitudinal Control

While flying an aircraft in a vertical plane, a good pilot simultaneously manipulates both throttle and elevator for better control of airspeed and altitude. The same is therefore true of an automatic controller. Although single-input longitudinal control is possible by either the elevator or the throttle, a multi-input control can remarkably improve closed-loop longitudinal performance and robustness. Furthermore, multi-input design may often provide the only realistic solution for controlling high-speed, high-performance aircraft such as a fighter aircraft. State-space design methods – such as linear optimal (LQG) control – are thus commonly employed for the multi-input, multi-output longitudinal controller, and form the topic of this section.

Since elevator and throttle servos are separately designed control units, they are seldom modified as part of longitudinal control design process. Instead, either the servo dynamics – being much faster than the corresponding modes to be controlled – is neglected in longitudinal controller design, or the servo states are removed from the feedback/feedforward paths of the longitudinal control system. In both cases, servo states do not form part of the longitudinal design plant. However, servo dynamics must be included in a final simulation of the closed-loop system.

Consider the longitudinal dynamics expressed in dimensional form as

$$\dot{\mathbf{x}} = \mathbf{A}\mathbf{x} + \mathbf{B}\mathbf{u}, \tag{3.159}$$

where the state vector is given by

$$\mathbf{x} = \begin{Bmatrix} u \\ \alpha \\ \theta \\ q \end{Bmatrix}, \tag{3.160}$$

the control input vector by

$$\mathbf{u} = \begin{Bmatrix} \delta_{\mathrm{E}} \\ \beta_{\mathrm{T}} \end{Bmatrix}, \tag{3.161}$$

and the state-space coefficient matrices are $\mathbf{A} = \mathbf{A}_{\mathrm{Long}}$ and $\mathbf{B} = \mathbf{B}_{\mathrm{Long}}$, given by equations (3.152) and (3.157), respectively. An optimal state feedback regulator with gain matrix, $\mathbf{K}$, is designed with the control law,

$$\mathbf{u} = -\mathbf{K}\mathbf{x}, \tag{3.162}$$

for minimizing the following quadratic, infinite-time objective function with symmetric cost parameter matrices, $(\mathbf{Q}, \mathbf{R})$:

$$\mathcal{J} = \frac{1}{2}\int_0^{\infty} \left[\mathbf{x}^T(\tau)\mathbf{Q}\mathbf{x}(\tau) + \mathbf{u}^T(\tau)\mathbf{R}\mathbf{u}(\tau)\right] \mathrm{d}\tau, \tag{3.163}$$

subject to the state equation constraint (3.159). The stabilizing steady-state solution for the feedback gain matrix is given by

$$\mathbf{K} = \mathbf{R}^{-1}\mathbf{B}^T\mathbf{S}, \tag{3.164}$$

where $\mathbf{S}$ is the constant, symmetric, positive semi-definite solution of the following algebraic Riccati equation (Chapter 2):

$$\mathbf{SA} + \mathbf{A}^T\mathbf{S} - \mathbf{SBR}^{-1}\mathbf{B}^T\mathbf{S} + \mathbf{Q} = \mathbf{0}. \tag{3.165}$$

Since the plant is controllable with elevator and throttle input, we can easily satisfy the sufficient conditions for the existence of a unique, positive semi-definite solution to the ARE with the symmetric cost matrices $\mathbf{R}$ (positive definite) and $\mathbf{Q}$ (positive semi-definite).

The output equation,

$$\mathbf{y} = \mathbf{Cx} + \mathbf{Du}, \tag{3.166}$$

is used to design a full-order observer with the state equation

$$\dot{\mathbf{x}}_o = (\mathbf{A} - \mathbf{LC})\,\mathbf{x}_o + (\mathbf{B} - \mathbf{LD})\,\mathbf{u} + \mathbf{Ly}, \tag{3.167}$$

where $\mathbf{x}_o(t)$ is the estimated state and $\mathbf{L}$ is the observer gain matrix, provided the plant $(\mathbf{A}, \mathbf{C})$ is observable (Appendix A). The plant $(\mathbf{A}, \mathbf{C})$ with airspeed and pitch rate measurement

$$\mathbf{C} = \begin{pmatrix} 1 & 0 & 0 & 0 \\ 0 & 0 & 0 & 1 \end{pmatrix} \tag{3.168}$$

has the observability test matrix

$$\mathbf{N} = \left[\mathbf{C}^T,\ \mathbf{A}^T\mathbf{C}^T,\ (\mathbf{A}^T)^2\mathbf{C}^T,\ (\mathbf{A}^T)^3\mathbf{C}^T\right], \tag{3.169}$$

of full rank, implying an observable plant with airspeed and pitch rate as outputs. We have the choice to design either a full-order or a reduced-order observer.

The closed-loop control system dynamics with a desired state, $\mathbf{x}_d(t)$, and linear feedforward/feedback control with output feedback,

$$\mathbf{u} = \mathbf{K}_d\mathbf{x}_d + \mathbf{K}\,(\mathbf{x}_d - \mathbf{x}_o), \tag{3.170}$$

is given by the state equation

$$\begin{Bmatrix} \dot{\mathbf{x}} \\ \dot{\mathbf{x}}_o \end{Bmatrix} = \begin{pmatrix} \mathbf{A} & -\mathbf{BK} \\ \mathbf{LC} & \mathbf{A} - \mathbf{BK} - \mathbf{LC} \end{pmatrix} \begin{Bmatrix} \mathbf{x} \\ \mathbf{x}_o \end{Bmatrix} + \begin{pmatrix} \mathbf{B}(\mathbf{K} + \mathbf{K}_d) \\ \mathbf{B}(\mathbf{K} + \mathbf{K}_d) \end{pmatrix} \mathbf{x}_d. \tag{3.171}$$

In longitudinal control applications, the desired state, $\mathbf{x}_d$, is usually a constant. The feedforward gain matrix, $\mathbf{K}_d$, is selected to ensure that the closed-loop error dynamics is independent of the desired state, thereby implying that

$$(\mathbf{A} + \mathbf{BK}_d)\,\mathbf{x}_d = \mathbf{0}. \tag{3.172}$$

With a proper choice of $\mathbf{K}_d$ satisfying equation (3.172), a suitable tracking controller can be designed.

3.8 Optimal Airspeed Control

Consider the design of an airspeed control system with both elevator and throttle inputs. A block diagram of a longitudinal control system with a full-order observer based on airspeed measurement is shown in Figure 3.27. Note that while engine and throttle servos are included in the aircraft plant, their dynamics is neglected in designing the airspeed control system because the servo frequencies are much higher than

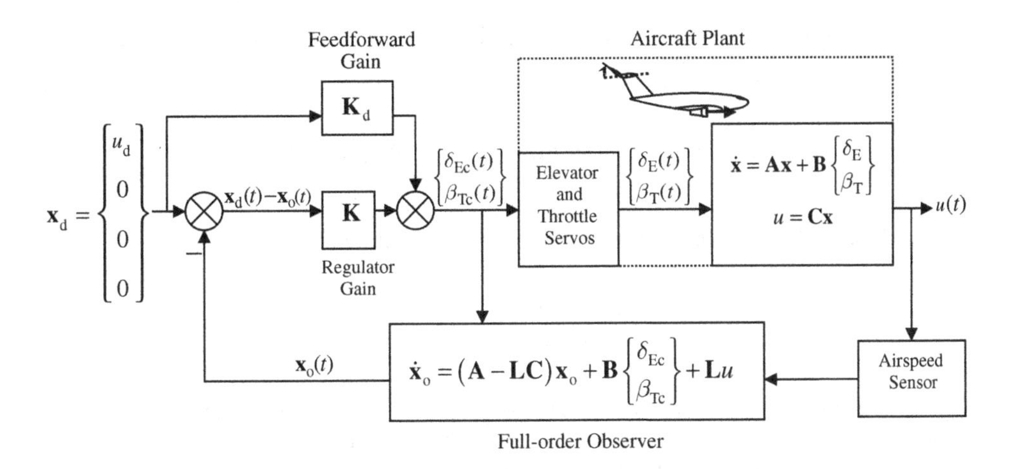

Figure 3.27 Block diagram of a multi-input airspeed control system with a full-order observer

that of the phugoid mode. For the same reason, airspeed sensor dynamics is also neglected in controller design.

To track a step airspeed command, the feedforward gain matrix, $\mathbf{K}_d$, must satisfy equation (3.164) with $\mathbf{f}_d(\mathbf{x}_d) = \mathbf{0}$,

$$(\mathbf{A} + \mathbf{B}\mathbf{K}_d)(u_d, 0, 0, 0)^T = \mathbf{0}, \tag{3.173}$$

which has no solution unless we assume $Z_T/m \simeq 0$, $M_T/m \simeq 0$ (which we generally can for a well-designed airplane). Furthermore, we assume $X_\delta/m \simeq 0$, in which case we can select

$$\mathbf{K}_d = \begin{pmatrix} -\frac{Z_u}{Z_\delta} & 0 & 0 & 0 \\ -\frac{X_u}{X_T} & 0 & 0 & 0 \end{pmatrix}. \tag{3.174}$$

Thus, a successful multi-input, step airspeed command tracking system must have a feedback/feedforward throttle and elevator control.

3.8.1 LQG/LTR Design Example

Consider the airplane of Example 3.3 in a steady climb with $\theta_e = 15^\circ$ and $U = 55$ m/s at standard sea level. The elevator control derivatives at standard sea level and reference airspeed $U = 55$ m/s are as follows:

$$\frac{X_\delta}{m} \simeq 0, \quad \frac{Z_\delta}{m} = 0.3 \text{ m/s}^2\text{/rad}, \quad \frac{M_\delta}{J_{yy}} = 0.1243 \text{ s}^{-2}.$$

The elevator servo is assumed to have the following approximate transfer function:

$$\frac{\delta_E(s)}{\delta_{Ec}(s)} = \frac{15}{s + 15}.$$

The airplane is powered by a single turboprop engine with the following parameters at standard sea level with reference speed $U = 55$ m/s:

$$\frac{X_T}{m} = 0.2452 \text{ m/s}^2/\text{deg}, \quad \frac{Z_T}{m} \simeq 0, \quad \frac{M_T}{J_{yy}} \simeq 0.$$

The engine is driven by a throttle servo of approximate transfer function,

$$\frac{\beta_T(s)}{\beta_{Tc}(s)} = \frac{5}{s+5}.$$

We will design an airspeed control system using both elevator and throttle inputs for the given flight condition, such that a step airspeed change of 1 m/s is obtained within a settling time not exceeding 15 s, the maximum elevator angle deflection required for the given airspeed change should be limited to $\pm 10°$, while the maximum throttle deflection should not exceed $\pm 0.5°$.

The longitudinal state coefficient matrices of the airplane in the given equilibrium flight condition are calculated as

$$\mathbf{A}_{\text{Long}} = \begin{pmatrix} -0.045 & 1.96 & -9.4757 & 0 \\ -0.0065455 & -1.9636 & -0.046164 & 1 \\ 0 & 0 & 0 & 1 \\ 0.0058909 & -6.8327 & 0.041548 & -2.9 \end{pmatrix},$$

$$\mathbf{B}_{\text{Long}} = \begin{pmatrix} 0 & 0.2452 \\ 0.3 & 0 \\ 0 & 0 \\ 0.1243 & 0 \end{pmatrix}.$$

For the given aircraft, we have

$$\mathbf{K}_d = \begin{pmatrix} 0.021667 & 0 & 0 & 0 \\ 0.183505 & 0 & 0 & 0 \end{pmatrix}.$$

An optimal feedback regulator and observer are to be designed by the linear quadratic Gaussian (LQG) approach for loop-transfer recovery (LTR) at either the plant's input, or its output (Chapter 2). As the plant is square, we can adopt either approach for LTR: LQG with recovery of return ratio at the plant input (denoted by LQGI) or at the plant output (LQGO). However, perfect LTR is not guaranteed for this plant, as it is non-minimum phase (i.e., it has a zero in the right-half s-plane). Successful LTR would require that the LQG compensator place its poles at plant zeros by inversion, thus resulting in an unstable control system for a non-minimum phase plant. Let us see whether LTR is achieved in the desired control system bandwidth of about 10 rad/s with adequate gain margins and roll-off at high frequencies.

LTR at Plant Input (LQGI)

Let us begin with the LQGI method, where

$$\mathbf{F} = \mathbf{B}, \quad \mathbf{S}_v = \mathbf{I}, \quad \mathbf{S}_w = \rho\mathbf{I},$$

and decrease the cost parameter, ρ, until the return ratio at plant input converges close to that of the full-state feedback regulator, $-\mathbf{K}(s\mathbf{I} - \mathbf{A})^{-1}\mathbf{B}$, over a desired frequency range. After some trial and error with the regulated step response, the optimal state feedback regulator cost parameters are chosen to be

$$\mathbf{Q} = 0.2\mathbf{I}, \quad \mathbf{R} = \begin{pmatrix} 4 & 0 \\ 0 & 1 \end{pmatrix},$$

for which the Control System Toolbox command *lqr.m* is used to solve the algebraic Riccati equation as follows:

```
>> a =[ -0.045          1.96         -9.4757     0;
        -0.0065455      -1.9636      -0.046164   1;
        0               0            0           1;
        0.0058909       -6.8327      0.041548    -2.9];

>> b = [0    0.2452; 0.3    0; 0      0; 0.1243    0];

>> [k,S,E]=lqr(a,b,0.2*eye(4),[4 0;0 1])

% Regulator gain matrix:
k =
        0.1588    2.6964    -5.4625    -0.8651
        0.1154    0.6118    -1.4384    -0.2233

% Algebraic Riccati solution:
S =
    0.4707     2.4952    -5.8660    -0.9108
    2.4952    41.4820   -84.2053   -13.3464
   -5.8660   -84.2053   173.5954    27.4462
   -0.9108   -13.3464    27.4462     4.3717

% Regulator poles (eigenvalues of a-b*k):
E =
        -2.4535 + 2.5776i
        -2.4535 - 2.5776i
        -0.3657 + 0.4130i
        -0.3657 - 0.4130i

>> c = [1 0 0 0; 0 0 0 1]; d=[0 0; 0 0];
>> sys=ss(a,b,c,d); % open-loop plant
>> pzmap(sys) % pole-zero map of open-loop plant
```

The regulator does not attempt to change the well-damped, short-period plant dynamics, but moves the lightly damped phugoid poles of the plant deeper into the left-half s-plane. The pole-zero map of the open-loop plant is depicted in Figure 3.28, showing zeros at the origin as well as in the right-half s-plane ($s = 14.5$). Clearly, an attempt to cancel the plant zeros by inversion for perfect LTR will lead to

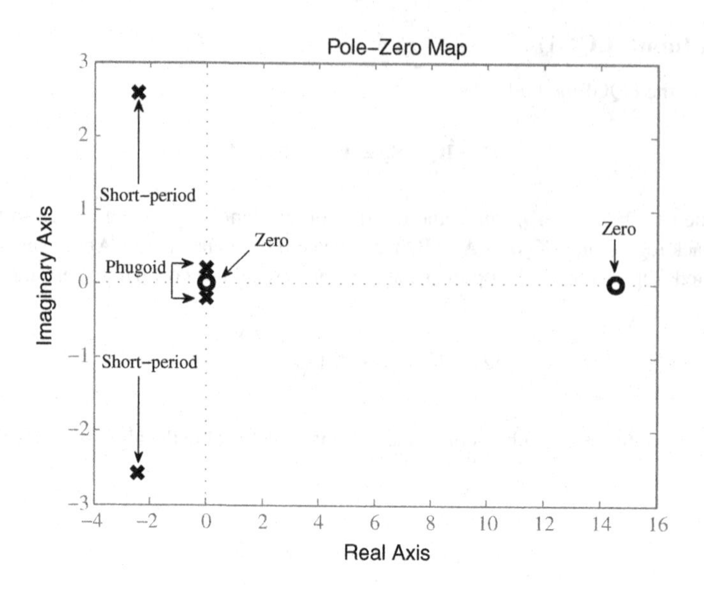

Figure 3.28 Pole-zero map of the open-loop plant

an unstable regulator. Thus, we cannot have perfect LTR for this plant. For LQGI Kalman filter design with $\rho = 10^{-8}$, we have the following Control System Toolbox statements:

```
>> Snu=eye(2); %Process noise PSD matrix
>> rho=1e-8; Sw=rho*eye(2); %Measurement noise PSD matrix
>> [L,S,E]=lqe(a,b,c,Snu,Sw,zeros(2,2)) %Kalman filter ARE

% Kalman filter gain matrix:
L =
       2452.6        -2.5331
       752.67        -2171.5
     -0.95992        0.71972
      -2.5331           1252

% Algebraic Riccati solution:
S =
  2.4526e-005  7.5267e-006 -9.5992e-009 -2.5331e-008
  7.5267e-006    0.0094566   5.227e-007 -2.1715e-005
 -9.5992e-009   5.227e-007  2.5919e-006  7.1972e-009
 -2.5331e-008 -2.1715e-005  7.1972e-009    1.252e-005

% Kalman filter poles (eigenvalues of a-L*c):
E =
        -2452
      -1242.9
      -14.528
   -0.0038565

>> Ac=a-b*k-L*c;
>> Bc=L;
>> Cc=-k;
>> Dc=d;
>> sysc=ss(Ac,Bc,Cc,Dc); % LQG compensator
```

```
>> sysKG=series(sysc,sys); % open-loop LQG transfer function
>> sysk=ss(a,b,-k,d); % return-ratio of regulator
>> w=logspace(-3,3);
% Singular values of return-ratios of LQGI and regulator
>> sigma(sysKG,w),hold on,sigma(sysk,w)
>> kd=[0.021667 0 0 0;0.183505 0 0 0]; %feedforward gain matrix
% Overall closed-loop system coefficients:
>> ACL=[a -b*k;L*c Ac];
>> BCL=[b*(k+kd);b*(k+kd)];

% Overall closed-loop system with desired airspeed change:
>> sysCL=ss(ACL,BCL(:,1),eye(size(ACL)),zeros(size(BCL(:,1))))
ACL =
            x1        x2         x3        x4         x5         x6        x7        x8
x1      -0.045      1.96     -9.476         0    -0.0283      -0.15    0.3527   0.05476
x2   -0.006546    -1.964   -0.04616         1   -0.04765    -0.8089     1.639    0.2595
x3           0         0          0         1          0          0         0         0
x4    0.005891    -6.833    0.04155      -2.9   -0.01974    -0.3352     0.679    0.1075
x5        2453         0          0    -2.533      -2453       1.81    -9.123     2.588
x6       752.7         0          0     -2171     -752.7     -2.773     1.593      2173
x7     -0.9599         0          0    0.7197     0.9599          0         0    0.2803
x8      -2.533         0          0      1252      2.519     -7.168    0.7205     -1255

BCL(:,1) =
            u1
   x1   0.0733
   x2  0.05415
   x3        0
   x4  0.02244
   x5   0.0733
   x6  0.05415
   x7        0
   x8  0.02244

% Unit step response to desired airspeed change:
>> [y,t]=step(sysCL);
% Closed-loop control inputs:
>> for i=1:size(y,1)
>>   u(:,i)=kd*[1 0 0 0]'+k*([1 0 0 0]'-y(i,5:8)');
>> end
```

Note that the Kalman filter has all real poles, the first three of which are very much deeper into the left-half s-plane than those of the regulator, while the last is a lightly damped pole close to the imaginary axis that attempts to provide an approximate integral action in the overall control system (by trying to cancel the plant zeros at the origin and $s = 14.5$). The singular value spectrum of the return ratio at plant input for $\rho = 10^{-6}$ and $\rho = 10^{-8}$ is compared with that of the full-state feedback regulator in Figure 3.29. Notice a good recovery with $\rho = 10^{-8}$ in the frequency range $2 \leq \omega \leq 1000$ rad/s, which decreases to $2 \leq \omega \leq 60$ rad/s for $\rho = 10^{-6}$. At smaller frequencies, the recovery of the return ratio is hampered by the absence of a pole at $s = 0$ in the plant. However, at larger frequencies, an increased roll-off assures robustness with high-frequency noise component. The step response to a unit change in airspeed for $\rho = 10^{-8}$ is plotted in Figure 3.30, while the associated control inputs are shown in Figure 3.31. Note the rapid decay of short-period transients as well as a suppression of the phugoid mode, resulting in a settling time of about 11 s, but a lack of integral action causes a steady-state airspeed error of 0.05 m/s (5%). As expected, the steady-state values of pitch angle and angle of attack, as well as the throttle and elevator

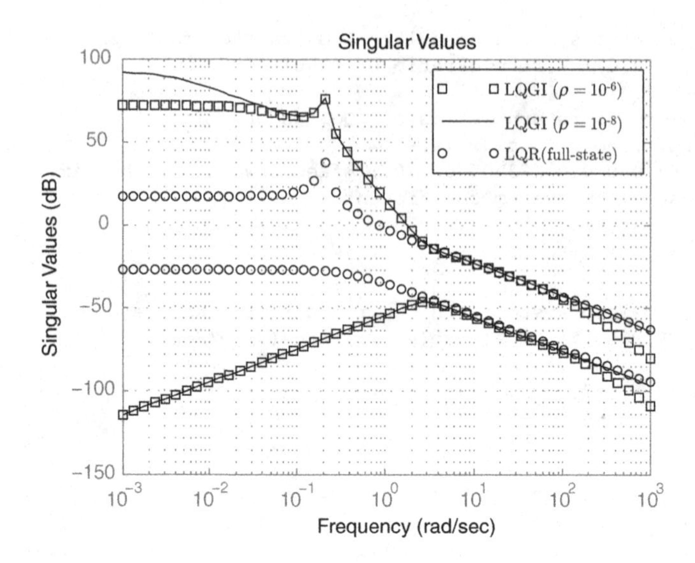

Figure 3.29 Singular values of return ratio at plant's input for two LQG compensated closed-loop systems ($\rho = 10^{-6}, 10^{-8}$) and their comparison with corresponding singular values for the full-state feedback LQR regulator

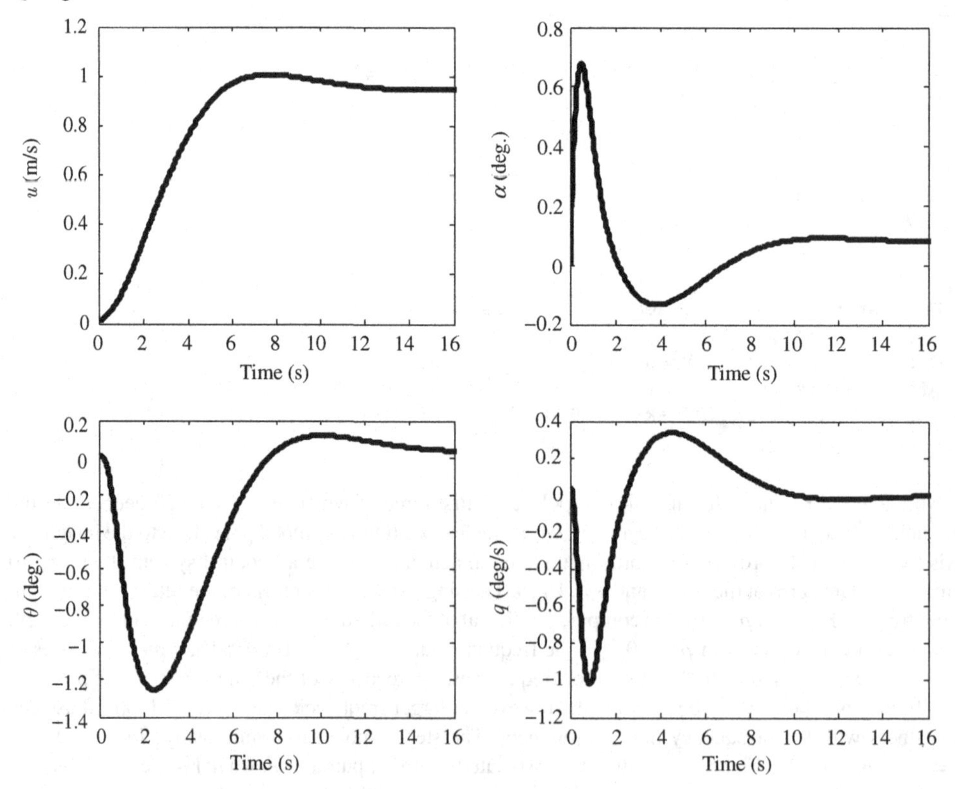

Figure 3.30 Closed-loop step response for a unit-step airspeed change for the LQG compensated closed-loop system with $\rho = 10^{-8}$

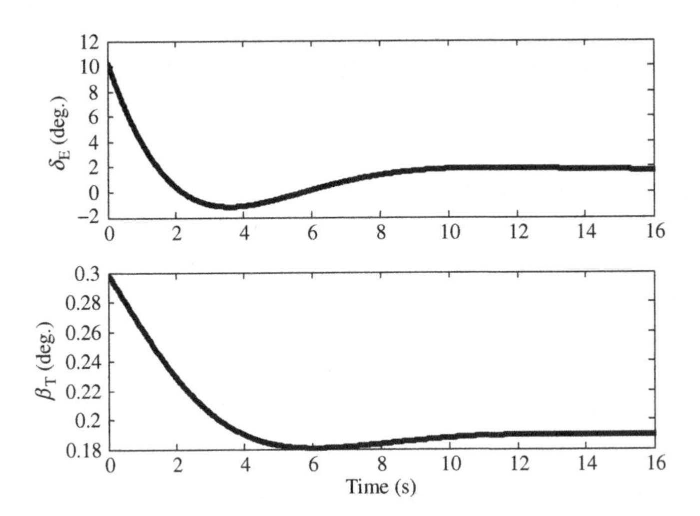

Figure 3.31 Closed-loop control inputs for a unit-step airspeed change for the LQG compensated closed-loop system with $\rho = 10^{-8}$

angles, are changed due to a change in the airspeed of climb. The required control input magnitudes (Figure 3.31) are seen to be within design requirements. However, an actuator-in-loop simulation needs to be carried out for a more accurate estimate of the settling time and control inputs.

LTR at Plant Output (LQGO)

For LQGO, we select a Kalman filter with

$$\mathbf{F} = \mathbf{B}, \quad \mathbf{S}_\nu = \mathbf{I}, \quad \mathbf{S}_w = \mathbf{I},$$

and iteratively design an optimal regulator by changing the cost parameter, ρ,

$$\mathbf{Q} = \mathbf{I}, \quad \mathbf{R} = \begin{pmatrix} 4 & 0 \\ 0 & 1 \end{pmatrix},$$

such that the return ratio of the LQG system at plant output is close to that of the Kalman filter, $-\mathbf{C}(s\mathbf{I} - \mathbf{A})^{-1}\mathbf{L}$, over a desired frequency range, and the desired performance objectives are achieved. The design calculations result in an acceptable closed-loop performance with $\rho = 4.7$, and the necessary steps are carried out by the following Control System Toolbox statements:

```
>> Snu=1e-8*eye(2); Sw=eye(2);
>> [L,S,E]=lqe(a,b,c,Snu,Sw,zeros(2,2))
L =
   6.0171e-005  2.5788e-007
  -6.1201e-008 -3.7937e-010
  -2.9857e-007  4.4617e-014
   2.5788e-007  1.4443e-009

S =
   6.0171e-005 -6.1201e-008 -2.9857e-007  2.5788e-007
  -6.1201e-008  2.3697e-010  1.2824e-010 -3.7937e-010
```

```
 -2.9857e-007  1.2824e-010  2.8662e-008  4.4617e-014
  2.5788e-007 -3.7937e-010  4.4617e-014  1.4443e-009

E =
       -2.4525 +      2.5776i
       -2.4525 -      2.5776i
    -0.0018694 +      0.2018i
    -0.0018694 -      0.2018i

>> rho=4.7;
>> [k,S,E]=lqr(a,b,eye(4),rho*[4 0;0 1])

k =
       0.16468        2.7422       -5.5638      -0.88122
       0.12133       0.63457       -1.4957      -0.23211

S =
        2.3257        12.163       -28.669       -4.4491
        12.163        198.33       -403.23       -63.917
       -28.669       -403.23        832.96        131.69
       -4.4491       -63.917        131.69        20.982

E =
       -2.4536 +      2.5776i
       -2.4536 -      2.5776i
      -0.37218 +      0.4185i
      -0.37218 -      0.4185i

>> sysL=ss(a,-L,c,d); %Return ratio of Kalman filter
>> Ac=a-b*k-L*c;
>> Bc=L;
>> Cc=-k;
>> Dc=d;
>> sysc=ss(Ac,Bc,Cc,Dc); % LQG compensator
>> sysGK=series(sys,sysc); %Return ratio at plant output
>> w=logspace(-3,3);
>> sigma(sysGK,w),hold on,sigma(sysL,w)


>> kd=[0.021667 0 0 0;0.183505 0 0 0];
>> ACL=[a -b*k;L*c Ac];
>> BCL=[b*(k+kd);b*(k+kd)];
>> sysCL=ss(ACL,BCL(:,1),eye(size(ACL)),zeros(size(BCL(:,1))))

ACL =
                 x1          x2          x3          x4          x5          x6          x7          x8
   x1        -0.045        1.96      -9.476           0    -0.02975     -0.1556      0.3667     0.05691
   x2     -0.006546      -1.964    -0.04616           1     -0.0494     -0.8227       1.669      0.2644
   x3             0           0           0           1           0           0           0           0
   x4      0.005891      -6.833     0.04155        -2.9    -0.02047     -0.3409      0.6916      0.1095
   x5     6.017e-005          0           0  2.579e-007    -0.07481       1.804      -9.109     0.05691
   x6      -6.12e-008         0           0 -3.794e-010    -0.05595      -2.786       1.623       1.264
   x7    -2.986e-007          0           0  4.462e-014  2.986e-007           0           0           1
   x8     2.579e-007          0           0  1.444e-009    -0.01458      -7.174      0.7331       -2.79
```

```
BCL(:,1) =
                 u1
    x1     0.07475
    x2      0.0559
    x3           0
    x4     0.02316
    x5     0.07475
    x6      0.0559
    x7           0
    x8     0.02316

>> [y,t]=step(sysCL);
>> for i=1:size(y,1)
>>     u(:,i)=kd*[1 0 0 0]'+k*([1 0 0 0]'-y(i,5:8)');
>> end
```

Note that a pair of regulator poles is virtually collocated with the pair of Kalman filter poles at the short-period poles of the plant, signifying that there is no controller action on the short-period dynamics (which are adequately damped), and all activity is concentrated on suppressing the lightly damped phugoid mode. The second pair of Kalman filter poles is quite close to the origin, indicating an attempt to provide approximate integral action by trying to cancel non-minimum phase plant zeros. The return ratio at the plant output (Figure 3.32) is approximately recovered in the low-frequency range ($0.1 \leq \omega \leq 2$) rad/s, and a further recovery is not possible because a reduction of ρ below 4.7 causes unacceptable closed-loop response. At high frequencies, the roll-off offered by the LQG compensator is much better than that of the Kalman filter alone, thereby implying a far greater high-frequency noise rejection. The step response to a unit change in airspeed for $\rho = 4.7$ is plotted in Figures 3.33 and 3.34, showing virtually no change in the closed-loop performance compared with that of the LQGI system designed earlier, except for a slight increase in the settling time of the transients.

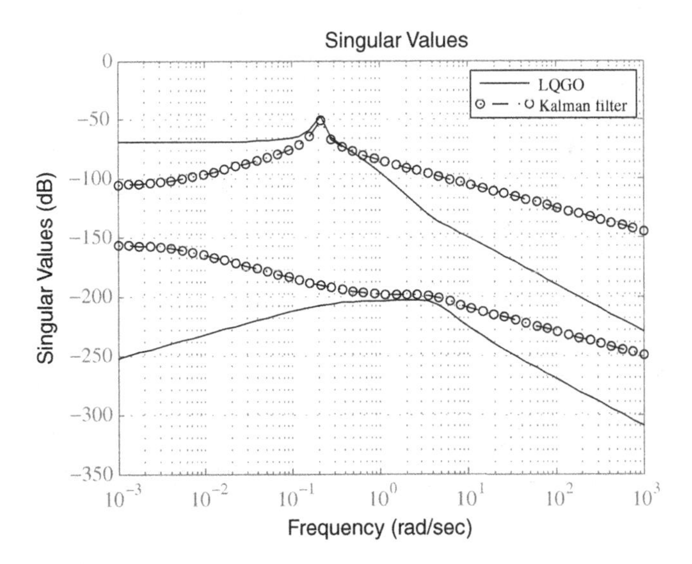

Figure 3.32 Singular values of return ratio at plant's output for LQG compensated closed-loop system ($\rho = 4.7$) and their comparison with corresponding singular values for the Kalman filter

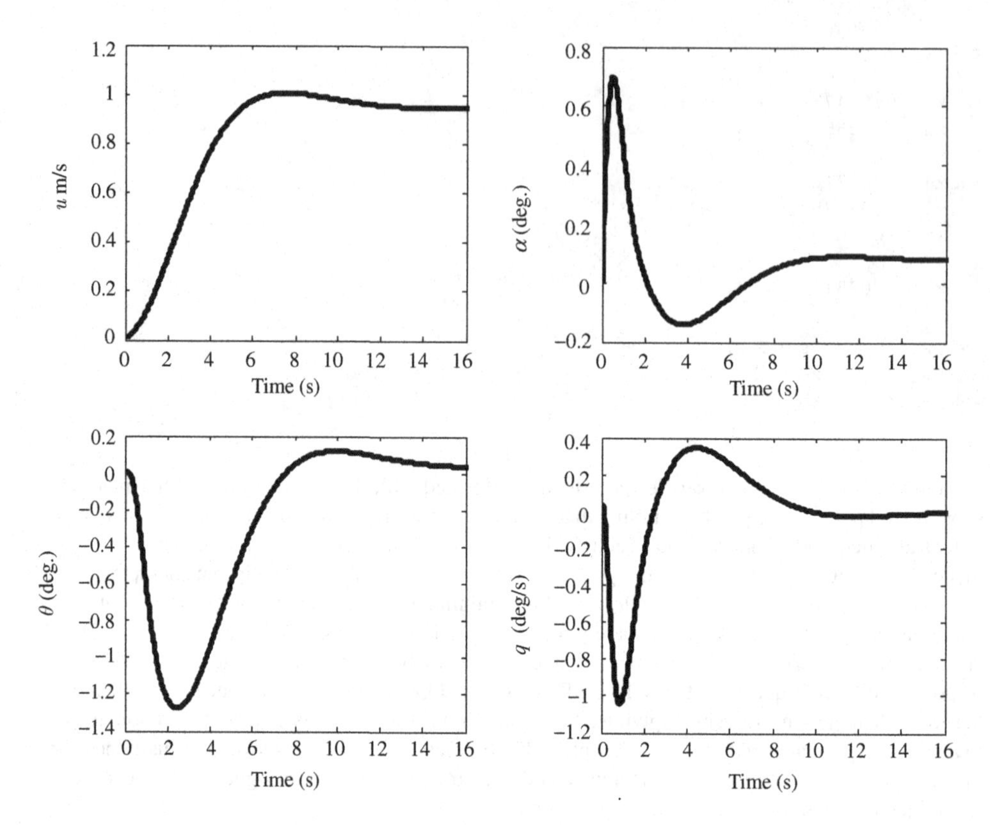

Figure 3.33 Closed-loop step response for a unit-step airspeed change for the LQG/LTR (output) compensated closed-loop system with $\rho = 4.7$

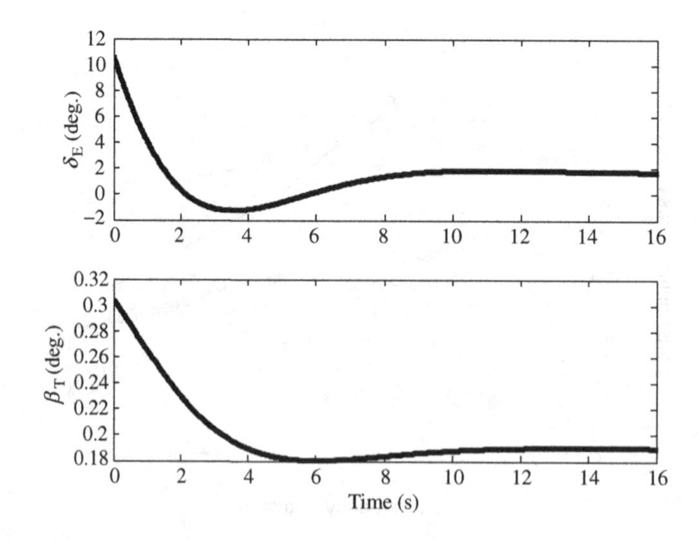

Figure 3.34 Closed-loop control inputs for a unit-step airspeed change for the LQG/LTR (output) compensated closed-loop system with $\rho = 4.7$

We now construct the closed-loop system with the first-order servos as follows:

```
>> wa=15; wb=5; % servo poles
% Plant dynamics matrix with servos:
>> Abar=[a b;zeros(1,4) -wa 0; zeros(1,4) 0 -wb]

Abar =
   -0.0450    1.9600   -9.4757         0         0    0.2452
   -0.0065   -1.9636   -0.0462    1.0000    0.3000         0
         0         0         0    1.0000         0         0
    0.0059   -6.8327    0.0415   -2.9000    0.1243         0
         0         0         0         0  -15.0000         0
         0         0         0         0         0   -5.0000

>> Bbar=[zeros(size(b)); wa 0; 0 wb] %Augmented plant's B matrix

Bbar =
     0     0
     0     0
     0     0
     0     0
    15     0
     0     5

>> Cbar=[c zeros(2,2)] %Augmented plant's C matrix

Cbar =
     1     0     0     0     0     0
     0     0     0     1     0     0

>> Ac=[Abar -Bbar*k; L*Cbar a-L*c-b*k];%Closed-loop dynamics matrix
>> damp(Ac) % Closed-loop system's eigenvalues

        Eigenvalue               Damping      Freq. (rad/s)
 -1.50e+001                     1.00e+000       1.50e+001
 -5.00e+000                     1.00e+000       5.00e+000
 -1.87e-003 + 2.02e-001i        9.26e-003       2.02e-001
 -1.87e-003 - 2.02e-001i        9.26e-003       2.02e-001
 -3.72e-001 + 4.18e-001i        6.65e-001       5.60e-001
 -3.72e-001 - 4.18e-001i        6.65e-001       5.60e-001
 -2.45e+000 + 2.58e+000i        6.89e-001       3.56e+000
 -2.45e+000 - 2.58e+000i        6.89e-001       3.56e+000
 -2.45e+000 + 2.58e+000i        6.89e-001       3.56e+000
 -2.45e+000 - 2.58e+000i        6.89e-001       3.56e+000

>> Bc=[Bbar*(k+kd); b*(k+kd)] % Closed-loop input coefficients matrix

Bc =
         0         0         0         0
         0         0         0         0
         0         0         0         0
         0         0         0         0
    2.7952   41.1330  -83.4570  -13.2184
    1.5242    3.1728   -7.4783   -1.1606
```

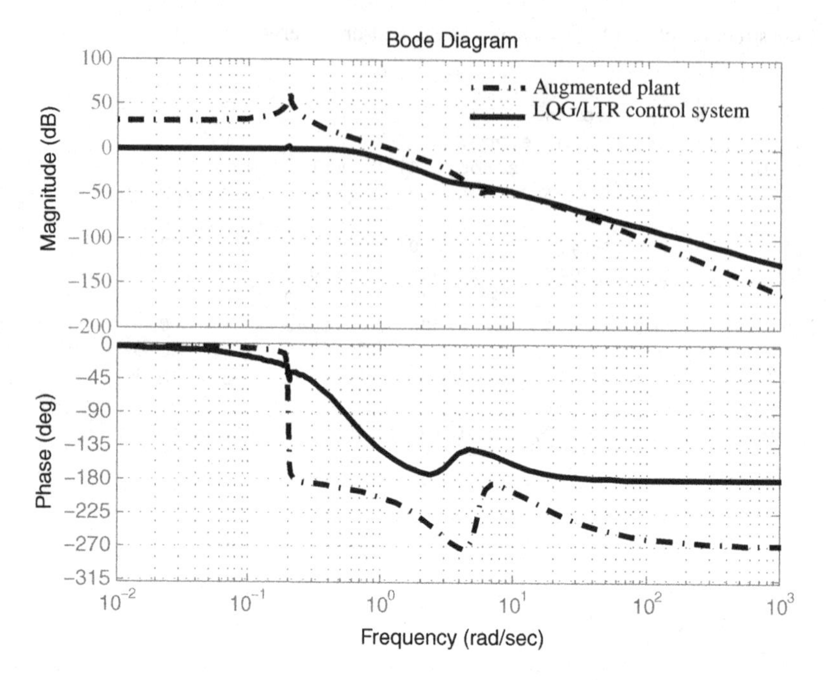

Figure 3.35 Closed-loop frequency response of the transfer function, $u(s)/u_d(s)$, augmented with servo dynamics, designed by the LQG/LTR (output) method with $\rho = 4.7$

```
0.0747     0.1556    -0.3667    -0.0569
0.0559     0.8227    -1.6691    -0.2644
     0          0          0          0
0.0232     0.3409    -0.6916    -0.1095
```

The eigenvalues of the closed-loop system indicate only minor changes in the regulator poles due to the addition of servo dynamics. However, the higher-frequency observer poles are moved closer to the dominant poles due to servo lag, which is fortunately a more optimal position from a control effort viewpoint. A shift in observer poles can be avoided by designing the observer based upon the augmented plant (rather than the plant without servos, as we have done here). Finally, the two first-order servos now appear as a single second-order system due to feedback. The closed-loop frequency response of $u(s)/u_d(s)$ is computed as follows and plotted in Figure 3.35, which shows an infinite gain margin and a phase margin of 135°, indicating excellent closed-loop robustness:

```
>> syscl=ss(Ac,Bc(:,1),[1 zeros(1,9)],0); bode(syscl),grid
```

Note the suppression of phugoid mode seen as 65 dB reduction in the peak magnitude at phugoid frequency ($\omega = 0.202$ rad/s). Thus we have designed a fairly robust, multi-variable airspeed control system using both elevator and throttle inputs.

3.8.2 H_∞ Design Example

We now carry out the design of the multi-input airspeed control system by the mixed-sensitivity H_∞ method (Chapter 2). After some trial and error, the following frequency weighting functions are selected for the design:

$$\mathbf{W}_1(s) = \begin{pmatrix} \frac{s+325}{500s+1} & 0 \\ 0 & \frac{s+500}{500s+1} \end{pmatrix},$$

$$\mathbf{W}_2(s) = \begin{pmatrix} \frac{1.5}{s+1} & 0 \\ 0 & \frac{1}{s+1} \end{pmatrix},$$

$$\mathbf{W}_3(s) = \begin{pmatrix} \frac{s+1}{s} & 0 \\ 0 & \frac{s+1}{s} \end{pmatrix}.$$

Bode plots of the selected frequency weights are shown in Figures 3.36 and 3.37. Note that the frequency weight, $\mathbf{W}_3(s)$, adds a pure integrator in the forward path, thereby providing integral action and nullifying the steady-state error due to step desired outputs. The iteration for maximum value of

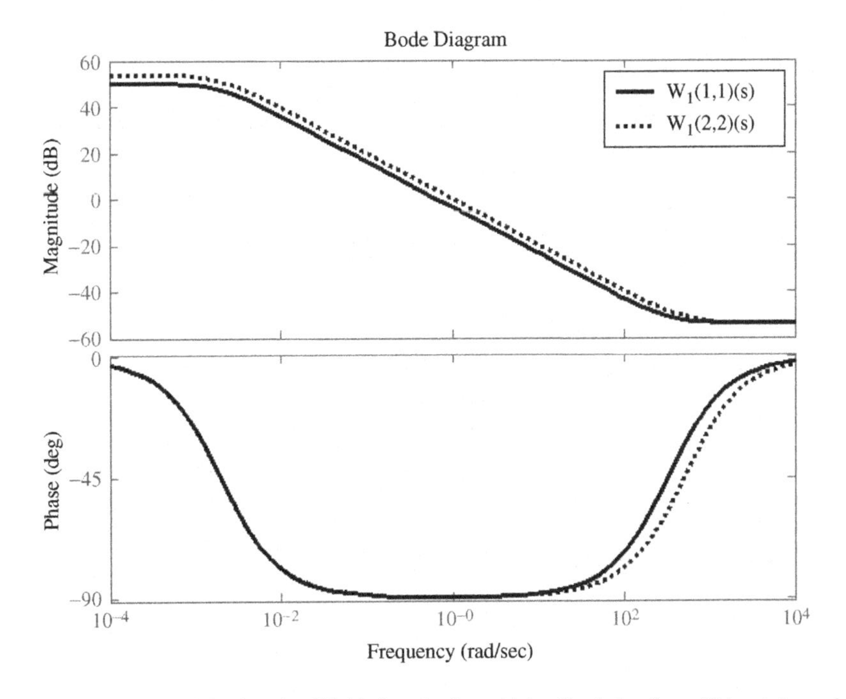

Figure 3.36 Frequency weighting function, $\mathbf{W}_1(s)$, for mixed-sensitivity H_∞ design for multi-input airspeed control system

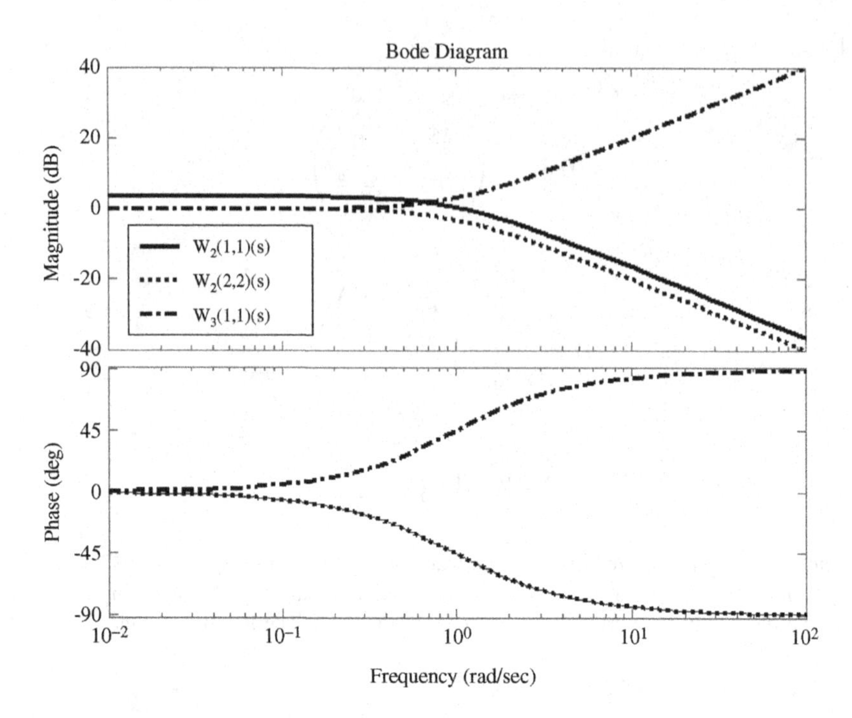

Figure 3.37 Frequency weighting functions, $\mathbf{W}_2(s)$, $\mathbf{W}_3(s)$, for mixed-sensitivity H_∞ design for multi-input airspeed control system

the parameter γ that minimizes the H_∞ norm of the mixed-sensitivity transfer matrix with a stabilizing controller is carried out using the Robust Control Toolbox function *hinfopt.m* in MATLAB as follows:

```
>> c1=500; w1=[1 0.65*c1;c1 1;1 c1;c1 1];
>> w2=[0 1.5;1 1;0 1;1 1];
>> w3=[1 1;0 1;1 1;0 1];
>> [A,B1,B2,C1,C2,D11,D12,D21,D22]=augtf(a,b,c,d,w1,w2,w3)

A =
   -0.0450    1.9600   -9.4757         0         0         0         0         0
   -0.0065   -1.9636   -0.0462    1.0000         0         0         0         0
         0         0         0    1.0000         0         0         0         0
    0.0059   -6.8327    0.0415   -2.9000         0         0         0         0
   -1.0000         0         0         0   -0.0020         0         0         0
         0         0         0   -1.0000         0   -0.0020         0         0
         0         0         0         0         0         0   -1.0000         0
         0         0         0         0         0         0         0   -1.0000

B1 =
     0     0
     0     0
     0     0
     0     0
     1     0
     0     1
```

```
     0     0
     0     0

B2 =
          0    0.2452
     0.3000         0
          0         0
     0.1243         0
          0         0
          0         0
     1.0000         0
          0    1.0000

C1 =
    -0.0020         0         0         0    0.6500         0         0         0
          0         0         0   -0.0020         0    1.0000         0         0
          0         0         0         0         0         0    1.5000         0
          0         0         0         0         0         0         0    1.0000
     0.9550    1.9600   -9.4757         0         0         0         0         0
     0.0059   -6.8327    0.0415   -1.9000         0         0         0         0

C2 =
    -1     0     0     0     0     0     0     0
     0     0     0    -1     0     0     0     0

D11 =
     0.0020         0
          0    0.0020
          0         0
          0         0
          0         0
          0         0

D12 =
          0         0
          0         0
          0         0
          0         0
          0    0.2452
     0.1243         0

D21 =
     1     0
     0     1

D22 =
     0     0
     0     0

>> [gamopt,acp,bcp,ccp,dcp,acl,bcl,ccl,dcl]=hinfopt(A,B1,B2,C1,C2,D11,D12,D21,D22);


<< H-Infinity Optimal Control Synthesis >>
```

```
 No     Gamma     D11<=1   P-Exist   P>=0   S-Exist   S>=0   lam(PS)<1    C.L.
--------------------------------------------------------------------------------
  1 1.0000e+000    OK        OK      FAIL      OK       OK        OK        UNST
  2 5.0000e-001    OK       FAIL      OK       OK       OK        OK        STAB
  3 2.5000e-001    OK       FAIL      OK       OK       OK        OK        STAB
  4 1.2500e-001    OK       FAIL      OK       OK       OK        OK        STAB
  5 6.2500e-002    OK       FAIL      OK       OK       OK        OK        STAB
  6 3.1250e-002    OK       FAIL      OK       OK       OK        OK        STAB
  7 1.5625e-002    OK       FAIL      OK       OK       OK        OK        STAB
  8 7.8125e-003    OK       FAIL      OK       OK       OK        OK        STAB
  9 3.9063e-003    OK       FAIL      OK       OK       OK        OK        STAB
 10 1.9531e-003    OK        OK       OK       OK       OK        OK        STAB
 11 2.9297e-003    OK       FAIL      OK       OK       OK        OK        STAB
 12 2.4414e-003    OK       FAIL      OK       OK       OK        OK        STAB
 13 2.1973e-003    OK       FAIL      OK       OK       OK        OK        STAB
 14 2.0752e-003    OK       FAIL      OK       OK       OK        OK        STAB
 15 2.0142e-003    OK       FAIL      OK       OK       OK        OK        STAB
 16 1.9836e-003    OK        OK       OK       OK       OK        OK        STAB
 17 1.9989e-003    OK        OK       OK       OK       OK        OK        STAB

      Iteration no. 17 is your best answer under the tolerance: 0.0100 .

gamopt =
    0.0020

acp =
-6.5219   -1.1383    2.2450   -5.9453    4.0852   -1.9169   -5.5450   16.6402
-0.6147   -0.3551    0.3317   -0.6881    1.1395    0.1429   -0.4164    2.1933
 1.3172    0.1089   -0.9974    3.3214   -0.3020    0.7945    3.1870   -5.2853
-3.9753   -0.3726    3.2651  -11.0567    0.4967   -2.1084  -10.5384   17.6255
 0.5436    0.0127   -0.3862    0.6458   -2.6694    2.4288    0.9722    1.7679
-1.3247    0.4237   -0.0445   -0.1729   -0.3704   -2.4083   -1.6401   -0.5119
-1.1377   -0.4669    0.4669   -1.7915   -0.2411   -0.0117   -3.0708    5.1019
 0.3643   -0.1537   -0.2912    0.7889   -0.1954    0.1286    0.2811   -2.4601

bcp =
   -0.0914    0.0272
    0.6939   -0.1028
    0.1064    0.9433
    0.0715    0.2801
    0.3264   -0.1113
    0.8142   -0.0406
   -0.5898   -0.0163
   -0.3461   -0.0063

ccp =
 6.6117    1.0342   -3.8010   11.7574   -3.1229    2.5270   11.5624  -25.0487
 3.2654    0.7274   -0.5817    0.3054   -4.0272    0.9972   -0.2564   -6.2503

dcp =
     0     0
     0     0
```

```
acl =

  Columns 1 through 12
   -0.0450    1.9600   -9.4757         0         0         0         0         0    0.8007    0.1784   -0.1426    0.0749
   -0.0065   -1.9636   -0.0462    1.0000         0         0         0         0    1.9835    0.3103   -1.1403    3.5272
         0         0         0    1.0000         0         0         0         0         0         0         0         0
    0.0059   -6.8327    0.0415   -2.9000         0         0         0         0    0.8218    0.1286   -0.4725    1.4614
   -1.0000         0         0         0   -0.0020         0         0         0         0         0         0         0
         0         0         0   -1.0000         0   -0.0020         0         0         0         0         0         0
         0         0         0         0         0         0   -1.0000         0    6.6117    1.0342   -3.8010   11.7574
         0         0         0         0         0         0         0   -1.0000    3.2654    0.7274   -0.5817    0.3054
    0.0914         0         0   -0.0272         0         0         0         0   -6.5219   -1.1383    2.2450   -5.9453
   -0.6939         0         0    0.1028         0         0         0         0   -0.6147   -0.3551    0.3317   -0.6881
   -0.1064         0         0   -0.9433         0         0         0         0    1.3172    0.1089   -0.9974    3.3214
   -0.0715         0         0   -0.2801         0         0         0         0   -3.9753   -0.3726    3.2651  -11.0567
   -0.3264         0         0    0.1113         0         0         0         0    0.5436    0.0127   -0.3862    0.6458
   -0.8142         0         0    0.0406         0         0         0         0   -1.3247    0.4237   -0.0445   -0.1729
    0.5898         0         0    0.0163         0         0         0         0   -1.1377   -0.4669    0.4669   -1.7915
    0.3461         0         0    0.0063         0         0         0         0    0.3643   -0.1537   -0.2912    0.7889

  Columns 13 through 16
   -0.9875    0.2445   -0.0629   -1.5326
   -0.9369    0.7581    3.4687   -7.5146
         0         0         0         0
   -0.3882    0.3141    1.4372   -3.1136
         0         0         0         0
         0         0         0         0
   -3.1229    2.5270   11.5624  -25.0487
   -4.0272    0.9972   -0.2564   -6.2503
    4.0852   -1.9169   -5.5450   16.6402
    1.1395    0.1429   -0.4164    2.1933
   -0.3020    0.7945    3.1870   -5.2853
    0.4967   -2.1084  -10.5384   17.6255
   -2.6694    2.4288    0.9722    1.7679
   -0.3704   -2.4083   -1.6401   -0.5119
   -0.2411   -0.0117   -3.0708    5.1019
   -0.1954    0.1286    0.2811   -2.4601

bcl =
         0         0
         0         0
         0         0
         0         0
    1.0000         0
         0    1.0000
         0         0
         0         0
   -0.0914    0.0272
    0.6939   -0.1028
    0.1064    0.9433
    0.0715    0.2801
    0.3264   -0.1113
    0.8142   -0.0406
   -0.5898   -0.0163
   -0.3461   -0.0063

ccl =

Columns 1 through 12
  -0.0000         0         0         0    0.0013         0         0         0         0         0         0         0
        0         0         0   -0.0000         0    0.0020         0         0         0         0         0         0
        0         0         0         0         0         0    0.0030         0         0         0         0         0
```

```
         0         0         0         0         0         0         0    0.0020         0         0         0         0
    0.0019    0.0039   -0.0189         0         0         0         0         0    0.0016    0.0004   -0.0003    0.0001
    0.0000   -0.0137    0.0001   -0.0038         0         0         0         0    0.0016    0.0003   -0.0009    0.0029


  Columns 13 through 16

         0         0         0         0
         0         0         0         0
         0         0         0         0
         0         0         0         0
   -0.0020    0.0005   -0.0001   -0.0031
   -0.0008    0.0006    0.0029   -0.0062

dcl =
  1.0e-005 *

    0.3998         0
         0    0.3998
         0         0
         0         0
         0         0
         0         0

>> syscl=ss(acl,bcl(:,1),ccl,dcl(:,1));
>> [y,t,x]=step(syscl,0:0.01:20);
>> sysc=ss(acp,bcp,ccp,dcp);
>> for i=1:size(y,1)
>>      dy(i,:)=[1 0]-y(i,5:6)/gamopt;
>> end
>> u=lsim(sysc,dy,t);
```

The closed-loop step response to unit airspeed change is plotted in Figures 3.38 and 3.39, showing a settling time of about 8 s and a zero steady-state error (due to integral action of $\mathbf{W}_3(s)$). This is a much better performance than the LQG/LTR compensators designed earlier for the same plant. However, the overall control system is now much larger in size, with 16 states, compared to 8 for the LQG/LTR system.

3.8.3 Altitude and Mach Control

A variation of the multi-input airspeed control system is that applied to maintain a constant indicated airspeed for low-subsonic airplanes, or a constant Mach number for high-speed aircraft, while cruising, climbing, or descending. A constant indicated airspeed, U_i, would imply a near-constant dynamic pressure, $\bar{q}$, irrespective of the altitude, h, according to

$$\bar{q} = \frac{1}{2}\rho(h)U(h)^2 = \frac{1}{2}\rho_0 U_i^2 = \text{const.}, \tag{3.175}$$

where ρ_0 is the sea level density. Similarly, for a constant Mach number, we have

$$\mathcal{M} = \frac{U}{\sqrt{\gamma \bar{R}\bar{T}(h)}} = \text{const.}, \tag{3.176}$$

where $\bar{T}$ is the outside air temperature and γ, $\bar{R}$ are constants (Tewari 2006). In both cases, the flight involves a steadily changing true airspeed, U, as the altitude (and thus ρ and $\bar{T}$) is varied. The autopilot

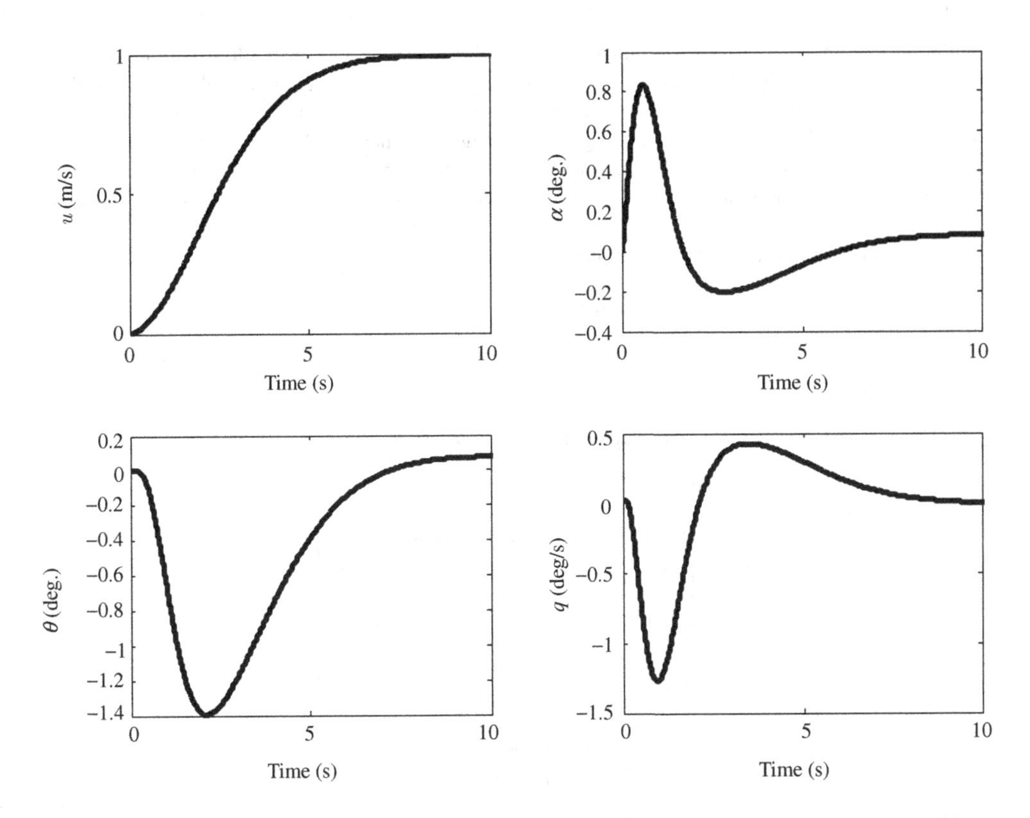

Figure 3.38 Closed-loop step response for unit airspeed change of mixed-sensitivity H_∞ controller for multi-input airspeed control system

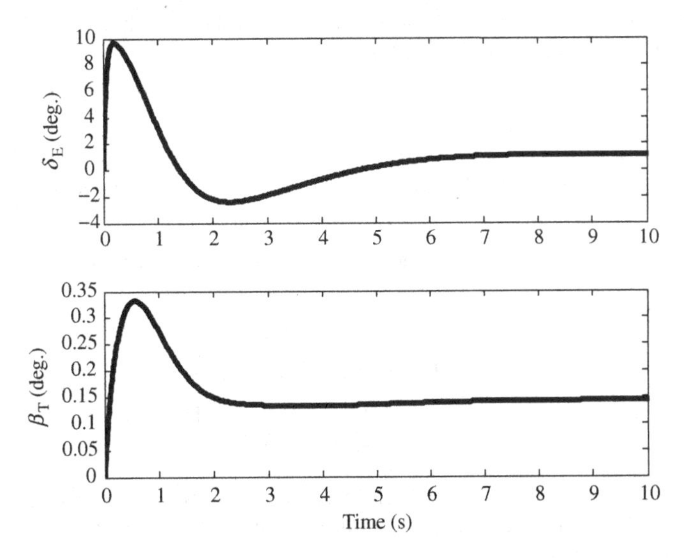

Figure 3.39 Closed-loop control inputs for unit airspeed change of mixed-sensitivity H_∞ controller for multi-input airspeed control system

provides the true airspeed reference command, $U(h)$, to the inner SAS loop, according to either equation (3.175) or equation (3.176). If the altitude is also held constant (as in navigational subsystem (b)) then indicated airspeed, true airspeed, and Mach number are simultaneously constant, barring small variations in atmospheric weather conditions of a long cruising flight. By having altitude as an additional state variable we add a pure integrator in the forward path, thereby enabling a zero steady-state error for step commands in airspeed and altitude. Thus it becomes easier to design an altitude and Mach (airspeed) hold autopilot by LQG techniques.

The additional state equation for the altitude, $h(t)$, is

$$\dot{h} = (U + u)\sin\phi = (U + u)\sin(\theta_e + \theta - \alpha), \tag{3.177}$$

which we write in terms of a small deviation, $\bar{h}(t)$, from a straight-line ($\theta_e =$ const.) equilibrium altitude profile, $h_e(t) = h_0 + Ut\sin\theta_e$, as

$$\dot{\bar{h}} = u\sin\theta_e + U(\theta - \alpha)\cos\theta_e, \tag{3.178}$$

resulting in the following augmentation of the longitudinal plant:

$$\begin{Bmatrix} \dot{\bar{h}} \\ \dot{u} \\ \dot{\alpha} \\ \dot{\theta} \\ \dot{q} \end{Bmatrix} = \left(\begin{array}{c|cccc} 0 & \sin\theta_e & -U\cos\theta_e & U\cos\theta_e & 0 \\ \hline 0 & & & & \\ 0 & & \mathbf{A}_{\text{Long}} & & \\ 0 & & & & \\ 0 & & & & \end{array}\right) \begin{Bmatrix} \bar{h} \\ u \\ \alpha \\ \theta \\ q \end{Bmatrix} + \left(\begin{array}{cc} 0 & 0 \\ \hline & \\ \multicolumn{2}{c}{\mathbf{B}_{\text{Long}}} \\ & \end{array}\right) \begin{Bmatrix} \beta_T \\ \delta_E \end{Bmatrix}. \tag{3.179}$$

Example 3.4 Let us design an airspeed and altitude hold system for a jet fighter aircraft with the following longitudinal dynamics data, cruising at Mach 0.6 at a standard altitude of 10.6349 km where the atmospheric density is 0.3809 kg/m^3 and the speed of sound is 296.654 m/s (corresponding to $U = 178$ m/s):

$$\mathbf{A}_{\text{Long}} = \begin{pmatrix} 4.3079 \times 10^{-7} & 0.00048733 & -9.6783 & 0 \\ -2.316 \times 10^{-7} & -0.0001771 & -0.0090011 & 0.99998 \\ 0 & 0 & 0 & 1 \\ 2.0538 \times 10^{-7} & -0.00023293 & 5.7156 \times 10^{-6} & -0.00064266 \end{pmatrix},$$

$$\mathbf{B}_{\text{Long}} = \begin{pmatrix} -0.0618 & 1.7818 \\ 0.0021016 & -0.0001 \\ 0 & 0 \\ 0.5581 & 0.039453 \end{pmatrix}.$$

Here, the airspeed is in meters per second while all angles and angular rates ($\alpha, \theta, q, \delta_E, \beta_T$) are in radians (or radians per second). Eigenvalues of $\mathbf{A}_{\text{Long}}$ show that the phugoid mode is unstable for the aircraft in the given flight condition.

We would like to design an airspeed and altitude hold autopilot with feedback of altitude and airspeed outputs, such that an initial altitude perturbation of ± 100 m is brought to zero in a settling time of 120 s, without exceeding an airspeed deviation of ±1 m/s, and elevator and throttle deflections of ±1°. An LQG compensator is designed for this purpose with the following parameters:

$$\mathbf{F} = \mathbf{B}, \quad \mathbf{S}_v = 0.01\mathbf{I}, \quad \mathbf{S}_w = \mathbf{I},$$

$$\mathbf{Q} = 10^{-5}\begin{pmatrix} 1 & 0 & 0 & 0 & 0 \\ 0 & 500 & 0 & 0 & 0 \\ 0 & 0 & 0 & 0 & 0 \\ 0 & 0 & 0 & 0 & 0 \\ 0 & 0 & 0 & 0 & 0 \end{pmatrix}, \quad \mathbf{R} = \mathbf{I}.$$

The resulting return ratio singular values and the closed-loop initial response are computed with the following MATLAB statements and plotted in Figures 3.40–4.42:

```
>> a = [ 0    0    -178.0000    178.0000 0
      0    0.0000  0.0005 -9.6783      0
      0    0.0000    -0.0002     -0.0090      1.0000
      0  0       0           0       1.0000
      0  0.0000  -0.0002         0.0000     -0.0006];
>> b = [  0        0
     -0.0618   1.7818
      0.0021     -0.0001
      0          0
      0.5581     0.0395];
>> c = [  1        0      0      0      0
      0      1      0      0      0];
>> d=zeros(2,2);
```

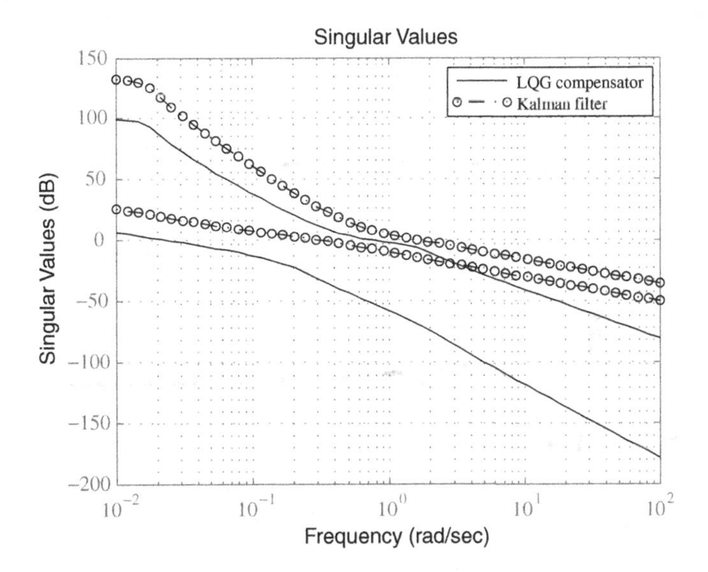

Figure 3.40 Singular values of return ratio at plant's output for LQG compensated altitude and Mach hold autopilot, and their comparison with corresponding singular values of the Kalman filter

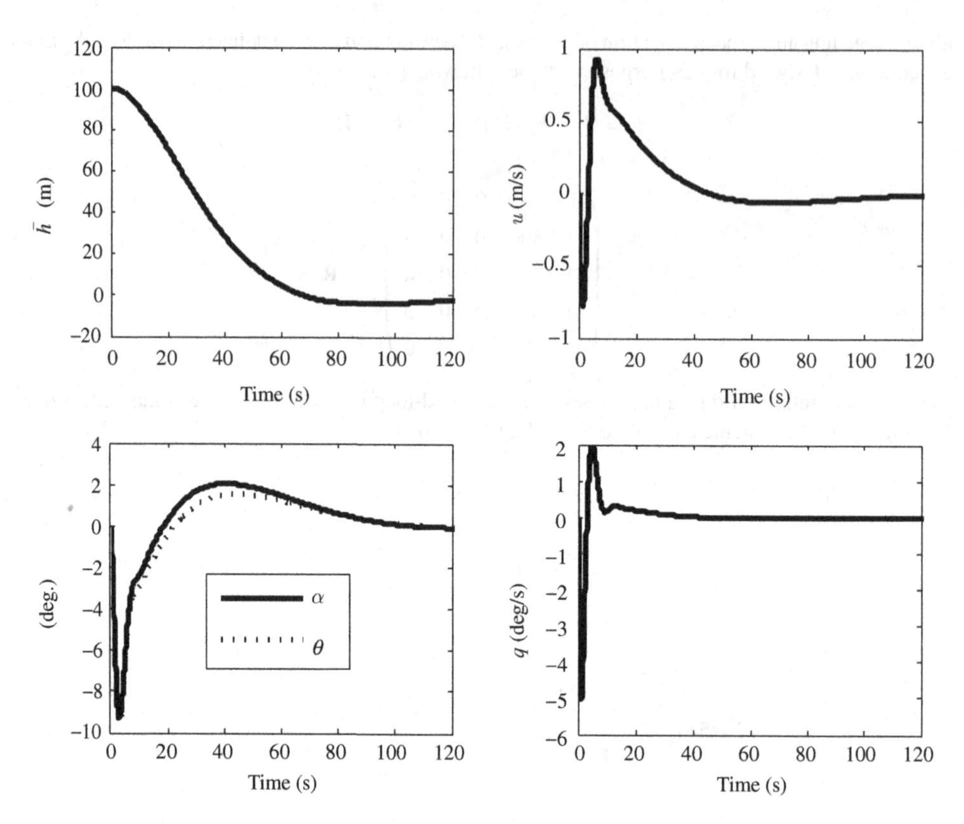

Figure 3.41 Closed-loop initial response for a 100 m altitude error for the LQG/LTR compensated altitude and Mach hold autopilot

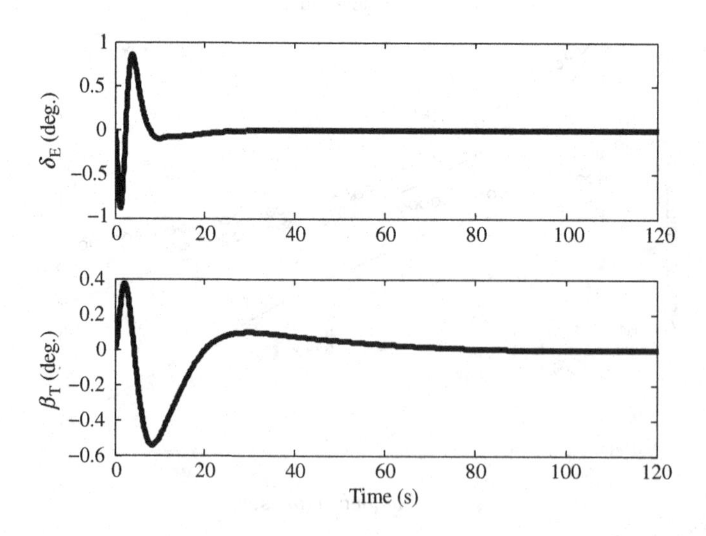

Figure 3.42 Closed-loop control inputs for an initial 100 m altitude error for the LQG/LTR altitude and Mach hold autopilot

```
>> Snu=1e-2*eye(2); Sw=eye(2);
>> [L,S,E]=lqe(a,b,c,Snu,Sw,zeros(2,2))

L =
    0.3156   -0.1104
   -0.1104    1.6347
   -0.0032   -0.1357
   -0.0029   -0.1370
   -0.0044   -0.0557

S =
    0.3156   -0.1104   -0.0032   -0.0029   -0.0044
   -0.1104    1.6347   -0.1357   -0.1370   -0.0557
   -0.0032   -0.1357    0.0172    0.0173    0.0094
   -0.0029   -0.1370    0.0173    0.0174    0.0094
   -0.0044   -0.0557    0.0094    0.0094    0.0076

E =
  -0.4115 + 0.7081i
  -0.4115 - 0.7081i
  -0.8165
  -0.1558 + 0.0994i
  -0.1558 - 0.0994i

>> [k,S,E]=lqr(a,b,.00001*[1 0 0 0 0;0 500 0 0 0;zeros(3,5)],eye(2))

k =
    0.0016   -0.0604   -9.5852   11.4367    2.6034
    0.0027    0.0368  -21.5892   21.4508    0.0786

S =
  1.0e+004 *

    0.0000    0.0000   -0.0002    0.0002    0.0000
    0.0000    0.0000   -0.0010    0.0010   -0.0000
   -0.0002   -0.0010    1.2067   -1.2092   -0.0064
    0.0002    0.0010   -1.2092    1.2121    0.0067
    0.0000   -0.0000   -0.0064    0.0067    0.0005

E =
  -0.3663 + 0.6258i
  -0.3663 - 0.6258i
  -0.7282
  -0.0237 + 0.0231i
  -0.0237 - 0.0231i

>> sys=ss(a,b,c,d); %open-loop plant
>> sysL=ss(a,-L,c,zeros(2,2));
>> Ac=a-b*k-L*c;
>> Bc=L;
>> Cc=-k;
>> Dc=zeros(2,2);
>> sysc=ss(Ac,Bc,Cc,Dc); % autopilot
>> sysGK=series(sys,sysc);
```

```
>> w=logspace(-2,2);
>> sigma(sysGK,w),hold on,sigma(sysL,w),hold on,grid
>> syss=series(sysc,sys);
>> sysCL=feedback(syss,eye(2)) %closed-loop system


ACL =
             x1         x2         x3         x4         x5            x6           x7         x8         x9
   x1         0          0       -178        178          0             0            0          0          0
   x2         0          0     0.0005     -9.678          0     -0.004709     -0.06933      37.88     -37.51
   x3         0          0    -0.0002     -0.009          1    -3.188e-006    0.0001305    0.01797   -0.02187
   x4         0          0          0          0          1             0            0          0          0
   x5         0          0    -0.0002          0    -0.0006     -0.001026      0.03224      6.202      -7.23
   x6   -0.3156     0.1104          0          0          0       -0.3156       0.1104       -178        178
   x7    0.1104     -1.635          0          0          0        0.1057       -1.704      37.88     -47.19
   x8  0.003209     0.1357          0          0          0      0.003206       0.1358    0.01777   -0.03087
   x9  0.002895      0.137          0          0          0      0.002895        0.137          0          0
   x10 0.004395    0.05566          0          0          0       0.00337       0.0879      6.202      -7.23


              x10
   x1           0
   x2     0.02093
   x3   -0.005459
   x4           0
   x5      -1.456
   x6           0
   x7     0.02093
   x8      0.9945
   x9           1
   x10     -1.457

BCL =
              u1         u2
   x1          0          0
   x2          0          0
   x3          0          0
   x4          0          0
   x5          0          0
   x6     0.3156    -0.1104
   x7    -0.1104      1.635
   x8  -0.003209    -0.1357
   x9  -0.002895     -0.137
   x10 -0.004395   -0.05566

CCL =
        x1   x2   x3   x4   x5   x6   x7   x8   x9  x10
   y1    1    0    0    0    0    0    0    0    0    0
   y2    0    1    0    0    0    0    0    0    0    0

DCL =
       u1  u2
   y1   0   0
   y2   0   0


>> [y,t,x]=initial(sysk,[100 0 0 0 0]');
>> u=lsim(sysc,y,t);
```

Figure 3.40 shows that the return ratios at the plant output are close to those of the Kalman filter in the desired control bandwidth of $0.01 \leq \omega \leq 5$ rad/s (although a perfect LTR is not obtained). There is an excellent roll-off at higher frequencies, indicating good noise rejection. The closed-loop initial response settles as required in about 120 s with the airspeed deviations kept within ± 1 m/s (Figure 3.41), and both the control angles not exceeding $\pm 1^\circ$. The maximum pitch and angle-of-attack deviations are limited to $\pm 10^\circ$ while the pitch rate is limited to $\pm 6^\circ$. The pitch rate quickly becomes small while the altitude and airspeed transients take longer to settle. The integral action promises zero steady-state error for step disturbances. If one would like to increase the robustness of the design by better LTR in the control bandwidth then an approximately tenfold increase in the control magnitudes must be tolerated. However, this will have little impact on the settling time of the transients.

A Mach and altitude hold autopilot is not the only application of simultaneous control of airspeed and altitude. A common feature of modern aircraft is the ability to land in bad weather, which requires automatic interception and tracking of a constant *glide-slope* radio signal transmitted by an instrument landing system. Such an automatic landing system is easily designed using the multi-variable approach presented in this section, and requires a precise geometric altitude measurement by onboard radar while maintaining a constant airspeed, right up to the point of touchdown on the runway. The only change required in the design presented above is to regard the initial altitude error, $\bar{h}$, as the reference altitude above the runway when intercepting the glide-slope signal. As seen above, the integral action provided by the controller will always ensure a smooth (zero pitch rate) landing.

3.9 Lateral-Directional Control Systems

Aircraft motion out of the plane of symmetry is termed lateral-directional dynamics and consists of roll (σ, P), yaw (Ψ, R), and sideslip (V). An aircraft's flight direction in the horizontal plane is governed by the lateral-directional dynamics. Therefore, lateral-directional stability and control are essential for aircraft navigation.

3.9.1 Lateral-Directional Plant

The lateral-directional dynamic plant is represented by the following governing equations derived from equations (3.17), (3.13), and (3.19):

$$\begin{aligned}
\dot{\sigma} &= P + Q\sin\sigma\tan\theta + R\cos\sigma\tan\theta, \\
\dot{\Psi} &= Q\sin\sigma\sec\theta + R\cos\sigma\sec\theta, \\
Y + mg\sin\sigma\cos\theta &= m(\dot{V} + RU - PW), \\
L &= J_{xx}\dot{P} + J_{xz}(\dot{R} + PQ) + \left(J_{zz} - J_{yy}\right)QR, \\
N &= J_{zz}\dot{R} + J_{xz}(\dot{P} - QR) + \left(J_{yy} - J_{xx}\right)PQ.
\end{aligned} \tag{3.180}$$

While the fifth-order nonlinear plant is affected by the longitudinal variables (θ, Q, U, W), if we introduce the small-perturbation approximation presented earlier for the longitudinal case, and express the heading (or yaw) angle by $\Psi = \psi_e + \psi$, we have the following linearized, lateral-directional plant:

$$\begin{aligned}
\dot{\sigma} &= p + r\tan\theta_e, \\
\dot{\psi} &= r\sec\theta_e, \\
Y + mg\sigma\cos\theta_e &= mU(\dot{\beta} + r), \\
L &= J_{xx}\dot{p} + J_{xz}\dot{r}, \\
N &= J_{zz}\dot{r} + J_{xz}\dot{p},
\end{aligned} \tag{3.181}$$

where σ, ψ, p, r are small perturbations in roll angle, yaw angle, roll rate, and yaw rate, respectively, from a straight-line, equilibrium flight condition, $\sigma_e = V_e = P_e = R_e = Q_e = 0$, $\psi_e = \text{const.}$, and β is the small *sideslip angle* given by

$$\beta = \tan^{-1} \frac{v}{U \cos \alpha} \simeq \frac{v}{U}.$$

The small sideslip perturbation is thus $v = V \simeq U\beta$.

In a manner similar to the longitudinal case, the lateral-directional aerodynamic force and moments without controls application can be expressed as linear functions of the perturbation variables as follows:

$$\begin{aligned} Y &= Y_\beta \beta + Y_{\dot\beta} \dot\beta + Y_p p + Y_r r, \\ L &= L_\beta \beta + L_{\dot\beta} \dot\beta + L_p p + L_r r, \\ N &= N_\beta \beta + N_{\dot\beta} \dot\beta + N_p p + N_r r, \end{aligned} \tag{3.182}$$

where the stability derivatives

$$Y_\beta = \left(\frac{\partial Y}{\partial \beta} \right)_e, \text{ etc.}$$

result from a first-order Taylor series expansion about the equilibrium flight condition. Due to longitudinal symmetry, the lateral-directional force and moments are independent of the longitudinal perturbation quantities, (θ, q, u, α). The β and $\dot\beta$ derivatives are primarily caused by the vertical tail and fuselage combination. They arise in quite the same way as the α and $\dot\alpha$ derivatives arise due to the wing-fuselage and horizontal tail combination. As we shall see later, the β derivatives influence the lateral-directional static stability. The derivatives L_p, N_r affect the damping of the dynamic modes (as in the case of M_q) and are thus called roll and yaw damping, respectively. Finally, the derivatives Y_p, L_r, N_p are called cross-coupling derivatives and represent the aerodynamic coupling between roll, yaw, and sideslipping motions. They are primarily due to the wing and the vertical tail.

In addition to the perturbation variables, we have the control inputs applied by the pilot/automatic controller in the form of aileron angle, δ_A, and rudder angle, δ_R. A positive aileron angle is one that creates a positive rolling moment, which is an upward deflection of the control surface on the right wing, and a downward deflection on the left, as shown in Figure 3.23. Similarly, a positive rudder deflection is one that produces a positive yawing moment, that is, a deflection to the right (Figure 3.23).

By substituting equation (3.182) into equation (3.181) and adding the control force and moments, we have the following state-space representation for lateral-directional dynamics:

$$\begin{Bmatrix} \dot\sigma \\ \dot\psi \\ \dot\beta \\ \dot p \\ \dot r \end{Bmatrix} = \mathbf{A}_{\mathrm{LD}} \begin{Bmatrix} \sigma \\ \psi \\ \beta \\ p \\ r \end{Bmatrix} + \mathbf{B}_{\mathrm{LD}} \begin{Bmatrix} \delta_A \\ \delta_R \end{Bmatrix}, \tag{3.183}$$

where

$$\mathbf{A}_{\mathrm{LD}} = \begin{bmatrix} 0 & 0 & 0 & 1 & \tan\theta_{\mathrm{e}} \\ 0 & 0 & 0 & 0 & \sec\theta_{\mathrm{e}} \\ \frac{g\cos\theta_{\mathrm{e}}}{U} & 0 & \frac{Y_\beta}{mU} & \frac{Y_p}{mU} & \frac{Y_r}{mU} - 1 \\ 0 & 0 & AL_\beta + BN_\beta & AL_p + BN_p & AL_r + BN_r \\ 0 & 0 & CN_\beta + BL_\beta & CN_p + BL_p & CN_r + BL_r \end{bmatrix}, \tag{3.184}$$

$$\mathbf{B}_{\mathrm{LD}} = \begin{bmatrix} 0 & 0 \\ 0 & 0 \\ \frac{Y_{\mathrm{A}}}{mU} & \frac{Y_{\mathrm{R}}}{mU} \\ AL_{\mathrm{A}} + BN_{\mathrm{A}} & AL_{\mathrm{R}} + BN_{\mathrm{R}} \\ CN_{\mathrm{A}} + BL_{\mathrm{A}} & CN_{\mathrm{R}} + BL_{\mathrm{R}} \end{bmatrix}, \tag{3.185}$$

with the inertial parameters given by

$$\begin{aligned} A &= \frac{J_{zz}}{J_{xx}J_{zz} - J_{xz}^2}, \\ B &= \frac{J_{xz}}{J_{xx}J_{zz} - J_{xz}^2}, \\ C &= \frac{J_{xx}}{J_{xx}J_{zz} - J_{xz}^2}. \end{aligned} \tag{3.186}$$

As for the longitudinal case, the lateral-directional modes can be found by examining the eigenvalues of $\mathbf{A}_{\mathrm{LD}}$. There are three distinct lateral-directional modes in a conventional aircraft: (a) a *pure rolling mode* that primarily involves rotation about *ox* and is usually well damped by the wings; (b) a long-period *spiral mode* largely consisting of a coordinated turn (roll and yaw) without much of a sideslip; and (c) a short-period *Dutch-roll mode* having yawing and sideslipping oscillations (in mutually opposite directions) with a small roll component. Control of the pure rolling mode is the easiest, requiring only the aileron input, whereas Dutch-roll damping can be simply increased by having a rudder in closed-loop with either the yaw rate, or the sideslip angle. However, the suppression of the spiral mode – essential for maintaining a desired heading (or bank) angle despite disturbances – requires a coordinated rudder and aileron combination (much like the suppression of the phugoid mode by elevator and throttle), thereby bringing us into the realm of multi-variable control design.

A general lateral-directional control system is depicted by the block diagram in Figure 3.43. The *stability augmentation system* (SAS) is the innermost loop requiring an increase of damping in the spiral and Dutch-roll modes. There is a tracking system called the *autopilot* in the middle that follows reference commands generated by a navigational system, while the outermost loop involves continuous adjustment of controller parameters with changing flight parameters using nonlinear mathematical laws called *gain scheduling*. A high-speed aircraft generally requires gain scheduling with both altitude and Mach number for efficient performance throughout the flight envelope.

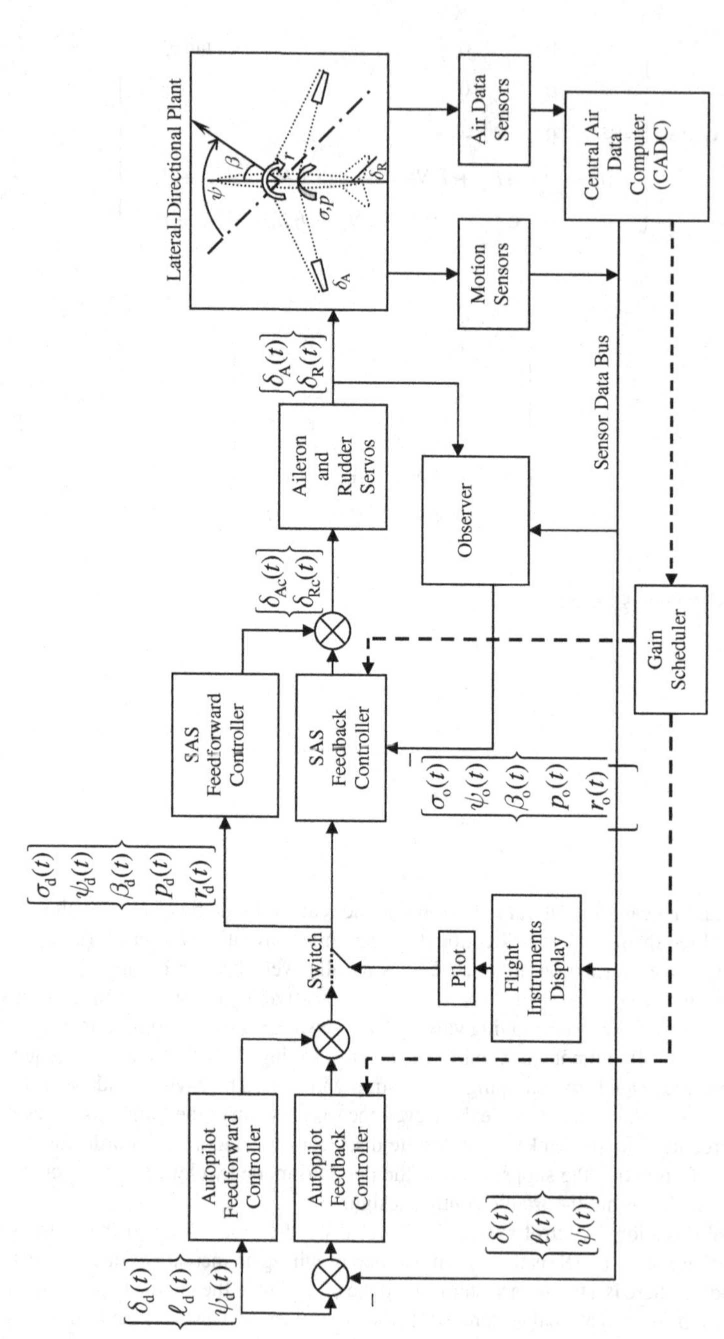

Figure 3.43 A general aircraft lateral-directional control system, including the inner stability augmentation system (SAS), the middle autopilot loop, and the outer, nonlinear gain scheduler for the SAS controller gains based upon flight parameters

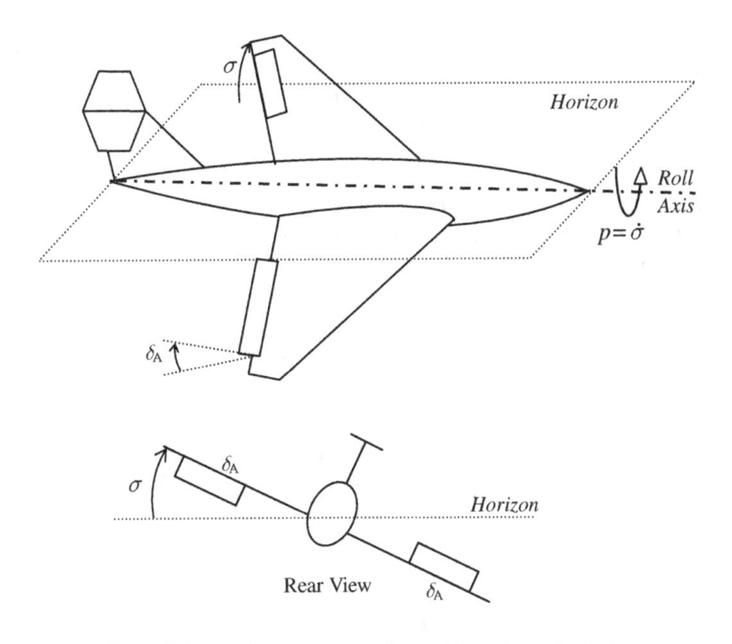

Figure 3.44 Rolling maneuver of aircraft/aerodynamic missile

3.9.2 Optimal Roll Control

The optimal control theory can be applied to terminal roll control of aircraft and missiles with aerodynamic control surfaces. Such an application involves isolating the pure rolling mode of the lateral-directional dynamics and posing the optimal control problem as a single-axis rotation with a fixed terminal time. The governing equation of motion for an aircraft in pure rolling mode with bank angle, $\sigma(t)$, roll rate, $p(t)$, and aileron deflection, $\delta_A(t)$,[9] (Figure 3.44) can be expressed as

$$\begin{aligned} J_{xx}\dot{p} &= L_p p + L_A \delta_A, \\ \dot{\sigma} &= p, \end{aligned} \tag{3.187}$$

which can be written in the following time-invariant state-space form with state vector, $\mathbf{x}(t) = [\sigma(t), p(t)]^T$, and control input, $u(t) = \delta_A(t)$:

$$\dot{\mathbf{x}} = \mathbf{f}(\mathbf{x}, u) = \mathbf{A}\mathbf{x} + \mathbf{B}u, \quad \mathbf{x}(0) = \mathbf{0}, \tag{3.188}$$

where

$$\mathbf{A} = \frac{\partial \mathbf{f}}{\partial \mathbf{x}} = \begin{pmatrix} 0 & 1 \\ 0 & \frac{L_p}{J_{xx}} \end{pmatrix}$$

[9] $\delta_A(t)$ can alternatively be regarded as fin deflection angle for an aerodynamically controlled missile, or the exhaust guide vane deflection of a rocket creating a rolling moment.

and

$$\mathbf{B} = \frac{\partial \mathbf{f}}{\partial u} = \begin{pmatrix} 0 \\ \frac{L_A}{J_{xx}} \end{pmatrix} u.$$

In order to simplify notation we define $a = L_p/J_{xx}$, $b = L_A/J_{xx}$.

We select $\varphi = Q[\sigma_f - p(t_f)]^2$, $L = Ru^2$, and impose the terminal roll rate constraint, $p(t_f) = 0$. From the Euler–Lagrange equations it follows that

$$\dot{\boldsymbol{\lambda}} = -H_{\mathbf{x}}^T = \begin{pmatrix} 0 \\ -\lambda_1 - a\lambda_2 \end{pmatrix}, \tag{3.189}$$

$$\boldsymbol{\lambda}(t_f) = \left(\frac{\partial \varphi}{\partial \mathbf{x}}\right)^T_{t=t_f} + \mathbf{G}^T \mu = \begin{Bmatrix} -2Q(\sigma_f - \sigma) \\ \mu \end{Bmatrix}, \tag{3.190}$$

and

$$H_u = 2Ru^* + b\lambda_2(t_f) = 0, \tag{3.191}$$

of which the closed-form solution yields

$$\begin{aligned} \lambda_1(t) &= \lambda_1(t_f) = -2Q[\sigma_f - \sigma(t_f)] \\ \lambda_2(t) &= \tfrac{1}{a}\left\{-\lambda_1(t_f)u_s(t_f - t) + [\lambda_1(t_f) + a\mu]\, e^{a(t_f - t)}\right\}, \end{aligned} \tag{3.192}$$

$$u^*(t) = -\frac{\lambda_2(t)}{2RJ} = -\frac{b}{2aR}\left\{-\lambda_1(t_f)u_s(t_f - t) + [\lambda_1(t_f) + a\mu]\, e^{a(t_f - t)}\right\}, \tag{3.193}$$

where $u_s(\tau)$ is the unit step function applied at $t = \tau$.

In order to apply the terminal constraint, we solve for the roll rate as follows:

$$\begin{aligned} p(t) &= \int_0^t e^{a(t-\tau)} b u^*(\tau) \mathrm{d}\tau \\ &= -\frac{b^2}{2a^2 R}\left\{\lambda_1(t_f)(1 - e^{at}) - \frac{1}{2}[\lambda_1(t_f) + a\mu]\left[e^{a(t_f - t)} - e^{a(t + t_f)}\right]\right\}, 0 \le t \le t_f. \end{aligned} \tag{3.194}$$

For $p(t_f) = 0$ we must have

$$\mu = \lambda_1(t_f)\frac{1 - 2e^{at_f} + e^{2at_f}}{a(1 - e^{2at_f})}, \tag{3.195}$$

substituting which into equation (3.194) and integrating produces the following expressions for the control input, roll rate, and bank angle:

$$u^*(t) = \frac{b\lambda_1(t_f)}{2aR}\left[u_s(t_f - t) - 2\frac{1 - e^{at_f}}{1 - e^{2at_f}} e^{a(t_f - t)}\right], \tag{3.196}$$

$$p(t) = -\frac{b^2\lambda_1(t_f)}{2a^2R}\left\{1 - e^{at} - \frac{1-e^{at_f}}{1-e^{2at_f}}\left[e^{a(t_f-t)} - e^{a(t+t_f)}\right]\right\}, \quad 0 \le t \le t_f, \tag{3.197}$$

$$\begin{aligned}\sigma(t) &= \int_0^t p(\tau)\mathrm{d}\tau \\ &= -\frac{b^2\lambda_1(t_f)}{2a^3R}\left[1 + at - e^{at} - \frac{1-e^{at_f}}{1-e^{2at_f}}\left(2e^{at_f} - e^{2at_f} - 1\right)\right], \quad 0 \le t \le t_f.\end{aligned} \tag{3.198}$$

Finally, substituting equation (3.192) and $t = t_f$ into equation (3.198) yields the final bank angle,

$$\sigma(t_f) = \sigma_f\left(\frac{D}{1+D}\right), \tag{3.199}$$

where

$$D = \frac{Qb^2}{a^3R\left(1-e^{2at_f}\right)}\left(2 + at_f - at_fe^{2at_f} + 2e^{2at_f} - 4e^{at_f}\right). \tag{3.200}$$

In order to make $\sigma(t_f) \approx \sigma_f$ one must choose Q, R such that $D \gg 1$.

Example 3.5 At a given flight condition, the roll characteristics of a fighter aircraft are given by $a = L_p/J_{xx} = -2\ \mathrm{s}^{-1}$ and $b = L_A/J_{xx} = 20\ \mathrm{s}^{-2}$. Beginning from zero initial condition, determine the aileron input required to achieve a steady bank angle of 10° in 0.5 s.

We select $Q = -a^3$, $R = b^{-2}$, which results in $D = 12\,123$ and $\sigma(t_f) = 9.9992°$. The optimal aileron input and the extremal trajectory are plotted in Figures 3.45 and 3.46, respectively. Note the maximum aileron deflection of ±12.2° at the ends of the control interval, where it must abruptly change from the

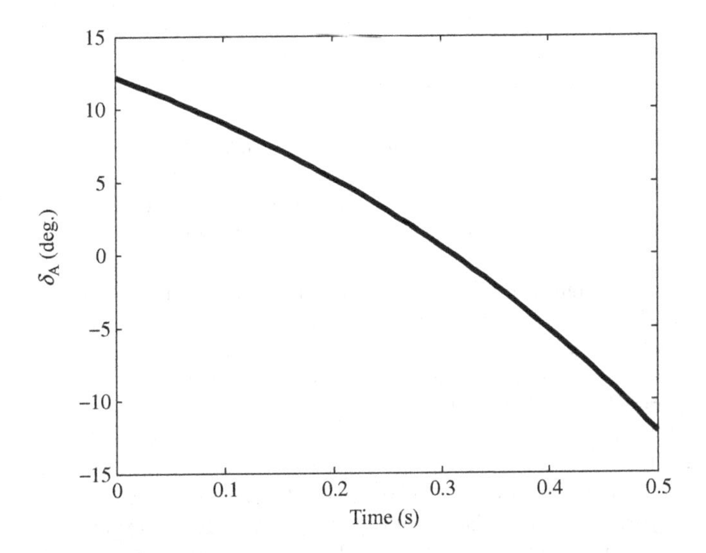

Figure 3.45 Optimal aileron input for banking maneuver of a fighter aircraft

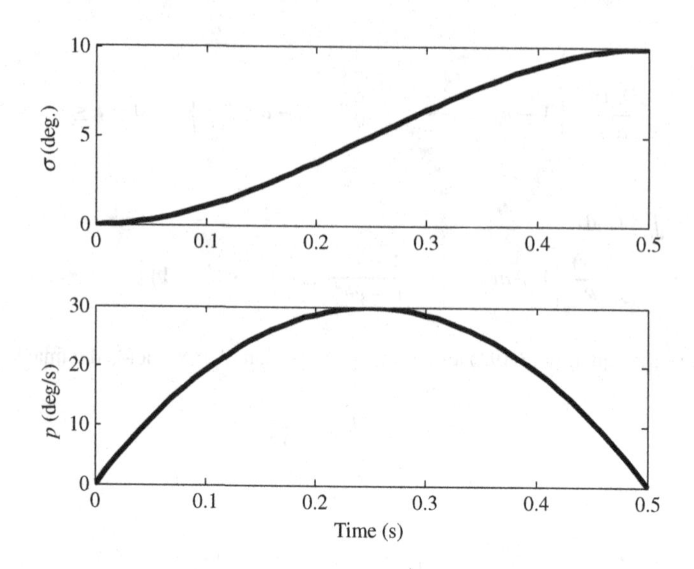

Figure 3.46 Optimal trajectory for banking maneuver of a fighter aircraft

null value at $t < 0$ and $t > t_f$. Also, note the maximum roll rate of 30 deg/s at the mid-point of the control interval.

3.9.3 Multi-variable Lateral-Directional Control: Heading-Hold Autopilot

While the pure rolling and Dutch-roll modes can be handled respectively by the aileron and rudder alone, the spiral mode requires a multi-variable approach due to the multiple degrees of freedom involved. Spiral mode control is necessary for good directional maneuverability, which involves pointing the aircraft at a desired heading angle, ψ_e, that is either set by the pilot or generated as the output of an inertial navigation system (navigational subsystem (a)). If the spiral mode is either unstable or poorly damped (as it is often likely to be), the aircraft will not maintain a constant heading in the presence of a lateral disturbance. Hence, most aircraft require an autopilot for holding a desired heading angle that is crucial for navigation. A heading-hold autopilot can be designed using a yaw rate, r, a rate-integrating (directional) gyro signal, ψ, and the aileron and rudder angle inputs, δ_A, δ_R. A multi-variable approach is necessary because the aileron deflection must be closely coordinated with rudder deflection in order to nullify any sideslip. We note that the lateral-directional plant has a pole at origin that enables us to design controllers without integral action (such as LQG) while ensuring zero steady-state error to step commands.

Example 3.6 Consider a fighter airplane climbing steadily with a constant flight-path angle of 30° and airspeed $U = 272$ m/s at an altitude of 5 km. At the given flight condition the airplane has the following parameters: $m = 6181$ kg; $J_{xx} = 1.5 \times 10^4$ kg.m^2, $J_{zz} = 5 \times 10^4$ kg.m^2, $J_{xz} = 456$ kg.m^2; $Y_\beta = -5.46 \times 10^5$ N/rad, $Y_p = 0$, $Y_r = 2.87 \times 10^5$ N.s/rad; $Y_A = 0$, $Y_R = -1.5 \times 10^5$ N/rad; $L_\beta = -7 \times 10^5$ N.m/rad, $L_p = -9.3 \times 10^4$ N.m.s/rad, $L_r = 1.02 \times 10^4$ N.m.s/rad; $L_A = 1.5 \times 10^6$ N.m/rad, $L_R = -805$ N.m/rad; $N_\beta = 1.1 \times 10^6$ N.m/rad, $N_p = -960$ N.m.s/rad, $N_r = -3.19 \times 10^4$ N.m.s/rad; $N_A = -1.06 \times 10^3$ N.m/rad, $N_R = 3.185 \times 10^4$ N.m/rad. At the given flight condition, design a heading-hold autopilot for the aircraft based on yaw rate and yaw angle outputs in order to achieve a zero steady-state heading deviation with a settling time of 10 s, in the presence of a lateral gust that causes an initial sideslip perturbation of $\beta(0) = 5°$. The maximum aileron and rudder deflections should not exceed ±25°. Consider first-order aileron and rudder servos of time constant 0.05 s for the closed-loop simulation.

With the given data, the lateral-directional dynamics matrix is

$$\mathbf{A}_{\mathrm{LD}} = \begin{bmatrix} 0 & 0 & 0 & 1 & 0.5774 \\ 0 & 0 & 0 & 0 & 1.1547 \\ 0.0312 & 0 & -0.3248 & 0 & -0.8293 \\ 0 & 0 & -46.0115 & -6.2023 & 0.6608 \\ 0 & 0 & 21.5518 & -0.0757 & -0.6311 \end{bmatrix},$$

whose eigenvalues are computed using Control System Toolbox as follows:

```
>> A =[   0            0            0       1.0000      0.5774
          0            0            0            0      1.1547
     0.0312            0      -0.3248            0     -0.8293
          0            0     -46.0115      -6.2023      0.6608
          0            0      21.5518      -0.0757     -0.6311];

>> damp(A)

          Eigenvalue                      Damping        Freq. (rad/s)
  1.72e-002                             -1.00e+000        1.72e-002
  0.00e+000                             -1.00e+000        0.00e+000
 -4.48e-001 + 4.28e+000i                 1.04e-001        4.31e+000
 -4.48e-001 - 4.28e+000i                 1.04e-001        4.31e+000
 -6.28e+000                              1.00e+000        6.28e+000
```

The pole associated with the pure rolling mode is $s = -6.28$, while the Dutch-roll mode is given by $\omega = 4.31$ rad/s, $\zeta = 0.104$, and the spiral mode is unstable with pole $s = 0.0172$. Finally, the plant pole at origin renders our task of feedback design much easier by providing integral action in the closed loop. We choose to design the bi-input, bi-output autopilot by output-weighted linear quadratic regulator (LQRY) and a Kalman filter as follows:

```
>> A = [ 0             0            0       1.0000      0.5774
         0             0            0            0      1.1547
    0.0312             0      -0.3248            0     -0.8293
         0             0     -46.0115      -6.2023      0.6608
         0             0      21.5518      -0.0757     -0.6311];

>> B = [ 0             0
         0             0
         0       -0.0892
  100.0271       -0.0343
    0.8910        0.6367];

>> C=[0 1 0 0 0; 0 0 0 0 1] % Heading angle and yaw rate outputs

% Regulator gain matrix by output weighted LQR:
>> K=lqry(A,B,C,zeros(2,2),[1 0;0 0.1],eye(2))
```

```
K =
    0.0848    0.9854    1.1700    0.0127    0.0046
   -0.0108   -0.1701   -0.4541   -0.0014    0.0448


>> Lp=lqr(A',C',10*eye(5),eye(2));L=Lp' % Kalman filter gain

L =
    0.2569    2.4205
    3.3518    0.8401
    0.0095    2.3001
   -0.3854  -12.0743
    0.8401    9.8886

% Augmented plant with aileron and rudder servos:
>> Abar=[A B;zeros(1,5) -20 0; zeros(1,5) 0 -20];
>> Bbar=[0 0 0 0 0 20 0;0 0 0 0 0 0 20]';
>> Cbar=[C zeros(2,2)];

>> Ac=[Abar  -Bbar*K;L*Cbar A-L*C-B*K] %Closed-loop dynamics matrix


Ac =
         0         0         0    1.0000    0.5774         0         0         0         0         0         0         0
         0         0         0         0    1.1547         0         0         0         0         0         0         0
    0.0312         0   -0.3248         0   -0.8293         0   -0.0892         0         0         0         0         0
         0         0  -46.0115   -6.2023    0.6608  100.0271   -0.0343         0         0         0         0         0
         0         0   21.5518   -0.0757   -0.6311    0.8910    0.6367         0         0         0         0         0
         0         0         0         0         0  -20.0000         0   -1.6969  -19.7089  -23.4003   -0.2538   -0.0912
         0         0         0         0         0         0  -20.0000    0.2162    3.4001    9.0803    0.0284   -0.8956
         0    0.2569         0         0    2.4205         0         0         0   -0.2569         0    1.0000   -1.8431
         0    3.3518         0         0    0.8401         0         0         0   -3.3518         0         0    0.3146
         0    0.0095         0         0    2.3001         0         0    0.0302   -0.0247   -0.3653   -0.0001   -3.1254
         0   -0.3854         0         0  -12.0743         0         0   -8.4874  -98.1914 -163.0602   -7.4719   12.2804
         0    0.8401         0         0    9.8886         0         0   -0.0687   -1.6099   20.7984   -0.0861  -10.5523


>> damp(Ac) % Closed-loop eigenvalues

        Eigenvalue                Damping      Freq. (rad/s)
 -2.04e+001                     1.00e+000       2.04e+001
 -5.32e+000 + 6.72e+000i        6.21e-001       8.57e+000
 -5.32e+000 - 6.72e+000i        6.21e-001       8.57e+000
 -5.03e-001 + 4.30e+000i        1.16e-001       4.33e+000
 -5.03e-001 - 4.30e+000i        1.16e-001       4.33e+000
 -3.25e-002                     1.00e+000       3.25e-002
 -5.75e-001 + 5.77e-001i        7.06e-001       8.15e-001
 -5.75e-001 - 5.77e-001i        7.06e-001       8.15e-001
 -3.34e+000                     1.00e+000       3.34e+000
 -6.10e+000                     1.00e+000       6.10e+000
 -6.28e+000                     1.00e+000       6.28e+000
 -2.00e+001                     1.00e+000       2.00e+001

% Closed-loop initial response to 5 deg. sideslip perturbation:
>> sys=ss(Ac,zeros(12,1),eye(12),zeros(12,1));
>> [y,t,x]=initial(sys,[0 0 5*pi/180 0 0 0 0 zeros(1,5)]',20);
>> u=-x*[K zeros(2,7)]'; % Commanded aileron & rudder deflections
```

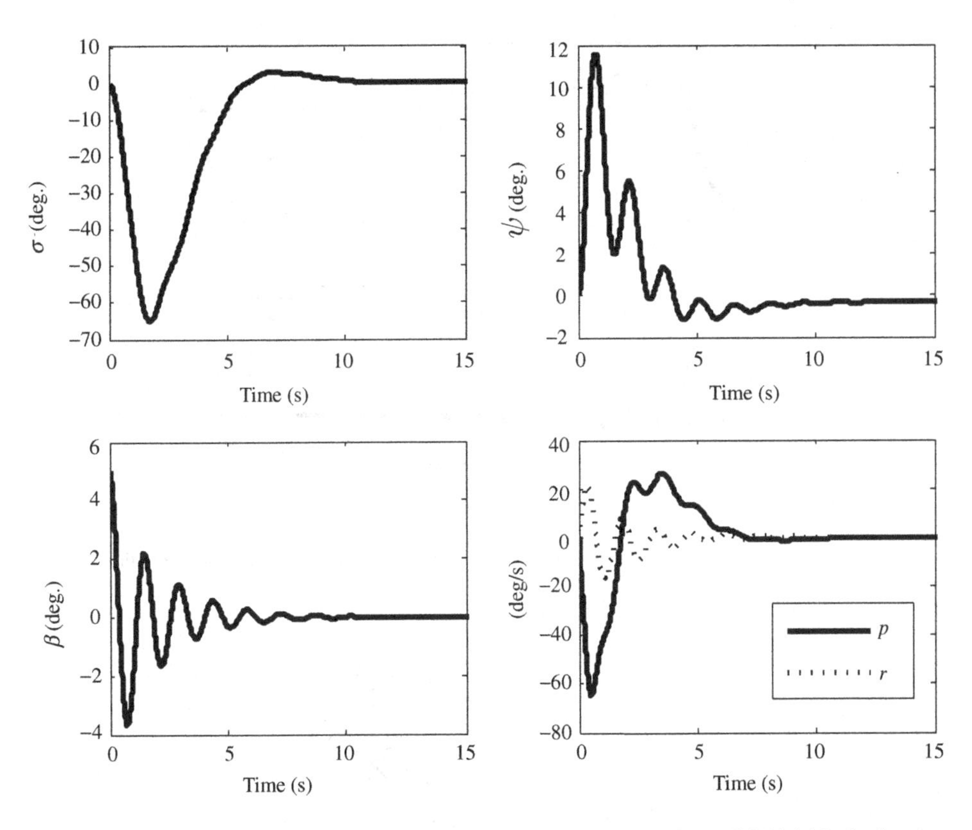

Figure 3.47 Closed-loop response of heading-hold autopilot for a fighter aircraft to a 5° initial sideslip disturbance

The closed-loop initial response is plotted in Figures 3.47 and 3.48 showing that all transients settle to zero in about 10 s, with the maximum bank angle of -65°, maximum yaw angle overshoot of 11.5°, and maximum aileron and rudder deflections of -5.85° and 2.3°, respectively. Clearly, the settling time can be reduced further by allowing larger aileron and rudder deflections. One can also improve the robustness of the design by employing either the LQG/LTR approach or the H_∞ method (Chapter 2).

3.10 Optimal Control of Inertia-Coupled Aircraft Rotation

In general aircraft motion, both aerodynamic force and torque vectors depend upon the six motion variables, $(v, \alpha, \beta, P, Q, R)$, thereby affecting the translational dynamics due to rotational motion, and vice versa. Hence, there is an intricate coupling between translational and rotational dynamics (called *six-degrees-of-freedom* motion) through aerodynamic forces and moments, to take which fully into account is a difficult task. However, some high-performance aircraft, such as fighter airplanes, are capable of rapid rotations involving all the three body axes. In such a maneuver, the time scale is small enough for the changes in the airspeed and altitude (and thus the dynamic pressure and Mach number) to be neglected. Furthermore, the instantaneous direction of the velocity vector, $\mathbf{v}$, can be regarded as fixed during the rapid attitude maneuver, which implies that the changes in the angle of attack, α, and sideslip angle, β, are only due to the change in the aircraft attitude. Consequently, during a rapid aircraft rotation one can take the stability axes, $(\mathbf{i}_e, \mathbf{j}_e, \mathbf{k}_e)$, to be essentially aligned with the instantaneously fixed wind

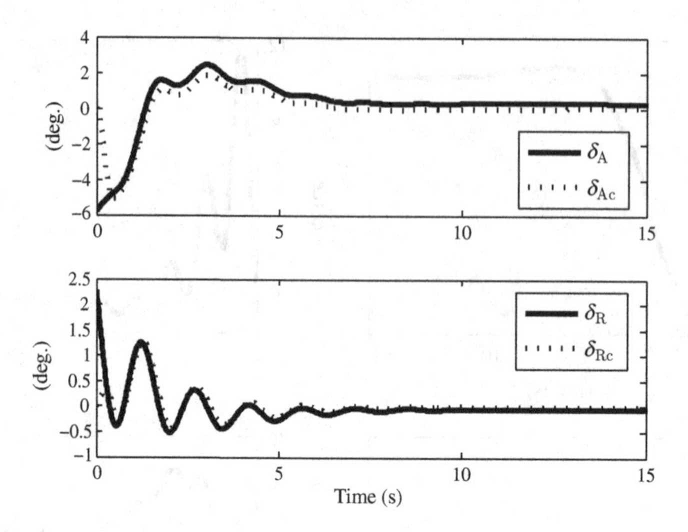

Figure 3.48 Closed-loop response of heading-hold autopilot for a fighter aircraft to a 5° initial sideslip disturbance (continued)

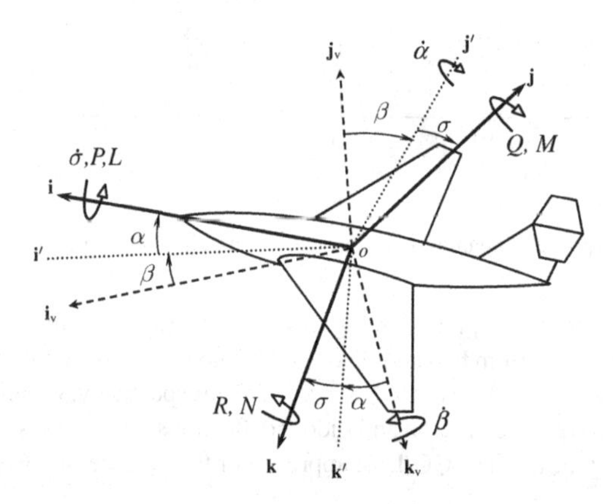

Figure 3.49 Coordinate frames for rapid rotation of a fighter aircraft

axes, $(\mathbf{i}_v, \mathbf{j}_v, \mathbf{k}_v)$, and used as a reference frame for measuring the aircraft attitude, (α, β, σ), as shown in Figure 3.49. However, now that the displaced attitude can be large, we refrain from using the traditional Euler angles that can have singularities during a large maneuver. Instead, we employ a non-singular attitude representation given by the attitude parameters vector, $\boldsymbol{\zeta}(t)$, such as the quaternion or the modified Rodrigues parameters (Tewari 2006).

The angular velocity, $\boldsymbol{\omega}$, and external torque, $\boldsymbol{\tau}$, vectors are resolved in the instantaneous body axes, $(\mathbf{i}, \mathbf{j}, \mathbf{k})$, as follows:

$$\begin{aligned}\boldsymbol{\omega} &= P\mathbf{i} + Q\mathbf{j} + R\mathbf{k} \\ \boldsymbol{\tau} &= L\mathbf{i} + M\mathbf{j} + N\mathbf{k}.\end{aligned} \tag{3.201}$$

The external torque can be divided into aerodynamic and control torques, wherein the aerodynamic part depends upon both the change in the attitude, (α, β), relative to the essentially constant velocity vector, and the angular velocity, (P, Q, R). Generally, the derivation of the aerodynamic torque for large and rapid changes in the airflow angles, (α, β), is a complex undertaking requiring *computational fluid dynamics*. For the sake of simplicity, we approach the problem of controlling large and rapid aircraft rotational dynamics simply by treating the external aerodynamic moments, (L, M, N), entirely as control inputs without considering precisely how they are physically generated. In this manner, we do not need to solve the complex aerodynamics problem *a priori*, but instead can develop the required aerodynamic forces and moments by post-processing the extremal trajectory and optimal control history.

The concepts developed in Chapter 2 for optimal terminal control can be applied to optimal control of an aircraft's rotational dynamics by taking into account aerodynamic moments and stability axes. Since the stability axis ox, being along an equilibrium flight direction, is not a principal axis, the product of inertia J_{xz} does not vanish, leading to inertia coupling among roll, pitch, and yaw motions. Hence, the rotational dynamics equations for an aircraft with reference to an equilibrium flight condition can be expressed as follows:

$$\begin{aligned} L &= J_{xx}\dot{P} + J_{xz}(\dot{R} + PQ) + (J_{zz} - J_{yy})QR, \\ M &= J_{yy}\dot{Q} + J_{xz}(R^2 - P^2) + (J_{xx} - J_{zz})PR, \\ N &= J_{zz}\dot{R} + J_{xz}(\dot{P} + QR) + (J_{yy} - J_{xx})PQ. \end{aligned} \tag{3.202}$$

Rapid, multi-axis rotation may be involved either in weapons pointing by fighter type aircraft, or in rapid recovery from unusual body rates and attitudes. The task of an optimal controller in such a case is to achieve a rapid change in the angular velocity and orientation of the aircraft's body-fixed stability axes. One can assume that the rotational dynamics is much faster than the translational dynamics, thereby allowing us to neglect linear velocity changes during the control interval. The rotational dynamics equations (3.202) in state-space form are thus

$$\begin{Bmatrix} \dot{P} \\ \dot{Q} \\ \dot{R} \end{Bmatrix} = \mathbf{f}(P, Q, R, L, M, N), \tag{3.203}$$

where

$$\mathbf{f} = \begin{Bmatrix} \frac{J_{xz}(J_{yy}-J_{xx}-J_{zz})}{J_{xx}J_{zz}-J_{xz}^2} PQ + \frac{J_{xz}^2+J_{zz}(J_{yy}-J_{zz})}{J_{xx}J_{zz}-J_{xz}^2} QR + \frac{J_{zz}}{J_{xx}J_{zz}-J_{xz}^2} L - \frac{J_{xz}}{J_{xx}J_{zz}-J_{xz}^2} N \\ \frac{J_{xz}}{J_{yy}}(P^2 - R^2) + \frac{J_{zz}-J_{xx}}{J_{yy}} PR + \frac{1}{J_{yy}} M \\ \frac{J_{xz}(J_{zz}-J_{yy}-J_{xx})}{J_{xx}J_{zz}-J_{xz}^2} QR + \frac{J_{xz}^2+J_{xx}(J_{xx}-J_{yy})}{J_{xx}J_{zz}-J_{xz}^2} PQ + \frac{J_{xx}}{J_{xx}J_{zz}-J_{xz}^2} N - \frac{J_{xz}}{J_{xx}J_{zz}-J_{xz}^2} L \end{Bmatrix}.$$

The optimal control problem for a rapid aircraft rotation with initial conditions,

$$\boldsymbol{\omega}(0) = \boldsymbol{\omega}_0, \quad \boldsymbol{\zeta}(0) = \boldsymbol{\zeta}_0, \tag{3.204}$$

and final condition,

$$\boldsymbol{\omega}(t_f) = \boldsymbol{\omega}_f, \tag{3.205}$$

for a fixed time, t_f, is posed with a zero terminal cost, $\varphi = 0$, and a quadratic control cost, $\mathbf{u}^T\mathbf{R}\mathbf{u}$, resulting in the following Hamiltonian:

$$H = \mathbf{u}^T\mathbf{R}\mathbf{u} + \boldsymbol{\lambda}^T(t)\mathbf{h}. \tag{3.206}$$

The necessary conditions for optimality yield the following extremal control input:

$$H_{\mathbf{u}} = L_{\mathbf{u}} + \boldsymbol{\lambda}^T\mathbf{B} = \mathbf{0}, \tag{3.207}$$

where

$$\mathbf{B} = \frac{\partial \mathbf{h}}{\partial \mathbf{u}} = \begin{Bmatrix} \mathbf{f}_{\mathbf{u}} \\ \mathbf{0} \end{Bmatrix} \tag{3.208}$$

and

$$\mathbf{f}_{\mathbf{u}} = \begin{pmatrix} \frac{J_{zz}}{J_{xx}J_{zz}-J_{xz}^2} & 0 & -\frac{J_{xz}}{J_{xx}J_{zz}-J_{xz}^2} \\ 0 & \frac{1}{J_{yy}} & 0 \\ -\frac{J_{xz}}{J_{xx}J_{zz}-J_{xz}^2} & 0 & \frac{J_{xx}}{J_{xx}J_{zz}-J_{xz}^2} \end{pmatrix}.$$

Taking $\mathbf{R} = \text{diag}(R_1, R_2, R_3)$, we have the following optimal control torque:

$$\mathbf{u}^* = \begin{Bmatrix} L^* \\ M^* \\ N^* \end{Bmatrix} = -\frac{1}{2}\mathbf{R}^{-1}\mathbf{f}_{\mathbf{u}}^T\boldsymbol{\lambda}. \tag{3.209}$$

Adopting the *quaternion* (Tewari 2006) as the kinematical set of attitude parameters, $\boldsymbol{\zeta} = (\mathbf{q}^T, q_4)^T)$, we have

$$\frac{d\{\mathbf{q}, q_4\}^T}{dt} = \frac{1}{2}\boldsymbol{\Omega}\{\mathbf{q}^T(t), q_4(t)\}^T \tag{3.210}$$

or

$$\mathbf{g}(\boldsymbol{\omega}, \boldsymbol{\zeta}) = \frac{1}{2}\boldsymbol{\Omega}(\boldsymbol{\omega})\boldsymbol{\zeta}, \tag{3.211}$$

where

$$\boldsymbol{\Omega} = \begin{pmatrix} 0 & \omega_z & -\omega_y & \omega_x \\ -\omega_z & 0 & \omega_x & \omega_y \\ \omega_y & -\omega_x & 0 & \omega_z \\ -\omega_x & -\omega_y & -\omega_z & 0 \end{pmatrix}. \tag{3.212}$$

Thus the co-state equations (to be solved for Lagrange multipliers) can be written as

$$\dot{\boldsymbol{\lambda}} = -\mathbf{A}^T\boldsymbol{\lambda}, \tag{3.213}$$

where

$$\mathbf{A} = \frac{\partial \mathbf{h}}{\partial \mathbf{x}} = \begin{pmatrix} \mathbf{f}_\omega & \mathbf{f}_\zeta \\ \mathbf{g}_\omega & \mathbf{g}_\zeta \end{pmatrix}, \tag{3.214}$$

$$\mathbf{f}_\omega \doteq \frac{\partial \mathbf{f}}{\partial \boldsymbol{\omega}} = \begin{pmatrix} j_1 Q & j_1 P + j_2 R & j_2 Q \\ 2 j_3 P + j_4 R & 0 & -2 j_3 R + j_4 P \\ j_5 Q & j_5 P + j_6 R & j_6 Q \end{pmatrix}, \tag{3.215}$$

$$\begin{aligned} j_1 &= \frac{J_{xz}(J_{yy} - J_{xx} - J_{zz})}{J_{xx}J_{zz} - J_{xz}^2}, \\ j_2 &= \frac{J_{xz}^2 + (J_{yy} - J_{zz})J_{zz}}{J_{xx}J_{zz} - J_{xz}^2}, \\ j_3 &= \frac{J_{xz}}{J_{yy}}, \quad j_4 = \frac{J_{zz} - J_{xx}}{J_{yy}}, \\ j_5 &= \frac{J_{xz}^2 + (J_{xx} - J_{yy})J_{xx}}{J_{xx}J_{zz} - J_{xz}^2}, \\ j_6 &= \frac{J_{xz}(J_{zz} - J_{yy} - J_{xx})}{J_{xx}J_{zz} - J_{xz}^2}, \end{aligned} \tag{3.216}$$

$$\mathbf{f}_\zeta \doteq \frac{\partial \mathbf{f}}{\partial \boldsymbol{\zeta}} = \mathbf{0}, \tag{3.217}$$

$$\mathbf{g}_\omega \doteq \frac{\partial \mathbf{g}}{\partial \boldsymbol{\omega}} = \frac{1}{2}\begin{pmatrix} q_4 & -q_3 & q_2 \\ q_3 & q_4 & -q_1 \\ -q_2 & q_1 & q_4 \\ -q_1 & -q_2 & -q_3 \end{pmatrix}, \tag{3.218}$$

$$\mathbf{g}_\zeta \doteq \frac{\partial \mathbf{g}}{\partial \zeta} = \frac{1}{2}\mathbf{\Omega}(\omega). \tag{3.219}$$

Thus, we can write the co-state equations as

$$\dot{\boldsymbol{\lambda}} = \begin{Bmatrix} \dot{\boldsymbol{\lambda}}_\omega \\ \dot{\boldsymbol{\lambda}}_\zeta \end{Bmatrix} = -\begin{Bmatrix} \mathbf{f}_\omega^T \boldsymbol{\lambda}_\omega + \mathbf{g}_\omega^T \boldsymbol{\lambda}_\zeta \\ \mathbf{g}_\zeta^T \boldsymbol{\lambda}_\zeta \end{Bmatrix}. \tag{3.220}$$

A necessary terminal boundary condition is imposed on the Lagrange's multipliers associated with the kinematical parameters, $\boldsymbol{\lambda}_\zeta$:

$$\boldsymbol{\lambda}_\zeta^T(t_f) = \left(\frac{\partial \Phi}{\partial \zeta}\right)_{t=t_f} = \mathbf{0}. \tag{3.221}$$

The set of state and co-state equations, along with the boundary conditions given by equations (3.204), (3.205), and (3.221), constitute the 2PBVP for optimal inertia-coupled maneuvers with fixed terminal time and a smooth control vector.

Example 3.7 Consider a fighter aircraft with the following inertia tensor in stability body axes:

$$\mathbf{J} = \begin{pmatrix} 15000 & 0 & 500 \\ 0 & 36000 & 0 \\ 500 & 0 & 50000 \end{pmatrix} \text{ kg.m}^2.$$

It is desired to bring the aircraft to rest from an initial orientation, $\mathbf{q} = \mathbf{0}$, and a large initial angular velocity, $\omega = (1.0, 0.3, -0.02)^T$ rad/s, in exactly 2 seconds, while minimizing the required control effort.

We solve the 2PBVP for this optimal control problem through MATLAB's inbuilt code *bvp4c.m* which is called by the program listed in Table 3.6. The differential equations and the boundary conditions required by *bvp4c.m* are specified by the respective codes *airrotode.m* and *airrotbc.m*, listed in Table 3.7. The control cost coefficient matrix is selected as $\mathbf{R} = \mathbf{J}^{-1}$ and the resulting extremal trajectory is plotted in Figures 3.50 and 3.51, and the optimal control torques in Figure 3.52. The rapidly rolling aircraft is brought to rest with a maximum rolling moment 14 kN.m, maximum pitching moment 7.4 kN.m, and maximum yawing moment 202 kN.m. These magnitudes include the aerodynamic moments generated due to a rotating aircraft, and can be represented in a non-dimensional form as follows:

$$C_l = \frac{L}{\frac{1}{2}\rho v^2 S b} = 0.0091, \quad C_m = \frac{M}{\frac{1}{2}\rho v^2 S \bar{c}} = 0.0275, \quad C_n = \frac{N}{\frac{1}{2}\rho v^2 S b} = 0.1318,$$

where the atmospheric sea-level density, $\rho = 1.225$ kg/m^3, and airspeed, $v = 100$ m/s, are taken for the reference flight condition of the aircraft with wing planform area, $S = 22$ m^2, mean aerodynamic chord, $\bar{c} = 2$ m, and wing span, $b = 11.37$ kg/m^3. Figure 3.51 shows that there is little change in the vehicle's attitude while the de-rotation is performed, thereby validating our original assumption of virtually constant linear velocity vector during the control interval.

Table 3.6

```
% Calling program for solving the 2PBVP for optimal aircraft
% rotational maneuver by collocation method using MATLAB's
% intrinsic code 'bvp4c.m'.
% Requires 'airrotode.m' and 'airrotbc.m'.
% (c) 2009 Ashish Tewari
%  dy/dx=f(y,x); a<=x<=b
%  y(x=a), y(x=b): Boundary conditions
%  y(1:3,1)=omega (rad/s) (Angular velocity vector)
%  y(4:7,1)=q_1,q_2,q_3,q_4 (Quaternion)
%  y(8:14,1)=lambda (Lagrange multipliers vector)

global tf; tf=2; % Terminal time (s)
global J; J=[15000 0 500;0 36000 0;500 0 50000];%(kg-m^2)
global wi; wi=[1 0.3 -0.02]'; % Initial ang. velocity (rad/s)
global Qi; Qi=[0 0 0 1]'; % Initial quaternion
global R; R=inv(J); % Control cost coefficients
% Collocation points & initial guess follow:
solinit = bvpinit(linspace(0,tf,5),[zeros(3,1);
                  [0.125 0 0 sqrt(1-.125^2)]'; zeros(7,1)]);
% 2PBVP Solution by collocation method:
options=bvpset('Nmax',100);
sol = bvp4c(@airrotode,@airrotbc,solinit,options);
x = linspace(0,tf); % Time vector (s)
y = deval(sol,x); % Solution state vector
plot(x,y(1:3,:)),xlabel('Time (s)'),ylabel('\omega (rad/s)')
figure
plot(x,y(4:7,:)),xlabel('Time (s)'),ylabel('q,q_4')
figure
jx=J(1,1);jy=J(2,2);jz=J(3,3);jxz=J(1,3);
Jxz=J(1,3);
D=jx*jz-jxz^2;
Fu=[jz/D 0 -jxz/D; 0 1/jy 0; -jxz/D 0 jx/D];
u=-0.5*inv(Rbar)*Fu'*y(8:10,:);
plot(x,u/1000),xlabel('Time (s)'),ylabel('u (x1000 N-m)')
```

3.11 Summary

Aircraft navigation involves following a specific flight path between two given horizontal locations in the presence of winds and planetary rotation, and can be carried out by dividing the problem into two parts: (a) maneuvering the aircraft in the horizontal plane by banking; and (b) modulating thrust and lift by throttle and angle of attack in order to maintain a quasi-level flight at a constant airspeed. Once a nominal trajectory is derived for subsystem (a) by solving an optimal two-point boundary value problem (2PBVP), it is provided as a reference to subsystem (b), which is essentially a tracking system subservient to the subsystem (a). The 2PBVP for (a) can be solved either subject to additional constraints, or by flying a great circle route which is the shortest distance between any two points on the planetary surface.

Aircraft dynamics generally involves small perturbations from an equilibrium flight condition. Such perturbations can be separated into those that are confined to the plane of symmetry (longitudinal dynamics) and those that occur outside it (lateral-directional dynamics). The aerodynamic and propulsive forces and moments are approximated by a first-order Taylor series expansion about the equilibrium condition, resulting in linear coefficients called stability and control derivatives, which can be rendered

Table 3.7

```
% Program for specifying governing ODEs expressed as
% state equations for the 2PBVP (to be called by 'bvp4c.m')
% (c) 2009 Ashish Tewari
function dydx=airrotode(x,y)
global tf;
global J;
global Rbar;
P=y(1);Q=y(2);R=y(3);
q1=y(4);q2=y(5);q3=y(6);q4=y(7);
jx=J(1,1);jy=J(2,2);jz=J(3,3);jxz=J(1,3);
Jxz=J(1,3);
D=jx*jz-jxz^2;
j1=jxz*(jy-jx-jz)/D;
j2=(jxz^2+jz*(jy-jz))/D;
j3=jxz/jy; j4=(jz-jx)/jy;
j5=(jxz^2+jx*(jx-jy))/D;
j6=jxz*(jz-jy-jx)/D;
Fw=[j1*Q j1*P+j2*R j2*Q; 2*j3*P+j4*R 0 j4*P-2*j3*R
    j5*Q j5*P+j6*R j6*Q];
Gw=0.5*[q4 -q3 q2; q3 q4 -q1;
    -q2 q1 q4; -q1 -q2 -q3];
Gq=0.5*[0 R -Q P; -R 0 P Q;
        Q -P 0 R; -P -Q -R 0];
Fu=[jz/D 0 -jxz/D; 0 1/jy 0; -jxz/D 0 jx/D];
if x<tf
    u=-0.5*inv(Rbar)*(Fu')*y(8:10,1);
else
    u=zeros(3,1);
end
dydx(1:3,1)=[j1*P*Q+j2*Q*R+(u(1,1)*jz-u(3,1)*jxz)/D;
             j3*(P^2-R^2)+j4*P*R+u(2,1)/jy;
             j5*P*Q+j6*Q*R+(u(3,1)*jx/D-u(1,1)*jxz)/D];
 dydx(4:7,1)=0.5*Gq*y(4:7,1);
 dydx(8:14,1)=-[Fw' Gw'; zeros(4,3) Gq']*y(8:14,1);

% Program for specifying boundary conditions for the 2PBVP.
% (To be called by 'bvp4c.m')
function res=airrotbc(ya,yb)
global wi;
global Qi;
res=[ya(1:3,1)-wi
    ya(4:7,1)-Qi
    yb(1:3,1)
    yb(11:14,1)];
```

non-dimensional by means of an appropriate reference area, length, and the flight dynamic pressure. Control of longitudinal dynamics is carried out by engine throttle and elevator. While single-variable longitudinal control is possible, a more robust design consists of both the inputs applied simultaneously, resulting in an active damping of both phugoid and short-period modes. A similar situation exists for controlling the lateral-directional modes by aileron and rudder inputs, mainly used for directional control by regulating sideslip and yaw rate. However, due to an inherent coupling of roll, yaw, and sideslip

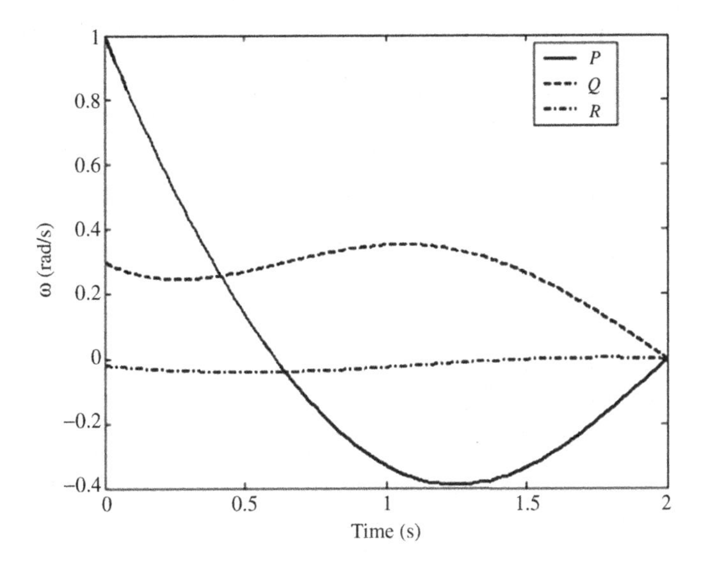

Figure 3.50 Extremal angular velocity for the multi-axis, rapid rotational maneuver of a fighter aircraft

motions, the aileron and rudder can be used individually as single inputs for controlling lateral-directional dynamics. Whereas the pure rolling and the Dutch-roll modes can be effectively controlled by single feedback loops, the spiral mode control (which is essential for a heading-hold autopilot) requires a multi-variable design approach. Large and rapid aircraft maneuvers require a nonlinear quaternion-based terminal controller that can be designed by the optimal (2PBVP) control techniques.

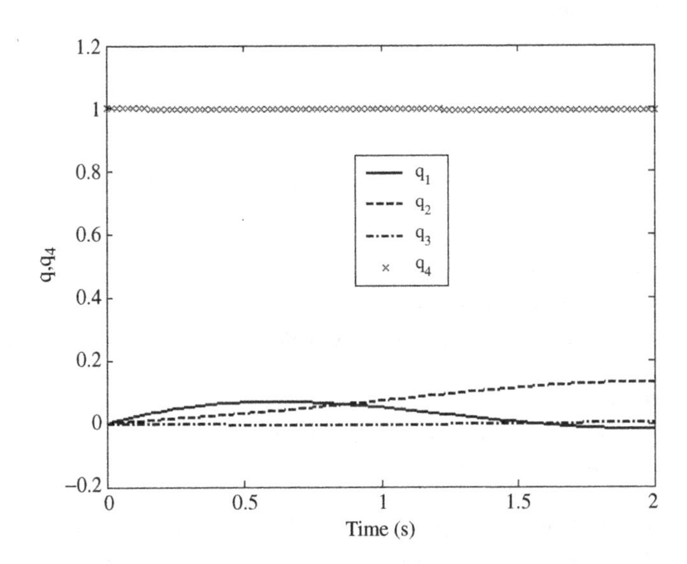

Figure 3.51 Extremal quaternion for the multi-axis, rapid rotational maneuver of a fighter aircraft

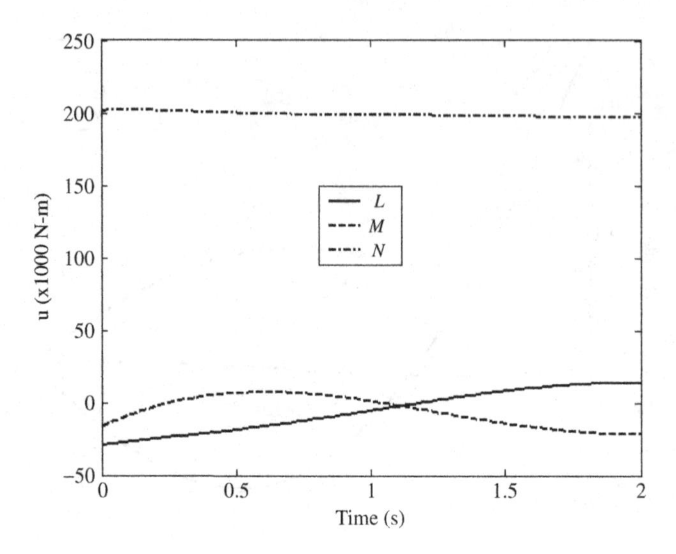

Figure 3.52 Optimal external torques for the multi-axis, rapid rotational maneuver of a fighter aircraft

Exercises

(1) Design an optimal cruise trajectory for an airplane flying in the Martian atmosphere at a constant altitude of 2 km, from $\delta = 20°$, $\ell = 0$ to $\delta = -30°$, $\ell = 20°$ with a constant airspeed, $v' = 120$ m/s, and initial velocity azimuth, $A_0 = 100°$. The flight has to be completed within 8 hr in the presence of a steady wind, $v_w = 10$ m/s, $A_w = 200°$. Assume a constant acceleration due to gravity, $g_0 = 3.71$ m/s^2, a constant surface radius, $R_0 = 3396.2$ km and a planetary rotation of period 24.6229 hr from west to east. The average atmospheric density at 2 km altitude can be computed from the following atmospheric model:

$$\rho = \frac{p}{191.8429\, T} \text{ (kg/m}^3\text{)},$$

where

$$T = 242 - 0.998\, h \text{ (K)}, \quad p = 699\, e^{-0.09h} \text{ (N/m}^2\text{)},$$

and h is in kilometers.

(2) Repeat Exercise 1 for a great circle route between the two points and plot the optimal bank angle history. What is the minimum flight time for the given data?

(3) Test the controllability of aircraft longitudinal dynamics plant with:
(a) elevator input alone;
(b) throttle input alone.

(4) Test the observability of the aircraft longitudinal dynamics plant with:
(a) pitch angle output alone;
(b) pitch rate output alone;
(c) normal acceleration output at sensor location, ℓ_z, from the center of mass along the ox axis.

(5) For safe aircraft landing in poor visibility conditions, it is necessary to have an autopilot that can follow a glideslope signal from a ground radio beacon in order to implement an instrument landing

system approach. After reaching a set height (called the decision height) above the runway, the pilot can either perform the landing manually or engage the automatic landing system to make a smooth touchdown at airspeed 10% above the stalling speed, that is, $h = 0$, $U = 1.1U_{\text{stall}}$ while making the sink rate, $-\dot{h} = -U \sin \Phi$, a small number (about 100 ft/min) that allows firm contact with the ground without bouncing. The plant for the instrument landing system approach must have the desired pitch angle, $\Theta_e = \Phi_e$, and the desired airspeed, $U = 1.1U_{\text{stall}}$, in the longitudinal coefficient matrices, **A**, **B**, which the pilot can set by a knob. Thus the autopilot is a set-point regulator for maintaining the state vector, $\mathbf{x} = (u, \Delta h, \alpha, \theta, q)^T = \mathbf{0}$, with the given equilibrium state, and can be designed in much the same way as in the case of altitude and airspeed hold autopilot. The observer can be based on pitch angle feedback from the glideslope sensor, airspeed measurement, and a radar altimeter provides feedback of the actual altitude above the runway. For the airplane of Example 3.3 and the LQG/LTR design example, design an automatic landing system for tracking a glideslope signal of $-3°$ down to the runway, with an initial approach speed, $U = 55$ m/s, and stalling speed at standard sea level, $U = 42.27$ m/s, using both elevator and throttle control inputs.

(6) Construct a Simulink® block diagram to simulate the multi-input airspeed control system of the LQG/LTR design example for a climb with a constant indicated airspeed, $U_i = 55$ m/s, beginning from the standard sea level to an altitude of 3 km. Use the following variation of atmospheric density in kg/m^3 with altitude (Tewari 2006):

$$\rho(h) = 1.225 \left(1 - \frac{0.0065h}{288.15}\right)^{4.258644}.$$

(7) Repeat Exercise 6 for a climb with a constant Mach number, $\mathcal{M} = 0.16$, beginning from the standard sea level to an altitude of 3 km. Use $\gamma = 1.4$, $\bar{R} = 287$ J kg^{-1} K^{-1}, and the following variation of atmospheric temperature in kelvin with altitude (Tewari 2006):

$$\bar{T}(h) = 288.15 - 0.0065h.$$

(8) Derive the feedforward gain equation (3.173) for an airspeed tracking system for a constant commanded airspeed, u_d (other perturbations being zero), and show that it does not have any solution unless $Z_T/m = M_T/m = 0$.

(9) Derive the lateral-directional control inputs for producing a steady, level, horizontal turn at a constant rate, r, without sideslip, beginning from a straight and level flight.

(10) It is often required to perform a steady-sideslip ($v =$ const.,$\beta =$ const.) maneuver in order to maintain a straight-line ground track in the presence of a cross-wind. What are the rudder and aileron inputs necessary for a steady sideslip, beginning from a straight and level flight?

(11) Aerobatic pilots often perform the *barrel-roll* maneuver, in which the airplane simultaneously rolls and sideslips in order to describe a cylinder in space with a horizontal axis (the inside of a horizontal barrel), while maintaining zero pitch and yaw rates. The roll and sideslip motions are synchronized such that $\beta = pR_0/U$, where R_0 is the constant radius of the barrel. Derive the equations of motion of the aircraft in a barrel-roll using the linearized lateral-directional dynamics model, and find the approximate transfer function between aileron and rudder inputs required for the maneuver.

(12) Test the observability of the aircraft lateral-directional dynamics plant with:
(a) bank angle output alone;
(b) yaw angle output alone;
(c) roll rate output alone;
(d) sideslip angle output alone;
(e) yaw rate output alone;
(f) lateral acceleration output, a_y, at sensor location, ℓ_y, from the center of mass along the ox axis.

(13) For an aircraft with the following lateral-directional dynamics in straight and level flight at altitude 11 km and airspeed 174 m/s, we have

$$\mathbf{A}_{\text{LD}} = \begin{pmatrix} 0 & 0 & 0 & 1 & 0 \\ 0 & 0 & 0 & 0 & 1 \\ 0.0732 & 0 & -0.1272 & 0 & -0.9909 \\ 0 & 0 & -2.0642 & -2.0936 & 0.4671 \\ 0 & 0 & 1.6298 & -0.0813 & -0.2698 \end{pmatrix},$$

$$\mathbf{B}_{\text{LD}} = \begin{pmatrix} 0 & 0 \\ 0 & 0 \\ 0 & -0.0316 \\ 4.7565 & -0.0830 \\ -0.0464 & 1.1867 \end{pmatrix}.$$

(a) Design a roll tracking system for achieving a desired bank angle in coordinated flight (i.e., with $\beta = 0$) by simultaneous aileron and rudder inputs and measurement of bank and sideslip angles.

(b) Use the code *airnav.m* (Table 3.5) to generate a nominal bank angle profile for flying the great circle route from New York JFK to London Heathrow in the presence of a steady wind of speed 20 m/s from the north-west.

(c) Construct a Simulink block diagram to simulate the closed-loop tracking system designed in part (a) in order to fly the nominal bank angle profile of part (b). Plot the aileron and rudder angle time histories for the flight.

(14) For the heading-hold autopilot designed in Example 3.6, find the smallest closed-loop settling time possible with the given initial condition, such that the allowable limits on aileron and rudder deflections are not exceeded. (*Hint: Try to increase the state-weighting terms in the design of the regulator, while keeping the observer gains unchanged.*)

(15) For the aircraft of Example 3.6, design an LQG/LTR yaw damper based on roll and yaw rate feedback and both aileron and rudder inputs such that the Dutch-roll mode damping ratio is doubled, while leaving its natural frequency unchanged. Simulate the closed-loop response due to a lateral gust that causes an initial sideslip perturbation of $\beta(0) = 5°$ at the given flight condition.

References

Åström, K.J. and Wittenmark, B. (1995) *Adaptive Control*. Addison-Wesley, Reading, MA.

Etkin, B. and Reid, L.D. (1995) *Dynamics of Flight: Stability and Control*. John Wiley & Sons, Inc., New York.

Hoak, D.E. *et al.* 1978: *USAF Stability and Control Datcom*. Air Force Flight Dynamics Laboratory, Wright-Patterson AFB.

Tewari, A. (2006) *Atmospheric and Space Flight Dynamics*. Birkhäuser, Boston.

Wertz, J.R. (ed.) (1978) *Spacecraft Attitude Determination and Control*. Kluwer Academic Publishers, Dordrecht.

4

Optimal Guidance of Rockets

4.1 Introduction

Rockets differ from aircraft and spacecraft due to the rapidly time-varying parameters of their equations of motion, which often requires special guidance and control design strategies. Furthermore, the fast response times required in both translation and rotation of rockets necessitate a much larger control loop bandwidth than that of either an aircraft or a spacecraft. The limited attitude maneuverability offered by thrust vector control is an additional aggravation for the rocket control engineer. Taken together with the nonlinear inertia-coupled rotational dynamics, static instability, and non-minimum phase behavior of rockets, these features make the control task a challenging one. However, there is a simplifying feature of rocket flight compared to aircraft, namely the lack of dependence of the essentially ballistic (zero-lift) trajectory on aerodynamic forces, which generally allows the lift and drag to be taken out of the controller design process.

We classify rockets as gravity-turn launch vehicles (including both satellite launch vehicles and long-range/intermediate-range ballistic missiles) and short-range missiles (including air-to-air, surface-to-air, and air-to-surface missiles). The flight of most rockets is confined to a vertical plane. This chapter is largely dedicated to guidance in a vertical plane. Other than discussing optimal guidance of short-range rockets in Section 4.2, we shall limit our treatment of rocket guidance to flight in a vertical plane.

Apart from the optimal trajectory generation that is crucial to rocket flight, the plant model for a tracking guidance system will be derived, and its stability, controllability, and observability examined. Then linear observer-based tracking guidance systems will be devised for gravity-turn rockets with the normal acceleration input. For attitude control, the gimballed nozzle and reaction jet actuator models will be developed, and separate roll, pitch, and yaw control systems will be designed and analyzed using multi-variable methods.

4.2 Optimal Terminal Guidance of Interceptors

We begin the study of rocket guidance with the general case of three-dimensional terminal guidance for interception of a maneuvering target. For our present purposes, we shall not distinguish between atmospheric and space flight and instead devise a method of generating reference trajectories for both. In doing so, the required acceleration inputs are generated either by aerodynamic means or by deflection of the rocket thrust (or both). Once the nominal trajectories are available, one can use them to derive control surface (or rocket nozzle) deflections by detailed aerodynamic and propulsive models that are discussed in the following sections. Since our focus is on translational motion, both target and interceptor

Advanced Control of Aircraft, Spacecraft and Rockets, Ashish Tewari.

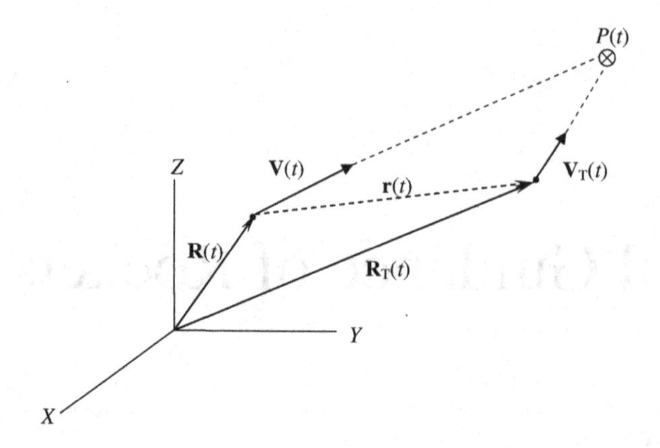

Figure 4.1 Geometry for three-dimensional interception of a maneuvering target with projected interception point, $P(t)$, and line-of-sight vector, $\mathbf{r}(t)$

are idealized as point masses, that is, their entire mass is assumed to be concentrated at their respective centers of mass.

The equations of motion of the interceptor and target in terms of their instantaneous position and velocity, $(\mathbf{R}, \mathbf{V})$ and $(\mathbf{R}_T, \mathbf{V}_T)$ respectively, measured in an inertial reference frame, XYZ (Figure 4.1), can be expressed as

$$\dot{\mathbf{R}} = \begin{Bmatrix} \dot{X} \\ \dot{Y} \\ \dot{Z} \end{Bmatrix} = \mathbf{V}, \tag{4.1}$$

$$\dot{\mathbf{R}}_T = \begin{Bmatrix} \dot{X}_T \\ \dot{Y}_T \\ \dot{Z}_T \end{Bmatrix} = \mathbf{V}_T, \tag{4.2}$$

$$\dot{\mathbf{V}} = \mathbf{u}, \tag{4.3}$$

$$\dot{\mathbf{V}}_T = \mathbf{u}_T, \tag{4.4}$$

where $\mathbf{u}(t) = [u_X(t), u_Y(t), u_Z(t)]^T$ is the acceleration input vector applied to the interceptor and $\mathbf{u}_T(t)$ is the target's acceleration vector assumed to be known perfectly at all times.

The interception problem is normally posed as guiding the interceptor such that the *line-of-sight* (LOS) vector,

$$\mathbf{r}(t) = \mathbf{R}_T(t) - \mathbf{R}(t),$$

vanishes at a given time, t_f, for a given set of initial conditions,

$$[\mathbf{R}(0), \mathbf{V}(0), \mathbf{R}_T(0), \mathbf{V}_T(0)].$$

Since only the relative motion between the target and the interceptor is important, we choose the state variables as the LOS vector and the *closure velocity* defined as

$$\mathbf{v}(t) = \mathbf{V}_{\mathrm{T}}(t) - \mathbf{V}(t).$$

Hence, a state-space representation of the relative motion is

$$\dot{\mathbf{x}} = \begin{pmatrix} \mathbf{0} & \mathbf{I} \\ \mathbf{0} & \mathbf{0} \end{pmatrix} \mathbf{x} + \begin{Bmatrix} \mathbf{0} \\ \mathbf{I} \end{Bmatrix} (\mathbf{u}_{\mathrm{T}} - \mathbf{u}), \tag{4.5}$$

where

$$\mathbf{x} = (\mathbf{r},\ \mathbf{v})^T. \tag{4.6}$$

The control task for the interceptor is to modify its velocity vector (by application of $\mathbf{u}(t)$) such that it moves toward a *projected interception point*, $P(t)$, as shown in Figure 4.1. Since the point $P(t)$ keeps on varying due to the maneuvering target, the control input, $\mathbf{u}(t)$, is determined from the solution of a boundary value problem with the terminal constraint, $\mathbf{r}(t_{\mathrm{f}}) = \mathbf{0}$.

We choose to pose the optimal control problem by the following choice of Hamiltonian:

$$H = \mathbf{u}^T\mathbf{R}\mathbf{u} + \boldsymbol{\lambda}^T(t)\mathbf{f}, \tag{4.7}$$

where $\mathbf{R}$ is symmetric and positive definite,

$$\mathbf{f} = \mathbf{A}\mathbf{x} + \mathbf{B}(\mathbf{u}_{\mathrm{T}} - \mathbf{u}), \tag{4.8}$$

and

$$\mathbf{A} = \begin{pmatrix} \mathbf{0} & \mathbf{I} \\ \mathbf{0} & \mathbf{0} \end{pmatrix}, \quad \mathbf{B} = \begin{Bmatrix} \mathbf{0} \\ \mathbf{I} \end{Bmatrix}. \tag{4.9}$$

The necessary conditions for optimality yield the extremal control input

$$H_{\mathbf{u}} = 2\mathbf{R}\mathbf{u}^T - \boldsymbol{\lambda}^T\mathbf{B} = \mathbf{0}, \tag{4.10}$$

leading to the optimal control input vector

$$\mathbf{u}^* = \frac{1}{2}\mathbf{R}^{-1}\mathbf{B}^T\boldsymbol{\lambda}. \tag{4.11}$$

A possible choice of controls cost parameters is for giving equal weighting to all the control input elements, that is, $\mathbf{R} = \rho\mathbf{I}$, where ρ is a positive real number. Thus, we have

$$\mathbf{u}^*(t) = \frac{1}{2\rho} \begin{Bmatrix} \lambda_4(t) \\ \lambda_5(t) \\ \lambda_6(t) \end{Bmatrix}, \tag{4.12}$$

where the Lagrange multipliers, $\boldsymbol{\lambda}(t)$, are determined from the solution of the following co-state equations:

$$\dot{\boldsymbol{\lambda}} = -\mathbf{A}^T\boldsymbol{\lambda} = \begin{Bmatrix} 0 \\ 0 \\ 0 \\ \lambda_1 \\ \lambda_2 \\ \lambda_3 \end{Bmatrix}. \tag{4.13}$$

The state equations (4.5) and the co-state equations (4.13) are to be integrated subject to the following boundary conditions:

$$\begin{aligned} \mathbf{r}(0) &= \mathbf{r}_0, \\ \mathbf{v}(0) &= \mathbf{v}_0, \\ \mathbf{r}(t_f) &= \mathbf{0}, \\ \lambda_4(t_f) &= 0, \\ \lambda_5(t_f) &= 0, \\ \lambda_6(t_f) &= 0. \end{aligned} \tag{4.14}$$

Clearly, the terminal boundary conditions on Lagrange multipliers yield the following closed-form solution:

$$\begin{aligned} \lambda_1(t) &= \text{const.}, \\ \lambda_2(t) &= \text{const.}, \\ \lambda_3(t) &= \text{const.}, \\ \lambda_4(t) &= -(t_f - t)\lambda_1, \\ \lambda_5(t) &= -(t_f - t)\lambda_2, \\ \lambda_6(t) &= -(t_f - t)\lambda_3, \end{aligned} \tag{4.15}$$

which, substituted into equation (4.12), yields

$$\mathbf{u}^* = -\frac{t_f - t}{2\rho}\begin{Bmatrix} \lambda_1 \\ \lambda_2 \\ \lambda_3 \end{Bmatrix}. \tag{4.16}$$

Note that the optimal control inputs vary linearly with $(t_f - t)$, called the *time to go*. Hence, the extremal trajectory can be obtained by direct integration of the state equations (4.5), into which the optimal control vector given by equation (4.16) has been substituted. Therefore, the two-point boundary value problem (2PBVP) consists of an iterative solution of the following initial value problem (IVP) for the extremal trajectory, $\mathbf{x}(t) = [\mathbf{r}^T(t), \mathbf{v}^T(t)]^T$:

$$\dot{\mathbf{x}} = \begin{pmatrix} \mathbf{0} & \mathbf{I} \\ \mathbf{0} & \mathbf{0} \end{pmatrix}\mathbf{x} + \begin{Bmatrix} \mathbf{0} \\ \mathbf{I} \end{Bmatrix}(\mathbf{u}_T - \mathbf{u}^*), \tag{4.17}$$

subject to the initial condition

$$\mathbf{r}(0) = \mathbf{r}_0, \quad \mathbf{v}(0) = \mathbf{v}_0, \tag{4.18}$$

and given the target acceleration, $\mathbf{u}_T(t)$, for the control interval $0 \leq t \leq t_f$. The IVP must be solved for various values of the constants, $(\lambda_1, \lambda_2, \lambda_3)$, until the terminal condition, $\mathbf{r}(t_f) = \mathbf{0}$, is satisfied.

Example 4.1 Consider a target moving with the following acceleration relative to a Cartesian frame fixed with respect to the ground-based radar station cum launcher of the surface-to-air interceptor missile:

$$\mathbf{u}_T(t) = \begin{Bmatrix} 1 \\ -2 \\ 0.1 \end{Bmatrix} \text{ m/s}^2.$$

The initial position and velocity of the target relative to the launcher are measured by the radar as follows:

$$\mathbf{R}_T(0) = \begin{Bmatrix} 500 \\ -600 \\ 500 \end{Bmatrix} \text{ m}, \quad \mathbf{V}_T(0) = \begin{Bmatrix} 100 \\ 100 \\ 10 \end{Bmatrix} \text{ m/s}.$$

The missile interceptor is powered by an engine with a propellant burn-out time of 50 s. If the interceptor is launched with an initial velocity of

$$\mathbf{V}(0) = \begin{Bmatrix} 150 \\ 40 \\ 5 \end{Bmatrix} \text{ m/s},$$

compute the optimal control inputs and the extremal trajectory for interception exactly at the missile burn-out time, $t_f = 50$ s.

In order to solve this problem, we write the MATLAB® code listed in Table 4.1 which utilizes the inbuilt solver, *bvp4c.m*, for 2PBVPs. The differential equations and boundary conditions required by *bvp4c.m* are specified by the respective codes *interceptode.m* and *interceptbc.m*, listed in Table 4.2. The results for $\rho = 1$ are plotted in Figures 4.2–4.5. Note the interception occurs with a zero position error in the given terminal time, with both interceptor and target simultaneously achieving the final position, $X(50) = 6750$, $Y(50) = 1900$, $u_Z(0) = 1125$ m. As expected, the largest acceleration input magnitudes are those at time $t = 0$, $u_X(0) = -0.9$, $u_Y(0) = -0.12$, $u_Z(0) = 1.05$ m/s^2, thereby yielding the constant Lagrange multipliers $\lambda_1 = 0.036$, $\lambda_2 = 0.0048$, and $\lambda_3 = -0.042$.

4.3 Non-planar Optimal Tracking System for Interceptors: 3DPN

In this section, we shall prove an important result, namely that the optimal tracking system for a general interception problem leads to *three-dimensional proportional navigation* (3DPN), which we briefly discussed in Chapter 1.

The job of the tracking guidance scheme is to provide closed-loop velocity commands that can minimize the relative distance between the target and the interceptor. Consider a missile with instantaneous inertial position, R(t), as shown in Figure 4.1. For intercepting a maneuvering target with inertial position and

Table 4.1

```
% Calling program for solving the 2PBVP for optimal missile
% guidance for interception of a maneuvering target by
% collocation method using MATLAB's intrinsic code 'bvp4c.m'.
% Requires 'interceptode.m' and 'interceptbc.m'.
% (c) 2009 Ashish Tewari
%  dy/dx=f(y,x); a<=x<=b
%  y(x=a), y(x=b): Boundary conditions
%  y(1:3,1)=omega (rad/s) (Angular velocity vector)
%  y(4:7,1)=q_1,q_2,q_3,q_4 (Quaternion)
%  y(8:14,1)=lambda (Lagrange multipliers vector)
global tf; tf=50; %Terminal time (s)
global r0; r0=[500 -600 500]'; %Initial relative position
global v0; v0=[-50 60 5]'; %Initial relative velocity
global R; R=1; %Control cost parameter
solinit = bvpinit(linspace(0,tf,50),[r0' v0' 0 0 0]);
sol = bvp4c(@interceptode,@interceptbc,solinit);
x = linspace(0,tf);
y = deval(sol,x);
plot(x,y(1:3,:)),xlabel('Time (s)'),ylabel('r (m)')
figure
V0=[100;100;10];uT=[1;-2;0.1];
rT=r0*ones(size(x))+V0*x+0.5*uT*(x.*x);
r=rT-y(1:3,:);
plot(x,rT,':',x,r),xlabel('Time (s)'),ylabel('(m)')
figure
plot(x,y(4:6,:)),xlabel('Time (s)'),ylabel('v (m/s)')
figure
u1=-y(7,:).*(tf-x)/(2*R);
u2=-y(8,:).*(tf-x)/(2*R);
u3=-y(9,:).*(tf-x)/(2*R);
plot(x,[u1; u2; u3]),xlabel('Time (s)'),ylabel('u (m/s^2)')
```

velocity, $[R_T(t), V_T(t)]$, the missile must be guided toward a projected interception point, $P(t)$, by giving it a velocity, $\mathbf{V}(t)$. Therefore, the velocity is the control input vector for the missile, whose kinematic equation of motion is simply

$$\frac{d\mathbf{R}}{dt} = \dot{\mathbf{R}} = \mathbf{V}. \tag{4.19}$$

How the velocity command is physically obeyed by the missile is presently not our concern, but should be taken into account while carrying out the closed-loop simulation. Short-range missiles (SRMs), such as surface-to-air, air-to-surface, and air-to-air missiles, are much more maneuverable than gravity-turn launch vehicles/long-range missiles, and thus require an entirely different guidance strategy. Furthermore, the SRM flight is not limited to the vertical plane. As opposed to liquid-fueled, long-range rockets, the SRMs are powered by solid-rocket motors that do not have any throttling capability. However, a few SRMs have a small second-stage ramjet engine for limited throttling and thus improved maneuverability in the terminal phase of flight.

The target's equations of motion are written as

$$\frac{d\mathbf{R}_T}{dt} = \dot{\mathbf{R}}_T = \mathbf{V}_T, \tag{4.20}$$

Table 4.2

```
% Program for specifying governing ODEs expressed as
% state equations for the 2PBVP (to be called by 'bvp4c.m')
% (c) 2009 Ashish Tewari
function dydx=interceptode(x,y)
global tf;
global R;
if x<tf
u=-y(7:9,1)*(tf-x)/(2*R);
else
    u=zeros(3,1);
end
vt=[100+x; 100-2*x; 10+0.1*x];
ut=[1 -2 0.1]';
dydx=[y(4:6,1)
       ut-u
       0
       0
       0];

% Program for specifying boundary conditions for the 2PBVP.
% (To be called by 'bvp4c.m')
function res=interceptbc(ya,yb)
global r0;
global v0;
res=[ya(1:3,1)-r0
      ya(4:6,1)-v0
      yb(1:3,1)];
```

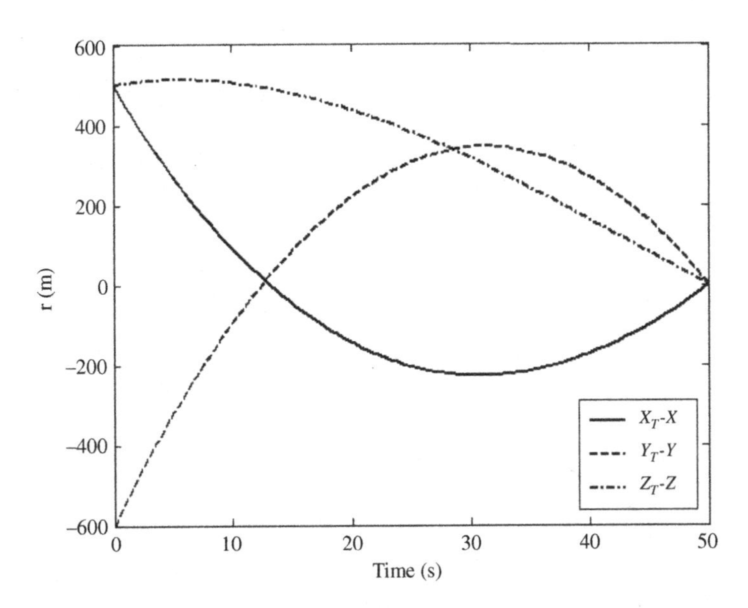

Figure 4.2 Extremal position of the target relative to the interceptor for the three-dimensional optimal interception problem

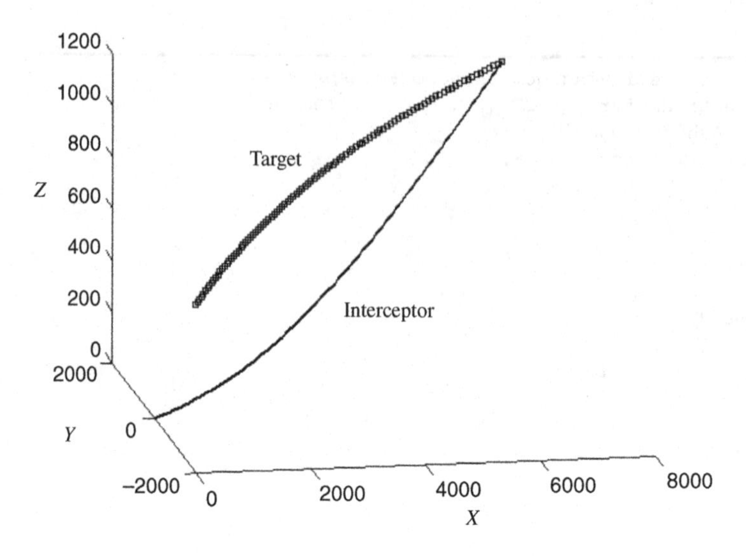

Figure 4.3 Extremal positions of the target and interceptor relative to the inertial radar station for the three-dimensional interception problem

$$\frac{d\mathbf{V}_T}{dt} = \dot{\mathbf{V}}_T = \mathbf{a}_T, \tag{4.21}$$

where $\mathbf{a}_T(t)$ is the target's acceleration. The motion of the target is assumed to be known by radar measurement of its position and velocity.

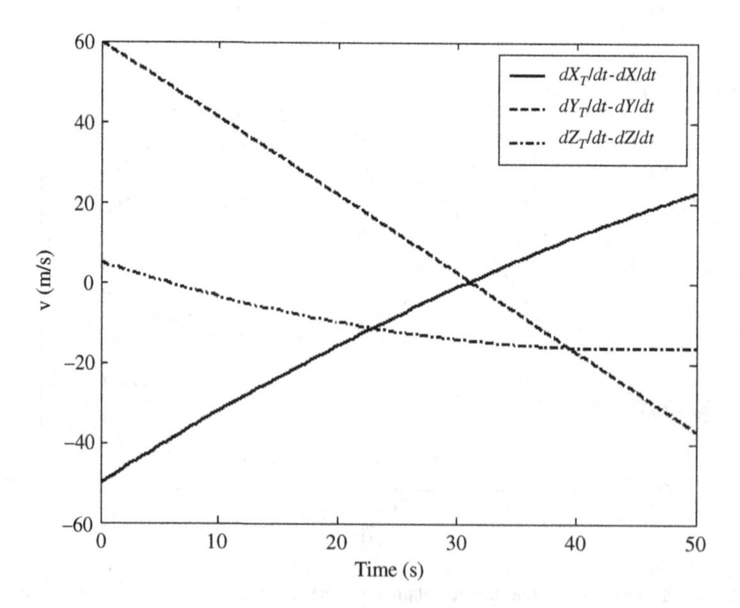

Figure 4.4 Extremal velocity of the target relative to the interceptor for the three-dimensional interception problem

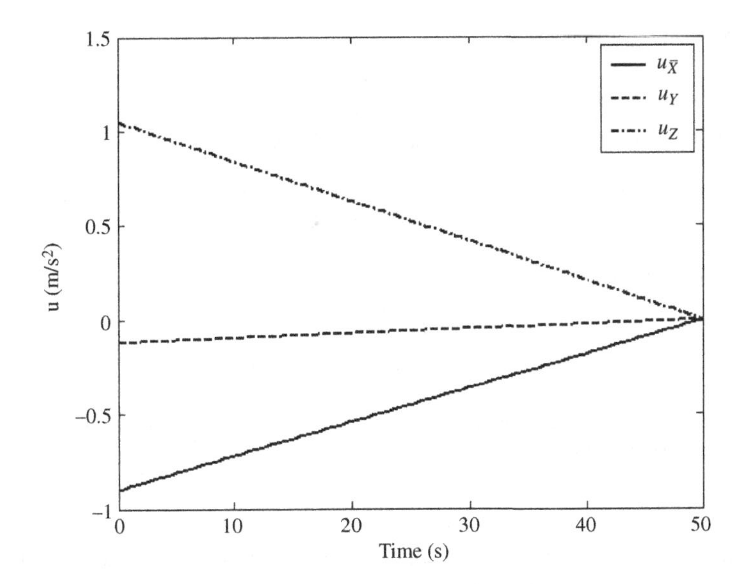

Figure 4.5 Optimal acceleration inputs to the interceptor for the three-dimensional interception problem

For a guidance system to be successful, the instantaneous position of the missile relative to the target,

$$\mathbf{r} = \mathbf{R} - \mathbf{R}_{\mathrm{T}}, \tag{4.22}$$

must be known with some precision. Such information can be provided by ground-based radar, airborne radar, or a radar/infrared/laser homing device mounted on the missile itself. By subtracting equation (4.20) from equation (4.19), we have the relative velocity vector given by

$$\mathbf{v} = \dot{\mathbf{r}} = \mathbf{V} - \mathbf{V}_{\mathrm{T}}. \tag{4.23}$$

In a state-space form, the equation of relative motion is thus given by

$$\dot{\mathbf{X}} = \mathbf{A}\mathbf{X} + \mathbf{B}\mathbf{u}, \tag{4.24}$$

where

$$\begin{aligned} \mathbf{X} &= \mathbf{r}, \\ \mathbf{u} &= \mathbf{v}, \\ \mathbf{A} &= \mathbf{0}, \\ \mathbf{B} &= \mathbf{I}. \end{aligned} \tag{4.25}$$

This is an important result showing that the exact relative motion plant is linear and time-invariant, irrespective of the magnitude of position and velocity errors, $(\mathbf{r}, \mathbf{v})$. Furthermore, the plant $(\mathbf{A}, \mathbf{B})$ is unconditionally controllable. Hence, one can directly apply a standard linear control methodology to the missile guidance problem.

Consider the full-state feedback, infinite-time, optimal linear quadratic regulator (LQR) approach (Chapter 2) with the state feedback control law

$$\mathbf{u} = -\mathbf{K}\mathbf{X}, \tag{4.26}$$

with the regulator gain matrix obeying

$$\mathbf{K} = \mathbf{R}^{-1}\mathbf{B}^T\mathbf{S}, \tag{4.27}$$

where $\mathbf{S}$ is the symmetric, positive semi-definite solution to the following algebraic Riccati equation (ARE):

$$\mathbf{SA} + \mathbf{A}^T\mathbf{S} - \mathbf{SBR}^{-1}\mathbf{B}^T\mathbf{S} + \mathbf{Q} = \mathbf{0}. \tag{4.28}$$

Selecting the cost parameters as

$$\begin{aligned} \mathbf{Q} &= c^2\mathbf{I}, \\ \mathbf{R} &= \mathbf{I}, \end{aligned} \tag{4.29}$$

where $c > 0$, the unique, positive definite solution to the ARE (and thus the regulator gain matrix) is obtained in closed form as follows:

$$\mathbf{K} = \mathbf{S} = c\mathbf{I}. \tag{4.30}$$

Hence, by adjusting a single, positive scalar variable, c, according to the missile's physical capabilities, one can achieve a desired accuracy by the guidance law

$$\mathbf{V} = \mathbf{V}_\mathrm{T} - c\mathbf{r}. \tag{4.31}$$

Equation (4.31), which adjusts the missile's velocity vector above that of the target in direct proportion to the relative distance vector, is called *three-dimensional proportional navigation*, and forms the core of all short-range missile guidance systems. One can represent the 3DPN guidance system by the block diagram of Figure 4.6, in which the target's velocity is the feedforward reference command, and the proportional feedback closes the loop, with constant scalar gain c. Apart from its simplicity, the unique feature of 3DPN guidance is that, being the solution to the infinite-time, optimal control problem with an exact plant, it is guaranteed to be successful for indefinitely large initial position error, provided the commands are faithfully followed by the missile. However, the plant dynamics is never as simple as the pure integrator shown in Figure 4.6.

If the missile plant has the dynamics of a pure integrator (i.e., a plant with pole at origin, $s = 0$), exemplified by the scalar case,

$$\frac{\mathrm{d}R}{\mathrm{d}t} = V, \tag{4.32}$$

where $V(t)$ is regarded as the control input and $R(t)$ the output, then the proportional navigation law,

$$V = V_\mathrm{T} - cr = V_\mathrm{T} - c(R - R_\mathrm{T}), \tag{4.33}$$

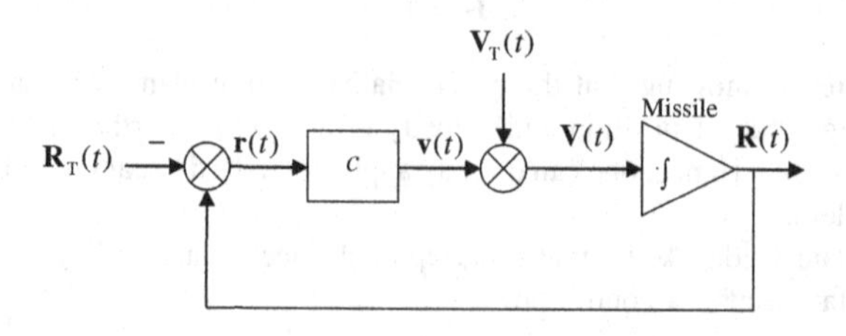

Figure 4.6 Block diagram of the ideal 3DPN system for interception of a maneuvering target

where $R_T(t)$ is the desired output, would converge to a zero miss distance for any value of the gain c. The basic reason for this behavior is the fact that a unit impulse input[1] applied to the plant given by equation (4.32), written as $V(t) = \delta(t)$, results in a unit step output, $R(t) = 1,\ t \geq 0$, which takes zero time to reach the desired displacement of unity. However, any physical object (such as a missile) has mass, m, that prevents it from changing its position instantaneously. Thus $V(t)$ is actually not the control input, but only a state variable that must also satisfy the kinetic equation,

$$m\frac{\mathrm{d}V}{\mathrm{d}t} = f, \tag{4.34}$$

where $f(t)$ is the force input applied to the plant. Hence, the plant transfer function, $R(s)/f(s)$, has a double pole at the origin, indicating an unstable system (Appendix B). Now a unit impulse input, $f(t) = \delta(t)$, results in a ramp output, $R(t) = t/m,\ t \geq 0$, scaled by the factor $1/m$, implying that it takes m seconds to reach the desired unit displacement. Such a delay in the response to a unit impulse input is termed a *lag* (or inertia). We may also have forces of gravity and friction (atmospheric drag), which can further complicate the situation. Clearly, the proportional navigation law can be physically implemented by taking its time derivative,

$$f = m[a_T - c(V - V_T)], \tag{4.35}$$

where $a_T(t)$ is the target's acceleration obtained by taking the time derivative of $V_T(t)$. Unfortunately, the control law of equation (4.35) only ensures that the difference between the speeds (not the positions) of the two objects is minimized, which is likely to produce a situation where the interceptor moves parallel to the target without intercepting it. This is true in any case if the proportional gain, c, is made very large.

We would like to improve the navigation law by taking into account the interceptor's inertia dynamics. As a reasonable approximation, we neglect drag in comparison with the thrust of the rocket engine. Furthermore, the interceptor's acceleration due to gravity is considered negligible compared to acceleration by rocket thrust. Thus we have the following dynamic equation of relative motion:

$$\frac{\mathrm{d}^2\mathbf{r}}{\mathrm{d}t^2} = \frac{\mathbf{f}_T}{m} - \mathbf{a}_T, \tag{4.36}$$

where $m(t)$ is the interceptor mass. We redefine the state vector to include velocity error, $\mathbf{v}(t)$, and change the control vector to be the excess of thrust acceleration input over target's acceleration:

$$\begin{aligned}
\mathbf{X} &= (\mathbf{r},\ \mathbf{v})^T, \\
\mathbf{u} &= \frac{\mathbf{f}_T}{m} - \mathbf{a}_T, \\
\mathbf{A} &= \left(\begin{array}{c|c} \mathbf{0} & \mathbf{I} \\ \hline \mathbf{0} & \mathbf{0} \end{array}\right), \\
\mathbf{B} &= \left(\begin{array}{c} \mathbf{0} \\ \hline \mathbf{I} \end{array}\right).
\end{aligned} \tag{4.37}$$

The optimal control law for the infinite-time regulator problem is now modified as follows:

$$\mathbf{u} = -\mathbf{R}^{-1}\mathbf{B}^T\mathbf{S}\mathbf{X}, \tag{4.38}$$

[1] An impulse is the fastest possible input one can apply to a system.

where $\mathbf{S}$ is the symmetric, positive semi-definite solution to the following algebraic Riccati equation (ARE):

$$\mathbf{SA} + \mathbf{A}^T\mathbf{S} - \mathbf{SBR}^{-1}\mathbf{B}^T\mathbf{S} + \mathbf{Q} = \mathbf{0}. \tag{4.39}$$

Selecting the cost parameters

$$\mathbf{Q} = \left(\begin{array}{c|c} c_r^2\mathbf{I} & \mathbf{0} \\ \hline \mathbf{0} & c_v^2\mathbf{I} \end{array} \right),$$

$$\mathbf{R} = \mathbf{I},$$

where $c_r > 0$, $c_v > 0$, we write the ARE solution matrix as

$$\mathbf{S} = \left(\begin{array}{c|c} \mathbf{S}_1 & \mathbf{S}_2 \\ \hline \mathbf{S}_2 & \mathbf{S}_3 \end{array} \right), \tag{4.40}$$

which, substituted into equation (4.39), yields

$$\begin{aligned} \mathbf{S}_1 &= c_r\sqrt{2c_r + c_v^2}\mathbf{I}, \\ \mathbf{S}_2 &= c_r\mathbf{I}, \\ \mathbf{S}_3 &= \sqrt{2c_r + c_v^2}\mathbf{I}. \end{aligned} \tag{4.41}$$

Thus, the optimal feedback gain matrix is given by

$$\mathbf{K} = -(\mathbf{S}_2,\ \mathbf{S}_3) = -\left(c_r\mathbf{I},\ \sqrt{2c_r + c_v^2}\mathbf{I}\right). \tag{4.42}$$

The 3DPN guidance system with inertia dynamics is schematically depicted in Figure 4.7, where the target's acceleration is the feedforward reference input and the missile position and velocity feedback loops ensure tracking of target's position and velocity, respectively. When applied to an actual interceptor, the 3DPN scheme will be successful provided the neglected effects of drag and gravity are indeed small. When such effects are included, the terminal position error is not exactly zero, leading to a minimum (closest) distance between the target and the interceptor, called the *miss distance*. The interception is regarded to be successful if the miss distance is less than a given value (usually determined by the explosive strength of the interceptor warhead). For example, a surface-to-air missile can have a miss distance of tens of meters for an interception to be successful, while an air-to-air missile (which has a

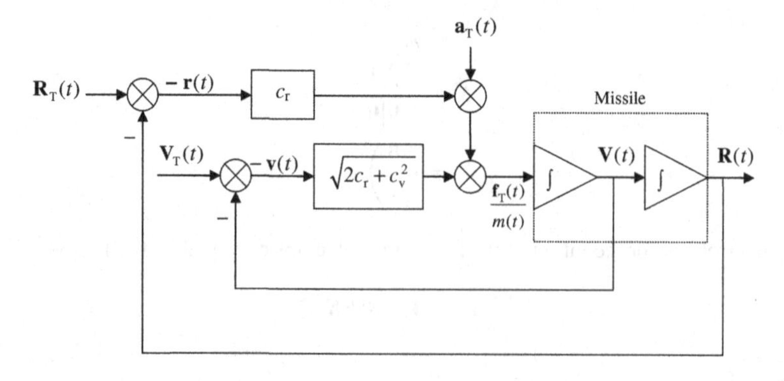

Figure 4.7 Block diagram of the 3DPN system including interceptor inertia

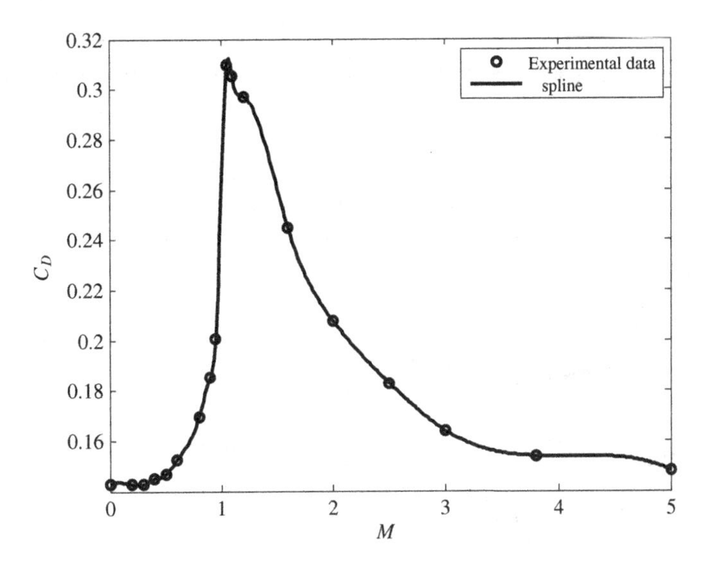

Figure 4.8 Variation of drag coefficient with Mach number of the surface-to-air missile

smaller warhead) must have a miss distance of only a few meters. In the following example, we will see how the interception accuracy is affected by the actual missile dynamics.

Example 4.2 Consider the interception of a maneuvering target by an advanced surface-to-air missile powered by a very responsive and high-thrust, hybrid rocket/ramjet engine. The missile has a burn-time of 20 s and a variation of its drag coefficient, C_D, at zero angle of attack – based upon a reference area, $S = 0.09$ m^2 – with Mach number, M, given by Figure 4.8. The atmospheric properties at a given altitude, Z – such as density, ρ, and speed of sound, a – that are required for the calculation of the drag,

$$D = \frac{1}{2}\rho V^2 S C_D,$$

are obtained from the standard atmosphere code, *atmosphere.m*, which is listed in Chapter 9 of Tewari (2006). The missile has initial mass of 600 kg and final burn-out mass of 80 kg, with a linear variation of mass with time. The target's acceleration time history is plotted in Figure 4.9 and the interceptor is launched at $t = 0$. Design a guidance system to achieve a miss distance of about 30 m with the given data.

We select, after some trial and error, the cost coefficients $c_r = 0.2$ and $c_v = 0.3$, implying $\sqrt{2c_r + c_v^2} = 0.7$. To simulate the flight of the interceptor by a fourth-order Runge–Kutta algorithm, we write the program listed in Table 4.3 which calls another program, *missile_state.m* (Table 4.4), for the state equations of the interceptor with mass and atmospheric drag taken into account. In order to post-process the computed closed-loop trajectory for acceleration inputs, drag, and thrust, we write code, *missile_state_u.m*, which is given in Table 4.5. The results of the simulation are plotted in Figures 4.10–4.14, showing that a miss distance of 30.17 m is achieved at approximately $t = 10$ s, with a maximum thrust requirement of 906 kN. While the required thrust may seem large for a conventional missile[2], it is well within the feasible *specific impulse* of 4000 s expected for a hybrid ramjet/rocket

[2] An existing missile closest in performance to the present example is the Russian SA-6 surface-to-air missile with a hybrid rocket/ramjet engine and a maximum flight speed of $M = 2.8$.

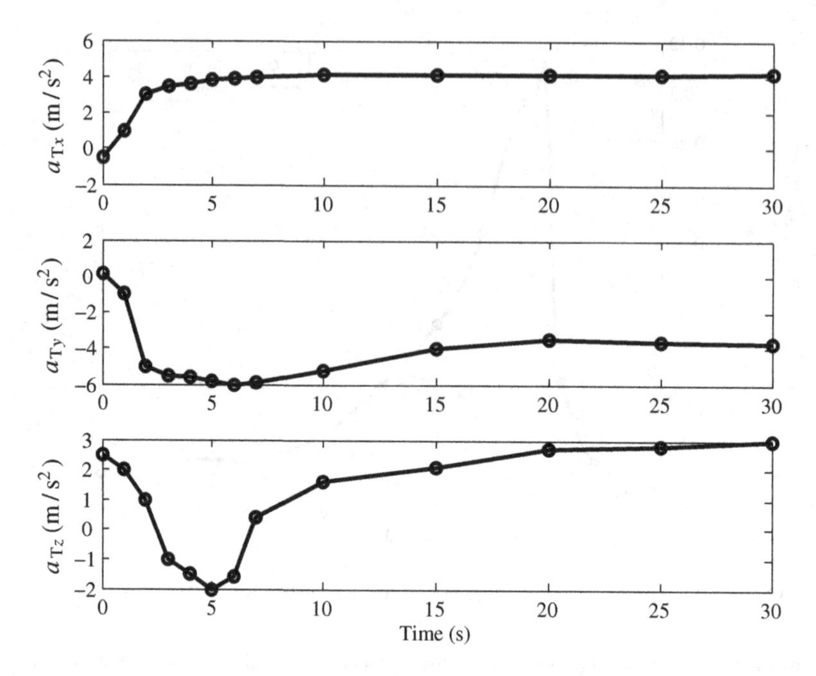

Figure 4.9 Acceleration history of a target for 3DPN interception by the surface-to-air missile

engine (Timnat 1990) in the Mach number range $2 \leq M \leq 5$. In the present case, the required specific impulse, I_{sp}, is calculated as follows:

$$I_{sp} = \frac{f_T}{\dot{m}g} = \frac{906013.4}{\frac{(600-80)}{30} \times 9.81} = 3552 \text{ s}.$$

The maximum speed achieved by the missile is 720 m/s ($M = 2.2$) at $t = 2.5$ s. Increasing c_r results in a smaller miss distance, but a larger thrust requirement.

4.4 Flight in a Vertical Plane

This section describes the translational motion of the center of mass of a flight vehicle in a vertical plane, $o_G y_G z_G$ (Figure 4.15), with radius $r(t)$, declination $\delta(t)$, speed $v(t)$, and flight-path angle $\phi(t)$. The control inputs generated by aerodynamic and propulsive means are the forward acceleration, $u_1(t)$, and the normal acceleration, $u_2(t)$. The duration of rocket flight is generally small enough for the effects of planetary rotation to be appreciable, apart from providing an initial velocity impulse to the vehicle. Therefore, the equations of motion, referred to a non-rotating, spherical planet of radius R_0, are written as follows (Chapter 3):

$$\dot{\delta} = \frac{v \cos \phi}{r}, \tag{4.43}$$

$$\dot{r} = v \sin \phi, \tag{4.44}$$

$$\dot{v} = u_1 - g \sin \phi, \tag{4.45}$$

Table 4.3

```
global T;T=[0 1 2 3 4 5 6 7 10 15 20 25 30]';
global aT;aT=[-0.5 1 3 3.5 3.6 3.8 3.9 4 4.11 4.12 4.13 4.15 4.2
              0.1 -1 -5 -5.5 -5.6 -5.8 -6.0 -5.9 -5.2 -4 -3.5 -3.6 -3.7
              2.5 2 1 -1 -1.5 -2 -1.6 0.4 1.6 2.1 2.7 2.8 3]';
global kr;kr=.2;
global kv;kv=.7;
global M; M = [0; 0.2; 0.3; 0.4; 0.5; 0.6; 0.8; 0.9; 0.95; 1.05; 1.1; 1.2;
               1.6; 2.0; 2.5; 3; 3.8; 5; 10; 99];
global CD;CD=[0.14264  0.14264  0.142728  0.145008  0.146688  0.152504 ...
              0.169688   0.185448   0.20044    0.309536 0.30512  0.297168 ...
              0.244784  0.207712  0.182912   0.163816   0.153904   0.1482 ...
              0.144952  0.144952];
global S; S =  0.09; % reference base area, m2
global tb; tb=20; % burn-time (s)
global m0; m0=600; % initial mass (kg)
global mf; mf=80; % burn-out mass (kg)
[t,x]=ode45('missile_state',[0 20],[1000 2000 3000 100 -50 10 0 0 0 0 0 0]');

n=size(t,1);
v=[];d=[];f=[];miss=[];
for i=1:n
      v(i,:)=norm(x(i,10:12));
end

[u,D,fT]=missile_state_u(t,x);

for i=1:n-1
         d(i,:)=norm(D(i,:)); % drag
         f(i,:)=norm(fT(:,i)); % thrust magnitude
         miss(i,:)=norm(x(i,7:9)-x(i,1:3)); % miss distance
end
```

$$v\dot{\phi} = u_2 + \left(\frac{v^2}{r} - g\right)\cos\phi. \tag{4.46}$$

The two acceleration inputs are related to the lift, $\mathcal{L}(t)$, drag, $\mathcal{D}(t)$, thrust, $f_T(t)$, thrust deflection angle, $\mu_1(t)$, and mass, $m(t)$ by the following equations:[3]

$$u_1 = \frac{f_T \cos\mu_1 - \mathcal{D}}{m}, \tag{4.47}$$

$$u_2 = \frac{f_T \sin\mu_1 + \mathcal{L}}{m}, \tag{4.48}$$

where the lift and drag are related to the angle of attack, $\alpha(t)$, and non-dimensional flow parameters – Mach number, M, Reynolds number, Re, and Knudsen number, Kn – through the non-dimensional lift

[3] The terminology for thrust deflection angles is explained in detail in Chapter 5. A deflection angle, μ_2, produces acceleration in a direction normal to the vertical plane, perpendicular to that created by the angle μ_1.

Table 4.4

```
function xdot=missile_state(t,x);
global T; % (nx1) vector of time points
global aT; % (nx3) matrix of target acceleration data
global kr;
global kv;
global M;
global CD;
global S;
global tb;
global m0;
global mf;
g=9.81;
rT=x(1:3,1);
vT=x(4:6,1);
rm=x(7:9,1);
vm=x(10:12,1);
ad=interp1(T,aT,t);
xdot(1:3,1)=vT;
xdot(4:6,1)=ad';
r=rm-rT;
v=vm-vT;
u=-kr*r+ad'-kv*v;
xdot(7:9,1)=vm;
m=m0-(m0-mf)*t/tb;
h=rm(3,1);
if h<0
    h=0;
    xdot(10:12,1)=zeros(3,1);
else
vel=norm(vm);
atmosp = atmosphere(h,vel,1.0);
rho = atmosp(2);
mach = atmosp(3);
Cd=interp1(M,CD,mach);
xdot(10:12,1)=u-[0 0 g]'-0.5*rho*norm(vm)*vm*S*Cd/m;
end
```

and drag coefficients, C_L and C_D, respectively, according to

$$\mathcal{L} = \frac{1}{2}\rho v^2 S C_L(\alpha, M, Re, Kn), \tag{4.49}$$

$$\mathcal{D} = \frac{1}{2}\rho v^2 S C_D(\alpha, M, Re, Kn). \tag{4.50}$$

For a rocket, even a moderate angle of attack can cause the transverse structural loads to quickly build up to destructive levels. Therefore, $\alpha(t) = 0$ (thus $\mathcal{L} = 0$) must be maintained by an attitude control system. Hence, $[f_T(t), \mu_1(t)]$ can be alternatively regarded as the control inputs, which will, however, necessitate the inclusion of the nonlinear constraint equations (4.47)–(4.50) in the control law derivation. Instead, we will derive control laws for $u_1(t)$ and $u_2(t)$, and then consider separately how these can be practically generated for a given vehicle. Such an approach of disregarding the aerodynamic forces in control derivation is valid for rockets and ballistic entry vehicles, because the drag – being the predominant

Table 4.5

```
function [u,D,fT]=missile_state_u(t,x);
global T; % (nx1) vector of time points
global aT; % (nx3) matrix of target acceleration data
global kr;
global kv;
global M;
global CD;
global S;
global tb;
global m0;
global mf;
g=9.81;
n=size(t,1);
u=[];D=[];fT=[];
for i=1:n-1
time=t(i,1);
timep1=t(i+1,1);
rT=x(i,1:3);
vT=x(i,4:6);
rm=x(i,7:9);
vm=x(i,10:12);
r=rm-rT;
v=vm-vT;
u(:,i)=-kr*r'-kv*v'+(x(i+1,10:12)-x(i,10:12))'/(timep1-time);
m=m0-(m0-mf)*time/tb;
atmosp = atmosphere(rm(1,3),norm(vm),1.0);
rho = atmosp(2);
mach = atmosp(3);
Cd=interp1(M,CD,mach);
drag=0.5*rho*norm(vm)*vm*S*Cd;
D(i,:)=drag;
fT(:,i)=-u(:,i)*m+[0 0 g]'+drag';
end
```

aerodynamic force – is essentially unaffected by control action in a zero-lift trajectory. Furthermore, the drag of a rocket with zero lift is quite small in comparison with its thrust. Thus, to all intents and purposes $u_1(t)$ and $u_2(t)$ are generated by thrust deflection, $[f_T(t), \mu_1(t)]$, while the angle of attack, $\alpha(t)$, is maintained near zero by a pitch attitude controller.

4.5 Optimal Terminal Guidance

Optimal control with end-point constraints can be applied to terminal guidance of rockets. For some simplified cases, well-known guidance laws are analytically derived. Other cases are posed as interesting boundary value problems that can be solved numerically. For the sake of simplicity, we confine our discussion to problems without interior constraints.

For the optimal guidance formulation, we will regard the radius, $r(t)$, declination, $\delta(t)$, speed, $v(t)$, and flight-path angle, $\phi(t)$, as state variables, and the forward acceleration input, $u_1(t)$, and normal acceleration input, $u_2(t)$, as the control inputs. For a spherical planet the acceleration due to gravity, g, is related to

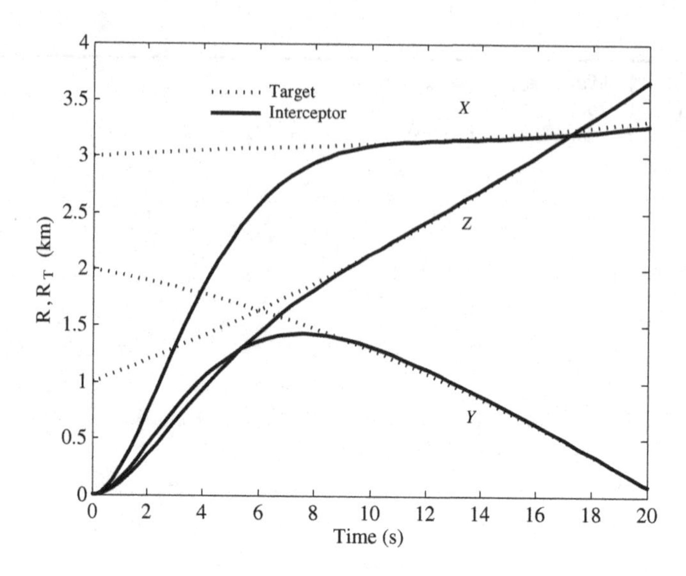

Figure 4.10 Position vectors of the target and the interceptor missile guided by a 3DPN system including interceptor inertia

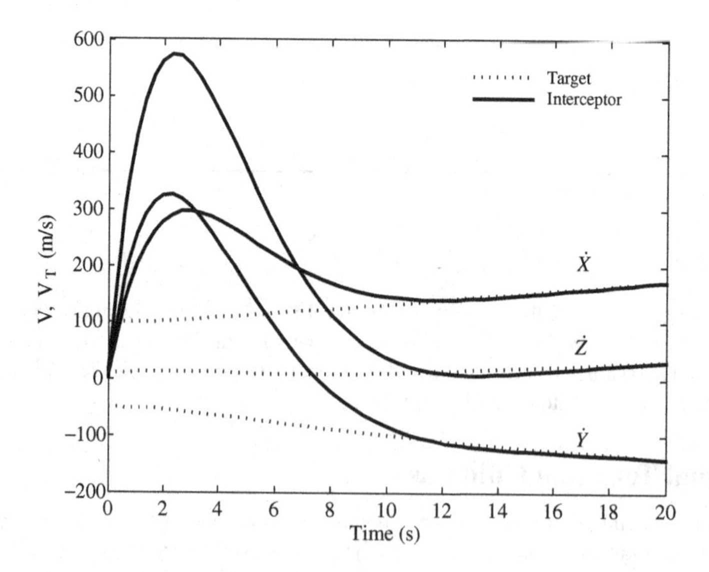

Figure 4.11 Velocity vectors of the target and the interceptor missile guided by a three-dimensional proportional navigation (3DPN) system including interceptor inertia

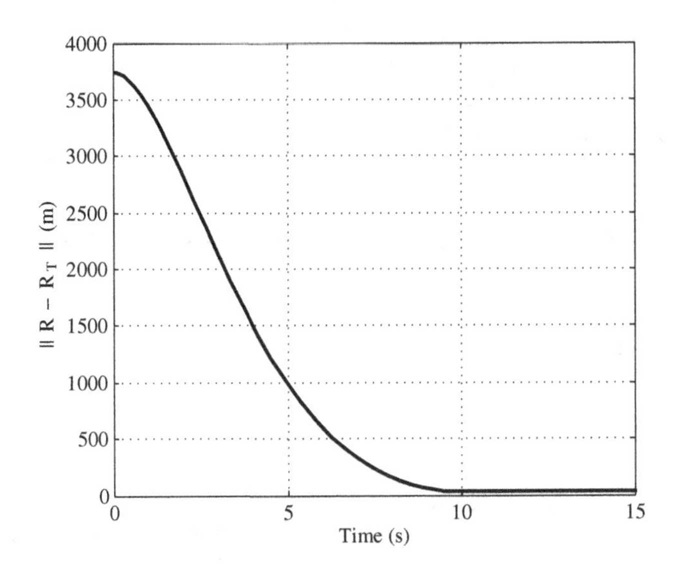

Figure 4.12 Distance between the target and the interceptor missile guided by a three-dimensional proportional navigation (3DPN) system including interceptor inertia

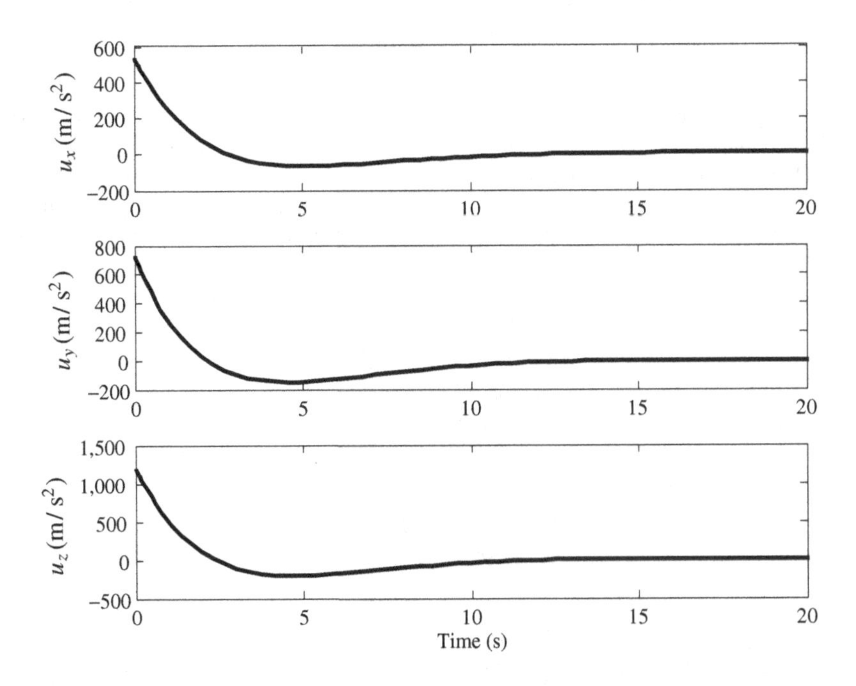

Figure 4.13 Thrust acceleration input components for interception by a three-dimensional proportional navigation (3DPN) system including interceptor inertia

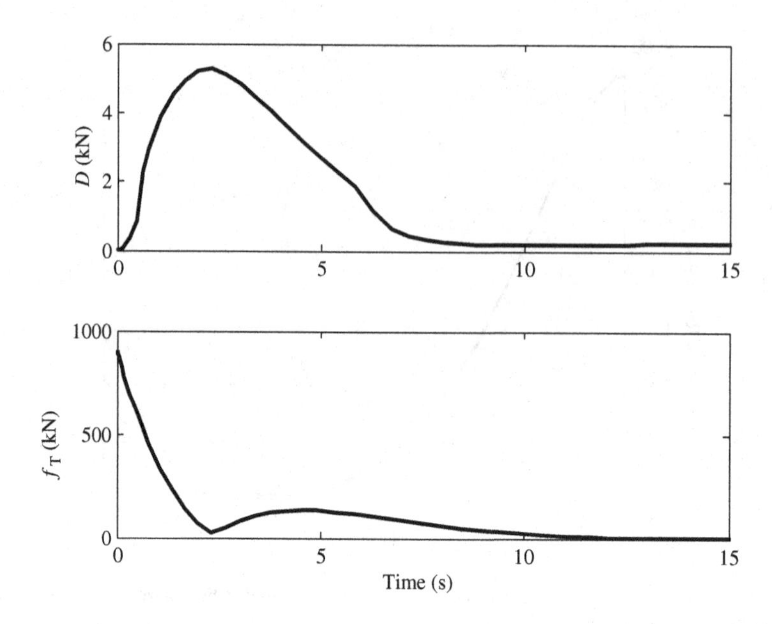

Figure 4.14 Drag and thrust magnitude for interception by a three-dimensional proportional navigation (3DPN) system including interceptor inertia

the radius by Newton's law of gravitation,

$$g = \frac{\mu}{r^2}, \tag{4.51}$$

where μ is the gravitational constant of the planet. One can directly substitute equation (4.51) into equations (4.45) and (4.46), thereby expressing the equations of motion entirely in terms of the four motion variables, (r, δ, v, ϕ). For the present, we impose no interior constraints on the two inputs, (u_1, u_2), whereas in practice they are generated by aerodynamic and propulsive forces that depend upon the motion variables. The forward acceleration input, u_1, is practically varied by changing the magnitude of the thrust,

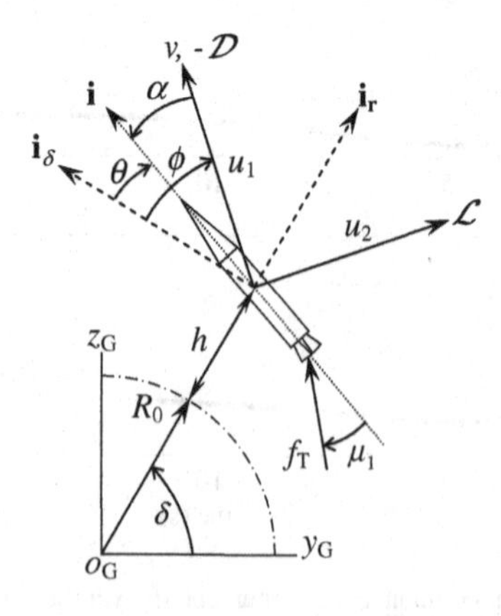

Figure 4.15 Flight of a rocket in a vertical plane

whereas the normal acceleration input, u_2, is controlled by varying the lift and/or the direction of the thrust relative to the velocity vector (called thrust vectoring). In this section, we assume that both the inputs are changed simultaneously and instantaneously as required, without modeling the mechanisms for doing so.

Consider optimization with quadratic cost and without terminal penalty, that is, $\varphi = 0$, $L = R_1 u_1^2 + R_2 u_2^2$ (Chapter 2). The Hamiltonian is thus $H = L + \boldsymbol{\lambda}^T \mathbf{f}$, where

$$\mathbf{f} = \begin{Bmatrix} \frac{v}{r}\cos\phi \\ v\sin\phi \\ u_1 - \frac{\mu}{r^2}\sin\phi \\ \frac{u_2}{v} + \left(\frac{v}{r} - \frac{\mu}{vr^2}\right)\cos\phi \end{Bmatrix}. \tag{4.52}$$

The necessary conditions for optimality yield the following extremal control input:

$$H_{\mathbf{u}} = L_{\mathbf{u}} + \boldsymbol{\lambda}^T \mathbf{B}, \tag{4.53}$$

where

$$\mathbf{B} = \frac{\partial \mathbf{f}}{\partial \mathbf{u}} = \begin{pmatrix} 0 & 0 \\ 0 & 0 \\ 1 & 0 \\ 0 & \frac{1}{v} \end{pmatrix},$$

which results in

$$2\begin{Bmatrix} R_1 u_1^* \\ R_2 u_2^* \end{Bmatrix} + \begin{Bmatrix} \lambda_3 \\ \frac{\lambda_4}{v} \end{Bmatrix} = \mathbf{0} \tag{4.54}$$

or

$$\begin{Bmatrix} u_1^* \\ u_2^* \end{Bmatrix} = -\begin{Bmatrix} \frac{\lambda_3}{2R_1} \\ \frac{\lambda_4}{2R_2 v} \end{Bmatrix}. \tag{4.55}$$

Substitution of the extremal control and trajectory yields the following set of governing equations for $\boldsymbol{\lambda}(t) = [\lambda_1(t), \lambda_2(t), \lambda_3(t), \lambda_4(t)]^T$:

$$\dot{\boldsymbol{\lambda}} = -\mathbf{A}^T \boldsymbol{\lambda}, \tag{4.56}$$

where

$$\mathbf{A} = \frac{\partial \mathbf{f}}{\partial \mathbf{x}} = \begin{pmatrix} 0 & -\frac{v\cos\phi}{r^2} & \frac{\cos\phi}{r} & -\frac{v\sin\phi}{r} \\ 0 & 0 & \sin\phi & v\cos\phi \\ 0 & \frac{2\mu\sin\phi}{r^3} & 0 & -\frac{\mu\cos\phi}{r^2} \\ 0 & \left(-\frac{v}{r^2} + \frac{2\mu}{vr^3}\right)\cos\phi & -\frac{u_2^*}{v^2} + \left(\frac{1}{r} + \frac{\mu}{v^2 r^2}\right)\cos\phi & -\left(\frac{v}{r} - \frac{\mu}{vr^2}\right)\sin\phi \end{pmatrix},$$

resulting in

$$\dot{\lambda}_1 = 0, \tag{4.57}$$

$$\dot{\lambda}_2 = \frac{v\cos\phi}{r^2}\lambda_1 - \frac{2\mu\sin\phi}{r^3}\lambda_3 - \lambda_4\left(\frac{2\mu}{vr^3} - \frac{v}{r^2}\right)\cos\phi, \tag{4.58}$$

$$\dot{\lambda}_3 = -\frac{\cos\phi}{r}\lambda_1 - (\sin\phi)\lambda_2 - \left[\frac{\lambda_4}{2R_2v^3} + \left(\frac{1}{r} + \frac{\mu}{v^2r^2}\right)\cos\phi\right]\lambda_4, \tag{4.59}$$

$$\dot{\lambda}_4 = \frac{v\sin\phi}{r}\lambda_1 - (v\cos\phi)\lambda_2 + \frac{\mu\cos\phi}{r^2}\lambda_3 + \lambda_4\left(\frac{v}{r} - \frac{\mu}{vr^2}\right)\sin\phi. \tag{4.60}$$

For an extremal trajectory, one must solve the 2PBVP comprising the state equations (4.43)–(4.46) along with the co-state equations (4.57)–(4.60), subject to the boundary conditions applied at $t = 0$ and $t = t_f$ specific to a particular flight application. We shall now consider some important flight applications of the present formulation.

4.6 Vertical Launch of a Rocket (Goddard's Problem)

The simplest case of rocket flight occurs for a vertical launch ($\phi = 90°$), which was studied by Robert Goddard for achieving the maximum possible altitude, $h(t) = r(t) - r_0$, above the planetary surface with a given propellant burn rate, $\beta(t)$. Considering a single-stage rocket with a constant exhaust speed, v_e, the ideal thrust can be simply expressed as

$$f_T = -\dot{m}v_e = \beta v_e. \tag{4.61}$$

In order to maintain vertical flight one must have $u_2 = 0$, which implies zero lift. Hence, the parasite drag coefficient, $C_{D0}(h, v)$ (based upon reference area, S), which depends upon altitude and airspeed due to prevailing flow parameters (e.g., Mach number, Reynolds number, Knudsen number), can be used to represent drag, and thus the forward acceleration, as follows:

$$u_1 = \frac{f_T - \mathcal{D}}{m} = \beta\frac{v_e}{m} - \frac{1}{2m}\rho v^2 SC_{D0}. \tag{4.62}$$

The following simple exponential atmospheric model can be proposed for vertical flight:

$$\rho = \rho_0 e^{-h/H}, \tag{4.63}$$

where the base density, ρ_0, and scale height, H, are constants. The equations of motion are finally written as

$$\dot{h} = \dot{r} = v, \tag{4.64}$$

$$\dot{v} = u_1 - g = \beta\frac{v_e}{m} - \frac{1}{2m}\rho_0 SC_{D0}e^{-h/H}v^2 - g_0\left(\frac{R_0}{R_0 + h}\right)^2, \tag{4.65}$$

$$\dot{m} = -\beta, \tag{4.66}$$

where g_0 is the acceleration due to gravity at planet's surface, $r = R_0$.

The optimal control problem for vertical launch (often referred to as *Goddard's rocket problem*) can be variously posed as maximization of either the final altitude or speed for given initial and final masses, and minimization of the initial mass for a given final mass (payload), given either a final altitude or a final speed. In any case, the final time, t_f, is considered fixed according to the burn program, resulting in

$$\beta = 0, \quad t > t_f. \tag{4.67}$$

Clearly, the burn rate can be regarded as the sole control input for Goddard's rocket problem,that is, $u(t) = \beta(t)$.

Consider optimization with a quadratic control cost and without any terminal penalty, that is, $H = Ru^2 + \boldsymbol{\lambda}^T \mathbf{f}$, where $R > 0$ and

$$\mathbf{f} = \begin{Bmatrix} v \\ \frac{v_e u}{m} - \frac{1}{2m}\rho_0 S C_{D0} e^{-h/H} v^2 - g_0 \left(\frac{R_0}{R_0 + h}\right)^2 \\ -u \end{Bmatrix}. \tag{4.68}$$

A necessary condition for optimality,

$$H_{\mathbf{u}} = L_{\mathbf{u}} + \boldsymbol{\lambda}^T \mathbf{B}, \tag{4.69}$$

where

$$\mathbf{B} = \frac{\partial \mathbf{f}}{\partial \mathbf{u}} = \begin{pmatrix} 0 \\ \frac{v_e}{m} \\ -1 \end{pmatrix},$$

yields the extremal control input

$$u^* = -\frac{v_e}{2Rm}\lambda_2 + \frac{1}{2R}\lambda_3. \tag{4.70}$$

The co-state equations for the optimal vertical launch can be expressed for the Lagrange multiplier vector, $\boldsymbol{\lambda}(t) = [\lambda_1(t), \lambda_2(t), \lambda_3(t)]^T$, as

$$\dot{\boldsymbol{\lambda}} = -\mathbf{A}^T \boldsymbol{\lambda}, \tag{4.71}$$

where

$$\mathbf{A} = \frac{\partial \mathbf{f}}{\partial \mathbf{x}} = \begin{pmatrix} 0 & 1 & 0 \\ a_{21} & a_{22} & a_{23} \\ 0 & 0 & 0 \end{pmatrix},$$

with

$$a_{21} = \frac{\rho_0 S e^{-h/H} v^2}{2m}\left[\frac{C_{D0}}{H} - \frac{\partial C_{D0}}{\partial h}\right] + g_0 \frac{R_0^2}{(R_0+h)^3},$$

$$a_{22} = -\frac{\rho_0 S e^{-h/H} v}{m}\left[C_{D0} + \frac{v}{2}\frac{\partial C_{D0}}{\partial v}\right],$$

$$a_{23} = -\frac{1}{m^2}\left[u^* v_e - \frac{\rho_0}{2} S C_{D0} e^{-h/H} v^2\right],$$

resulting in

$$\dot{\lambda}_1 = -\left[\frac{\rho_0 S e^{-h/H} v^2}{2m}\left(\frac{C_{D0}}{H} - \frac{\partial C_{D0}}{\partial h}\right) + g_0 \frac{R_0^2}{(R_0+h)^3}\right]\lambda_2, \tag{4.72}$$

$$\dot{\lambda}_2 = -\lambda_1 + \frac{\rho_0 S e^{-h/H} v}{m}\left(C_{D0} + \frac{v}{2}\frac{\partial C_{D0}}{\partial v}\right)\lambda_2, \tag{4.73}$$

$$\dot{\lambda}_3 = \frac{1}{m^2}\left(u^* v_e - \frac{\rho_0}{2} S C_{D0} e^{-h/H} v^2\right)\lambda_2. \tag{4.74}$$

The representative boundary conditions of Goddard's rocket problem could be posed as follows:

$$h(0) = 0, \quad \mathbf{v}(0) = \mathbf{v}_0, \quad m(0) = m_0,$$
$$m(t_f) = m_f, \quad \lambda_1(t_f) = 0, \quad \lambda_2(t_f) = 0.$$

Clearly, the terminal boundary conditions on the first two Lagrange multipliers yield the following closed-form solution:

$$\lambda_1(t) = 0, \quad \lambda_2(t) = 0, \tag{4.75}$$

which, substituted into equation (4.74), produces

$$\lambda_3(t) = \text{const.} \tag{4.76}$$

Thus, the optimal control input is the constant

$$u^* = \frac{1}{2R}\lambda_3. \tag{4.77}$$

The value of the constant λ_3 is either determined from the solution to the 2PBVP, or directly from state equation (4.66), as follows:

$$u^* = \frac{1}{2R}\lambda_3 = \frac{m_0 - m_f}{t_f}. \tag{4.78}$$

Therefore, Goddard's rocket problem possesses a closed-form solution for the constant, optimal control input (which is also the maximum possible value for a given final time, t_f). The extremal trajectory is obtained by merely solving the IVP associated with the state equations and state initial conditions.

Table 4.6

```
% Calling program for solving the 2PBVP for Goddard's
% rocket problem  by collocation method using MATLAB's
% intrinsic code 'bvp4c.m'.
% Requires 'goddardode.m' and 'goddardbc.m'.
% (c) 2009 Ashish Tewari
%  dy/dx=f(y,x); a<=x<=b
%  y(x=a), y(x=b): Boundary conditions
%  y(1,1)=altitude (m)
%  y(2,1)=speed (m/s)
%  y(3,1)=mass (kg)
%  y(4:6,1)=lambda (Lagrange multipliers vector)
global tf;tf=100; %terminal time
global R; R=1;

% Initial guess and 2PBVP solution follow:
solinit = bvpinit(linspace(0,tf,5),[100e3 0 2000 0 0 0]);
sol = bvp4c(@goddardode,@goddardbc,solinit);
x = linspace(0,tf);
y = deval(sol,x);
r0=6378.14e3;
g=9.81*r0^2./(r0+y(1,:)).^2;
ve=9.81*300;
% Optimal control:
u=0.5*(y(6,:)-ve*y(5,:)./y(3,:))/R;
```

Example 4.3 Consider the vertical launch of a rocket of initial mass, $m_0 = 20\,000$ kg, initial launch speed, $v_0 = 1$ m/s, final (payload) mass, $m_0 = 2000$ kg, and burn-out time, t_f. The rocket can be assumed to be single-stage (i.e., all the propellant is exhausted in one go) with a specific impulse, $v_e/g_0 = 300$ s, and a constant drag coefficient, $C_{D0} = 0.2$, based upon a reference area, $S = 2$ m^2. The Earth's atmosphere is assumed to have the following exponential model:

$$\rho_0 = 1.752 \text{ kg/m}^3, \quad H = 6700 \text{ m}. \tag{4.79}$$

Simulate the extremal trajectory of the rocket for three specific values of the burn-out time, $t_f = 30, 60, 100$ s.

Although it is not necessary to do so, we shall formally solve this problem as a 2PBVP with MATLAB's inbuilt solver, *bvp4c.m*, using the calling program listed in Table 4.6. The differential equations and the boundary conditions required by *bvp4c.m* are specified by the respective codes *goddardode.m* and *goddardbc.m*, listed in Table 4.7. The results are plotted in Figures 4.16–4.19.

4.7 Gravity-Turn Trajectory of Launch Vehicles

The ascent of a rocket launch vehicle from the planetary surface ($h = 0$) through the atmosphere to either a circular orbit ($h = h_f =$ const.; Figure 4.20), or a prescribed non-circular orbit [$h(t_f) = h_f$, $v(t_f) = v_f$, $\phi(t_f) = \phi_f$], in a fixed terminal time, t_f, usually follows a *gravity-turn trajectory* wherein the flight path curves naturally under the influence of gravity, without requiring any normal acceleration input ($u_2(t) = 0$). An attitude control system stabilizes the vehicle such that zero normal acceleration is maintained at all times. We shall demonstrate how such a trajectory can be generated as the extremal solution of the planar optimal guidance problem.

Table 4.7

```
% Program for specifying governing ODEs expressed as
% state equations for the Goddard rocket problem
% 2PBVP (to be called by 'bvp4c.m')
% (c) 2009 Ashish Tewari
function dydx=goddardode(x,y)
global tf;
global R;
g0=9.81;
r0=6378.14e3;
cd0=0.2;
rho0=1.752; H=6.7e3;
S=2;
ve=g0*300;
h=y(1);v=y(2);m=y(3);
lam1=y(4);lam2=y(5);lam3=y(6);
rho=rho0*exp(-h/H);
if m<2000
    u=0;
else
    u=0.5*(lam3-lam2*ve/m)/R;
end
dydx=[v
      u*ve/m-0.5*rho*v^2*S*cd0/m-g0*(r0/(r0+h))^2
     -u
     -(0.5*rho*v^2*S*cd0/(m*H)+g0*r0^2/(r0+h)^3)*lam2
     -lam1+rho*v*S*cd0*lam2/m
     (u*ve-0.5*rho*v^2*S*cd0)*lam2/m^2];

% Program for specifying boundary conditions for the
% Goddard rocket problem 2PBVP.
% (To be called by 'bvp4c.m')
function res=goddardbc(ya,yb)
res=[ya(1)
     ya(2)-1
     ya(3)-20000
     yb(3)-2000
     yb(4)
     yb(5)];
```

4.7.1 *Launch to Circular Orbit: Modulated Acceleration*

Consider a launch beginning from a planet's surface to a circular orbit of radius r_f with a fixed terminal time t_f. An attitude control system stabilizes the vehicle such that zero normal acceleration is maintained at all times ($u_2(t) = 0$), which requires $\lambda_4(t) = 0$. Furthermore, the final declination, $\delta(t_f)$, is generally unimportant for a launch vehicle, and is therefore unconstrained, which requires $\lambda_1(t_f) = 0$ (Chapter 2). Hence, equation (4.57) produces $\lambda_1(t) = 0$ which, substituted into equations (4.58)–(4.60), simplifies them considerably:

$$\lambda_1(t) = 0, \tag{4.80}$$

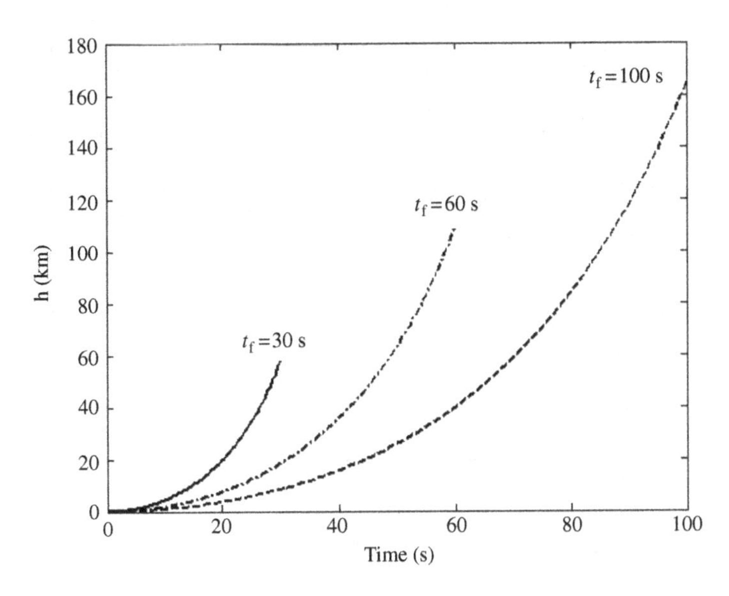

Figure 4.16 Extremal altitude of a vertically launched rocket for three different values of burn-out time, $t_f = 30, 60, 100$ s

$$\dot{\lambda}_2 = -\frac{2\mu \sin\phi}{r^3}\lambda_3, \tag{4.81}$$

$$\dot{\lambda}_3 = -(\sin\phi)\lambda_2, \tag{4.82}$$

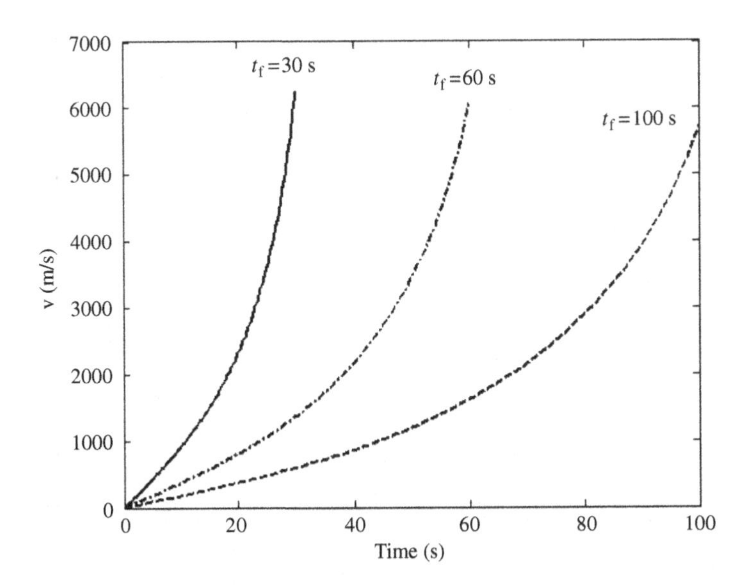

Figure 4.17 Extremal speed of a vertically launched rocket for three different values of burn-out time, $t_f = 30, 60, 100$ s

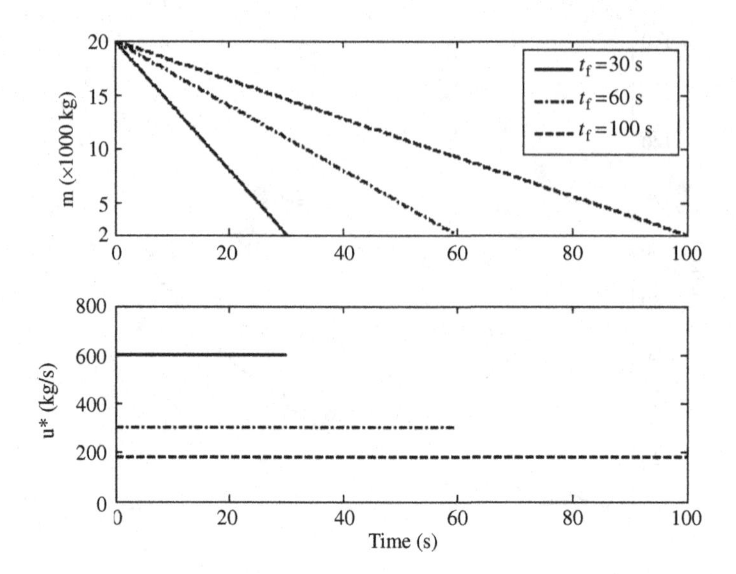

Figure 4.18 Extremal mass and optimal burn rate of a vertically launched rocket for three different values of burn-out time, $t_f = 30, 60, 100$ s

$$\lambda_4(t) = 0. \tag{4.83}$$

The boundary value problem for gravity-turn trajectory concerns a simultaneous integration of equations (4.81) and (4.82) along with the equations of motion

$$\dot{r} = v \sin\phi, \tag{4.84}$$

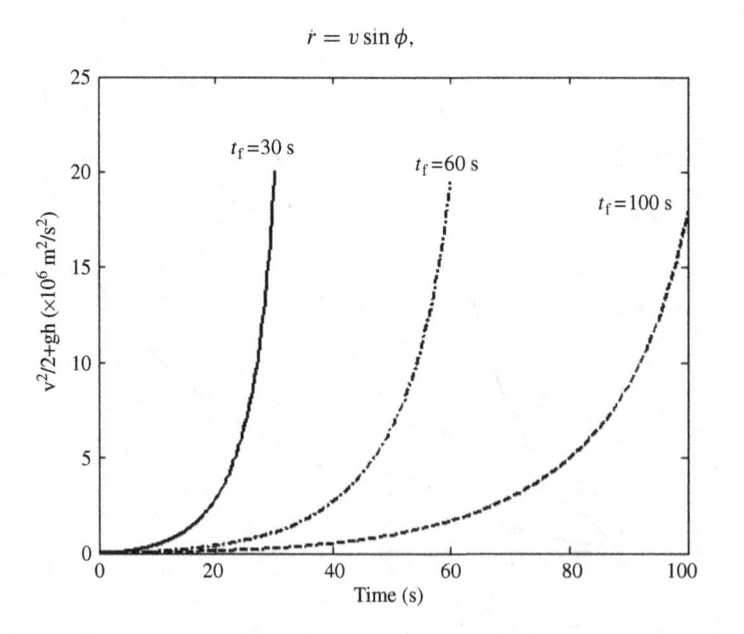

Figure 4.19 Extremal total specific energy of a vertically launched rocket for three different values of burn-out time, $t_f = 30, 60, 100$ s

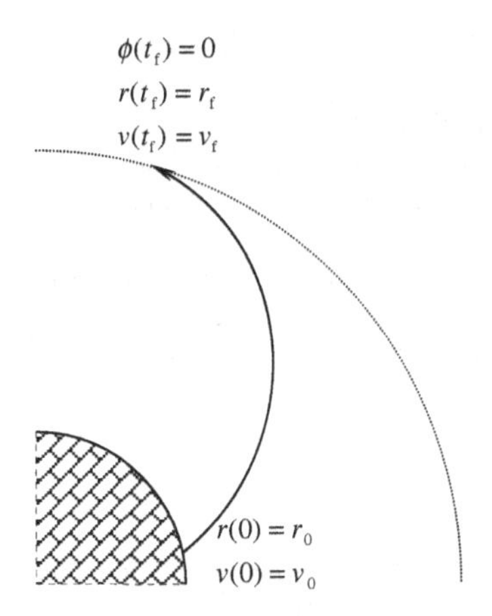

Figure 4.20 Gravity-turn trajectory for launch from the surface to a circular orbit of radius r_f

$$\dot{v} = u_1 - g \sin \phi, \tag{4.85}$$

$$v\dot{\phi} = \left(\frac{v^2}{r} - g\right) \cos \phi, \tag{4.86}$$

subject to the boundary conditions

$$r(0) = r_0, \tag{4.87}$$

$$v(0) = v_0, \tag{4.88}$$

$$r(t_f) = r_f, \tag{4.89}$$

$$\phi(t_f) = 0, \tag{4.90}$$

$$v(t_f) = \sqrt{\frac{\mu}{r_f}}. \tag{4.91}$$

The time of flight, t_f, is not know *a priori*, but must be selected depending upon the maximum available thrust.

The required optimal control input is the forward acceleration given by

$$u_1^*(t) = -\frac{\lambda_3(t)}{2R_1}. \tag{4.92}$$

This acceleration is the excess of thrust of the rocket engine above the atmospheric drag (which varies with speed and atmospheric density) per unit mass (a function of time). If the orbital speed is reached with zero acceleration ($\dot{v} = 0$) at $t = t_f$, the terminal time coincides with the condition $u_1^*(t_f) = 0$ (which implies $\lambda_3(t_f) = 0$) and is equal to the time when rocket propellant is exhausted (called *burn-out time*), thereby producing zero thrust simultaneously as the vehicle crosses the tangible atmosphere into space (zero atmospheric drag). In such a case, the trajectory is carefully timed such that burn-out occurs exactly at $t = t_f$.

It is clear from equation (4.86) that the trajectory curvature under the influence of gravity begins at a low speed immediately after launch, and ends with the final orbital speed, for which $v_f = \sqrt{g_f r_f} = \sqrt{\frac{\mu}{r_f}}$. The time of flight must be carefully selected such that $\phi(t_f) = 0$ for the circular orbit is reached simultaneously with $r(t_f) = r_f$ and $v(t_f) = v_f$.

An attitude control system must provide an initial tilt in the flight path by pitching the vehicle from the vertical in the desired orbital plane, and thereafter must maintain a zero normal acceleration, u_2, which requires a zero angle of attack, α (Figure 4.15). Such a maneuver is generally known as the *pitch program*, in which the pitch angle,

$$\theta = \phi - \alpha, \tag{4.93}$$

is carefully controlled such that $\alpha(t) = 0$. Therefore, the pitch control system must maintain a varying pitch rate through the flight according to equation (4.86):

$$\dot{\theta} = \dot{\phi} = \frac{1}{r}\left(v - \frac{\mu}{rv}\right)\cos\phi. \tag{4.94}$$

Example 4.4 Consider the launch of a rocket from the Earth's surface ($\mu = 398\,600.4$ km^3/s^2, $r_0 = 6378.14$ km) to a circular orbit of 300 km altitude. Immediately after launch and as soon as the vehicle leaves the launch pad, an initial speed, $v(0) = 1$ m/s, is assumed and a slight tilt from the vertical is provided in the desired direction for achieving the necessary orbital inclination.

The extremal trajectory is computed for three different values of the terminal time, $t_f = 600, 800, 1200$ s, using MATLAB's intrinsic code *bvp4c.m* which solves the 2PBVP by a collocation method (Chapter 2). The necessary MATLAB code for this purpose is listed in Table 4.8, and the results are plotted in Figures 4.21–4.24. The 2PBVP is solved subject to the following boundary conditions imposed

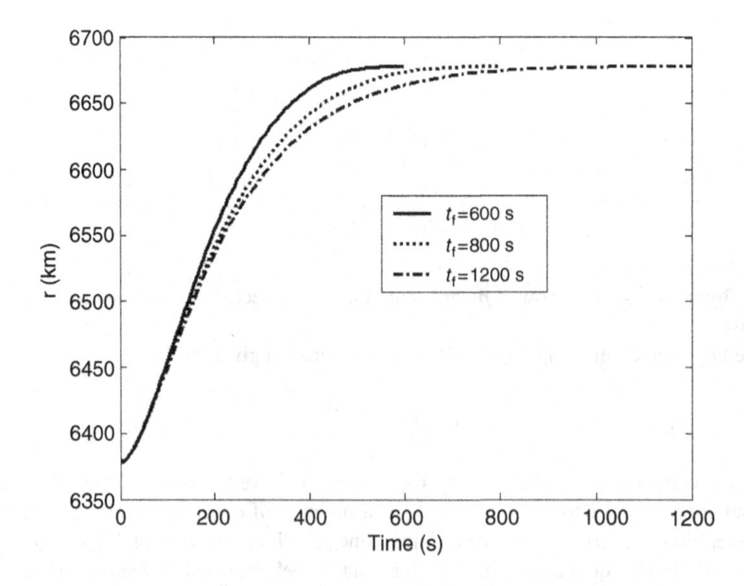

Figure 4.21 Radius in gravity-turn trajectory for launch from the Earth's surface to a circular orbit of 300 km altitude

Table 4.8

```
% Calling program for solving the gravity turn 2PBVP
% by collocation method by MATLAB's intrinsic code 'bvp4c.m'.
% Requires 'fiveode.m' and 'fivebc.m'.
% (c) 2009 Ashish Tewari
%  dy/dx=f(y,x); a<=x<=b
%  y(x=a), y(x=b): Boundary conditions
%  y(1,1)=r (km)
%  y(2,1)=v (km/s)
%  y(3,1)= phi (rad)
%  y(4,1)=lambda_2 (Lagrange multiplier of r)
%  y(5,1)=lambda_3 (Lagrange multiplier of v)
global mu; mu=398600.4; % Grav. const. (km^3/s^2)
global r0; r0=6378.14; % Planetary radius (km)
global R;   R=1; % Control cost coefficient
tf=600; % Terminal time (s)
% Collocation points & initial guess follow:
solinit = bvpinit(linspace(0,tf,5),[6680 8 0.1 0 0]);
% 2PBVP Solution by collocation method:
sol = bvp4c(@fiveode,@fivebc,solinit);
x = linspace(0,tf); % Time vector (s)
y = deval(sol,x); % Solution state vector

% Program for specifying governing ODEs expressed as
% state equations for the 2PBVP (to be called by 'bvp4c.m')
function dydx=fiveode(x,y)
global mu;
global r0;
global R;
% State equations follow:
dydx=[y(2)*sin(y(3))
    -0.5*y(5)/R-mu*sin(y(3))/y(1)^2
    (y(2)-mu/(y(1)*y(2)))*cos(y(3))/y(1)
    -2*mu*y(5)*sin(y(3))/y(1)^3
    -y(4)*sin(y(3))];

% Program for specifying boundary conditions for the 2PBVP.
% (To be called by 'bvp4c.m')
function res=fivebc(ya,yb)
global mu;
global r0;
% Boundary conditions follow:
res=[ya(1)-r0
     ya(2)-0.001
     yb(1)-r0-300
     yb(2)-sqrt(mu/(r0+300))
     yb(3)];
```

through the listed subroutine, *fivebc.m*:

$$r(0) = 6378.14\ \text{km}, \quad v(0) = 0.001\ \text{km/s},$$

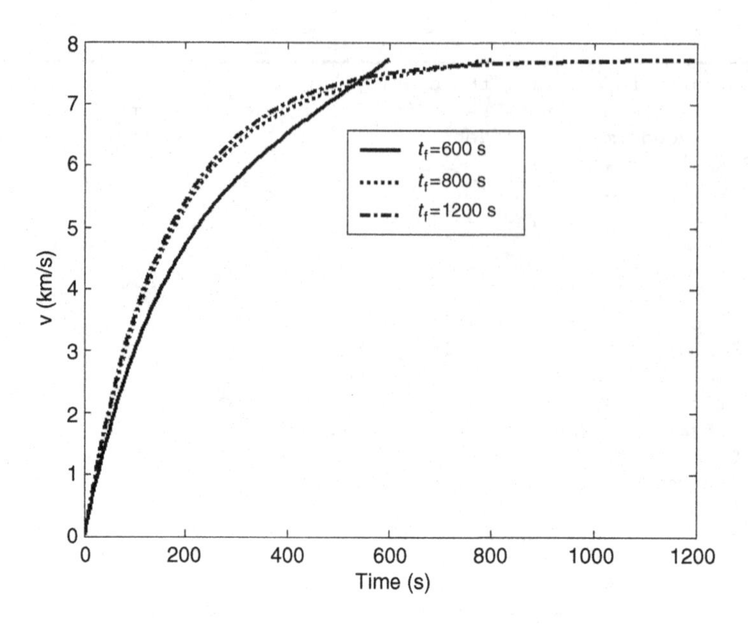

Figure 4.22 Speed in gravity-turn trajectory for launch from the Earth's surface to a circular orbit of 300 km altitude

$$r(t_f) = r_f = 6678.14 \text{ km}, \quad v(t_f) = \sqrt{\frac{\mu}{r_f}} = 7.7258 \text{ km/s}, \quad \phi(t_f) = 0.$$

Note that a smaller final time for launch requires a larger optimal control input.

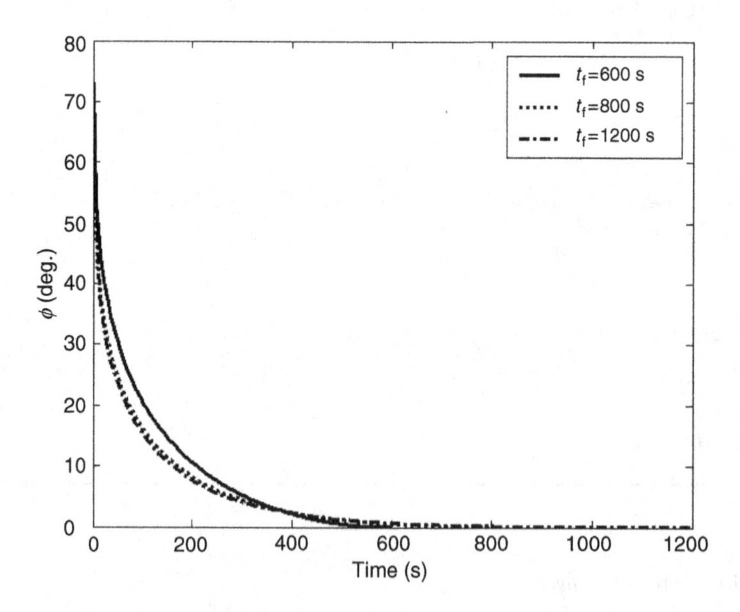

Figure 4.23 Flight-path angle in gravity-turn trajectory for launch from the Earth's surface to a circular orbit of 300 km altitude

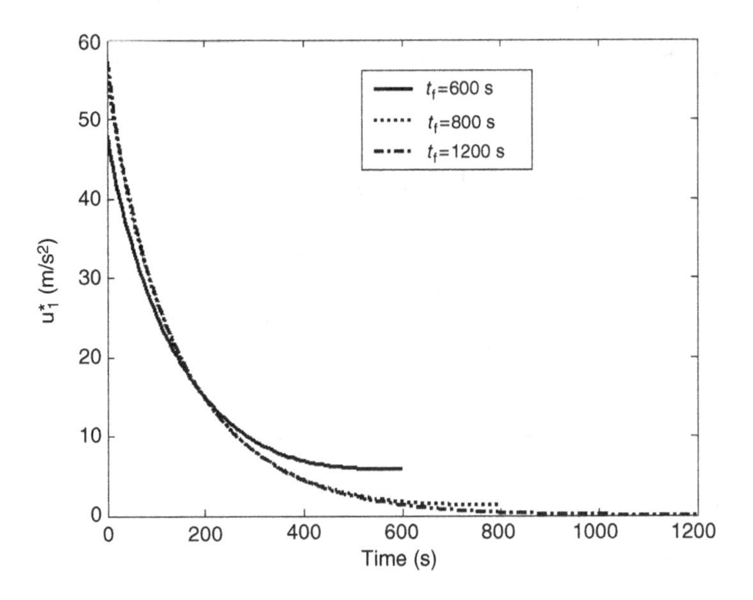

Figure 4.24 Forward acceleration input in gravity-turn trajectory for launch from the Earth's surface to a circular orbit of 300 km altitude

4.7.2 *Launch to Circular Orbit: Constant Acceleration*

A common application of launch trajectories is with a constant forward acceleration input, $u_1(t) = \bar{u}$. Such a trajectory is based upon the assumption that the thrust magnitude is constantly adjusted with time such that a constant forward acceleration is maintained throughout the ascent. The equations of motion and boundary conditions on state variables remain unchanged from equations (4.84)–(4.91). However, owing to the constant input, the co-state equations are modified. The optimal control problem is now posed with a terminal penalty on the final time and a quadratic control magnitude cost,

$$\varphi = t_f^2, \quad L = R\bar{u}^2.$$

The necessary conditions for optimality,

$$H_{\mathbf{u}} = L_{\mathbf{u}} + \boldsymbol{\lambda}^T \mathbf{B} = \mathbf{0},$$

$$\dot{\boldsymbol{\lambda}} = -\mathbf{A}^T \boldsymbol{\lambda},$$

$$\left(\frac{\partial \varphi}{\partial t} + L + \boldsymbol{\lambda}^T \mathbf{f} \right)_{t=t_f} = 0,$$

yield the following:

$$\lambda_3 = -2R\bar{u} = \text{const.}, \tag{4.95}$$

$$\dot{\lambda}_1 = 0, \tag{4.96}$$

$$\dot{\lambda}_3 = 0, \tag{4.97}$$

$$\lambda_4 = \frac{r^2 \lambda_2 \tan\phi}{r + \mu v^2}, \tag{4.98}$$

$$\dot{\lambda}_2 = \left[-\lambda_2 \left(\frac{\frac{2\mu}{vr} - v}{r + \mu v^2} \right) + 4R\bar{u}\frac{\mu}{r^3} \right] \sin\phi, \tag{4.99}$$

and

$$\begin{aligned} 2t_f + R\bar{u}^2 &+ \lambda_2(t_f)v(t_f)\sin\phi(t_f) \\ &+ \lambda_3(t_f)\left[\bar{u} - \frac{\mu}{r^2(t_f)}\sin\phi(t_f)\right] \\ &+ \lambda_4(t_f)\left[\frac{v(t_f)}{r(t_f)} - \frac{\mu}{r^2(t_f)v(t_f)}\right]\cos\phi(t_f) = 0. \end{aligned} \tag{4.100}$$

For a launch to a circular orbit, we have

$$\phi(t_f) = 0, \quad v(t_f) = \sqrt{\frac{\mu}{r_f}},$$

which, substituted into equation (4.100), yields

$$\bar{u} = \sqrt{\frac{2t_f}{R}}. \tag{4.101}$$

Clearly in this case, the co-state equations are unnecessary for determining the required control input, $\bar{u}$. Therefore, the 2PBVP consists of the state equations and boundary conditions on state variables, equations (4.84)–(4.91), with the constant control input selected by equation (4.100).

For a given terminal time, t_f, the final state, $r(t_f)$, $v(t_f)$, $\phi(t_f)$, is very sensitive to the selected control cost, R, which must be carefully selected by trial and error to achieve the desired orbit.

Example 4.5 Reconsider the launch of a rocket from the Earth's surface to a desired circular orbit of 300 km altitude, as stated in Example 4.4, but now with a constant forward acceleration input.

The extremal trajectory is computed for three different values of the control cost, $R = 3.1138 \times 10^6$, 3.6633×10^6, 3.938×10^6, with terminal time $t_f = 562$ s, using MATLAB's intrinsic 2PBVP solver, *bvp4c.m*. The necessary MATLAB code for this purpose is listed in Table 4.9, and the results are plotted in Figures 4.25–4.27. The 2PBVP is solved subject to the following boundary conditions imposed through the listed subroutine, *constubc.m*:

$$r(0) = 6378.14 \text{ km}, \quad v(0) = 0.001 \text{ km/s}, \quad \phi(t_f) = 0.$$

It is evident that only in the case of $R = 3.6633 \times 10^6$ is the desired circular orbit achieved with $r(t_f) = 6678.14$ km and $v(t_f) = 7.7258$ km/s. The other two values of control cost produce either a lower or a higher orbit, but at a later time, $t > t_f$.

4.8 Launch of Ballistic Missiles

The approach of guiding a gravity-turn rocket can be extended to a ballistic missile, whose re-entry crucially depends upon a specific terminal condition reached at the end of the *boost phase* (i.e., at the

Table 4.9

```
% Calling program for solving the gravity turn,
% constant fwd acceleration 2PBVP by collocation
% method using MATLAB's intrinsic code 'bvp4c.m'.
% Requires 'constuode.m' and 'constubc.m'.
% (c) 2009 Ashish Tewari
%  dy/dx=f(y,x); a<=x<=b
%  y(x=a), y(x=b): Boundary conditions
%  y(1,1)=r (km)
%  y(2,1)=v (km/s)
%  y(3,1)=phi (rad)
global mu; mu=398600.4; % Grav. const. (km^3/s^2)
global r0; r0=6378.14; % Planetary radius (km)
global R; R=3.6633e6; % Control cost coefficient
global tf; tf=562; % Terminal time (s)
% Collocation points & initial guess follow:
solinit = bvpinit(linspace(0,tf,5),[6680 8 0.1]);
% 2PBVP Solution by collocation method:
sol = bvp4c(@constuode,@constubc,solinit);
x = linspace(0,tf); % Time vector (s)
y = deval(sol,x); % Solution state vector

% Program for specifying governing state equations
% for the 2PBVP (to be called by 'bvp4c.m')
function dydx=constuode(x,y)
global tf;
global mu;
global r0;
global R;
if y(1)>=r0
if x<tf
u=sqrt(2*tf/R);
else
    u=0;
end
% State equations follow:
dydx=[y(2)*sin(y(3))
      u-mu*sin(y(3))/y(1)^2
      (y(2)-mu/(y(1)*y(2)))*cos(y(3))/y(1)];
else
    dydx=[0 0 0]';
end

% Program for specifying boundary conditions for the 2PBVP.
% (To be called by 'bvp4c.m')
function res=constubc(ya,yb)
global mu;
global r0;
% Boundary conditions follow:
res=[ya(1)-r0
     ya(2)-0.001
     yb(3)];
```

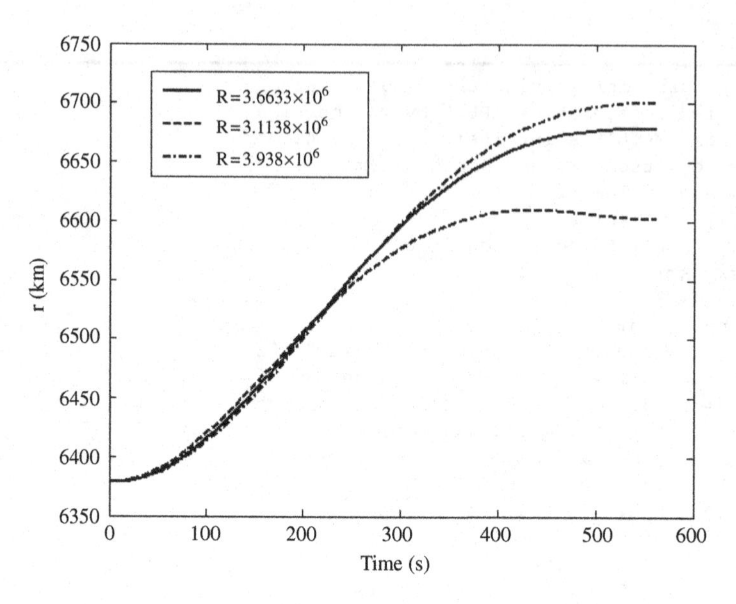

Figure 4.25 Radius in constant forward acceleration, gravity-turn trajectory for launch from the Earth's surface to a circular orbit

point of engine burn-out, $t = t_f$). However, the problem is much more difficult compared to a satellite launch vehicle in that the range traveled by the vehicle, given by the declination measured from the launch point, $\delta(t)$, is now important. The terminal condition is generally specified as a set of declination, altitude, inertial speed, and inertial flight-path angle, $(\delta_f, h_f, v_f, \phi_f)$, to be reached at $t = t_f$, as shown in

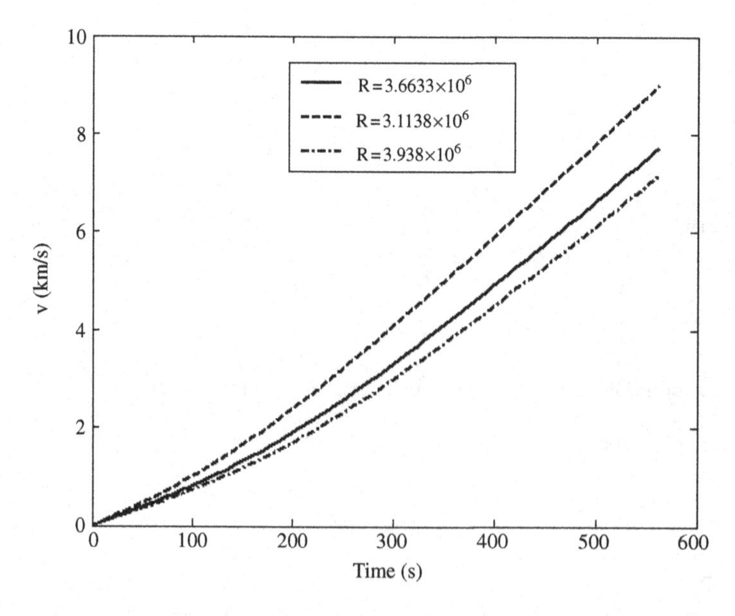

Figure 4.26 Speed in constant forward acceleration, gravity-turn trajectory for launch from the Earth's surface to a circular orbit

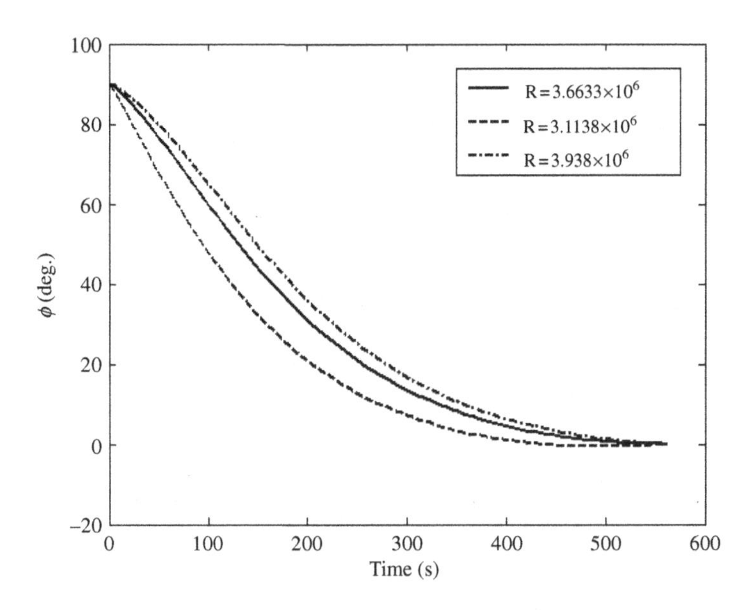

Figure 4.27 Flight-path angle in constant forward acceleration, gravity-turn trajectory for launch from the Earth's surface to a circular orbit

Figure 4.28. The trajectory followed by the ballistic missile warhead for $t \geq t_f$ is determined purely by Kepler's equations of motion for space flight (Chapter 6), which can be integrated backward in time – starting from the required entry conditions – in order to yield the necessary declination, altitude, speed, and flight-path angle at the end of the boost phase. Clearly, this is a separate 2PBVP, which we shall defer until Chapter 6.

The equations of motion and the co-state equations that must now be solved are the following:

$$\dot{\delta} = \frac{v \cos\phi}{r}, \tag{4.102}$$

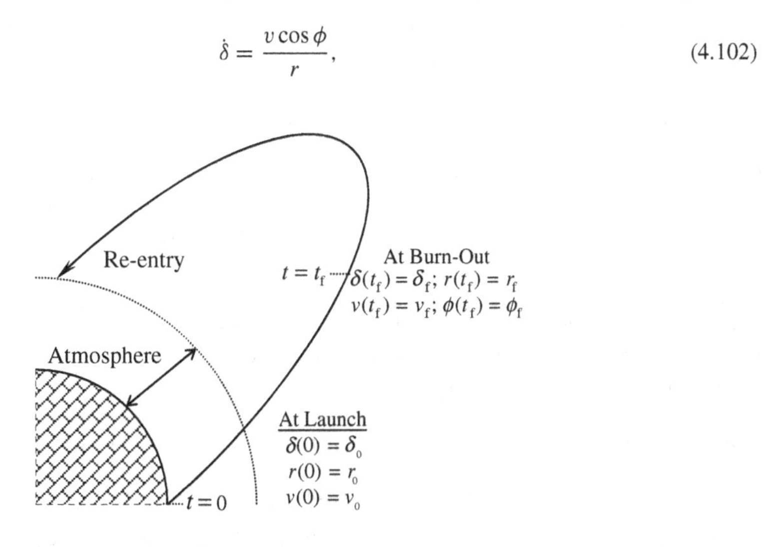

Figure 4.28 Launch of a ballistic missile from the surface to a given terminal condition in the space at the end of the boost phase, $(\delta_f, h_f, v_f, \phi_f)$

$$\dot{r} = v \sin\phi, \tag{4.103}$$

$$\dot{v} = u_1 - g\sin\phi, \tag{4.104}$$

$$v\dot{\phi} = u_2 + \left(\frac{v^2}{r} - g\right)\cos\phi, \tag{4.105}$$

$$\dot{\lambda}_1 = 0, \tag{4.106}$$

$$\dot{\lambda}_2 = \frac{v\cos\phi}{r^2}\lambda_1 - \frac{2\mu\sin\phi}{r^3}\lambda_3 - \lambda_4\left(\frac{2\mu}{vr^3} - \frac{v}{r^2}\right)\cos\phi, \tag{4.107}$$

$$\dot{\lambda}_3 = -\frac{\cos\phi}{r}\lambda_1 - (\sin\phi)\lambda_2 - \left[\frac{\lambda_4}{2R_2v^3} + \left(\frac{1}{r} + \frac{\mu}{v^2r^2}\right)\cos\phi\right]\lambda_4, \tag{4.108}$$

$$\dot{\lambda}_4 = \frac{v\sin\phi}{r}\lambda_1 - (v\cos\phi)\lambda_2 + \frac{\mu\cos\phi}{r^2}\lambda_3 + \lambda_4\left(\frac{v}{r} - \frac{\mu}{vr^2}\right)\sin\phi, \tag{4.109}$$

which include the optimal control inputs given by

$$\begin{aligned} u_1^* &= \frac{\lambda_3}{2R_1}, \\ u_2^* &= \frac{\lambda_4}{2R_2v}. \end{aligned} \tag{4.110}$$

We shall consider two cases of the ballistic missile trajectories, namely those that follow the gravity-turn approach (i.e., $u_2 = \lambda_4 = 0$) with a modulated forward acceleration, and those having a modulation of both forward and normal acceleration components, u_1, u_2. In each case, the boundary conditions on state and co-state variables are posed differently.

4.8.1 Gravity-Turn with Modulated Forward Acceleration

We begin with the solution of optimal, ballistic missile flight with the gravity-turn case and a modulated forward acceleration, $u_1(t)$, and note that the additional demand for meeting a final declination could increase the thrust requirement, when compared to a launch of a satellite to a circular orbit by the same rocket. The initial condition imposed on declination, speed, and altitude, and the final condition on declination, speed, altitude, and flight-path angle provide seven boundary conditions for the four state variables, $\delta(t)$, $h(t)$, $v(t)$, $\phi(t)$, and three co-state variables, $\lambda_1(t)$, $\lambda_2(t)$, $\lambda_3(t)$. The fourth co-state variable vanishes ($\lambda_4(t) = 0$) because of the demand of zero normal acceleration input, $u_2(t) = 0$, for a gravity-turn flight.

Example 4.6 Consider the launch of a ballistic missile from the Earth's surface with initial condition

$$r(0) = 6378.14 \text{ km}, \quad v(0) = 0.001 \text{ km/s},$$

such that at the point of rocket burn-out the following terminal conditions are reached:

$$\delta(t_f) = 30^\circ, \quad r(t_f) = 6578.14 \text{ km},$$

$$v(t_f) = \sqrt{\frac{\mu}{r_f}} = 9 \text{ km/s}, \quad \phi(t_f) = 5°.$$

To solve the 2PBVP concerned, we wrote the code listed in Table 4.10 which utilizes MATLAB's inbuilt solver, *bvp4c.m*. The differential equations and the boundary conditions required by *bvp4c.m* are specified by the respective codes *ballisticode.m* and *ballisticbc.m*, also listed in Table 4.10. The results for $t_f = 500$ s are plotted in Figures 4.29 and 4.30. The optimal acceleration varies from 0.088 km/s^2 immediately after launch to -0.0083 km/s^2 at burn-out, which requires a rather precise thrust modulation.

4.8.2 Modulated Forward and Normal Acceleration

The most difficult problem of flight in the vertical plane is controlling the declination, speed, altitude, and flight-path angle by both forward and normal acceleration inputs. The fourth co-state variable, $\lambda_4(t)$, is now non-zero and contributes to the optimal normal acceleration. For the four state equations $[\delta(t), h(t), v(t), \phi(t)]$ and four co-state $[\lambda_1(t), \lambda_2(t), \lambda_3(t), \lambda_4(t)]$ equations, we have the initial condition imposed on declination, speed, and altitude, as well as the final condition on declination, speed, altitude, and flight-path angle, which add up to only seven boundary conditions. For the remaining boundary condition, we have the possibility of restraining the initial flight-path angle by the normal acceleration input, which results in an instability due to the small initial speed (as we shall see later), thereby producing an ill-posed boundary value problem. A better choice would be to impose zero initial normal acceleration

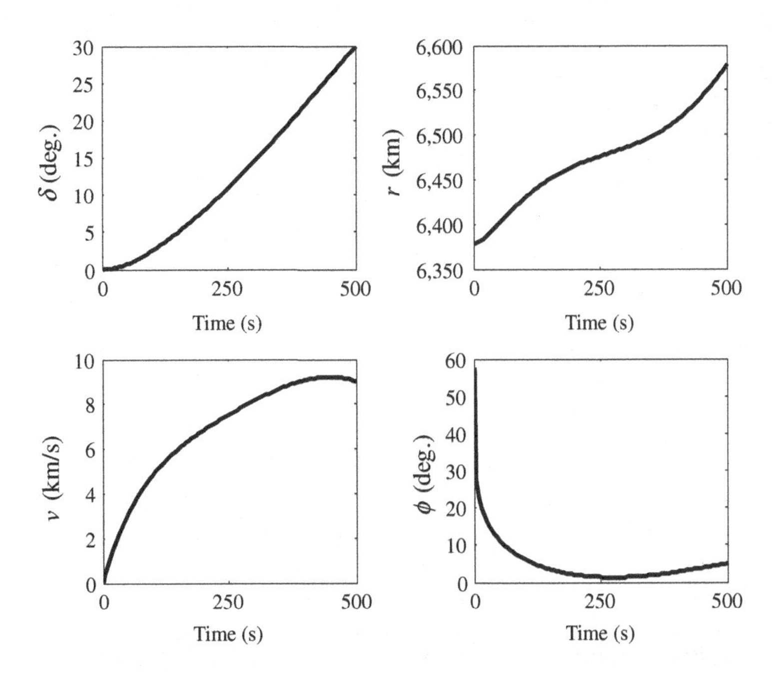

Figure 4.29 Extremal trajectory of gravity-turn flight of a ballistic missile from the Earth's surface to $\delta(t_f) = 30°$, $h(t_f) = 200$ km, $v(t_f) = 9$ km/s, and $\phi(t_f) = 5°$, with $t_f = 500$ s

Table 4.10

```
% Calling program for solving the ballistic,
% gravity turn with modulated fwd acceleration
% 2PBVP by collocation method using MATLAB's
% intrinsic code 'bvp4c.m'.
% Requires 'ballisticode.m' and 'ballisticbc.m'.
% (c) 2010 Ashish Tewari
%  dy/dx=f(y,x); a<=x<=b
%  y(x=a), y(x=b): Boundary conditions
%  y(1,1)=delta (rad)
%  y(2,1)=r (km)
%  y(3,1)=v (km/s)
%  y(4,1)=phi (rad)
%  y(5:7,1)=Lagrange multipliers
global R; R=1;
global tf; tf=500;
dtr=pi/180;
init=[30*dtr 6580 8 5*dtr 0 0 0];
solinit = bvpinit(linspace(0,tf,5),init);
sol = bvp4c(@ballisticode,@ballisticbc,solinit);
x = linspace(0,tf);
y = deval(sol,x);
u=-0.5*y(7,:)/R;

% Program for specifying governing state equations
% for the 2PBVP (to be called by 'bvp4c.m')
function dydx=ballisticode(x,y)
global R;
mu=398600.4;
r0=6378.14;
dydx=[y(3)*cos(y(4))/y(2)
   y(3)*sin(y(4))
   -0.5*y(7)/R-mu*sin(y(4))/y(2)^2
   (y(3)-mu/(y(2)*y(3)))*cos(y(4))/y(2)
   0
   y(5)*y(3)*cos(y(4))/y(2)^2-2*mu*y(7)*sin(y(4))/y(2)^3
   -y(5)*cos(y(4))/y(2)-y(6)*sin(y(4))];

% Program for specifying boundary conditions for the 2PBVP.
% (To be called by 'bvp4c.m')
function res=ballisticbc(ya,yb)
mu=398600.4;
r0=6378.14;
dtr=pi/180;
res=[ya(1)
    ya(2)-r0
    ya(3)-0.001
    yb(1)-30*dtr
    yb(2)-r0-200
    yb(3)-9
    yb(4)-5*dtr];
```

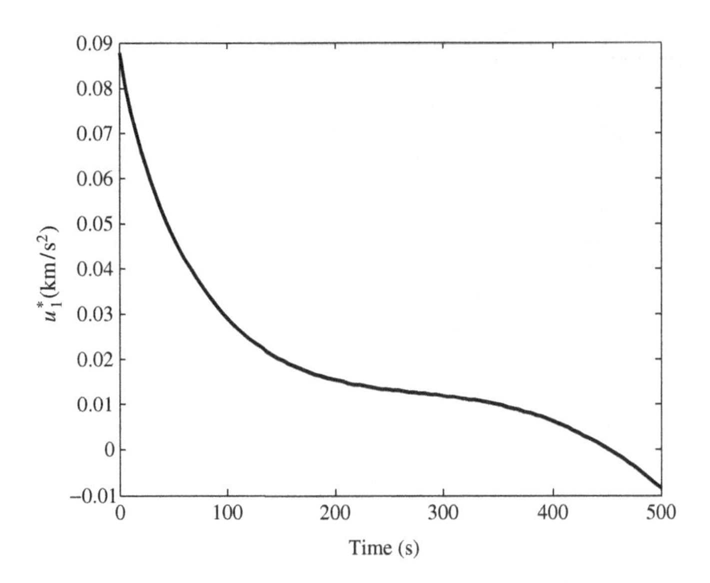

Figure 4.30 Optimal forward acceleration input of gravity-turn flight of a ballistic missile from the Earth's surface to $\delta(t_f) = 30°$, $h(t_f) = 200$ km, $v(t_f) = 9$ km/s, and $\phi(t_f) = 5°$, with $t_f = 500$ s

by having $\lambda_4(0) = 0$. The complete set of boundary conditions is thus the following:

$$\begin{aligned}
\delta(0) &= \delta_0, \\
h(0) &= h_0, \\
v(0) &= v_0, \\
\lambda_4(0) &= 0, \\
\delta(t_f) &= \delta_f, \\
h(t_f) &= h_f, \\
v(t_f) &= v_f, \\
\phi(t_f) &= \phi_f.
\end{aligned} \tag{4.111}$$

Example 4.7 We repeat the problem of Example 4.6 with both forward and normal acceleration inputs. For the control cost parameters, we select $R_1 = 1$ and vary R_2 in order to produce trajectories with different magnitudes of the ratio u_2/u_1. To solve the 2PBVP concerned, we wrote the code listed in Table 4.11 which utilizes MATLAB's inbuilt solver, *bvp4c.m*. The differential equations and the boundary conditions required by *bvp4c.m* are specified by the respective codes *normballisticode.m* and *normballisticbc.m* listed in Table 4.12. The results for $t_f = 500$ s are plotted in Figures 4.31–4.35 for $R_2 = 7, 10, 20$. In all cases, the normal acceleration input is seen to be an order of magnitude smaller than the forward acceleration, which is good for structural load considerations of a liquid fuelled rocket. The case for $R_2 = 7$ barely manages to leave ground (Figures 4.32 and 4.34) while $R_2 = 20$ has the largest normal acceleration input. The declination, speed, and forward acceleration of all the three cases are closely matched.

Table 4.11

```
% Calling program for solving the 2PBVP for
% flight with modulated fwd and normal
% acceleration inputs by collocation method
% using MATLAB's intrinsic code 'bvp4c.m'.
% Requires 'normballisticode.m' and
% 'normballisticbc.m'.
% (c) 2010 Ashish Tewari
%  dy/dx=f(y,x); a<=x<=b
%  y(x=a), y(x=b): Boundary conditions
%  y(1,1)=delta (rad)
%  y(2,1)=r (km)
%  y(3,1)=v (km/s)
%  y(4,1)=phi (rad)
%  y(5:8,1)=Lagrange multipliers
global R1; R1=1;
global R2; R2=7;
global tf; tf=500;
dtr=pi/180;
init=[30*dtr 6580 8 5*dtr 0 0 0 0];
solinit = bvpinit(linspace(0,tf,5),init);
sol = bvp4c(@normballisticode,@normballisticbc,solinit);
x = linspace(0,tf);
y = deval(sol,x);
u1=-0.5*y(7,:)/R1;
u2=-0.5*y(8,:)./(R2*y(3,:));
```

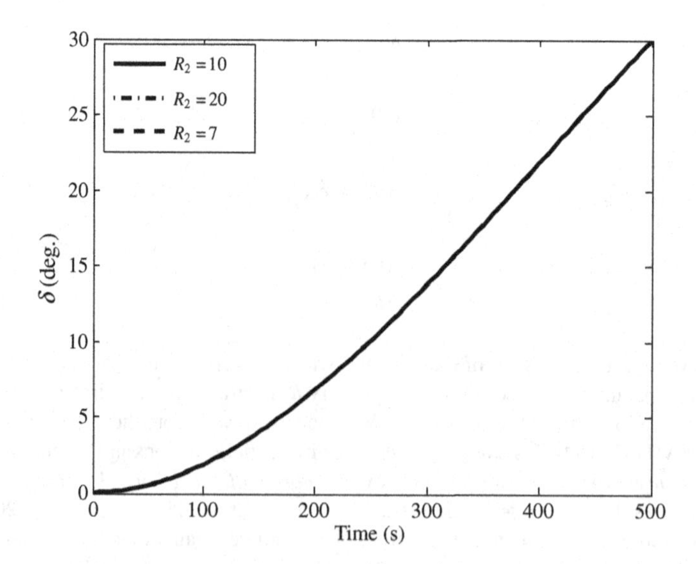

Figure 4.31 Extremal declination of a ballistic missile with modulated forward and normal acceleration inputs for three different values of cost parameter, $R_2 = 7, 10, 20$

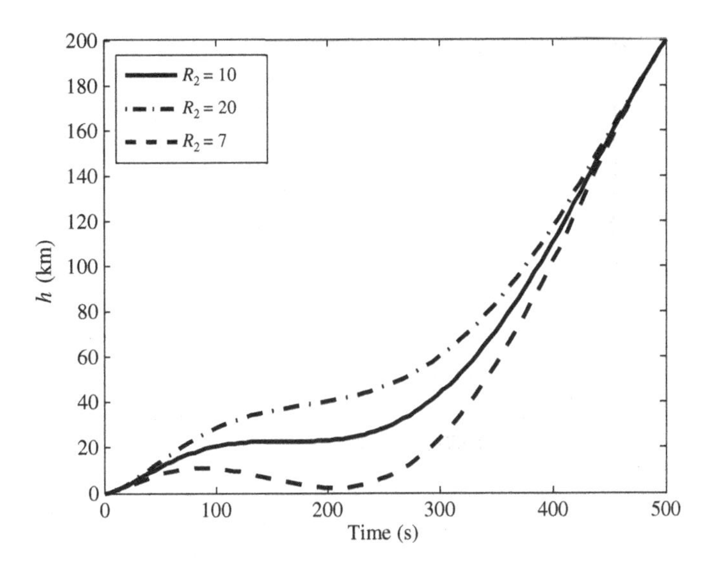

Figure 4.32 Extremal altitude of a ballistic missile with modulated forward and normal acceleration inputs for three different values of cost parameter, $R_2 = 7, 10, 20$

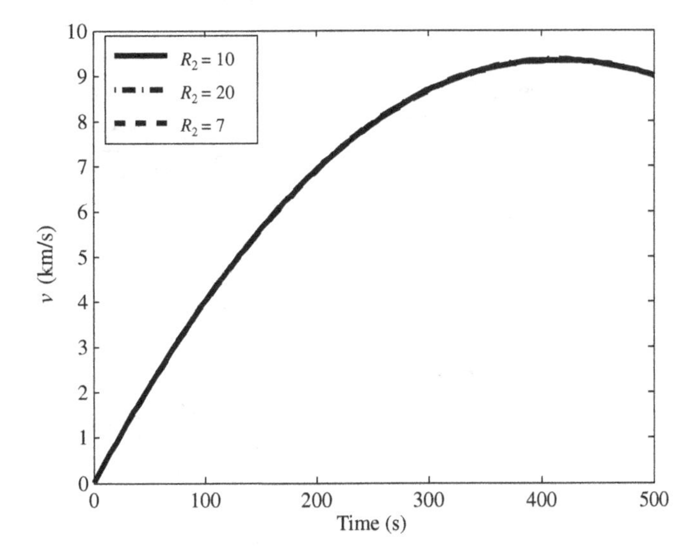

Figure 4.33 Extremal speed of a ballistic missile with modulated forward and normal acceleration inputs for three different values of cost parameter, $R_2 = 7, 10, 20$

4.9 Planar Tracking Guidance System

There is need for a tracking guidance system in order to maintain the vehicle on the desired (optimal) trajectory despite disturbances. The disturbances could be either internal or external called the process noise, $\mathbf{v}(t)$, or measurement noise, $\mathbf{w}(t)$. An example of process noise is the inability of the

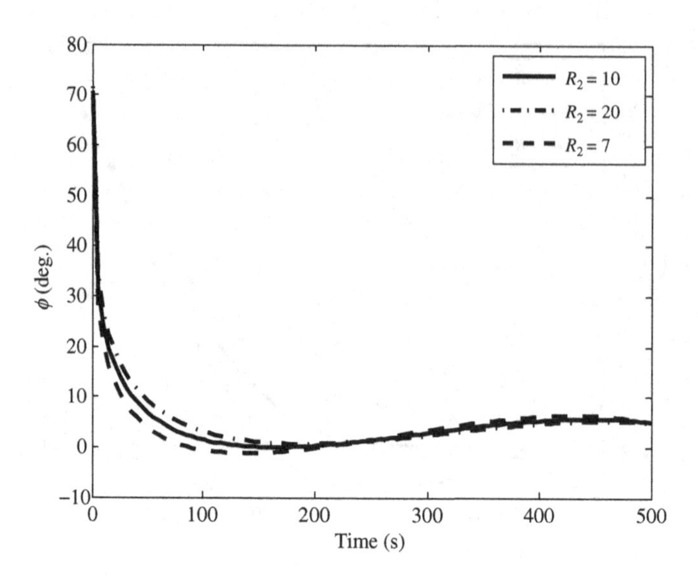

Figure 4.34 Extremal flight-path angle of a ballistic missile with modulated forward and normal acceleration inputs for three different values of cost parameter, $R_2 = 7, 10, 20$

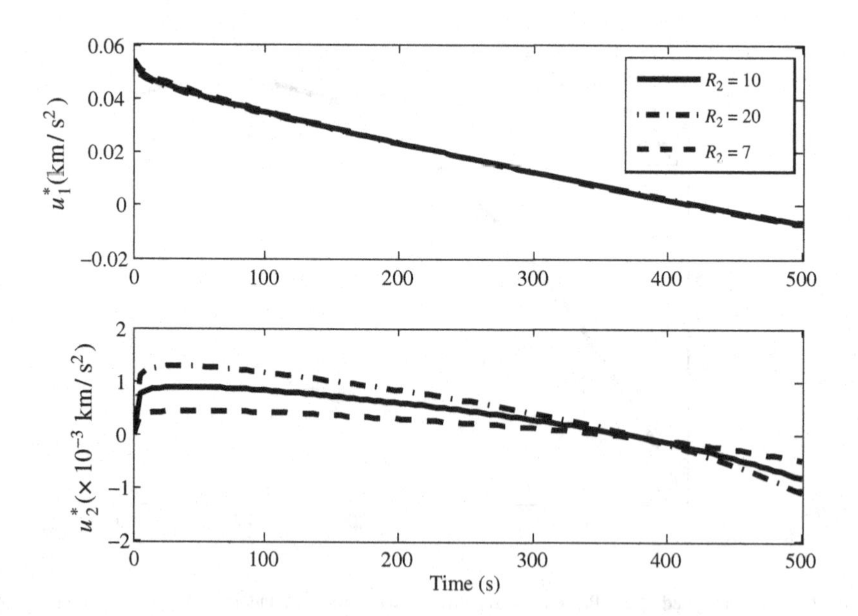

Figure 4.35 Extremal acceleration inputs of a ballistic missile with modulated forward and normal acceleration inputs for three different values of cost parameter, $R_2 = 7, 10, 20$

rocket to generate the forward and normal acceleration inputs precisely as required. The effects of the stochastic noise can be minimized by having a closed-loop control system that drives deviations from a nominal trajectory, $[h_n(t), \delta_n(t), v_n(t), \phi_n(t)]$, based upon measurement and feedback of output variables, as shown in Figure 4.36. Equations (4.43)–(4.46) describe the plant for the planar guidance problem, that

Table 4.12

```
% Program for specifying governing state equations
% of 2PBVP for flight with modulated fwd and normal
% acceleration inputs (to be called by 'bvp4c.m')
function dydx=normballisticode(x,y)
global R1;
global R2;
mu=398600.4;
dydx=[y(3)*cos(y(4))/y(2)
      y(3)*sin(y(4))
      -0.5*y(7)/R1-mu*sin(y(4))/y(2)^2
      -0.5*y(8)/(R2*y(3))+(y(3)- ...
       mu/(y(2)*y(3)))*cos(y(4))/y(2)
      0
      y(5)*y(3)*cos(y(4))/y(2)^2- ...
      2*mu*sin(y(4))*y(7)/y(2)^3- ...
      (2*mu/(y(2)*y(3))-y(3))*y(8)*cos(y(4))/y(2)^2
      -y(5)*cos(y(4))/y(2)-y(6)*sin(y(4))- ...
      y(8)*(0.5*y(8)/(R2*y(3)^3)+ ...
      (1+mu/(y(3)^2*y(2)))*cos(y(4))/y(2))
      y(5)*y(3)*sin(y(4))/y(2)-y(6)*y(3)*cos(y(4))+ ...
      mu*cos(y(4))*y(7)/y(2)^2+(y(3)/y(2)- ...
      mu/(y(2)^2*y(3)))*sin(y(4))*y(8)];

% Program for specifying boundary conditions for the 2PBVP.
% (To be called by 'bvp4c.m')
function res=normballisticbc(ya,yb)
mu=398600.4;
r0=6378.14;
dtr=pi/180;
res=[ya(1)
    ya(2)-r0
    ya(3)-0.001
    ya(8)
    yb(1)-30*dtr
    yb(2)-r0-200
    yb(3)-9
    yb(4)-5*dtr];
```

is, the problem of maneuvering the center of mass in a vertical plane, and are expressed in state-space form as follows:

$$\dot{\mathbf{x}} = \mathbf{f}(\mathbf{x}, \mathbf{u}), \quad \mathbf{x}(0) = \mathbf{x}_0. \tag{4.112}$$

The planar guidance systems track a nominal trajectory,

$$\mathbf{x}_n(t) = [h_n(t), \delta_n(t), v_n(t), \phi_n(t)]^T,$$

which is the solution to

$$\dot{\mathbf{x}}_n = \mathbf{f}(\mathbf{x}_n, \mathbf{u}_n), \quad \mathbf{x}_n(0) = \mathbf{x}_{n0}. \tag{4.113}$$

where the nominal control, $\mathbf{u}_n(t) = [u_{n1}(t), u_{n2}(t)]^T$, is obtained from the solution of an optimal boundary value problem with end-point and path constraints on the state and input variables (Chapter 2). The plant's

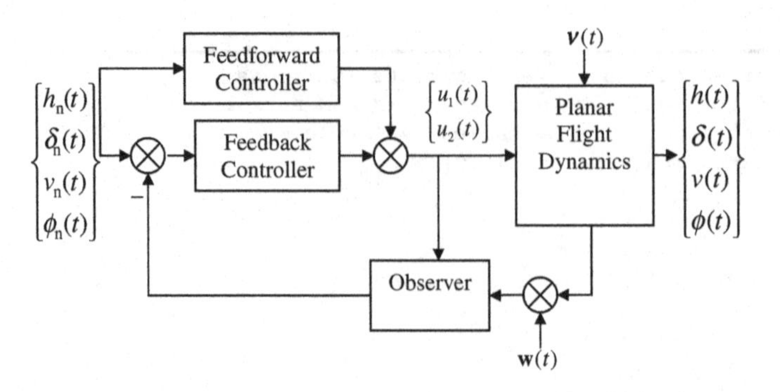

Figure 4.36 Planar guidance system for a rocket in a vertical plane

dynamic state equation (4.112) is expanded in a Taylor series about the nominal trajectory, $\mathbf{x}_n(t)$, resulting in the following linear approximation:

$$\Delta\dot{\mathbf{x}} = \left(\frac{\partial \mathbf{f}}{\partial \mathbf{x}}\right)_{\mathbf{x}=\mathbf{x}_n} \Delta\mathbf{x} + \left(\frac{\partial \mathbf{f}}{\partial \mathbf{u}}\right)_{\mathbf{x}=\mathbf{x}_n} \Delta\mathbf{u}, \quad \Delta\mathbf{x}(0) = \Delta\mathbf{x}_0, \tag{4.114}$$

where

$$\Delta\mathbf{x}(t) = \mathbf{x}(t) - \mathbf{x}_n(t) = [\Delta h(t), \Delta\delta(t), \Delta v(t), \Delta\phi(t)]^T, \tag{4.115}$$

$$\Delta\mathbf{u}(t) = \mathbf{u}(t) - \mathbf{u}_n(t) = [\Delta u_1(t), \Delta u_2(t)]^T. \tag{4.116}$$

Since a successful tracking guidance system should always maintain the trajectory in the vicinity of the nominal one, only first-order state and control variations need be considered in the derivation of the optimal control input. Consequently, the plant's *Jacobian* matrices,

$$\mathbf{A} = \left(\frac{\partial \mathbf{f}}{\partial \mathbf{x}}\right)_{\mathbf{x}=\mathbf{x}_n}, \quad \mathbf{B} = \left(\frac{\partial \mathbf{f}}{\partial \mathbf{u}}\right)_{\mathbf{x}=\mathbf{x}_n},$$

evaluated at $\mathbf{x} = \mathbf{x}_n$, are primarily important.

For a planar flight, the vector functional, $\mathbf{f}(.)$, is given by

$$\mathbf{f} = \left\{ \begin{array}{c} v \sin\phi \\ \dfrac{v\cos\phi}{r_0 + h} \\ u_1 - \dfrac{g_0 r_0^2}{(r_0+h)^2}\sin\phi \\ \dfrac{u_2}{v} + \left[\dfrac{v}{r_0+h} - \dfrac{g_0 r_0^2}{v(r_0+h)^2}\right]\cos\phi \end{array} \right\}, \tag{4.117}$$

which yields

$$\mathbf{A} = \begin{pmatrix} 0 & 0 & a_{13} & a_{14} \\ a_{21} & 0 & a_{23} & a_{24} \\ a_{31} & 0 & 0 & a_{34} \\ a_{41} & 0 & a_{43} & a_{44} \end{pmatrix}, \tag{4.118}$$

where

$$a_{13} = \sin\phi_n, \tag{4.119}$$

$$a_{14} = v_n \cos\phi_n, \tag{4.120}$$

$$a_{21} = -\frac{v_n \cos\phi_n}{(r_0 + h_n)^2}, \tag{4.121}$$

$$a_{23} = \frac{\cos\phi_n}{r_0 + h_n} \tag{4.122}$$

$$a_{24} = -\frac{v_n \sin\phi_n}{r_0 + h_n}, \tag{4.123}$$

$$a_{31} = \frac{2g_0 r_0^2 \sin\phi_n}{(r_0 + h_n)^3}, \tag{4.124}$$

$$a_{34} = -\frac{g_0 r_0^2 \cos\phi_n}{(r_0 + h_n)^2}, \tag{4.125}$$

$$a_{41} = \left[-\frac{v_n}{(r_0 + h_n)^2} + \frac{2g_0 r_0^2}{v_n (r_0 + h_n)^3}\right] \cos\phi_n, \tag{4.126}$$

$$a_{43} = -\frac{u_{n2}}{v_n^2} + \left[\frac{1}{r_0 + h_n} + \frac{g_0 r_0^2}{v_n^2 (r_0 + h_n)^2}\right] \cos\phi_n, \tag{4.127}$$

$$a_{44} = -\left[\frac{v_n}{r_0 + h_n} - \frac{g_0 r_0^2}{v_n (r_0 + h_n)^2}\right] \sin\phi_n, \tag{4.128}$$

and

$$\mathbf{B} = \begin{pmatrix} 0 & 0 \\ 0 & 0 \\ 1 & 0 \\ 0 & \frac{1}{v_n} \end{pmatrix}. \tag{4.129}$$

4.9.1 *Stability, Controllability, and Observability*

Since the nominal trajectory and control vary with time, $\mathbf{x}_n(t)$, $\mathbf{u}_n(t)$, the linearized plant, $[\mathbf{A}(t), \mathbf{B}(t)]$, is a time-varying one. Therefore, any tracking guidance law based upon the linear, time-varying plant must also be a function of time. However, before a stabilizing guidance law can be found, it is necessary to examine the stability, controllability, and observability of the linearized plant, all of which depend upon the nominal trajectory. The stability of the planar guidance plant is analyzed by examining the eigenvalues of the Jacobian on the nominal trajectory with the following characteristic equation:

$$|\lambda\mathbf{I} - \mathbf{A}| = \begin{vmatrix} \lambda & 0 & -a_{13} & -a_{14} \\ -a_{21} & \lambda & -a_{23} & -a_{24} \\ -a_{31} & 0 & \lambda & -a_{34} \\ -a_{41} & 0 & -a_{43} & (\lambda - a_{44}) \end{vmatrix} = 0. \tag{4.130}$$

The plant eigenvalues are thus given by

$$\begin{aligned}
&\lambda = 0, \\
&\lambda^3 - a_{44}\lambda^2 - (a_{13}a_{31} + a_{14}a_{41} + a_{34}a_{43})\,\lambda \\
&\qquad + a_{13}\,(a_{31}a_{44} - a_{34}a_{41}) - a_{14}a_{31}a_{43} = 0.
\end{aligned} \tag{4.131}$$

The eigenvalue $\lambda = 0$ is due to the downrange (δ) dynamics which does not contribute to changes in h, v, and ϕ, and is thus regarded as being uncoupled from them. Due to the presence of an eigenvalue at the origin, the plant is never asymptotically stable along the nominal trajectory. For stability, all the remaining eigenvalues (i.e., roots of the cubic polynomial in equation (4.131)) must lie strictly in the left-half s-plane, which requires

$$\begin{aligned}
a_{44} &< 0, \\
(a_{13}a_{31} + a_{14}a_{41} + a_{34}a_{43}) &< 0, \\
a_{13}\,(a_{31}a_{44} - a_{34}a_{41}) &> a_{14}a_{31}a_{43}.
\end{aligned} \tag{4.132}$$

The first inequality of equation (4.132) requires

$$\left[\frac{v_n}{r_0 + h_n} - \frac{g_0 r_0^2}{v_n (r_0 + h_n)^2}\right] \sin\phi_n > 0, \tag{4.133}$$

which for a climbing trajectory ($\sin\phi_n > 0$) translates into the requirement

$$v_n^2 > \frac{g_0 r_0^2}{r_0 + h_n}, \tag{4.134}$$

implying that the speed should be greater than that of a circular orbit of radius $r_0 + h_n$. Clearly, this stability criterion is never met for sub-orbital rockets as well as launch vehicles designed to put spacecraft into circular orbits. For a spacecraft in an eccentric orbit of semi-major axis a (Chapter 6), the criterion is locally satisfied at altitudes for which $h_n < a - r_0$, and is globally satisfied for only escape trajectories. Thus, most rockets have an inherently unstable translational dynamics and require some kind of guidance system for operation in presence of disturbances.

The main purpose of a rocket guidance system is to stabilize the vehicle's translational dynamics in a closed loop. However, controllability (or at least stabilizability) of the plant is necessary for a guidance system to be successful. For controllability of the fourth-order guidance plant, we must have the following test matrix of full rank:

$$\mathbf{P} = \left[\mathbf{B},\ \ \mathbf{AB},\ \ \mathbf{A}^2\mathbf{B},\ \ \mathbf{A}^3\mathbf{B}\right]. \tag{4.135}$$

For a non-zero normal acceleration input, Δu_2, the plant can be shown to be always controllable (rank$(P) = 4$). However, if it is required to guide the rocket using only Δu_1 by throttling of the engines, we have

$$\mathbf{B} = \begin{pmatrix} 0 \\ 0 \\ 1 \\ 0 \end{pmatrix}. \tag{4.136}$$

Thus at the initially low speeds shortly after launch, we have $a_{14} = a_{21} = a_{24} = 0$, which makes

$$\mathbf{P} = \begin{pmatrix} 0 & a_{13} & 0 & a_{13}(a_{13}a_{31} + a_{34}a_{43}) \\ 0 & a_{23} & 0 & a_{23}(a_{13}a_{31} + a_{34}a_{43}) \\ 1 & 0 & (a_{13}a_{31} + a_{34}a_{43}) & a_{34}(a_{13}a_{41} + a_{43}a_{44}) \\ 0 & a_{43} & (a_{13}a_{41} + a_{43}a_{43}) & [a_{44}(a_{13}a_{41} + a_{43}a_{44}) \\ & & & +a_{43}(a_{13}a_{31} + a_{34}a_{43})] \end{pmatrix} \tag{4.137}$$

singular (rank$(P) = 3$), thereby implying uncontrollability. Furthermore, since the uncontrollable mode is also unstable, we have an unstabilizable plant at $v \simeq 0$ without normal acceleration input. In such a case, the only way the flight path can be kept straight during the initial portion of a launch is by having a guide-rail support from the rocket launch-tower.

The sufficient condition for observability of the fourth-order guidance plant, with output given by

$$\mathbf{y} = \mathbf{Cx}, \tag{4.138}$$

is that the following observability test matrix must have full rank:

$$\mathbf{N} = \left[\mathbf{C}^T, \ \ \mathbf{A}^T\mathbf{C}^T, \ \ (\mathbf{A}^T)^2\mathbf{C}^T, \ \ (\mathbf{A}^T)^3\mathbf{C}^T\right]. \tag{4.139}$$

For a practical implementation, the measured outputs should be physical quantities, such as the flight state variables, h, δ, v, ϕ. For a rocket, the flight variables can be measured using onboard sensors, or by a ground based radar. It can be easily shown that the plant is unobservable with any single state variable taken as the output, except $y(t) = \delta(t)$. However, with $y(t) = \delta(t)$, the test matrix $\mathbf{N}$ has a very large condition number, indicating only a weakly observable plant. Thus it is more practical to use at least two outputs, one of which is $\delta(t)$. For example, the plant is unconditionally observable with $y_1 = h$ and $y_2 = \delta$, that is, with

$$\mathbf{C} = \begin{pmatrix} 1 & 0 & 0 & 0 \\ 0 & 1 & 0 & 0 \end{pmatrix}. \tag{4.140}$$

4.9.2 Nominal Plant for Tracking Gravity-Turn Trajectory

The most common planar tracking plant for a rocket is for a gravity-turn trajectory. We have seen above how such a trajectory is generated as an optimal solution of the 2PBVP with the boundary conditions specified at both the ends of the control interval, for example,

$$h(0) = 0, \tag{4.141}$$

$$v(0) = v_0, \tag{4.142}$$

$$h(t_f) = h_f, \tag{4.143}$$

$$\phi(t_f) = \phi_f, \tag{4.144}$$

$$v(t_f) = v_f. \tag{4.145}$$

Here we present a nominal gravity-turn trajectory example with modulated forward acceleration, $u_1(t)$, for launch to a circular orbit, which will be later used to design closed-loop tracking systems.

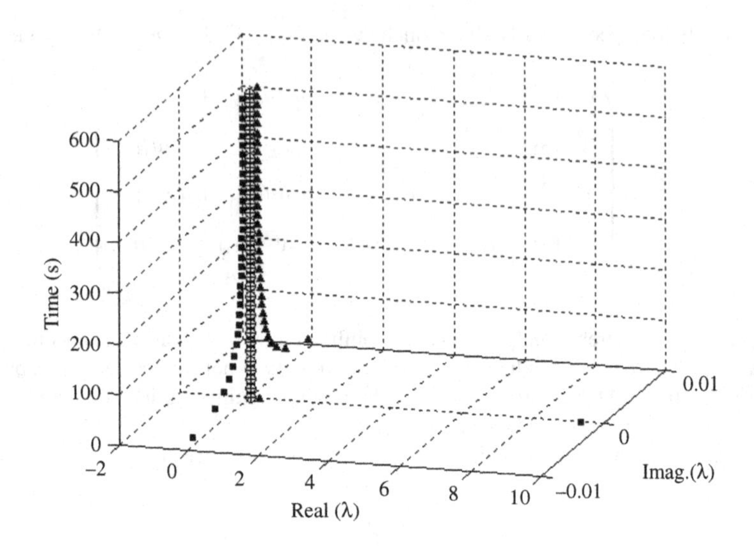

Figure 4.37 Guidance plant eigenvalues along the nominal gravity-turn trajectory of a launch vehicle

Example 4.8 Consider the launch of a rocket from the Earth's surface ($g_0 = 9.798275$ m/s^2, $r_0 = 6378.14$ km) to a 300 km high circular orbit, with an initial speed of $v_0 = 0.001$ km/s. The total flight time is $t_f = 600$ s. The nominal gravity-turn trajectory and the modulated control input obtained from the numerical solution of the optimal 2PBVP are tabulated in Table 4.13. The required orbital speed,

$$v(t_f) = \sqrt{\frac{g_0 r_0^2}{r_0 + h_f}} = 7.72576 \text{ km/s},$$

is seen to be reached with a zero flight-path angle at final altitude $h_f = 300$ km, just prior to burn-out in space, that is, $u_1(t_f) = 0$.

The eigenvalues of the plant and its controllability with only the forward acceleration control input are examined in Figures 4.37 and 4.38, respectively. As expected, the plant is unstable, with the positive real eigenvalues (circles in Figure 4.37) diminishing in magnitude as the speed increases.[4] One eigenvalue – corresponding to the uncoupled δ mode – always remains at the origin (marked by a + sign in Figure 4.37). The imaginary parts of the complex conjugate poles (solid squares and triangles in Figure 4.37) diminish with the increasing speed, thus exhibiting a coalescing tendency of the complex poles. At very low speeds shortly after launch, both of the complex eigenvalues migrate into the right-half plane, but remain in the left-half plane at higher speeds. Also as expected, and as seen in Figure 4.38, with only the forward acceleration input the plant is uncontrollable (unstabilizable) at launch, but becomes controllable as the flight speed builds up. The condition number of $\mathbf{P}$ in the controllable case is quite large throughout the trajectory, indicating a weakly controllable plant due to the δ mode.

[4] The real part of positive real eigenvalues is not visible for $t > 0$ in Figure 4.37 due to scale, where two poles (+ and ∘) appear to be on the imaginary axis at each instant. At burn-out ($t = t_f$), when the speed equals that of the circular orbit, the real part becomes zero, implying a double pole at origin (unstable).

Table 4.13 Nominal gravity-turn trajectory and control input

t (s)	h_n(km)	v_n(km/s)	ϕ_n(deg.)	u_{n1} (m/s^2)
0	0	0.0010	80.6050	48.0037
6.0606	0.5999	0.2336	53.5197	45.8082
12.1212	2.2035	0.4599	47.0522	43.8710
18.1818	4.6311	0.6785	42.8007	42.0906
24.2424	7.7533	0.8895	39.5566	40.4314
30.3030	11.4663	1.0932	36.9036	38.8732
36.3636	15.6834	1.2897	34.6449	37.4023
42.4242	20.3307	1.4795	32.6717	36.0089
48.4848	25.3440	1.6626	30.9150	34.6854
54.5455	30.6669	1.8394	29.3302	33.4256
60.6061	36.2501	2.0102	27.8855	32.2244
66.6667	42.0494	2.1752	26.5577	31.0776
72.7273	48.0260	2.3346	25.3291	29.9815
78.7879	54.1448	2.4886	24.1863	28.9328
84.8485	60.3749	2.6375	23.1185	27.9286
90.9091	66.6881	2.7815	22.1167	26.9664
96.9697	73.0603	2.9206	21.1741	26.0440
103.0303	79.4686	3.0553	20.2845	25.1593
109.0909	85.8927	3.1855	19.4427	24.3102
115.1515	92.3156	3.3115	18.6447	23.4952
121.2121	98.7203	3.4334	17.8865	22.7125
127.2727	105.0929	3.5515	17.1649	21.9607
133.3333	111.4209	3.6658	16.4773	21.2385
139.3939	117.6923	3.7764	15.8209	20.5445
145.4545	123.8973	3.8836	15.1936	19.8776
151.5152	130.0268	3.9875	14.5934	19.2366
157.5758	136.0723	4.0881	14.0186	18.6204
163.6364	142.0274	4.1856	13.4675	18.0281
169.6970	147.8863	4.2802	12.9388	17.4588
175.7576	153.6429	4.3719	12.4310	16.9116
181.8182	159.2922	4.4608	11.9430	16.3856
187.8788	164.8306	4.5471	11.4737	15.8800
193.9394	170.2548	4.6309	11.0220	15.3941
200.0000	175.5617	4.7122	10.5872	14.9272
206.0606	180.7485	4.7912	10.1682	14.4785
212.1212	185.8133	4.8679	9.7642	14.0475
218.1818	190.7549	4.9425	9.3748	13.6336
224.2424	195.5718	5.0150	8.9990	13.2361
230.3030	200.2629	5.0855	8.6362	12.8544
236.3636	204.8276	5.1540	8.2859	12.4882
242.4242	209.2660	5.2208	7.9476	12.1367
248.4848	213.5777	5.2858	7.6207	11.7996
254.5455	217.7628	5.3491	7.3048	11.4763
260.6061	221.8217	5.4108	6.9993	11.1665
266.6667	225.7550	5.4709	6.7040	10.8696
272.7273	229.5636	5.5296	6.4183	10.5853
278.7879	233.2480	5.5869	6.1420	10.3131
284.8485	236.8095	5.6428	5.8747	10.0526
290.9091	240.2491	5.6975	5.6160	9.8035
296.9697	243.5682	5.7509	5.3657	9.5655

Table 4.13 *(Continued)*

t (s)	h_n(km)	v_n(km/s)	ϕ_n(deg.)	u_{n1} (m/s^2)
303.0303	246.7681	5.8031	5.1234	9.3380
309.0909	249.8502	5.8543	4.8889	9.1209
315.1515	252.8161	5.9043	4.6620	8.9138
321.2121	255.6675	5.9534	4.4424	8.7164
327.2727	258.4062	6.0015	4.2299	8.5283
333.3333	261.0339	6.0487	4.0243	8.3493
339.3939	263.5522	6.0950	3.8253	8.1790
345.4545	265.9632	6.1405	3.6328	8.0172
351.5152	268.2689	6.1853	3.4466	7.8637
357.5758	270.4712	6.2293	3.2666	7.7181
363.6364	272.5721	6.2726	3.0925	7.5801
369.6970	274.5737	6.3153	2.9243	7.4496
375.7576	276.4780	6.3574	2.7617	7.3262
381.8182	278.2872	6.3989	2.6048	7.2098
387.8788	280.0036	6.4398	2.4533	7.1000
393.9394	281.6292	6.4803	2.3071	6.9968
400.0000	283.1664	6.5202	2.1661	6.8997
406.0606	284.6172	6.5598	2.0302	6.8087
412.1212	285.9840	6.5989	1.8994	6.7234
418.1818	287.2690	6.6377	1.7734	6.6437
424.2424	288.4746	6.6761	1.6523	6.5694
430.3030	289.6031	6.7142	1.5360	6.5002
436.3636	290.6570	6.7520	1.4244	6.4360
442.4242	291.6384	6.7895	1.3173	6.3765
448.4848	292.5497	6.8268	1.2148	6.3215
454.5455	293.3935	6.8638	1.1168	6.2709
460.6061	294.1722	6.9007	1.0232	6.2245
466.6667	294.8881	6.9374	0.9340	6.1821
472.7273	295.5437	6.9739	0.8491	6.1434
478.7879	296.1416	7.0102	0.7685	6.1083
484.8485	296.6843	7.0465	0.6921	6.0766
490.9091	297.1742	7.0826	0.6199	6.0481
496.9697	297.6139	7.1186	0.5518	6.0227
503.0303	298.0060	7.1545	0.4879	6.0002
509.0909	298.3532	7.1904	0.4280	5.9803
515.1515	298.6579	7.2262	0.3721	5.9629
521.2121	298.9229	7.2620	0.3203	5.9479
527.2727	299.1509	7.2977	0.2724	5.9351
533.3333	299.3444	7.3334	0.2285	5.9242
539.3939	299.5063	7.3691	0.1885	5.9152
545.4545	299.6392	7.4048	0.1524	5.9078
551.5152	299.7460	7.4404	0.1203	5.9019
557.5758	299.8294	7.4761	0.0919	5.8973
563.6364	299.8923	7.5117	0.0674	5.8939
569.6970	299.9375	7.5474	0.0467	5.8914
575.7576	299.9679	7.5830	0.0299	5.8897
581.8182	299.9864	7.6187	0.0168	5.8887
587.8788	299.9960	7.6544	0.0074	5.8882
593.9394	299.9995	7.6901	0.0019	5.8880
600.0000	300.0000	7.7258	0	5.8880

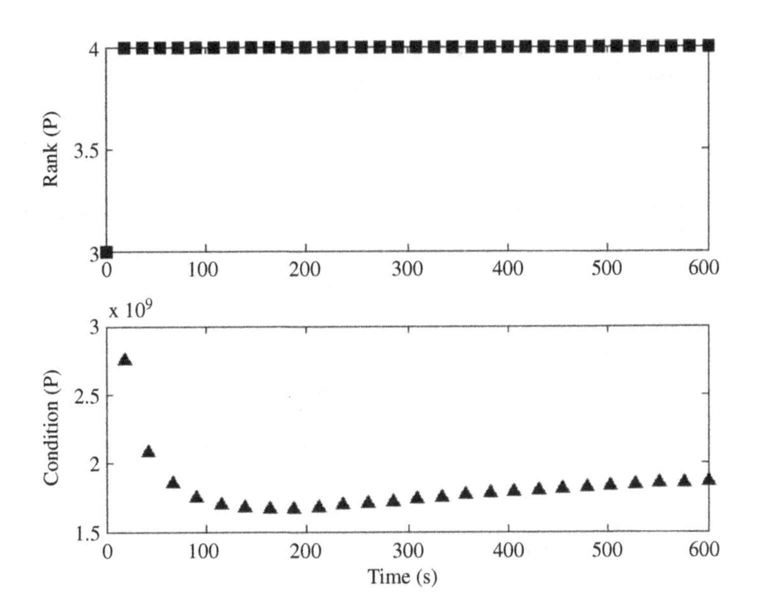

Figure 4.38 Rank and condition number of controllability test matrix with only forward acceleration input along the nominal gravity-turn trajectory of a launch vehicle

4.10 Robust and Adaptive Guidance

As discussed above, the equations of motion (equation (4.112)) of a planar flight are essentially nonlinear, but can be linearized about a reference (nominal) trajectory. In such a case, a linear feedback control law could be designed in order to keep the small deviations from nominal from growing large. Such a stabilizing controller, however, requires that its gains, $\mathbf{K}(t)$, should be updated in real time by solving a time-varying control problem. While it may be possible to do so for a slowly varying plant (such as a spacecraft), the rapidly changing trajectory of a rocket precludes real-time computation of controller gains by even the simplest approach – a linear time-invariant (LTI) approximation at every time instant. Instead, a way must be found for varying the gains with the changing flight variables according to pre-selected functions called *adaptation laws*. Thus, even a linear feedback control system becomes essentially nonlinear when controller gains are adapted to changing trajectory.

The simple approach of stabilizing a time-varying plant by treating it as if it were frozen at a given instant is likely to succeed if the mathematical character of the linearized plant does not change drastically with time and the feedback gains are not too large (implying large control inputs at any time). In other words, if the plant matrices, $\mathbf{A}(t)$, $\mathbf{B}(t)$, remain bounded and smoothly varying with time, one can have confidence in applying the results of an LTI stability analysis for the time-varying system. Generally, caution must be exercised in interpreting stability from the instantly frozen system eigenvalues, since a rapidly time-varying system could be unstable even though all its eigenvalues always lie in the left-half complex plane.[5] However, since the linearized rocket guidance plant after launch ($v > 0$) does not exhibit any drastic discontinuity, singularity, or unbounded behavior with time, we can expect a locally LTI (or quasi-steady) approximation to be valid for $t > 0$.

[5] An example of a linear, time-varying, unstable system with all eigenvalues always in the left-half plane is that given by the differential equations, $\dot{x}_1 = -ax_1 + e^{bt}x_2$, $\dot{x}_2 = -cx_2$, where a, b, c are positive, real constants.

The nominal control, $\mathbf{u}_n(t)$, is the nonlinear feedforward control, pre-selected to achieve the nominal trajectory. It is either prescribed as the solution of an optimal boundary-value problem, or assumed to be solvable from the nominal state equation (4.113), by *dynamic inversion* as follows:

$$\mathbf{u}_n = \mathbf{g}(\mathbf{x}_n, \dot{\mathbf{x}}_n). \quad (4.146)$$

Assuming small deviations from a nominal flight path, a linear, state feedback guidance law is sufficient to maintain the vehicle on the nominal trajectory at all times,

$$\Delta\mathbf{u} = -\mathbf{K}(t)\Delta\mathbf{x}, \quad (4.147)$$

or

$$\mathbf{u}(t) = \mathbf{u}_n(t) + \Delta\mathbf{u} = \mathbf{u}_n(t) + \mathbf{K}(t)[\mathbf{x}_n(t) - \mathbf{x}(t)], \quad (4.148)$$

where $\mathbf{K}(t)$ is the time-varying feedback gain matrix that may be computed at each instant by an LTI method, such as linear optimal control (Chapter 2). However, with an observer-based feedback, the feedback control law changes to

$$\Delta\mathbf{u} = \mathbf{K}\left(\mathbf{x}_n - \mathbf{x}_o\right) = -\mathbf{K}\left(\Delta\mathbf{x} + \mathbf{e}_o\right), \quad (4.149)$$

where the estimated state, $\mathbf{x}_o(t)$, and estimation error, $\mathbf{e}_o(t) = \mathbf{x}(t) - \mathbf{x}_o(t)$, are based upon the measurement of output variables, $\mathbf{y}(t)$, that are generally a linear combination of the state and control variables according to the output equation,

$$\mathbf{y}(t) = \mathbf{C}(t)\mathbf{x}(t) + \mathbf{D}(t)\mathbf{u}(t). \quad (4.150)$$

It is often advantageous to use a linear quadratic Gaussian (LQG) compensator for feedback control due to its good robustness properties, which we shall do here. The plant linearized about the nominal trajectory is described by

$$\Delta\dot{\mathbf{x}} = \mathbf{A}\Delta\mathbf{x} + \mathbf{B}\Delta\mathbf{u} + \mathbf{F}\boldsymbol{\nu}, \quad \Delta\mathbf{x}(0) = \Delta\mathbf{x}_0, \quad (4.151)$$

$$\Delta\mathbf{y} = \mathbf{C}\Delta\mathbf{x} + \mathbf{D}\Delta\mathbf{u} + \mathbf{w}, \quad (4.152)$$

where $\mathbf{y}(t) = \mathbf{y}_n(t) + \Delta\mathbf{y}(t)$ is the measured output, $\boldsymbol{\nu}(t)$ the process noise, and $\mathbf{w}(t)$ the measurement noise. A full-order observer designed as a steady-state Kalman filter[6] at each time instant with given plant matrices, $\mathbf{A}, \mathbf{B}, \mathbf{F}, \mathbf{C}, \mathbf{D}$, white noise with process noise spectral density matrix, $\mathbf{S}_\nu$, measurement noise spectral density matrix, $\mathbf{S}_\mathbf{w}$, and cross-spectral density matrix between process and measurement noise, $\mathbf{S}_{\mathbf{w}\nu}$, has the state equation

$$\dot{\mathbf{x}}_o = (\mathbf{A} - \mathbf{L}\mathbf{C})\,\mathbf{x}_o + \mathbf{B}\Delta\mathbf{u} + \mathbf{L}\Delta\mathbf{y}. \quad (4.153)$$

The steady-state Kalman filter gain matrix, $\mathbf{L}$, is given by

$$\mathbf{L} = \left(\mathbf{P}\mathbf{C}^T + \mathbf{F}\mathbf{S}_{\mathbf{w}\nu}\right)\mathbf{S}_\mathbf{w}^{-1}, \quad (4.154)$$

where $\mathbf{P}$ is the optimal covariance matrix, the unique, symmetric, and positive semi-definite solution to the following steady-state, algebraic Riccati equation:

$$\mathbf{0} = \mathbf{A}_G\mathbf{P} + \mathbf{P}\mathbf{A}_G^{\mathrm{T}} - \mathbf{P}\mathbf{C}^{\mathrm{T}}\mathbf{S}_\mathbf{w}^{-1}\mathbf{C}\mathbf{P} + \mathbf{F}\mathbf{S}_G\mathbf{F}^{\mathrm{T}}, \quad (4.155)$$

[6] The extension to a reduced-order observer is easily carried out (Appendix A); thus there is no loss of generality.

and

$$\mathbf{A}_G = \mathbf{A} - \mathbf{F}\Psi\mathbf{W}^{-1}\mathbf{C},$$
$$\mathbf{S}_G = \mathbf{S}_\nu - \mathbf{S}_{\mathbf{w}\nu}\mathbf{S}_\mathbf{w}^{-1}\mathbf{S}_{\mathbf{w}\nu}^T. \tag{4.156}$$

When the process and measurement white noise are uncorrelated (as they usually are), we have $\mathbf{S}_{\mathbf{w}\nu} = \mathbf{0}$ and a considerable simplification takes place in the Kalman filter derivation. A sufficient condition for the existence of a unique, positive semi-definite solution to equation (4.155) is that the plant $(\mathbf{A}, \mathbf{C})$ is observable, $\mathbf{S}_G$ is positive semi-definite, and $\mathbf{S}_\mathbf{w}$ is positive definite. Generally, the matrices $\mathbf{S}_\nu$, $\mathbf{S}_\mathbf{w}$, $\mathbf{S}_{\mathbf{w}\nu}$ are treated as design parameters for achieving a desired robustness through loop-transfer recovery (LTR; Chapter 2). Alternative design techniques, such as H_∞, can also achieve robustness objectives in a manner similar to the LQG/LTR method.

Having selected a method for computing the feedback controller gains at every time instant, we are left with the problem of deriving empirical adaptation laws for changing the gains with measured output variables, such as

$$\mathbf{K} = \mathbf{G}_\mathbf{K}(\mathbf{y}),$$
$$\mathbf{L} = \mathbf{G}_\mathbf{L}(\mathbf{y}). \tag{4.157}$$

A block diagram of an adaptive guidance system is shown in Figure 4.39, where the adaptation loops are nonlinear feedback from output variables, $\mathbf{y}(t)$, to the controller and observer gain matrices, $\mathbf{K}(t)$, $\mathbf{L}(t)$, respectively. Even though an adaptive, nonlinear control theory has been devised (Åström and Wittenmark 1995; Slotine and Li 1991), there is no uniform procedure in existence that can be generally applied across the board, and a designer has to resort to trial and error for a given system. We will consider an adaptation method called *gain scheduling* that is commonly applied to flight control wherein pre-determined functions of control system gains with certain flight parameters are stored onboard a flight control computer. At every instant, the concerned flight variables – such as the altitude and flight speed – are sensed either by onboard sensors, or fed through telemetry by a ground-based radar station, and are finally used to determine the control system gains.

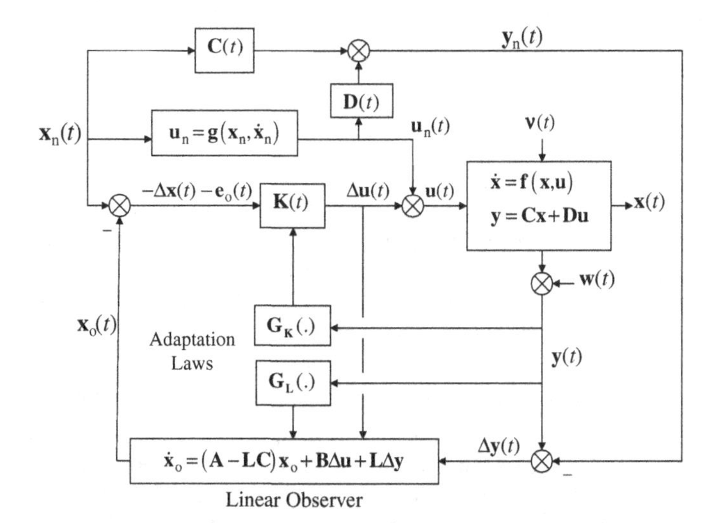

Figure 4.39 Closed-loop guidance system with nonlinear feedforward control, $\mathbf{u}_n(t)$, an observer-based linear feedback control, $\Delta u = \mathbf{K}(t)[\Delta x(t) - \mathbf{e}_o(t)]$, and nonlinear adaptation loops, $(\mathbf{K}, \mathbf{L}) = [\mathbf{G}_\mathbf{K}(\mathbf{y}), \mathbf{G}_\mathbf{L}(\mathbf{y})]$

In order to derive the adaptation laws for a rocket, we first solve the steady-state, infinite-time, optimal control problem for the regulator and Kalman filter, essentially treating the plant as if it were frozen in time at every instant along the nominal trajectory. Then the time-varying regulator and Kalman filter gains, $\mathbf{K}(t)$, $\mathbf{L}(t)$, are plotted with time and their functional behavior with a pre-eminent flight variable (e.g., the flight speed, v) is derived by curve-fitting, resulting in approximate schedules for varying gain matrices, $\mathbf{K}(v)$, $\mathbf{L}(v)$. The following applications illustrate the gain scheduling approach for the rocket guidance problem.

4.11 Guidance with State Feedback

Consider a gravity-turn launch to a circular orbit. The most common guidance task in such a case is to regulate the altitude, speed, and flight-path angle. A separate attitude control system maintains the vehicle's longitudinal axis along the flight direction. The guidance system should be capable of mitigating initial errors in (h, v, ϕ) from the nominal trajectory such that the terminal conditions are precisely met, thereby leaving the final orbit unaffected. However, it is not important to have a specific terminal downrange angle in the final circular orbit, where phasing orbital maneuvers can be performed subsequent to orbital injection in order to carry out a rendezvous mission. In fact, a brief *coasting* (or zero thrust) phase usually follows burn-out, in which final orbital and attitude corrections (including spin-up of the payload if necessary) are performed before releasing the satellite into orbit.

4.11.1 Guidance with Normal Acceleration Input

Rocket engines are quite finicky in the sense that their thrust cannot be controlled with great precision. Solid propellant rockets have no mechanism for throttling or even shutting down the engines once ignited. Liquid propellant engines can be throttled up to some extent by regulating the supply of propellants through valves. However, a propellant flow rate below a certain limit cannot sustain combustion, resulting in flame-out (zero thrust). Even when throttling is possible, the engine response is highly nonlinear. Due to difficulty in throttling a large rocket engine, most launch vehicles employ guidance solely through the normal acceleration control input, $\Delta u_2(t)$, which is easily created by gimballing the rocket nozzle along an axis normal to the plane of flight (called *pitch axis*) by a small deflection angle, μ_1, as shown in Figure 4.15. In the absence of a significant lift as the attitude controller maintains a zero angle of attack, the normal acceleration is produced entirely by μ_1, and is given by

$$u_2 = T \sin \mu_1 \simeq T\mu_1, \tag{4.158}$$

which enables a linear plant if μ_1 is taken to be the control input. The rocket thrust, f_T, is always assumed to be nominal as it is independent of flight-path perturbations. Having normal acceleration (rather than forward acceleration) as the control input is also advantageous from the propulsive efficiency viewpoint, because operation at off-nominal thrust magnitudes is avoided.

Since the nonlinear equations of motion (equation (4.112)) of a planar flight are linearized about a reference (nominal) trajectory, a linear feedback control law,

$$\Delta u_2 = -\mathbf{K}(t)\Delta \mathbf{x}, \tag{4.159}$$

can be designed in order to keep the small deviations from nominal from growing large. For such a controller to be stabilizing requires that its gain matrix, $\mathbf{K}(t)$, should be updated in real time by solving a time-varying control problem. However, it is much more practical to select a constant set of gains that are capable of stabilizing the trajectory in the presence of the largest expected disturbances.

The guidance system based upon regulation of $\Delta h, \Delta v, \Delta\phi$ does not require the measurement of the downrange angle, δ, thus one can work with a simpler guidance plant, $\Delta \mathbf{x} = (\Delta h, \Delta v, \Delta\phi)^T$, that is always controllable with the normal acceleration input, $\Delta u = \Delta u_2$, and has the following state-space coefficient matrices:

$$\mathbf{A} = \begin{pmatrix} 0 & a_{13} & a_{14} \\ a_{31} & 0 & a_{34} \\ a_{41} & a_{43} & a_{44} \end{pmatrix}, \quad \mathbf{B} = \begin{pmatrix} 0 \\ 0 \\ \frac{1}{v} \end{pmatrix}, \tag{4.160}$$

where the coefficients a_{ij} are the same as those in equation (4.119). With a full-state feedback law, $u_2(t) = -\mathbf{K}(t)[\Delta h, \Delta v, \Delta\phi]$, where

$$\mathbf{K}(t) = \left[k_{\rm h}(t),\ k_{\rm v}(t),\ k_{\phi}(t)\right], \tag{4.161}$$

one can use a standard linear design approach with fixed gains, as demonstrated in the following example.

Example 4.9 For the gravity-turn trajectory of Example 4.8, design a control system for regulating the altitude, speed, and flight-path angle using full-state feedback and normal acceleration control input. The gimballing actuator has a saturation limit of $|\Delta u_2| \leq 5$ m/s^2 and a deadband of ± 0.4 m/s^2 (i.e., the actuator does not respond to commanded normal acceleration less than 0.4 m/s^2 in magnitude). Design a state feedback regulator for meeting the terminal conditions ($t = 600$ s) of circular orbit injection with an error not exceeding ± 50 m altitude and ± 10 m/s speed in the presence of an initial flight-path error of $\Delta\phi = 0.1$ rad at $t = 0$.

For a full-state feedback controller based upon regulation of Δh, Δv, $\Delta\phi$, the nominal trajectory variables, $h_{\rm n}(t)$, $v_{\rm n}(t)$, $\phi_{\rm n}(t)$, $u_{\rm n1}(t)$, specified in Tables 4.13 and 4.14, are converted into radius, rn (km), speed, vn (km/s), flight-path angle, phin (rad), and forward acceleration input, un1 km/s^2, and stored in the MATLAB workspace as the row vectors, along with the row vector, tn, of time instants of the same column dimension. After some preliminary trials, we select the following constant feedback gains:

$$k_{\rm h} = 0.01\ {\rm s}^{-2}, \quad k_{\rm v} = 0.1\ {\rm s}^{-1}, \quad k_{\phi} = 1.6\ {\rm km/s^2/rad}.$$

The simulation of the actual nonlinear time-varying system with constant feedback gains is carried out using the Simulink® model shown in Figure 4.40, with the nonlinear subsystem block named *Rocket*

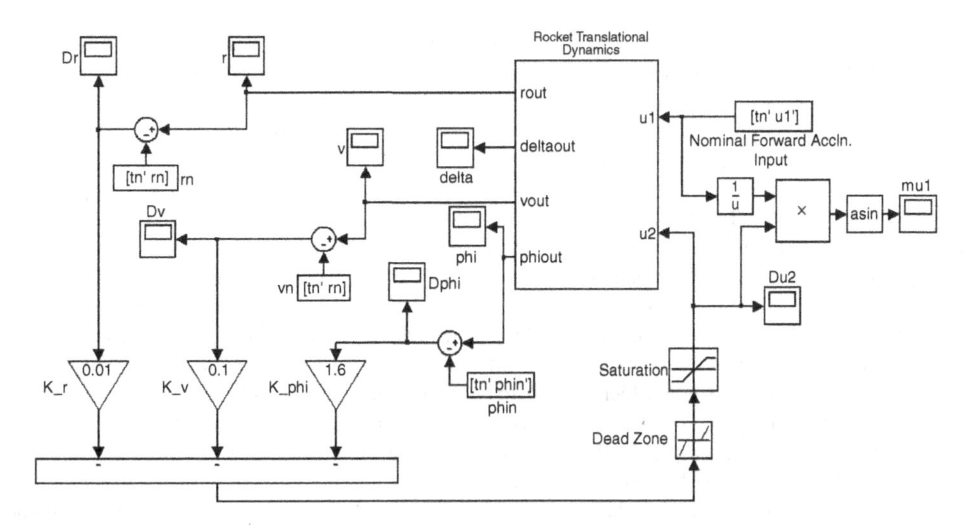

Figure 4.40 Simulink block diagram for simulating closed-loop initial response of state feedback rocket guidance system with nonlinear rocket plant

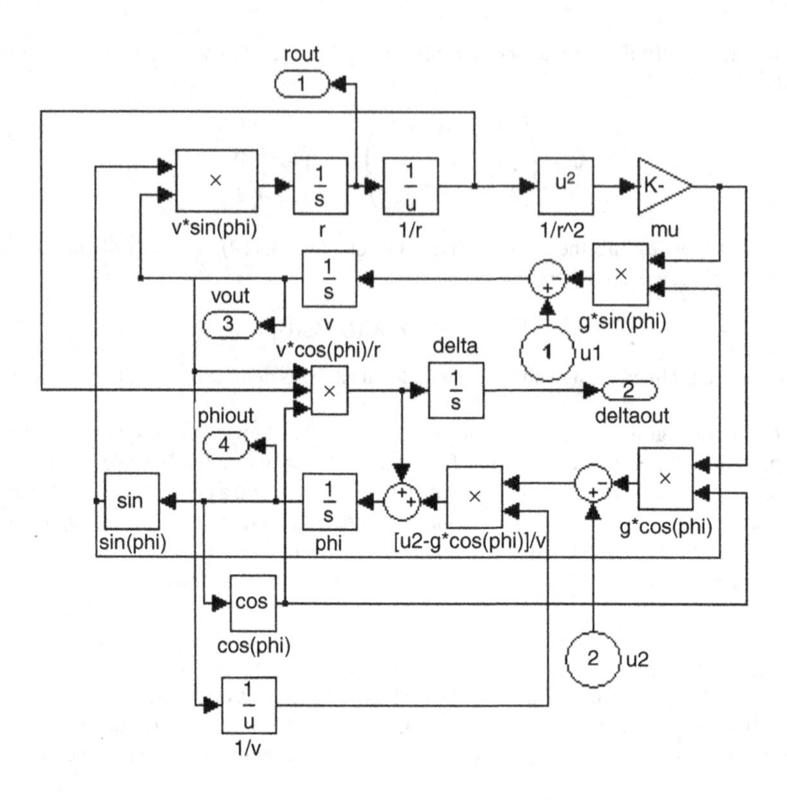

Figure 4.41 Simulink *Rocket Translational Dynamics* subsystem block for simulating nonlinear rocket translational dynamics in vertical plane

Translational Dynamics depicted in Figure 4.41. The simulation blocks use the radius, $r = r_0 + h$, instead of the altitude, h. The initial conditions of $r = 6378.14$ km, $v = 0.001$ km/s, and $\phi = 1.4068 + 0.1$ rad applied at $t = 0$ are specified in the respective integrator blocks of the *Rocket Translational Dynamics* subsystem (Figure 4.41). Note that although only h, v, ϕ are being controlled, the completely nonlinear, vertical plane rocket dynamics – including the downrange angle, δ – is being simulated in the *Rocket Translational Dynamics* subsystem. The results of the simulation carried out with 100 data points of the nominal trajectory (Tables 4.13 and 4.14) and a fourth-order Runge–Kutta scheme are plotted in Figures 4.42–4.45. The initial error of 0.1 rad (5.7°) in the flight-path angle – and the attendant undershoot in flight speed – is successfully seen to be attenuated in about 50 s (Figure 4.43), without affecting the desired terminal conditions in any way. Figure 4.43 shows that the terminal altitude and speed errors are only 3 m and −3.8 m/s respectively, while the terminal flight-path angle error is virtually zero. The gimballing actuator is initially saturated with input, $\Delta u_2 = -5$ m/s^2 (Figure 4.44), but quickly settles down to a near-zero value, which appears to be acceptable given the rather large initial flight-path error. Also plotted in Figure 4.44 is the simulated thrust deflection angle, μ_1, which initially deflects to −6.4° for correcting the flight-path angle, and thereafter has small amplitude (±1°) damped oscillations whenever flight-path and speed errors occur. Figure 4.45 focusses on the simulated closed-loop flight-path angle error in the first minute of flight, showing the corrective control action applied by the regulator. The success of the fixed gain regulator in guiding a time-varying trajectory is thus demonstrated by this example.

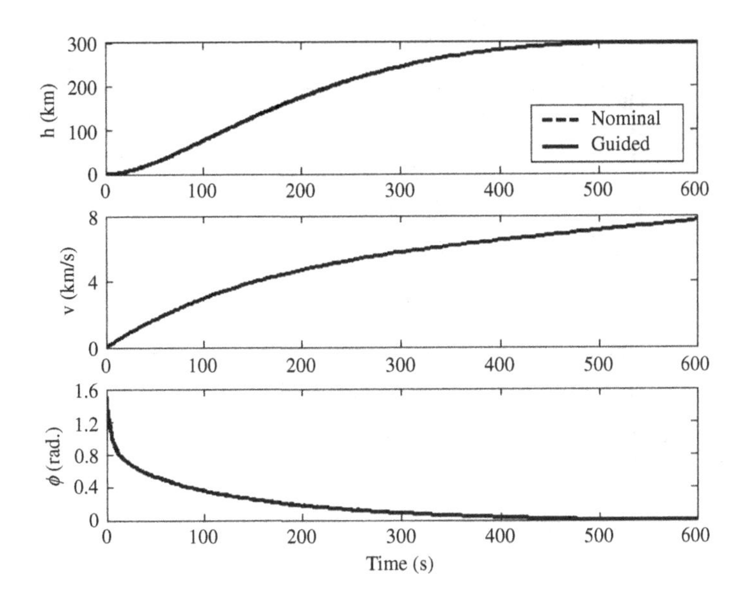

Figure 4.42 Nominal (open-loop) and initially perturbed (closed-loop) trajectories with state feedback rocket guidance system for gravity-turn launch to a circular orbit

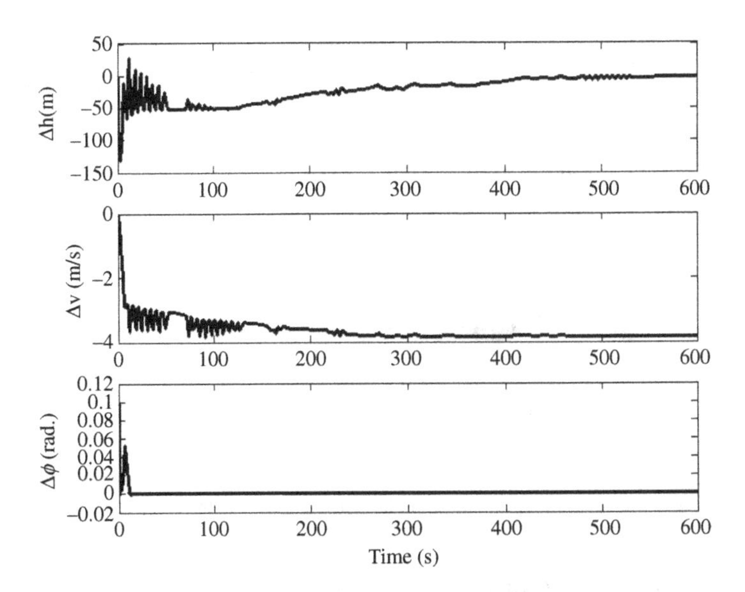

Figure 4.43 Closed-loop simulated trajectory errors with initial flight-path perturbation using state feedback rocket guidance for gravity-turn launch to a circular orbit

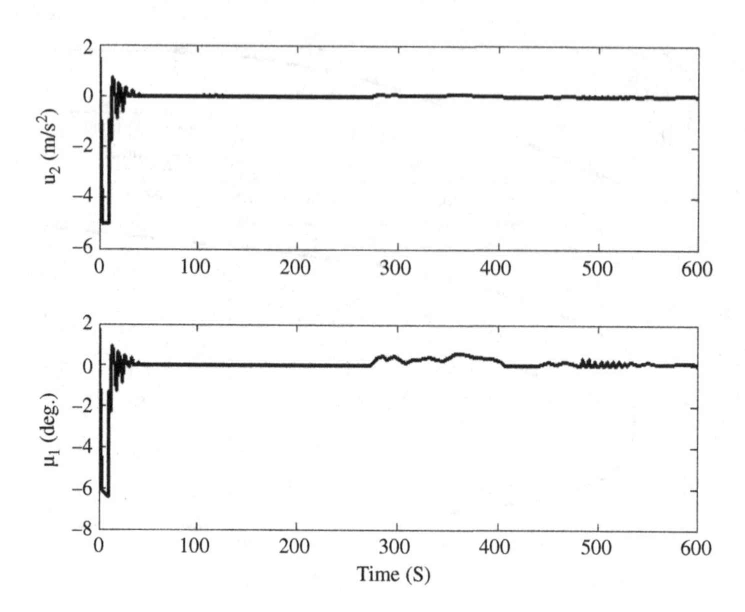

Figure 4.44 Normal acceleration input and thrust deflection angle with state feedback rocket guidance for gravity-turn launch to a circular orbit

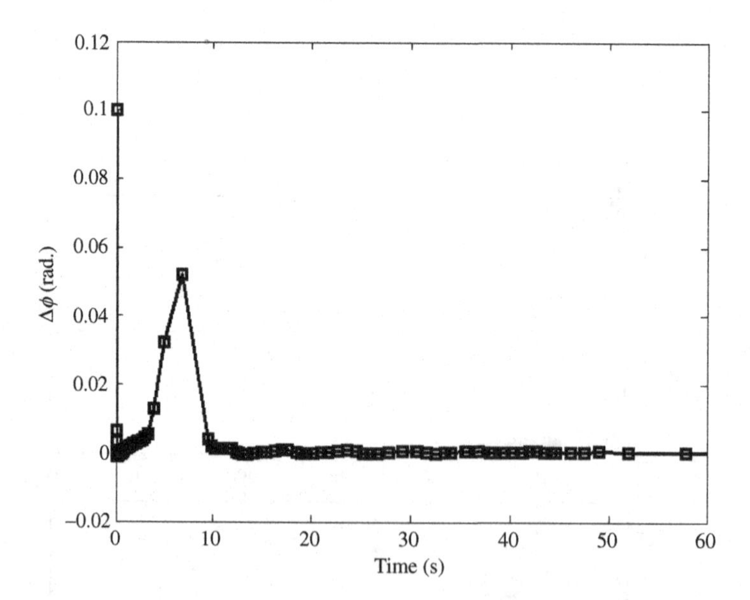

Figure 4.45 Closed-loop simulated flight-path angle error in the first minute of flight

4.12 Observer-Based Guidance of Gravity-Turn Launch Vehicle

While the examples of the previous section demonstrated the use of full-state feedback controllers for successfully guiding a launch vehicle, it is not necessary that all the state variables be available for instantaneous measurement. Instead, the state feedback regulator can be replaced by an observer-based output feedback compensator. It should be noted that the third-order guidance plant without downrange dynamics, $\delta(t)$, is observable with a single output variable: either the speed error, Δv, or the altitude error, Δh.[7] Thus, theoretically, a single-output observer based upon either the speed or the altitude measurement can be designed. However, a single-output observer is unlikely to produce an acceptable closed-loop performance and robustness due to poor conditioning of the observability test matrix, $\mathbf{N}$, which leads to a dominant observer pole with a very small real part. Instead, a bi-output observer would prove to be much better from both performance and robustness viewpoints.

4.12.1 Altitude-Based Observer with Normal Acceleration Input

Consider the single output to be the altitude error, $y = \Delta h$, with the partitioned state vector, $\mathbf{x} = (\mathbf{x}_1, \mathbf{x}_2)$, where $\mathbf{x}_1 = \Delta h$, $\mathbf{x}_2 = (\Delta v, \Delta\phi)$, and the following state-space coefficient matrices:

$$\mathbf{A} = \left(\begin{array}{c|cc} 0 & a_{13} & a_{14} \\ \hline a_{31} & 0 & a_{34} \\ a_{41} & a_{43} & a_{44} \end{array}\right), \quad \mathbf{B} = \left(\begin{array}{c} 0 \\ \hline 0 \\ \frac{1}{v_n} \end{array}\right). \tag{4.162}$$

The output equation, $\mathbf{y} = \mathbf{x}_1$, and the partitioning of $\mathbf{A}$, $\mathbf{B}$ reveal the following sub-matrices required for the reduced-order observer:

$$\mathbf{C} = 1, \quad \mathbf{A}_{11} = 0, \quad \mathbf{A}_{12} = (a_{13} \;\; a_{14}), \quad \mathbf{B}_1 = 0, \tag{4.163}$$

$$\mathbf{A}_{21} = \begin{pmatrix} a_{31} \\ a_{41} \end{pmatrix}, \quad \mathbf{A}_{22} = \begin{pmatrix} 0 & a_{34} \\ a_{43} & a_{44} \end{pmatrix}, \quad \mathbf{B}_2 = \begin{pmatrix} 0 \\ \frac{1}{v_n} \end{pmatrix}. \tag{4.164}$$

The state equation of the reduced-order observer with estimated state, $\mathbf{x}_{o2} = \mathbf{L}\mathbf{y} + \mathbf{z}$, and gain matrix, $\mathbf{L} = (L_1, L_2)^T$, is the following (Chapter 2):

$$\dot{\mathbf{z}} = \mathbf{F}\mathbf{z} + \mathbf{H}\mathbf{u} + \mathbf{G}\mathbf{y}, \tag{4.165}$$

where

$$\mathbf{F} = \mathbf{A}_{22} - \mathbf{L}\mathbf{C}\mathbf{A}_{12} = \begin{pmatrix} -L_1 a_{13} & a_{34} - L_1 a_{14} \\ a_{43} - L_2 a_{13} & a_{44} - L_2 a_{14} \end{pmatrix}, \quad \mathbf{H} = \mathbf{B}_2 - \mathbf{L}\mathbf{C}\mathbf{B}_1 = \begin{pmatrix} 0 \\ \frac{1}{v_n} \end{pmatrix}, \tag{4.166}$$

and

$$\mathbf{G} = \mathbf{F}\mathbf{L} + (\mathbf{A}_{21} - \mathbf{L}\mathbf{C}\mathbf{A}_{11})\mathbf{C}^{-1} = \begin{pmatrix} a_{31} - L_1^2 a_{13} + L_2 a_{34} - L_1 L_2 a_{14} \\ a_{41} + L_1 a_{43} - L_1 L_2 a_{13} + L_2 a_{44} - L_2^2 a_{14} \end{pmatrix}. \tag{4.167}$$

The poles of the second-order observer can be readily determined in closed form from the desired observer characteristic equation

$$\begin{aligned} |\lambda \mathbf{I} - \mathbf{F}| &= \lambda^2 + (L_1 a_{13} + L_2 a_{14} - a_{44})\lambda \\ &\quad + L_1(a_{14}a_{43} - a_{13}a_{44}) + L_2 a_{34} a_{13} - a_{34} a_{43}. \end{aligned} \tag{4.168}$$

[7] It can be easily shown that the third-order plant with $y = \Delta\phi$ becomes unobservable for $\phi_n = 0, \pm\pi/2$, which may occur at some point in the nominal trajectory (e.g., for a launch to circular orbit, $\phi_n(t_f) = 0$). Hence, flight-path angle measurement cannot be employed for a single-output observer.

For selected observer natural frequency, ω, and damping ratio, ζ, the observer gains are thus

$$
\begin{aligned}
L_1 &= \frac{\omega^2 + a_{34}a_{43} - \frac{a_{13}a_{34}}{a_{14}}(a_{44} + 2\zeta\omega)}{a_{14}a_{43} - a_{13}a_{44} - \frac{a_{13}^2 a_{34}}{a_{14}}}, \\
L_2 &= \frac{2\zeta\omega + a_{44} - L_1 a_{13}}{a_{14}}.
\end{aligned}
\tag{4.169}
$$

Selection of the second-order observer's characteristics, (ω, ζ), forms the crux of observer design. Care must be exercised in selecting the observer poles for the rocket guidance problem with thrust acceleration input, because a real part of too small a magnitude may cause the observer to interact with the dominant closed-loop dynamics, thereby amplifying the transient response as well as increasing the settling time. On the other hand, observer poles too deep into the left-half plane would require a large controller activity that can either exceed the thrust limitations of the rocket or destabilize the time-varying system. Thus, a balance must be struck between the speed of response and observer gain magnitudes.

Example 4.10 Design a reduced-order observer for the gravity-turn trajectory guidance system of Example 4.9 using the altitude error, Δh, as the only output. Replace the full-state feedback controller of Example 4.9 with the observer-based output feedback compensator and simulate the closed-loop response to the given initial trajectory perturbation.

Ideally, a likely observer performance criterion would be a well-damped step response with settling time a fraction of that of the state feedback system. However, when applied to rocket guidance problem, one must also consider the transient control input limits that may be exceeded and the time-varying system destabilized if the observer gains are either too large, or too small. Given the regulator of Example 4.9 with a closed-loop settling time of about 50 s, we initially select the second-order observer's natural frequency, $\omega = 0.1\sqrt{2}$ rad/s and a critical damping ratio of $\zeta = 1/\sqrt{2}$, which corresponds to an observer settling time of about 40 s. The observer gains (equation (4.169)) are plotted in Figure 4.46, while the observer poles computed along the nominal trajectory are constant at $-0.1 \pm 0.1i$.

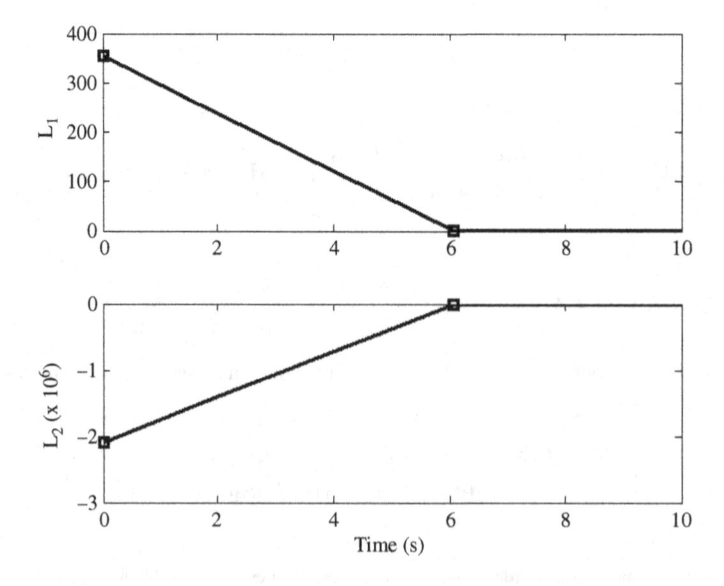

Figure 4.46 Reduced-order observer gains with altitude measurement as the sole output, for the gravity-turn launch to a circular orbit

Note the large observer gain magnitudes at $t = 0$, which quickly reduce to almost zero at the next time instant ($t = 6.06$ s). This is due to an ill-conditioned observability test matrix at the initially small speed. Any attempt to use the large observer gains shown in Figure 4.46 is certain to destabilize the time-varying closed-loop system. Hence, we take $L_1 = L_2 = 0$ for the entire trajectory, which implies the following state equation for the estimated state vector:

$$\dot{\mathbf{x}}_{o2} = \mathbf{A}_{22}\mathbf{x}_{o2} + \mathbf{B}_2\mathbf{u} + \mathbf{A}_{21}\mathbf{y}. \tag{4.170}$$

As the observer is initially unstable, it is necessary to reduce the regulator gain, k_ϕ, from 1.6 to 0.65, which would otherwise amplify the initial error, $\Delta\phi$. In order to compensate for the loss of closed-loop performance due to a reduction in k_ϕ magnitude, we increase the altitude feedback gain, k_h, from 0.01 to 0.2. Thus the constant regulator gains are revised to the following values:

$$k_h = 0.2 \text{ s}^{-2}, \quad k_v = 0.1 \text{ s}^{-1}, \quad k_\phi = 0.65 \text{ km/s}^2\text{/rad},$$

which are substituted into the control law for the compensated system,

$$\Delta u_2 = -k_h \Delta h - \begin{pmatrix} k_v, & k_\phi \end{pmatrix} \mathbf{x}_{o2}.$$

Figure 4.47 shows the Simulink block diagram used for simulating the closed-loop compensated guidance system with a fourth-order Runge–Kutta method. The *Rocket Coefficients*, *Observer Coefficients*, and *Adaptive Observer* subsystem blocks required in the simulation are given in Figures 4.48, 4.49, and 4.50, respectively. The nominal trajectory and forward acceleration input (Tables 4.13 and 4.14)

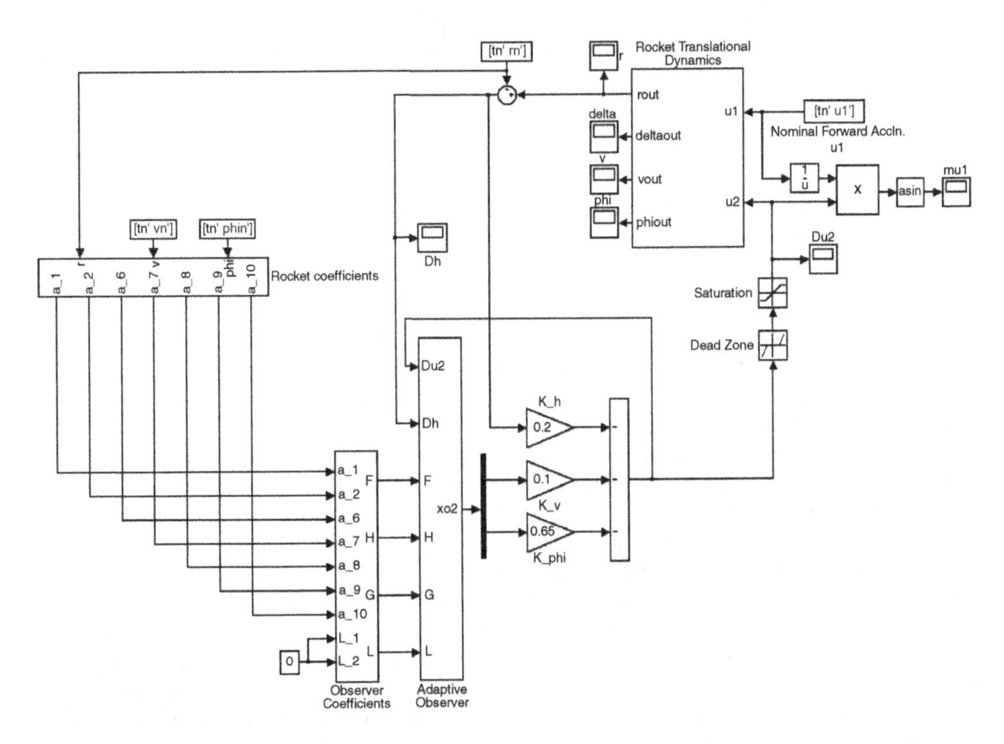

Figure 4.47 Simulink block diagram for simulating closed-loop initial response of nonlinear rocket guidance system with reduced-order, single-output observer

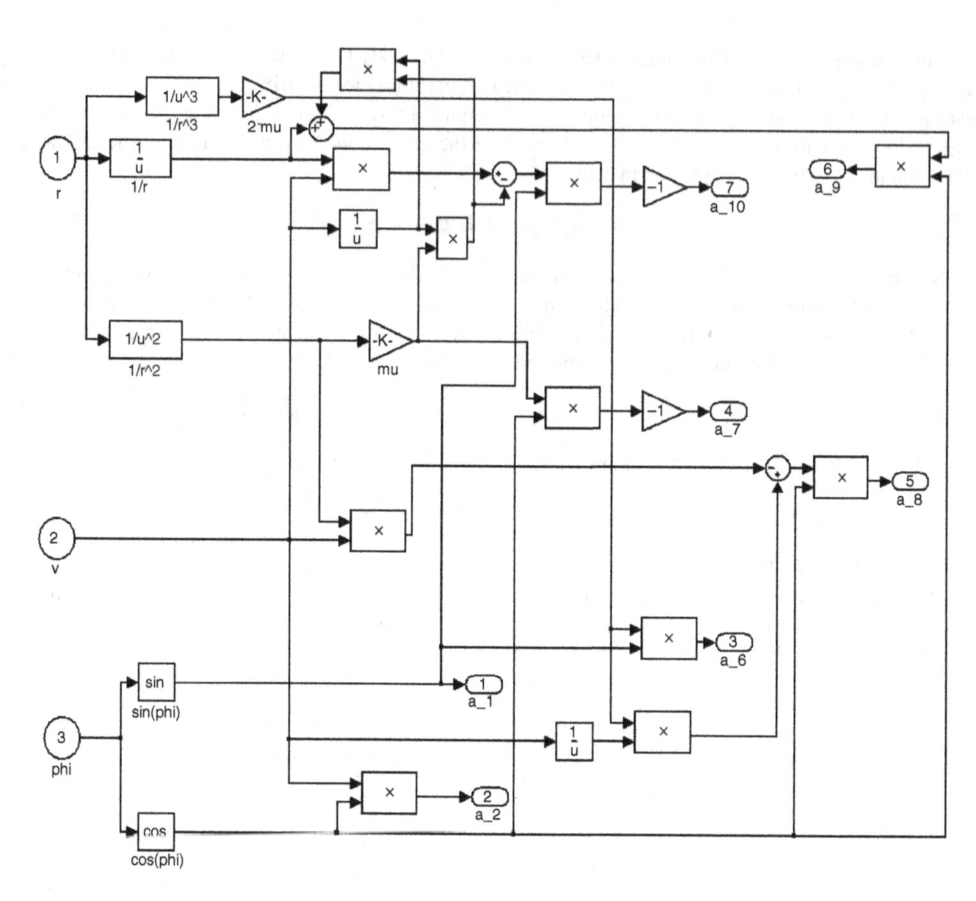

Figure 4.48 Simulink *Rocket Coefficients* subsystem block for computing linear observer coefficient matrices required by the block diagram of Figure 4.47

are stored *a priori* in the MATLAB workspace as global variables tn, rn, vn, phin, un1. The results of the simulation for initial trajectory error, $\Delta\phi(0) = 0.1$ rad, are plotted in Figures 4.51–4.53. Figure 4.52 indicates an excess altitude in the initial launch phase due to a starting error (5.7°) in the flight-path angle, which causes an initial undershoot in the flight speed. By having an altitude-based observer, the trajectory error is corrected by the end of powered flight, as shown in Figure 4.51. The small eccentricity in the initial orbit due to a slightly negative flight-path angle at $t = 600$ s can be rectified by a final circularization maneuver (Chapter 6). Figure 4.53 shows the closed-loop control activity for the given initial flight-path disturbance. Control saturation occurs during the first 200 s of flight, for which a maximum gimbal angle of $\pm 17°$ is required. This should be compared with only $-6.4°$ angle deflection necessary for the full-state feedback case (Figure 4.44). Furthermore, the control activity was largely confined to the first 20 s of flight for the full-state feedback regulator, rather than the entire duration of powered flight required here. Thus as expected, the addition of a single-output observer significantly degrades the closed-loop response. In addition, there is an increased sensitivity to feedback gain values, thereby implying a reduction in the stability margin. The present example illustrates the limitations of a guidance system based upon a single-output observer.

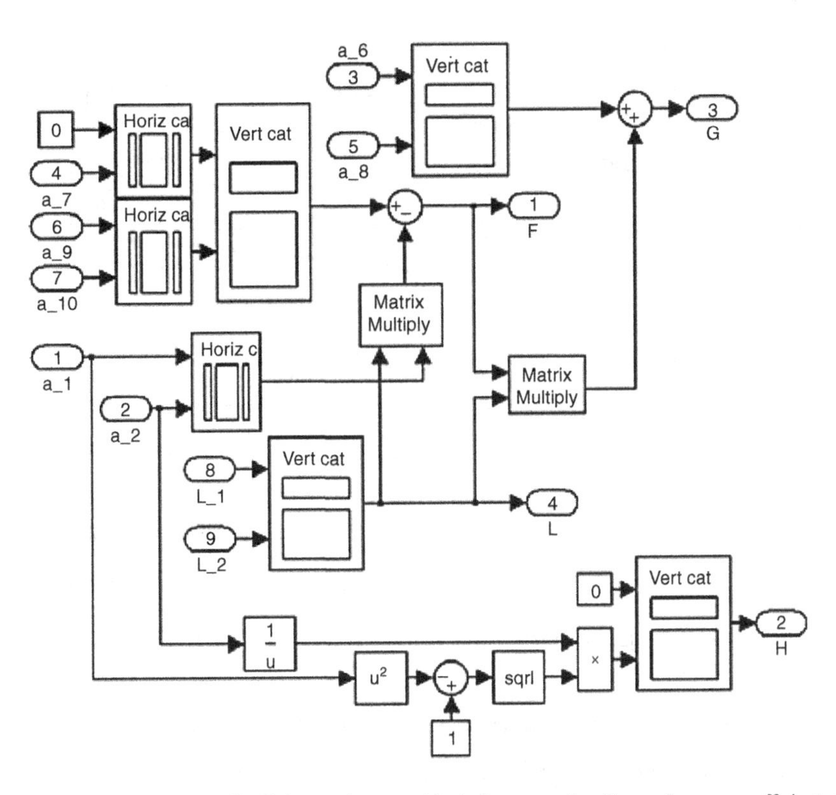

Figure 4.49 Simulink *Observer Coefficients* subsystem block for computing linear observer coefficient matrices required by the block diagram of Figure 4.47

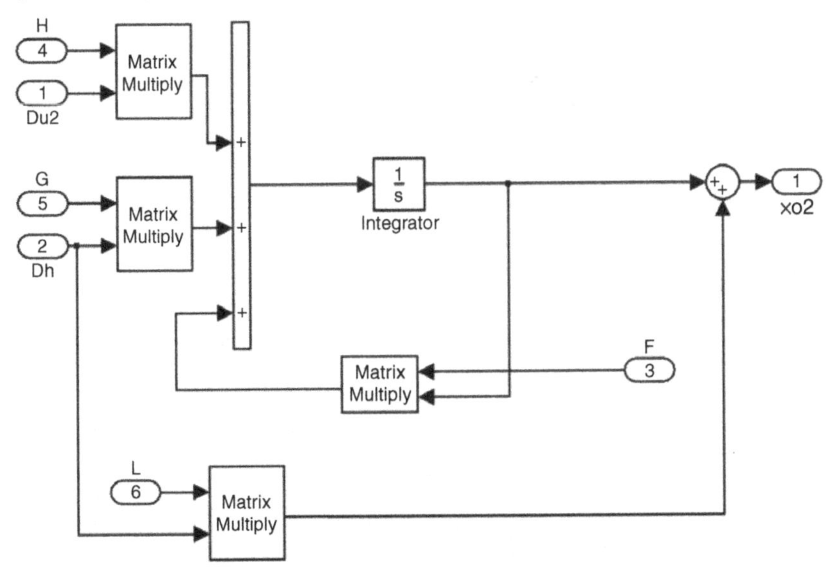

Figure 4.50 Simulink *Adaptive Observer* subsystem block for adaptive observer state equations required by the block diagram of Figure 4.47

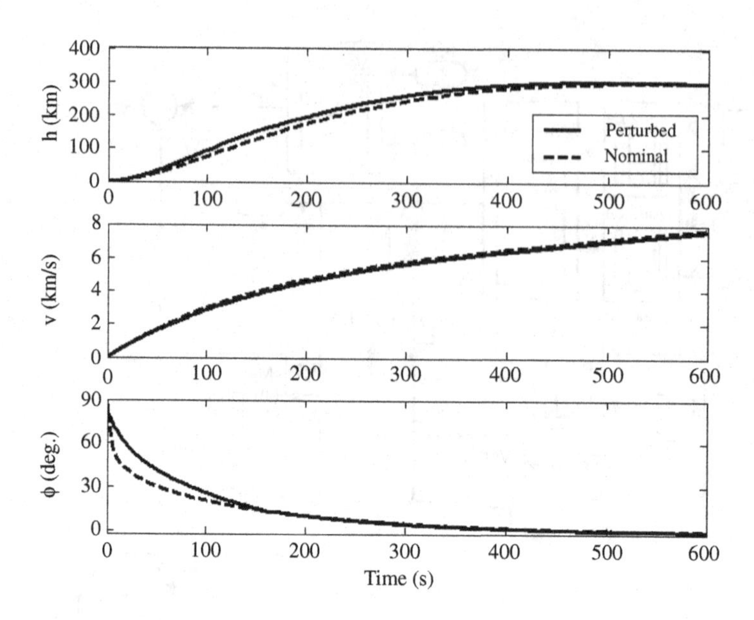

Figure 4.51 Closed-loop simulated trajectory of rocket guidance system with reduced-order, single-output observer

4.12.2 *Bi-output Observer with Normal Acceleration Input*

Single-output observers are inadequate for guiding a rocket in its ascent to orbit, as shown in Example 4.10. Let us now examine the fate of an observer that employs both altitude and speed

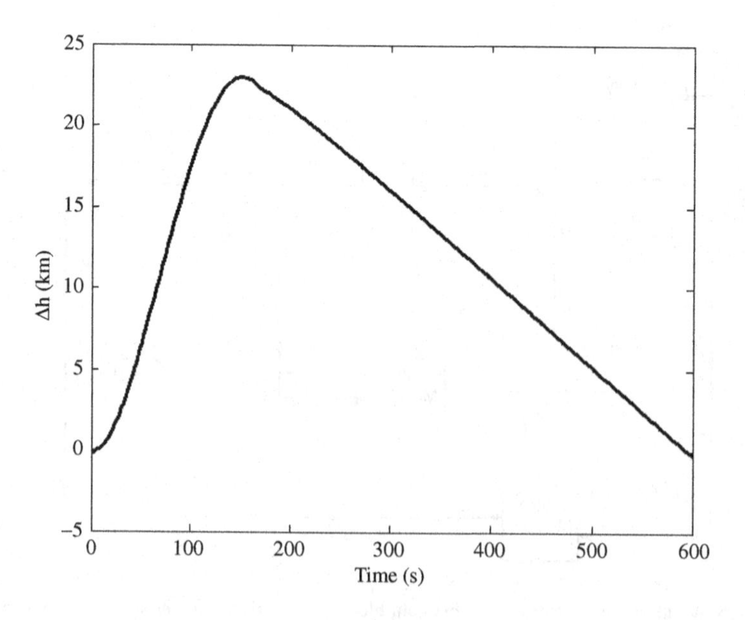

Figure 4.52 Closed-loop altitude error of rocket guidance system with reduced-order, single-output observer

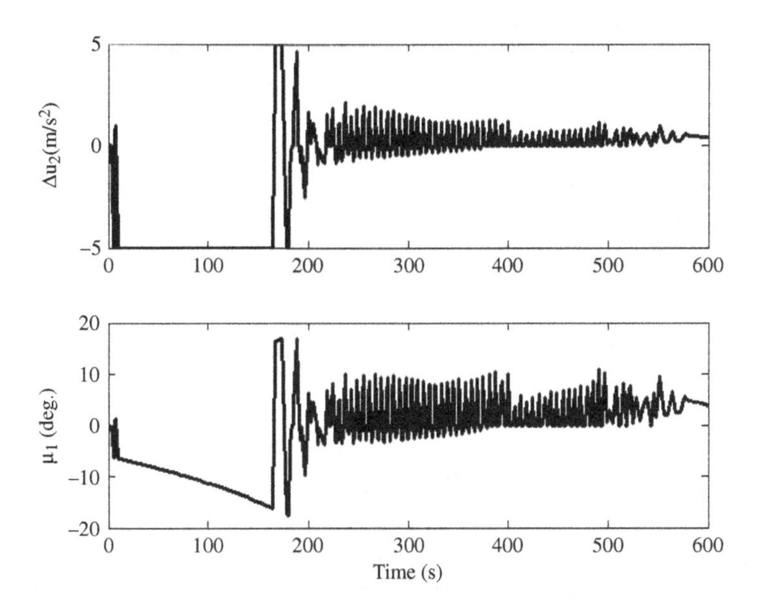

Figure 4.53 Closed-loop normal acceleration and gimbal angle input of rocket guidance system with reduced-order, single-output observer

measurements, $\mathbf{y} = \mathbf{x}_1 = (\Delta h, \Delta v)^T$, for which we have $\mathbf{x}_2 = \Delta\phi$, and the following partitioned coefficient matrices:

$$\mathbf{A} = \begin{pmatrix} 0 & a_{13} & | & a_{14} \\ a_{31} & 0 & | & a_{34} \\ \hline a_{41} & a_{43} & | & a_{44} \end{pmatrix}, \quad \mathbf{B} = \begin{pmatrix} 0 \\ 0 \\ \hline \frac{1}{v_n} \end{pmatrix}. \tag{4.171}$$

The sub-matrices required for the reduced-order observer are thus

$$\mathbf{C} = \mathbf{I}, \quad \mathbf{A}_{11} = \begin{pmatrix} 0 & a_{13} \\ a_{31} & 0 \end{pmatrix}, \quad \mathbf{A}_{12} = \begin{pmatrix} a_{14} \\ a_{34} \end{pmatrix}, \quad \mathbf{B}_1 = \begin{pmatrix} 0 \\ 0 \end{pmatrix}, \tag{4.172}$$

$$\mathbf{A}_{21} = (a_{41} \quad a_{43}), \quad \mathbf{A}_{22} = a_{44}, \quad \mathbf{B}_2 = \frac{1}{v_n}. \tag{4.173}$$

The state equation of the single-order observer with estimated state, $x_{o2} = \mathbf{L}\mathbf{y} + z$, and gain matrix, $\mathbf{L} = (L_1, L_2)$, is

$$\dot{z} = Fz + Hu + \mathbf{G}\mathbf{y}, \tag{4.174}$$

where

$$F = \mathbf{A}_{22} - \mathbf{LCA}_{12} = a_{44} - L_1 a_{14} - L_2 a_{34}, \quad H = \mathbf{B}_2 - \mathbf{LCB}_1 = \frac{1}{v_n}, \tag{4.175}$$

and

$$\begin{aligned} \mathbf{G} &= \mathbf{FL} + (\mathbf{A}_{21} - \mathbf{LCA}_{11})\,\mathbf{C}^{-1} \\ &= (FL_1 + a_{41} - L_2 a_{31} \quad FL_2 + a_{43} - L_1 a_{13}). \end{aligned} \tag{4.176}$$

The pole of the observer, $\lambda = F$, is determined from the desired settling-time, t_s, by

$$F \simeq -\frac{4}{t_s}, \tag{4.177}$$

and alternative choices for the observer gains are either

$$\begin{aligned} L_1 &= 0, \\ L_2 &= \frac{a_{44} + 4/t_s}{a_{34}} = -\frac{\left(\frac{\mu}{vr^2} - \frac{v}{r}\right)\sin\phi + \frac{4}{t_s}}{\frac{\mu}{r^2}\cos\phi} \end{aligned} \tag{4.178}$$

or

$$\begin{aligned} L_2 &= 0, \\ L_1 &= \frac{a_{44} + 4/t_s}{a_{14}} = -\frac{\left(\frac{\mu}{vr^2} - \frac{v}{r}\right)\sin\phi + \frac{4}{t_s}}{v\cos\phi}. \end{aligned} \tag{4.179}$$

The two choices are equivalent, and produce an identical closed-loop response.

The main advantage of the bi-output observer over the single-output one is that the former is much more robust than the latter due to a better conditioning of the observability test matrix. Therefore, we expect that the bi-output observer can be successfully implemented in the actual guidance system.

Example 4.11 Redesign the observer-based compensator for the gravity-turn trajectory guidance system of Example 4.10 using the altitude and speed errors, Δh, Δv, as the two outputs. Simulate the closed-loop response to the initial trajectory perturbation given in Example 4.9. Model the actuator nonlinearities of Example 4.9 more realistically with limits applied directly to the gimbal angle, μ_1, as a saturation limit of $|\mu_1| \leq 15°$ and a deadband of $\pm 0.5°$ (i.e., the gimbal actuator does not respond to commanded deflection less than $\pm 0.5°$, and has maximum deflection $\pm 15°$).

Selecting the observer pole as $F = -1$, we have

$$L_1 = 0, \quad L_2 = \frac{a_{44} + 1}{a_{34}},$$

$$G = (a_{41} - L_2 a_{31} \quad a_{43} - 4L_2), \quad H = \frac{1}{v_n}.$$

The observer gain, L_2, plotted along the nominal trajectory, is shown in Figure 4.54. Evidently, except for the singularity at $v = 0$, the observer gain is much smaller in magnitude compared to the gains of Example 4.10 (Figure 4.46), even for an observer pole much deeper into the left-half s-plane. This is due to the fact that the condition number of the observability test matrix, **N**, with the two outputs, Δh, Δv, for the third-order rocket guidance plant is several orders of magnitude smaller than that with a single output.

Since the observer is stable, we will directly utilize the adaptive regulator gains of Example 4.9 in order that the usual degradation of performance with observer-based feedback is avoided. Hence, for closing the loop of the compensated system, we have

$$\Delta u_2 = -0.01\Delta h - 0.1\Delta v - 1.6 x_{o2}.$$

The simulation of the gravity-turn guidance system with the bi-output observer is carried out by a fourth-order Runge–Kutta method using the MATLAB code, *rocket_obs.m* listed in Table 4.14. Separate code,

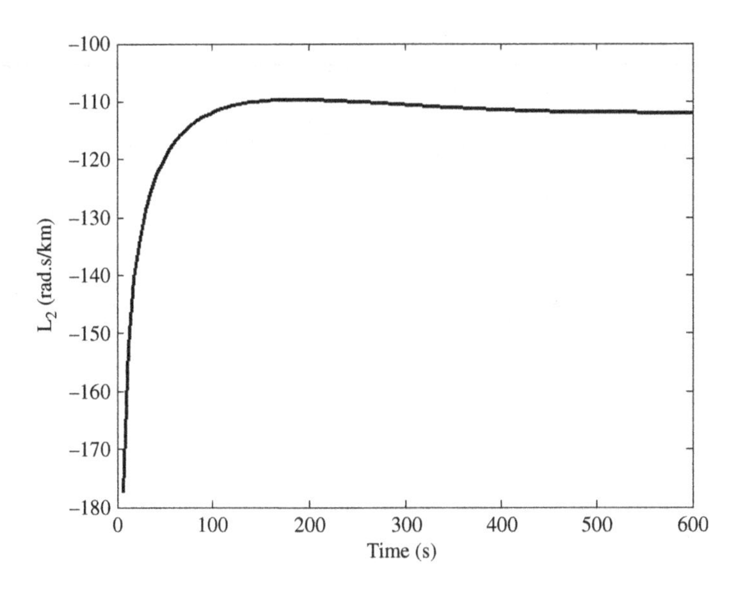

Figure 4.54 Observer gain for a bi-output, speed- and altitude-based, reduced-order observer for a rocket guidance system

Table 4.14

```
% Program 'rocket_obs.m' for generating the closed-loop
% state equations for a launch vehicle to circular orbit while
% guided by a normal acceleration controlled, bi-output
% reduced-order compensator based upon measurement and feedback of
% speed and altitude.
% (c) 2010 Ashish Tewari
% To be used by Runge-Kutta solver 'ode45.m'
% t: current time (s); x(1,1): current radius; x(2,1): current speed
% x(3,1): current flight-path angle; x(4,1): reduced-order observer state.
%
function xdot=rocket_obs(t,x)
global tn;   %Nominal time listed in the calling program as a row vector.
global rn;   %Nominal radius listed in the calling program as a row vector.
global vn;   %Nominal speed listed in the calling program as a row vector.
global phin; %Nominal flight-path angle listed in calling
             %program as row vector.
global un;   %Nominal fwd accln. input listed in calling
             %program as row vector.
dtr=pi/180;
mu=398600.4; %Earth's gravitational constant (km^3/s^2)
r=interp1(tn,rn,t); %Interpolation for nominal radius at current time.
v=interp1(tn,vn,t); %Interpolation for nominal speed at current time.
phi=interp1(tn,phin,t); %Interpolation for nominal f.p.a. at current time.
u=interp1(tn,un,t); %Interpolation for nominal input at current time.

% Linearized plant's state coefficient matrix:
A=[0 sin(phi) v*cos(phi);
```

(Continued)

Table 4.14 (*Continued*)

```
2*mu*sin(phi)/r^3    0   -mu*cos(phi)/r^2;
(-v/r^2+2*mu/(v*r^3))*cos(phi) ...
(1/r+mu/(v^2*r^2))*cos(phi) -(v/r-mu/(v*r^2))*sin(phi)];

% Regulator gains follow:
k=[0.01 0.1 1.6];

% Observer coefficients follow:
a1=A(1,2);a2=A(1,3);a6=A(2,1);a7=A(2,3);a8=A(3,1);a9=A(3,2);a10=A(3,3);
F=-1;
L2=(a10-F)/a7;
G=F*[0 L2]+[a8 a9]-[L2*a6 0];
H=1/v;

% Trajectory error variables:
dr=x(1,1)-r;
dv=x(2,1)-v;
dphi=x(3,1)-phi;

% Estimated trajectory error variables:
z=x(4,1);
xo2=L2*dv+z;

% Feedback control input:
du=-k(1,2)*dv-k(1,1)*dr-k(1,3)*xo2;
mu1=asin(du/u)/dtr; %Gimbal angle
% Gimbal actuator's saturation and deadzone follow:
if abs(mu1)>15
    du=abs(u)*sin(15*dtr)*sign(du);
end
if abs(mu1)<0.5
    du=0;
end

%Closed-loop state equations
xdot(1,1)=x(2,1)*sin(x(3,1));
xdot(2,1)=u-mu*sin(x(3,1))/x(1,1)^2;
xdot(3,1)=du/x(2,1)+(x(2,1)/x(1,1)-mu/(x(1,1)^2*x(2,1)))*cos(x(3,1));
xdot(4,1)=F*z+H*du+G*[dr dv]'; %Observer state equation
```

rocket_obs_u.m (Table 4.15), was written to compute the closed-loop control input by post-processing the integration results of *rocket_obs.m*. The simulation results are plotted in Figures 4.55–4.57, showing an almost exact launch even in the presence of a large initial angular error, even with the gimbal angle deflection limited by saturation and deadband nonlinearities. Virtually no difference between the perturbed and nominal trajectories is visible in Figure 4.55, while Figure 4.56 shows that a prompt control action does not allow the altitude and speed errors to grow beyond 240 m and -2.5 m/s, respectively, while bringing the flight-path angle error to less than $\pm 0.1^\circ$ in about 200 s. The gimbal drive is seen in Figure 4.57 to be saturated for almost the entire trajectory, which means an excellent use of control effort. The best feature of the bi-output compensator is its large robustness with respect to variations in controller gains, k_h, k_v, k_ϕ, L_2, allowing it to have a practical application.

Table 4.15

```
% Program 'rocket_obs_u.m' for generating the closed-loop
% gimbal angle input for a launch vehicle to circular orbit while
% guided by a normal acceleration controlled, bi-output
% reduced-order compensator based upon measurement and feedback of
% speed and altitude.
% (c) 2010 Ashish Tewari
% T: time vector (s); x: Runge-Kutta solution of 'rocket_obs.m'.
% x(:,1): radius vector (km); x(:,2): speed vector (km/s)
% x(:,3): flight-path angle vector (rad.); x(:,4): reduced-order
% observer state vector.
%
function u=rocket_obs_u(T,x)
global tn;   %Nominal time listed in the calling program as a row vector.
global rn;   %Nominal radius listed in the calling program as a row vector.
global vn;   %Nominal speed listed in the calling program as a row vector.
global phin; %Nominal flight-path angle listed in the calling
             %program as a row vector.
global un;   %Nominal forward acceleration input listed in the calling
             %program as a row vector.
mu=398600.4; %Earth's gravitational constant (km^3/s^2)
dtr=pi/180;
n=size(T,1)
for i=1:n
t=T(i,1);
r=interp1(tn,rn,t);
v=interp1(tn,vn,t);
phi=interp1(tn,phin,t);
U=interp1(tn,un,t);
A=[0 sin(phi) v*cos(phi);
2*mu*sin(phi)/r^3   0  -mu*cos(phi)/r^2;
(-v/r^2+2*mu/(v*r^3))*cos(phi)  (1/r+mu/(v^2*r^2))*cos(phi) ...
-(v/r-mu/(v*r^2))*sin(phi)];
k=[0.01 0.1 1.6];
a1=A(1,2);a2=A(1,3);a6=A(2,1);a7=A(2,3);a8=A(3,1);a9=A(3,2);a10=A(3,3);
F=-1;
L2=(a10-F)/a7;
G=F*[0 L2]+[a8 a9]-[L2*a6 0];
H=1/v;
dr=x(i,1)-r;
dv=x(i,2)-v;
dphi=x(i,3)-phi;
z=x(i,4);
xo2=L2*dv+z;
du=-k(1,2)*dv-k(1,1)*dr-k(1,3)*xo2;
mu1=asin(du/U)/dtr;
if abs(mu1)>15
    du=abs(U)*sin(15*dtr)*sign(du);
end
if abs(mu1)<0.5
    du=0;
end
u(i,1)=asin(du/U);
end
```

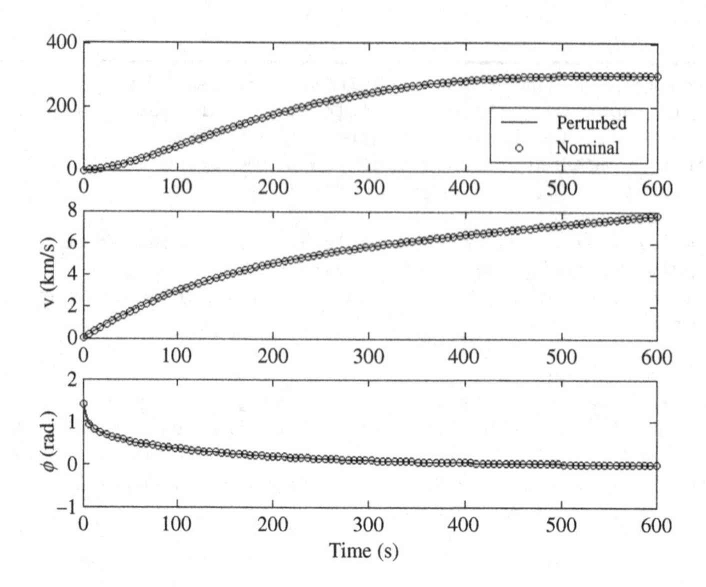

Figure 4.55 Closed-loop compensated trajectory of a guided launch vehicle with a bi-output reduced-order observer-based compensator

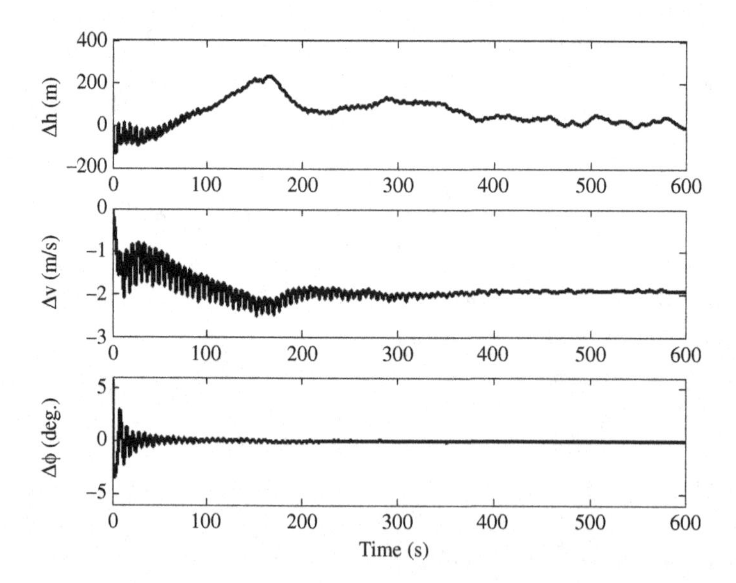

Figure 4.56 Closed-loop trajectory errors of a guided launch vehicle with a bi-output reduced-order observer-based compensator

4.13 Mass and Atmospheric Drag Modeling

In order to build a realistic closed-loop guidance simulation model, one must account for atmospheric drag and vehicle mass variation. A launch vehicle consists of several stages, each of which has its own drag, mass, and thrust characteristics. While the details of an atmospheric model, aerodynamic behavior,

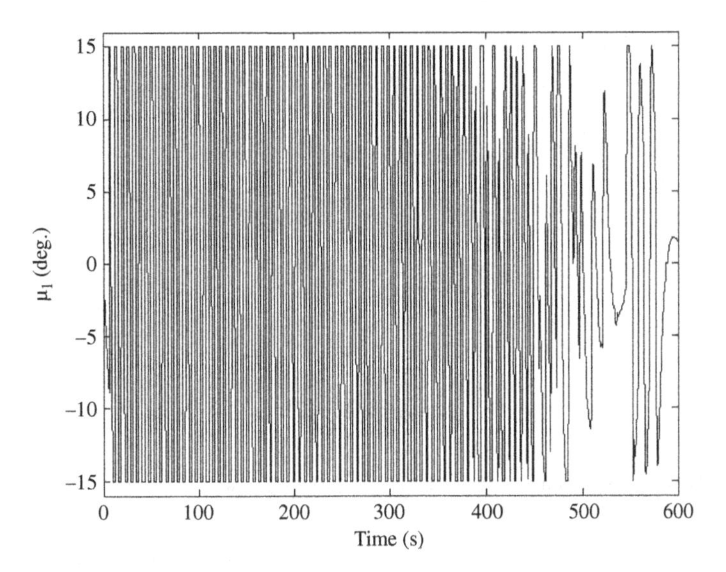

Figure 4.57 Closed-loop gimbal angle deflection input of a guided launch vehicle with a bi-output reduced-order observer-based compensator

and rocket propulsion can be found in Tewari (2006), we will briefly list the features necessary for building a closed-loop simulation.

The drag force of a particular stage is given by

$$D = \frac{1}{2}\rho v^2 S C_D(\alpha, M, Re, Kn). \tag{4.180}$$

The variation of density, ρ, and other atmospheric properties with the altitude, h, is modeled by an appropriate standard atmospheric model, such as that given in Chapter 9 of Tewari (2006). Since the vehicle quickly crosses the laminar flow regime after launch, variations with the Reynolds number, Re, are commonly neglected within the continuum phase of the flight, and a constant, average turbulent Reynolds number can be employed. Furthermore, an attitude control system (discussed below) always maintains a zero angle of attack ($\alpha = 0$) to minimize the transverse loads on the vehicle's shell-like structure. Therefore, the drag coefficient, C_D, is approximated as a function only of the Mach number, M, and Knudsen number, Kn, both of which are carefully computed using the flight parameters, (h, v), and a standard atmospheric model.

We model the drag coefficient (referred to the base area, S) of a stage taken to be a sphere-cone-cylinder (Figure 4.58) as follows:

$$\begin{aligned}
C_D &= C_{Dc}(M), \quad Kn < Kn_c, \\
C_D &= C_{Dfm}, \quad Kn > Kn_f, \\
C_D &= C_{Dc} + (C_{Dfm} - C_{Dc})\left[\frac{1}{3}\log_{10}(A\, Kn) + B\right], \quad Kn_c < Kn < Kn_f,
\end{aligned} \tag{4.181}$$

where $C_{Dc}(M)$ is the drag coefficient in the continuum flow limit as a specific function of Mach number depending only upon the geometry of the stage. By the hypersonic Mach number independence principle, C_{Dc} becomes invariant with M after a critical Mach number, $M > M_c$. The flight Knudsen number,

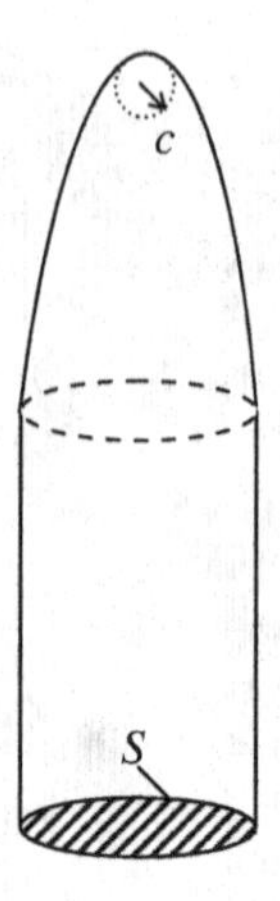

Figure 4.58 Geometry of a rocket stage, showing base area, S, and nose radius, c

$Kn = \lambda/c$, where λ is the molecular mean free-path and c is the nose radius of the spherical stagnation region (Figure 4.58).

In the free-molecular limit, $(Kn > Kn_{\text{f}})$, the drag coefficient, $C_{D\text{fm}}$, for a sphere-cone forebody is given by the cold-wall approximation (Schaaf and Chambre 1958) as

$$C_{D\text{fm}} = 1.75 + \frac{\sqrt{\pi}}{2s}, \tag{4.182}$$

with $s = \frac{v}{\sqrt{2RT}}$ denoting the molecular speed ratio. The constants, (A, B), of the logarithmic interpolation in the transition regime (called the *bridging relation*), as well as the limiting Knudsen numbers, $(Kn_{\text{c}}, Kn_{\text{f}})$, depend exclusively upon the vehicle's geometry, and are selected in order to have a smooth transition between the continuum and the free-molecular flow regimes.

The mass of the ith rocket stage can be assumed to vary linearly from an initial mass, m_{0i}, to a final mass, $m_{\text{f}i}$, for the given burn time, $t_{\text{b}i}$:

$$m(t) = m_{0i} - \left(\frac{m_{0i} - m_{\text{f}i}}{t_{\text{b}i}} \right) t. \tag{4.183}$$

The initial mass of the next stage is obtained after subtracting the structural mass of the previous stage, $m_{\text{s}i}$, from its final mass according to

$$m_{0(i+1)} = m_{\text{f}i} - m_{\text{s}i}. \tag{4.184}$$

Finally, the thrust required is computed by adding the drag deceleration, D/m, to the required forward acceleration input, u_1:

$$f_{\text{T}}(t) = m(t)u_1(t) + D(t). \tag{4.185}$$

In this process, we neglect the slight decrease in forward acceleration due to gimballing, because the gimbal angle is generally small. The latter approximation is valid for a rocket, but not for any other atmospheric flight vehicle. This is due to the fact that the drag – which is usually several times less than the thrust – does not vary significantly from its value along the zero-lift (ballistic) nominal trajectory even for the off-nominal case, since the angles of attack and sideslip are always maintained near zero by an attitude control system. Therefore, drag can be regarded as being outside the feedback loop.

Table 4.16 Continuum drag coefficient for the rocket stages and payload

M	C_{Dc} (Stage I)	C_{Dc} (Stage II)	C_{Dc} (Payload)
0	0.4755	0.1783	0.0594
0.2000	0.4755	0.1783	0.0594
0.3000	0.4758	0.1784	0.0595
0.4000	0.4834	0.1813	0.0604
0.5000	0.4890	0.1834	0.0611
0.6000	0.5083	0.1906	0.0635
0.8000	0.5656	0.2121	0.0707
0.9000	0.6182	0.2318	0.0773
0.9500	0.6681	0.2505	0.0835
1.0500	1.0318	0.3869	0.1290
1.1000	1.0171	0.3814	0.1271
1.2000	0.9906	0.3715	0.1238
1.6000	0.8160	0.3060	0.1020
2.0000	0.6924	0.2596	0.0865
2.5000	0.6097	0.2286	0.0762
3.0000	0.5461	0.2048	0.0683
3.8000	0.5130	0.1924	0.0641
5.0000	0.4940	0.1853	0.0617
10.0000	0.4832	0.1812	0.0604
99.0000	0.4832	0.1812	0.0604

Example 4.12 Simulate the drag, mass, and thrust required for the nominal gravity-turn trajectory of Example 4.8 using the following two-stage rocket data:

$$S = 4\ \text{m}^2, \quad c = 0.5\ \text{m},$$

$$Kn_c = 0.0146241, \quad Kn_f = 1000 Kn_c, \quad A = 2, \quad B = 0.5113,$$

$$m_{01} = 30\,856.5\ \text{kg}, \quad m_{s1} = 1581.6\ \text{kg}, \quad t_{b1} = 250\ \text{s},$$

$$m_{02} = 8262.4\ \text{kg}, \quad m_{s2} = 395.6\ \text{kg}, \quad t_{b2} = 350\ \text{s}.$$

The variation of the continuum drag coefficient, C_{Dc}, with Mach number, M, for the first and second stages as well as the final payload are given in Table 4.16.

We begin by coding the computation steps for drag and mass in a MATLAB program called *rocket_drag_mass.m*, which is listed in Table 4.17. This code calls a 1976 US standard atmosphere based subroutine called *atmosphere.m* that is listed in Tewari (2006), and can also be downloaded from http://home.iitk.ac.in/~ashtew/index_files/atmosphere.m.

Alternatively, the reader can build his/her own atmospheric properties code according to a standard atmospheric model. The results of the code run for the nominal gravity-turn trajectory given in Example 4.8 are plotted in Figure 4.59. Note that the thrust is always several times larger than the drag.

Example 4.13 Taking into account the atmospheric drag and mass computed in Example 4.12, simulate the closed-loop response of the bi-output, reduced-order compensator designed in Example 4.11 for guidance with normal acceleration input and the given gimbal actuator limits.

We construct a Simulink model, depicted in Figure 4.60, whose *Rocket Translational Dynamics* and *Observer Coefficients* subsystem blocks are depicted in Figures 4.61 and 4.62, respectively. The *Rocket*

Table 4.17

```
% Program 'rocket_drag_mass.m' for calculating the mass and drag of a rocket
% with specified stage data, as a function of time, radius, and
% relative speed.
% (c) 2010 Ashish Tewari
% Requires the standard atmosphere code 'atmosphere.m'.
function [D,m]=rocket_drag_mass(t,r,v)
% r: radius (km)
% v: relative speed (km/s)
% m: mass (kg)
% D: drag (N)
global c; %Nose radius (m) specified in the calling program.
global S; %Base area (m^2)  specified in the calling program.
global m01; %I-stage initial mass (kg) specified in the calling program.
global m02; %II-stage initial mass (kg) specified in the calling program.
global mL; %Payload mass (kg) specified in the calling program.
global ms1 %I-stage structural mass (kg) specified in the calling program.
global ms2 %II-stage structural mass (kg) specified in the calling program.
global tb1; %I-stage burn time (s) specified in the calling program.
global tb2; %II-stage burn time (s) specified in the calling program.
global M;   %Mach numbers at which CDc is tabulated in the calling program.
global CDc1; %I-stage CDc values specified in the calling program.
global CDc2; %II-stage CDc values specified in the calling program.
global CDcL; %Payload CDc values specified in the calling program.
global Gamma;%Specific-heat ratio (1.41 for air) specified in the
             %calling program.
mu=398600.4;
v=v*1000;
alt=(r-6378.14)*1000;
atmosp = atmosphere(alt,v,c);
rho = atmosp(2);
Qinf = 0.5*rho*v^2;
mach = atmosp(3);
Kn=atmosp(4);
s = mach*sqrt(Gamma/2);
CDFM=1.75+sqrt(pi)/(2*s);
if t<=tb1
    m=m01-(m01-m02-ms1)*t/tb1;
    CDC=interp1(M,CDc1,mach);
elseif t<=(tb1+tb2)
    m=m02-(m02-mL-ms2)*(t-tb1)/tb2;
    CDC=interp1(M,CDc2,mach);
else
     m=mL;
     CDC=interp1(M,CDcL,mach);
end
iflow=atmosp(6);
if iflow==2
    CD=CDC;
elseif iflow==1
    CD=CDFM;
else
    CD = CDC + (CDFM - CDC)*(log10(2*Kn)/3+0.5113);
end
D=Qinf*S*CD;
```

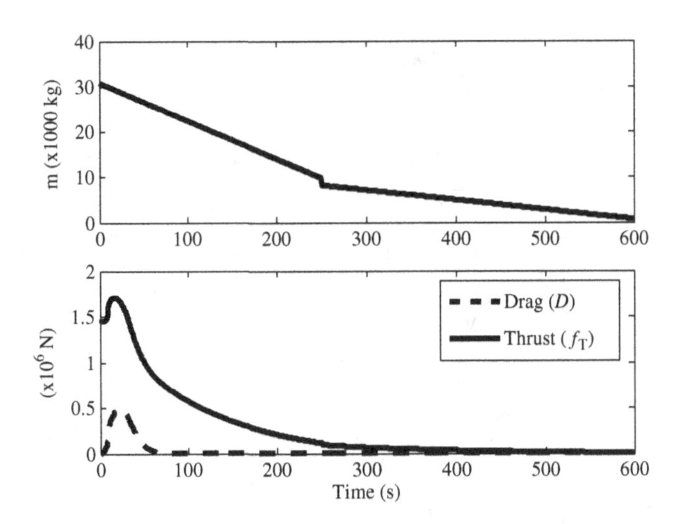

Figure 4.59 Mass, atmospheric drag, and required thrust of a launch vehicle along the nominal, gravity-turn trajectory

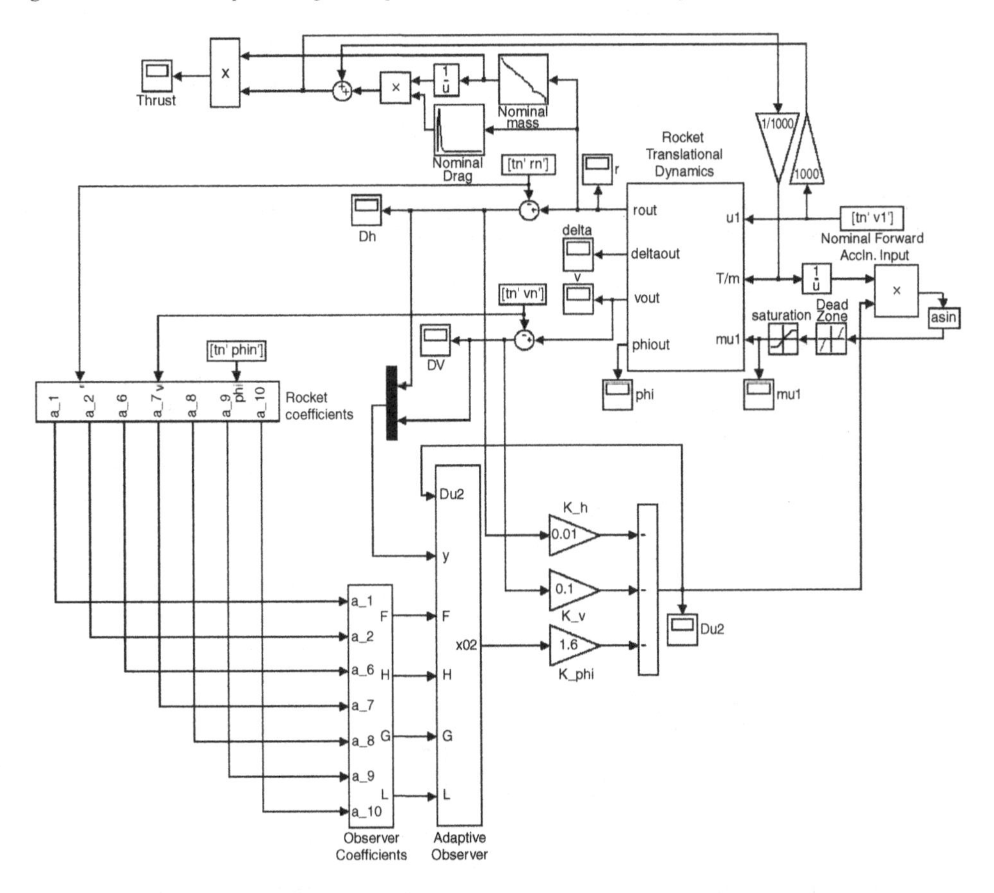

Figure 4.60 Simulink block diagram for simulating closed-loop initial response of nonlinear rocket guidance system with a bi-output observer as well as mass and drag taken into account

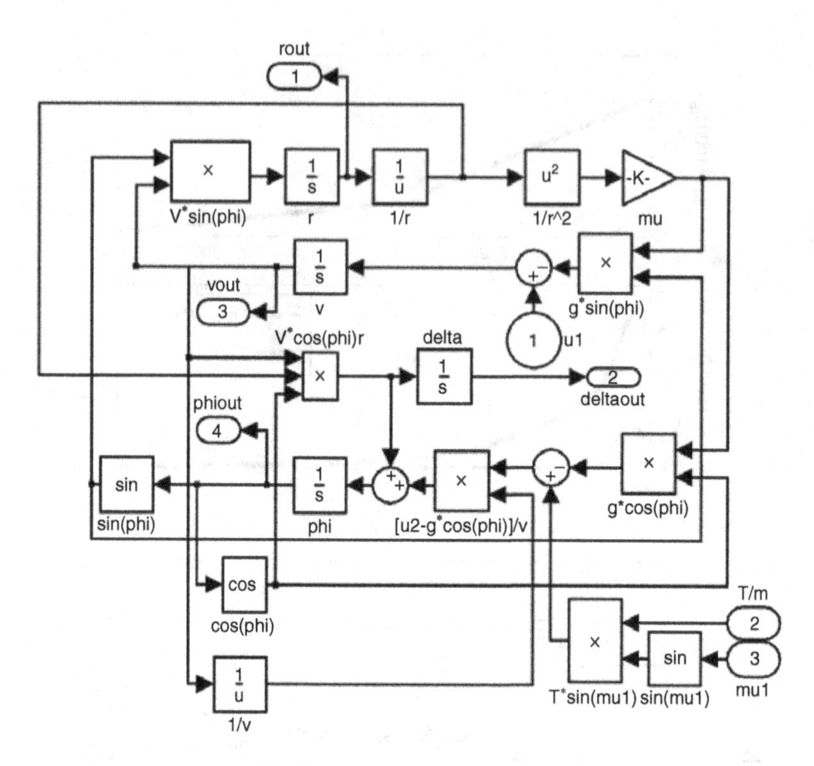

Figure 4.61 Simulink *Rocket Translational Dynamics* subsystem block for computing nonlinear rocket translational dynamics plant with gimbal input, as required by the block diagram of Figure 4.60

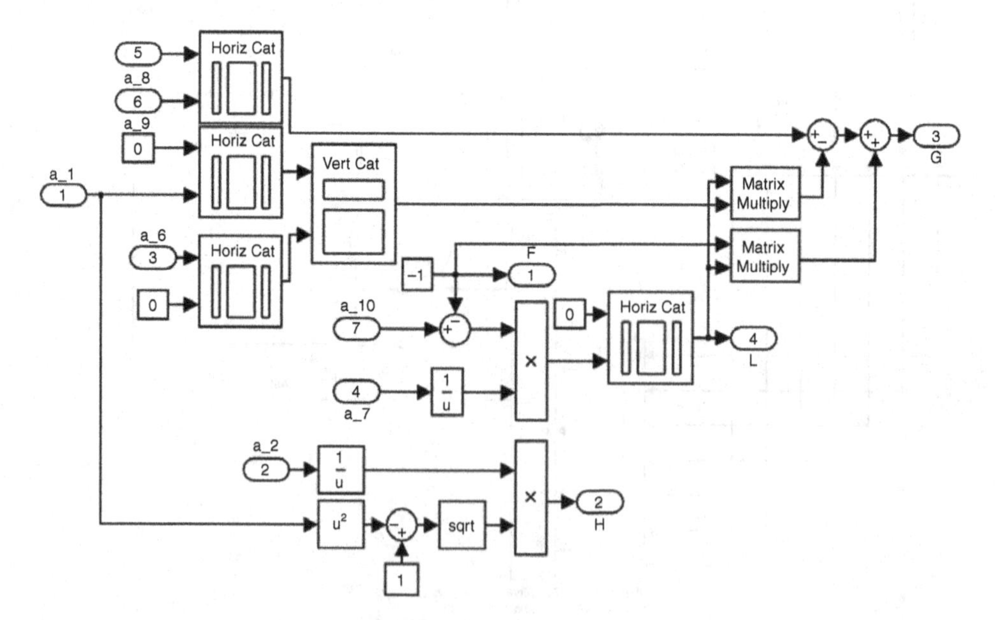

Figure 4.62 Simulink *Observer Coefficients* subsystem block for computing bi-output observer coefficient matrices required by the block diagram of Figure 4.60

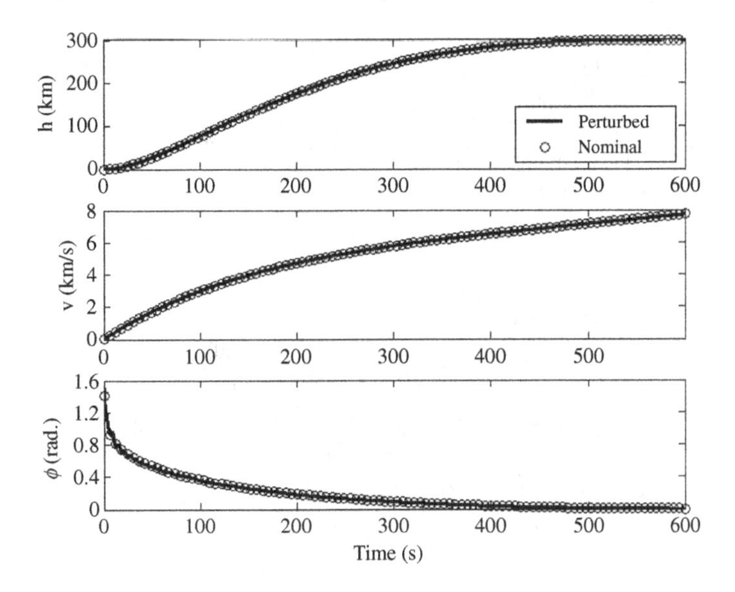

Figure 4.63 Closed-loop simulated trajectory of nonlinear rocket guidance system with a bi-output observer as well as mass and drag taken into account

Coefficients and *Adaptive Observer* subsystems are the same as those of Figures 4.48 and 4.50, respectively, with the plant output vector, $\mathbf{y}$, in place of Δh. Before beginning the simulation, the nominal trajectory is stored in the MATLAB workspace row vectors tn, rn, vn, phin, and the nominal forward acceleration input in the vector u1. Furthermore, the mass and drag computed by the code *rocket_drag_mass.m*

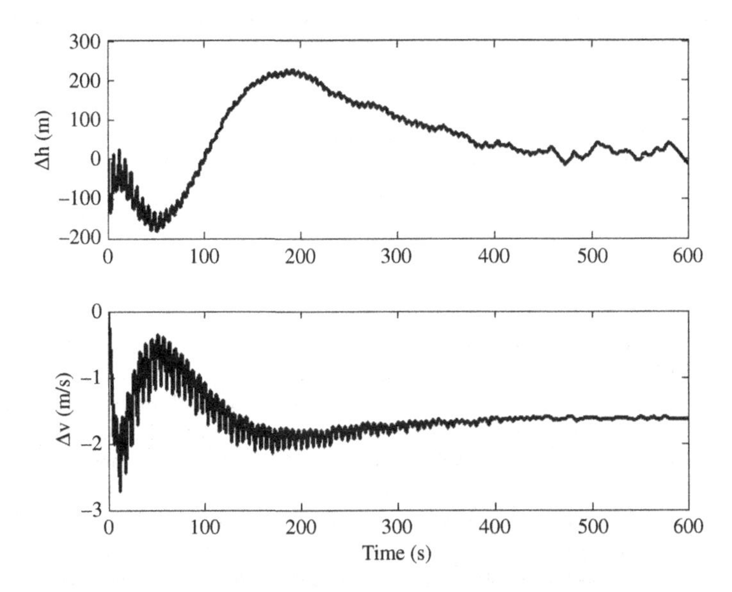

Figure 4.64 Closed-loop altitude and speed errors of nonlinear rocket guidance system with a bi-output observer as well as mass and drag taken into account

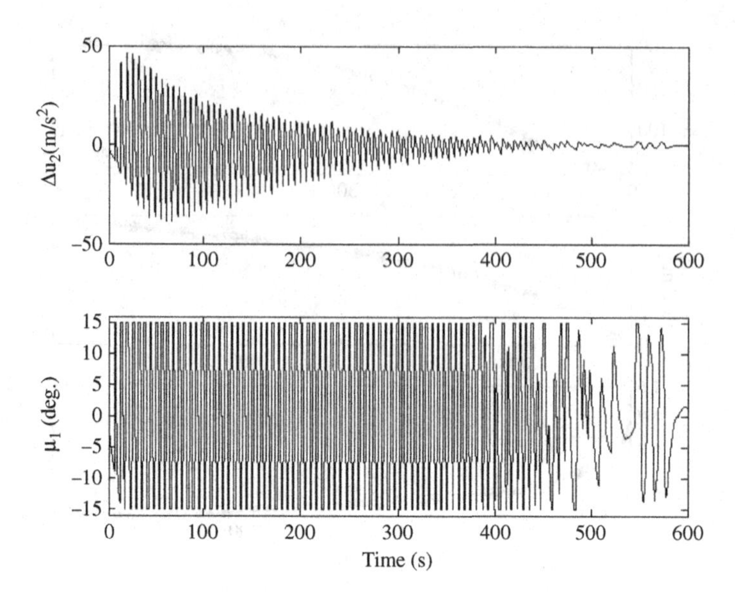

Figure 4.65 Closed-loop normal acceleration command and gimbal angle control input for nonlinear rocket guidance system with a bi-output observer as well as mass and drag taken into account

(Table 4.17) along the nominal trajectory are stored as column vectors m and D, respectively. Note the presence of table look-up blocks for nominal mass and drag in Figure 4.60. The results of the simulation carried out by the fourth-order Runge–Kutta method with a relative tolerance of 10^{-5} are plotted in Figures 4.63–4.65. There is virtually no difference in the closed-loop trajectory compared to that observed in Example 4.11, even though mass and atmospheric drag are now being accounted for. However, a small amplitude, high-frequency oscillation is now superimposed on the altitude and speed error plots shown in Figure 4.64, without affecting the net error magnitudes observed in Figure 4.56. Such an oscillation can be regarded as process noise due to neglecting mass and drag in the plant model used for control law derivation.

Figure 4.65 shows that gimbal activity is unchanged from that observed in Figure 4.57, because the high-frequency process noise component has no effect on the control input due to the actuator deadzone. If a larger process noise were present due to other modeling errors, a lag compensator could be added to insulate the feedback loop from its effects.

4.14 Summary

Nominal trajectories of various rocket flight problems can be derived by the optimal, two-point boundary value problem (2PBVP) of control cost minimization. These include optimal guidance of short-range interceptors such as surface-to-air missiles, which can be shown to result in three-dimensional proportional navigation. Flight of most rockets is largely confined to the vertical (longitudinal) plane and is usually ballistic in nature (zero lift and sideforce) in order to avoid excessive transverse loads. Since the atmospheric drag is generally quite small in comparison with the rocket's thrust, guidance and control systems can be designed by neglecting drag as a first-order approximation. The vertical flight is posed as a special 2PBVP called Goddard's rocket problem. For the general case, the guidance system must maintain the vehicle on a special ballistic trajectory called the gravity-turn, which allows the flight-path angle to change naturally from a near vertical to horizontal in a given time, under the sole influence of gravity. Small deviations from the nominal trajectory are minimized by a normal acceleration control input, produced by thrust deflection (pitch gimbal angle). Linear feedback regulators and observers can be designed by either pole-placement or LQR methods for the linearized trajectory error plant. A typical rocket is a time-varying plant, due to not only the varying speed and altitude of the nominal trajectory but also the changing mass. Therefore, rocket guidance and control systems must have gains that adapt themselves to the varying plant parameters.

Exercises

(1) Examine the controllability of the rocket guidance plant for flight in a vertical plane with only the normal acceleration input, $u_2(t)$. What (if any) is the difference compared to the controllability of the plant with only the forward acceleration input, $u_1(t)$? Include the very low speed near $t = 0$, as well as the nearly orbital speed close to $t = t_f$, in your analysis.

(2) Using the data given in Example 4.8, validate the results of Exercise 1 by plotting rank and condition number of the controllability test matrix, **P**, against time.

(3) Demonstrate that the complete rocket guidance plant for flight in a vertical plane is unconditionally unobservable with any of the flight variables, $h(t)$, $v(t)$, $\phi(t)$, used either singly, or jointly in any combination thereof, as the measured outputs.

(4) Show that the third-order guidance plant without downrange dynamics, $\delta(t)$, becomes unobservable with $\Delta\phi$ as the single output variable whenever $\phi_n = 0$ or $\phi_n = \pm\pi/2$.

(5) Repeat the reduced-order design of Example 4.10 for a gravity-turn trajectory with $\Delta v(t)$ as the sole output variable. Is the closed-loop, time-varying system with your observer and the regulator designed in Example 4.9 stable? Why?

(6) Repeat the reduced-order design of Example 4.11 using the equivalent observer gain choice, $L_2 = 0$. Suitably modify the code *rocket_obs.m* (Table 4.14) for your design and carry out the closed-loop simulation using the Runge–Kutta solver, *ode45.m*. Also compute the control input by post-processing the simulated state vector. Do you observe any difference in your simulation compared to that reported in Example 4.11?

References

Åström, K.J. and Wittenmark, B. (1995) *Adaptive Control*. Addison-Wesley, Reading, MA.

Schaaf, S.A. and Chambre, P.L. (1958) *Flow of Rarefied Gases*. Princeton University Press, Princeton, NJ.

Slotine, J.-J.E. and Li, W. (1991) *Applied Nonlinear Control*. Prentice Hall, Englewood Cliffs, NJ.

Tewari, A. (2006) *Atmospheric and Space Flight Dynamics*. Birkhäuser, Boston.

Timnat, Y.M. (1990) Recent developments in ramjets, ducted rockets, and scramjets. *Progress in Aerospace Sciences* **27**, 201–235.

5

Attitude Control of Rockets

5.1 Introduction

It is crucial for a launch vehicle to be always aligned with the velocity vector due to aerodynamic load considerations. The thin structure of a vehicle filled with liquid propellants is much like an elongated aluminum can of soda where any significant transverse acceleration due to lift (or sideforce) can prove to be destructive. For this reason, all launch vehicles are equipped with an attitude control system for making the vehicle's orientation follow the flight-path curvature such that angle of attack and sideslip angle essentially remain zero. In order to have longitudinal (pitch–yaw) stability at lower altitudes, launch vehicles are generally fitted with aerodynamic fins near the base. If made movable, the fins can also apply limited control torques in the denser region of the atmosphere (as in the case of the German V-2 rocket of the Second World War). However, since a launch vehicle operates at a wide range of speeds and altitudes, one cannot rely upon aerodynamic moments generated by movable fins to control the vehicle throughout its trajectory. Instead, the attitude control torque is generated by a combination of thrust-vectoring (i.e., changing the direction of the thrust vector), cold gas reaction jets, and/or small (vernier) rockets.

Since there is no chance of confusion with the temperature in this chapter, we will use the symbol T for the thrust of rocket engine (instead of f_T used earlier).

5.2 Attitude Control Plant

The creation of attitude control torque by thrust-vectoring requires a balancing of the transverse forces. Thus, a minimum of two equal and opposite control forces are necessary in order to form a pure couple. While a short-range missile operating entirely inside the lower atmosphere can have the control force produced by an aerodynamic fin to oppose that produced by vectored thrust, this is not feasible for a gravity-turn launch vehicle. In a launch vehicle, thrust-vectoring can be mainly achieved in three distinct ways:[1] (a) gimballing of a single main engine balanced by reaction jets or vernier rockets; (b) gimballing of two or more main engines; and (c) differential thrust from gimballed nozzles. Some launch vehicles (such as the Space Shuttle) employ (a) and (b) alternatively in different stages. The concept of differential thrust for attitude control has not yet been employed due to practical limitations.

[1] Some short-range rockets – such as the German V-2 and the Russian SCUD – alternatively employ graphite vanes at the nozzle exit in order to generate limited control moments by rotation and deflection of the exhaust gases.

Advanced Control of Aircraft, Spacecraft and Rockets, Ashish Tewari.

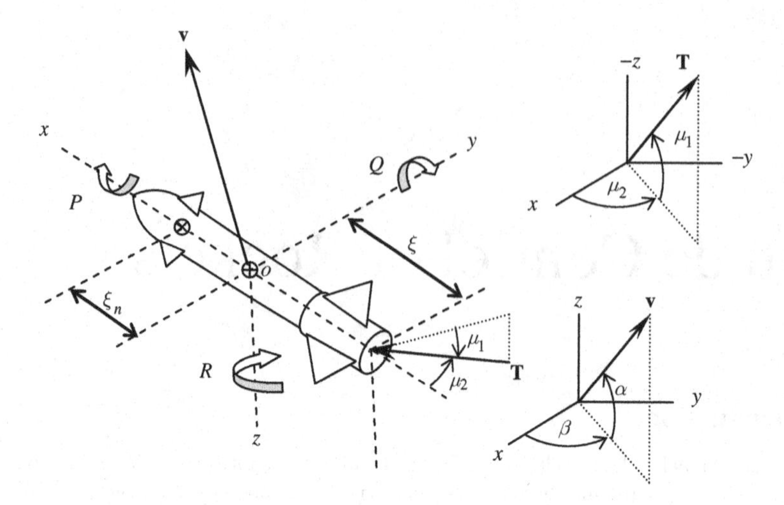

Figure 5.1 Attitude dynamic variables of a rocket given by translational angles, α, β, rotational body rates, P, Q, R, and thrust deflection angles, μ_1, μ_2

Consider a rocket with small thrust deflection angles μ_1 and μ_2 measured from the body axes as shown in Figure 5.1. For a generic thrust-deflection model without either force balance or thrust differential, we have the following control forces and moments:

$$\begin{aligned}
F_y &= -T\cos\mu_1\sin\mu_2 \simeq -T\mu_2,\\
F_z &= -T\sin\mu_1 \simeq -T\mu_1,\\
\tau_y &= -T\xi\sin\mu_1 \simeq -T\xi\mu_1,\\
\tau_z &= T\xi\cos\mu_1\sin\mu_2 \simeq T\xi\mu_2,
\end{aligned} \tag{5.1}$$

while the expression for rolling moment, τ_x, specifically depends upon whether the reaction jet or a gimballing system is employed.

Rotation of the vehicle about a reference orientation gives rise to attitude dynamics. Consider a reference frame, $ox_n y_n z_n$, such that the axis ox_n – represented by the unit vector $\mathbf{i}_n$ – is along the nominal flight direction, and oy_n ($\mathbf{j}_n$) is normal to the plane of flight. The third axis, oz_n ($\mathbf{k}_n$) completes the right-handed triad, $\mathbf{k}_n = \mathbf{i}_n \times \mathbf{j}_n$. Such a frame describes the nominal attitude of the vehicle's body axes along the nominal trajectory, and has an angular velocity

$$\omega_n = Q_n \mathbf{j}_n, \tag{5.2}$$

where, in the nominal case (i.e., when the desired gravity-turn trajectory (h_n, v_n, ϕ_n) is faithfully followed by the guided vehicle), we have the following desired pitch rate signal generated by the guidance system (Chapter 4):

$$Q_n = \dot{\phi}_n = \left[\frac{v_n}{r_0 + h_n} - \frac{g_0 r_0^2}{v_n(r_0 + h_n)}\right]\cos\phi_n. \tag{5.3}$$

We assume the time scale of speed variations is much larger than that of attitude dynamics; thus, $v \simeq v_n$ is taken for the attitude plant.

The orientation of the body-fixed frame, $oxyz$, relative to the instantaneous nominal attitude, $ox_n y_n z_n$, can be represented by the aircraft 3-2-1 Euler's angles (ψ, θ, φ) – see Chapter 3 – which are considered to be small yaw, pitch, and roll angular perturbations, respectively. The net angular velocity of the perturbed body-fixed frame, $oxyz$, relative to a ground reference frame (assumed stationary) is then given by

$$\boldsymbol{\omega} = \dot{\varphi}\mathbf{i} + \dot{\theta}\mathbf{j}' + \dot{\psi}\mathbf{k}_n + Q_n\mathbf{j}_n, \tag{5.4}$$

where the unit vectors, $\mathbf{j}'$, $\mathbf{j}_n$, $\mathbf{k}_n$, are obtained from the 3-2-1 rotation matrix (Chapter 3) with small-angle approximation as follows:

$$\begin{aligned} \mathbf{j}' &\simeq \mathbf{j} - \varphi\mathbf{k}, \\ \mathbf{j}_n &\simeq \psi\mathbf{i} + \mathbf{j} - \varphi\mathbf{k}, \\ \mathbf{k}_n &\simeq -\theta\mathbf{i}. \end{aligned} \tag{5.5}$$

Therefore, the rotational kinematics equations for the rocket are determined by the net roll, pitch, and yaw rates, P, Q, R, respectively, as shown in Figure 5.1, and given by the following:

$$\begin{aligned} \dot{\varphi} &= P - Q_n\psi \\ \dot{\theta} &= Q - Q_n = q \\ \dot{\psi} &= R + Q_n\varphi. \end{aligned} \tag{5.6}$$

The decoupling of pitch from roll–yaw kinematics is evident. The pitch perturbation solution is given by

$$\theta(t) = \int_0^t [Q(\tau) - Q_n(\tau)]\mathrm{d}\tau, \tag{5.7}$$

while the roll–yaw kinematics obeys the following linear time-varying state equation:

$$\begin{Bmatrix} \dot{\varphi} \\ \dot{\psi} \end{Bmatrix} = \begin{pmatrix} 0 & -Q_n \\ Q_n & 0 \end{pmatrix} \begin{Bmatrix} \varphi \\ \psi \end{Bmatrix} + \begin{pmatrix} 1 & 0 \\ 0 & 1 \end{pmatrix} \begin{Bmatrix} P \\ R \end{Bmatrix}. \tag{5.8}$$

Clearly, the attitude kinematics is driven by the roll, pitch, and yaw rates, which, in turn, are governed by rotational kinetics. The task of the attitude control system is to drive the angular errors, φ, θ, ψ, to zero as quickly as possible by suitably modifying the body rates through the application of control inputs.

The nominal pitch rate, Q_n, is synchronized with the rate of change of the flight-path angle (equation (5.3)) in order to maintain the longitudinal axis, ox, along the instantaneous flight direction. However, as soon as ox has a misalignment with the nominal attitude, the off-nominal normal acceleration causes a sideways translation of the vehicle, which is described using the angle of attack, α, and angle of sideslip, β, as follows (Figure 5.1):

$$\mathbf{v} = v(\cos\alpha\cos\beta\mathbf{i} + \cos\alpha\sin\beta\mathbf{j} + \sin\alpha\mathbf{k}) \tag{5.9}$$

or, assuming small perturbations, α, β,

$$\mathbf{v} \simeq v(\mathbf{i} + \beta\mathbf{j} + \alpha\mathbf{k}). \tag{5.10}$$

The acceleration produced by the misalignment of the velocity vector is given by

$$\begin{aligned} \dot{\mathbf{v}} + \boldsymbol{\omega} \times \mathbf{v} &\simeq \dot{v}\mathbf{i} + v(\dot{\beta}\mathbf{j} + \dot{\alpha}\mathbf{k}) \\ &\quad + v(P\mathbf{i} + Q\mathbf{j} + R\mathbf{k}) \times (\mathbf{i} + \beta\mathbf{j} + \alpha\mathbf{k}) \\ &\simeq \left[(\dot{v} + v\alpha Q_n)\mathbf{i} + v(\dot{\beta} + R)\mathbf{j} + v(\dot{\alpha} - Q)\mathbf{k}\right]. \end{aligned} \tag{5.11}$$

The normal acceleration is thus

$$\mathbf{a}_n = a_y\mathbf{j} + a_z\mathbf{k} = v\left[\left(\dot{\beta} + R\right)\mathbf{j} + (\dot{\alpha} - Q)\mathbf{k}\right]. \tag{5.12}$$

Clearly, there is a separation of longitudinal and lateral acceleration terms. The pitch and yaw rate perturbations, q, R, as well as translations, α, β, thus give rise to off-nominal normal acceleration terms given by

$$\begin{aligned} v(\dot{\alpha} - q) &= vQ_\text{n} + mv\dot{\phi} = v(Q_\text{n} - \dot{\phi}_\text{n} + \Delta\dot{\phi}) = v\Delta\dot{\phi}, \\ v(\dot{\beta} + R) &= v\Delta\dot{\psi} \end{aligned} \tag{5.13}$$

If we take into account the aerodynamic and off-nominal control forces, F_y, ΔF_z, the equations for translational perturbation about the nominal trajectory can be expressed as follows:

$$\begin{aligned} mv(\dot{\alpha} - q) &= \Delta F_z + Z_\alpha \alpha, \\ mv(\dot{\beta} + R) &= F_y + Z_\alpha \beta. \end{aligned} \tag{5.14}$$

Here, we have accounted for the pitch-yaw aerodynamic symmetry enjoyed by most launch vehicles and ballistic missiles, wherein $Z_\alpha = Y_\beta$; furthermore, owing to the small size of the stabilizing lifting-surfaces (fins), the stability derivatives representing aerodynamic lag in lift and sideforce, as well as forces due to pitch and yaw rates, are negligible. It is to be noted that explicit gravity and centripetal acceleration due to flight-path curvature terms do not occur in the rocket's α and β dynamics model, because – in contrast with aircraft control systems – the nominal gravitational and centripetal accelerations are included in the terms vQ and vR by virtue of equations (5.6) and (5.3).

The rotational kinetic equations of motion for a launch vehicle with pitch–yaw inertial symmetry ($J_{yy} = J_{zz}$), using the standard terminology for aerodynamic stability derivatives (Chapter 3) are written as follows:

$$\begin{aligned} J_{xx}\dot{P} &= \tau_x + L_p P, \\ J_{yy}\dot{Q} + PR(J_{xx} - J_{yy}) &= \tau_y + M_q Q + M_\alpha \alpha, \\ J_{yy}\dot{R} + PQ(J_{yy} - J_{xx}) &= \tau_z + M_q R - M_\alpha \beta, \end{aligned} \tag{5.15}$$

where $M_q = N_r$ and $M_\alpha = -N_\beta$ are assumed due to aerodynamic symmetry.

Note that the roll kinetics,

$$J_{xx}\dot{P} = \tau_x + L_p P, \tag{5.16}$$

is decoupled from pitch and yaw kinetics as well as α, β dynamics, but pitch and yaw have a mutual inertial coupling owing to a non-zero roll rate, P. Furthermore, pitch–yaw rotations have kinematic and aerodynamic coupling with α, β translations. The success of an attitude control system for a launch vehicle depends upon having a dedicated roll control loop in order to quickly nullify even a small roll rate, thereby removing the inertial coupling from pitch and yaw dynamics. In such a case, we have a linearized pitch dynamics plant given by

$$\begin{aligned} mv(\dot{\alpha} - q) &= \Delta F_z + Z_\alpha \alpha, \\ J_{yy}\dot{q} &= \Delta\tau_y + M_q q + M_\alpha \alpha, \end{aligned} \tag{5.17}$$

where $\Delta\tau_y$ is the off-nominal control pitching moment. Similarly, the yaw dynamics plant for a small yaw rate, R, and sideslip angle, β, is

$$\begin{aligned} mv(\dot{\beta} + R) &= F_y + Z_\alpha \beta, \\ J_{yy}\dot{R} &= \tau_z + M_q R - M_\alpha \beta, \end{aligned} \tag{5.18}$$

where τ_z is the off-nominal control yawing moment. Note the similarity in the pitch and yaw dynamics, which are seen to have identical coefficients if one replaces R by $-R$. Therefore, we can use identical feedback gains for both pitch and yaw loops.

The decoupling of roll, pitch, and yaw dynamics indicates that one can design separate roll, pitch, and yaw control loops for a launch vehicle. The roll controller must have the highest bandwidth (speed) for quickly reducing the roll rate, while the pitch and yaw controllers should maintain the vehicle's attitude along the nominal trajectory.

5.3 Closed-Loop Attitude Control

A thrust-vectored vehicle is statically unstable in pitch and yaw ($M_\alpha > 0$) and is also unstable in roll when unstabilized by aerodynamic fins (or when operating beyond the sensible atmosphere). Thus feedback stabilization is crucial for all launch vehicles. The outputs for an attitude control system are traditionally provided by a fully gimballed, gyro stabilized platform called an *inertial measurement unit* (IMU) that has pendulums, rate gyros, and accelerometers arranged strategically to pick up angular and rate deviations in roll, pitch, and yaw and forms the integral part of an inertial navigation system. However, modern digital electronics has enabled the replacement of the gyro stabilized platform by *strap-on IMU* units rigidly attached to the vehicle for better resolution, ruggedness, and reduced weight. Such strap-on units have fewer moving parts, and mainly rely upon accelerometers, tuning forks, oscillating crystals, and laser optics (*ring laser gyros*). Detection of angular and rate errors is carried out by comparing the sensed attitude with that generated by the guidance system for a nominal trajectory that is stored onboard. The attitude command output of the guidance system is provided as either a slaving command to the gyro stabilized platform, or a reference input to a strap-on IMU. In the earliest guided rocket vehicles such as the German V-2 rocket, the pitch reference was provided by a simple clockwork mechanism called the *pitch program*. With the availability of digital computers, it is now routine to compute roll and pitch attitude in real time for a nominal trajectory stored on a chip, which can also be updated by telemetry signals from the ground.

With the available attitude reference and sensor data, it is the job of the attitude controller to provide commands for the thrust gimbal angle actuators and/or reaction gas jets in order to nullify the attitude errors, even in the presence of sensor noise and modeling uncertainties. Due to the decoupling of roll dynamics from pitch and yaw, as well as pitch and yaw symmetry, it is possible to have separate control loops for roll, pitch, and yaw. A block diagram of a generic guidance and attitude control system for a rocket is depicted in Figure 5.2. Note the bus for guidance command signals generated by the closed-loop guidance system, to be used as reference for the roll, pitch, and yaw control loops, as well as the engine gimbal servos and thrust cut-off mechanism. Thus the attitude control system is subservient to the guidance system.

The guidance system provides nominal pitch and roll reference for the attitude control plant. However, whenever trajectory corrections are required, the guidance system also generates normal acceleration commands, $\Delta u_2(t)$, $\Delta u_3(t)$,

$$\Delta \mathbf{u} = \Delta u_2 \mathbf{k} + \Delta u_3 \mathbf{j}, \tag{5.19}$$

that are fed forward to the gimbal actuators, as shown in Figure 5.2. As we have seen earlier, $\Delta u_2(t)$ is necessary for guiding the vehicle in a vertical plane. Similarly, the out-of-plane acceleration, $\Delta u_3(t)$, is necessary when a change of plane (i.e., $\psi \neq 0$) is required.

5.4 Roll Control System

The roll control system of a launch vehicle is quite similar to that of a reaction jet controlled, single-axis spacecraft (Chapter 7), although some aerodynamic damping in roll, $L_p < 0$, may be available due to stabilizing fins, in which case a single-variable controller can be designed in a manner quite similar to that of aircraft (Chapter 3). Since roll control of aircraft and spacecraft is quite extensively covered in Chapters 4 and 7, respectively, we shall refrain from considering the very similar approach for rockets

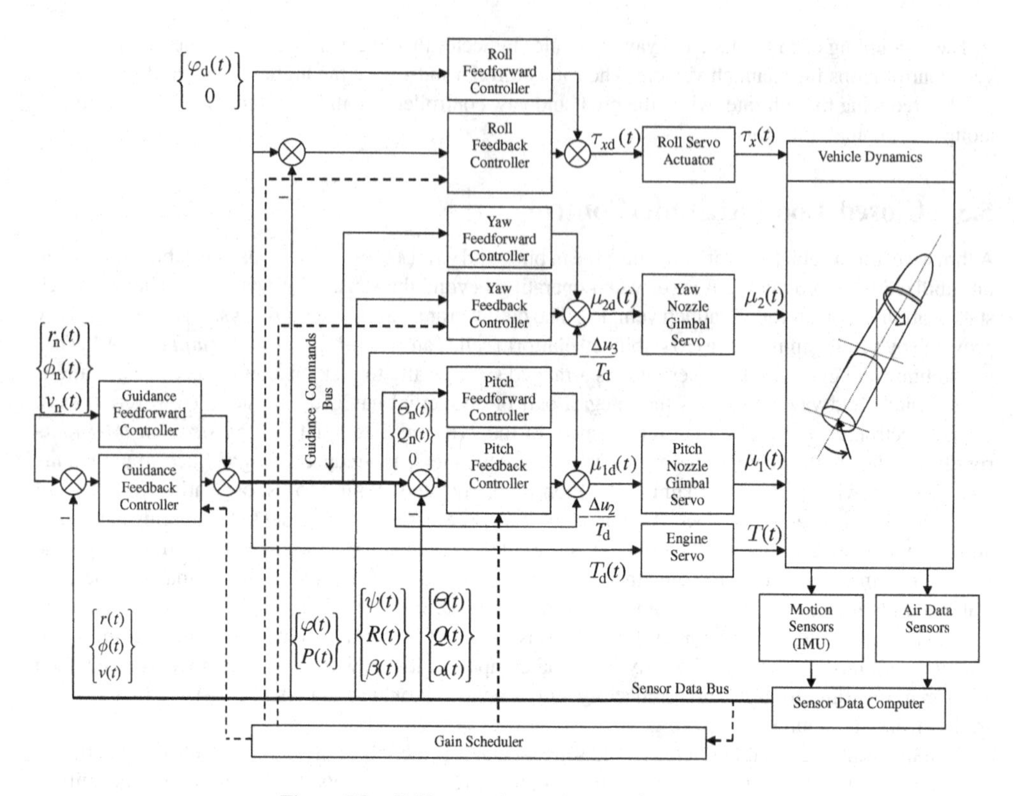

Figure 5.2 Guidance and control system for a rocket

here. Suffice it to say that a single-input, single-output roll controller, using either roll angle or roll rate feedback, can be designed by the classical transfer function approach, which quickly brings a small roll rate perturbation to zero, thereby maintaining decoupled pitch and yaw dynamics.

5.5 Pitch Control of Rockets

Pitch and yaw control of a launch vehicle is essential for maintaining a zero angle of attack and a zero sideslip angle at all times; this is achieved by rotating the vehicle so as to align the longitudinal axis, ox, along the flight path, which is naturally curved by the influence of gravity. By having a roll controller that maintains a zero roll rate despite disturbances, we have a decoupled pitch and yaw motion that can be separately controlled. Furthermore, due to pitch and yaw symmetry in geometry, mass distribution, and the identical actuating systems for the yaw and pitch gimbal angles, μ_1, μ_2, respectively, we can have identical feedback gains for both pitch and yaw loops. Beginning with the pitch control system, we consider motion in the longitudinal plane (Figure 4.15), for which the kinematical relationship between the angle of attack, α, the pitch angle, Θ, and the flight-path angle, ϕ, is

$$\Theta = \phi - \alpha. \tag{5.20}$$

5.5.1 Pitch Program

In order to maintain $\alpha(t) = 0$, we must have $\Theta(t) = \phi(t)$ at all times. Therefore, the output, $\phi(t)$, of the guidance system provides a reference command, $\Theta_n(t)$, for the pitch control system. For a nominal trajectory, $\Theta_n(t) = \phi_n(t)$ or $Q_n(t) = \dot{\phi}_n(t)$ (equation (5.3)). Such a variation of the desired pitch angle according to a known function of time is called the *pitch program* and is usually stored onboard as a data file in the flight control computer. The earliest pitch program was devised for the German V-2 rocket (also used later in some Russian SCUD missile variants) in the form of a simple, pre-set clockwork mechanism that tilted a gyro stabilized program with time in order to provide a pitch reference for targeting the short-range missile. The pitch program should, however, be based upon the actual – rather than the nominal – trajectory in order to accurately regulate the angle of attack. Thus, the pitch controller is fed the nominal pitch rate command, $Q_n(t)$, based upon currently sensed trajectory variables, as well as a normal acceleration guidance command, $\Delta u_2(t)$, generated by the closed-loop guidance system for correcting trajectory errors (Chapter 4).

For rotating the vehicle so as to achieve the nominal pitch rate given by equation (5.3), the nominal pitching moment input can be calculated as follows:

$$\tau_{yn} = J_{yy}\dot{Q}_n - M_q Q_n = \left(J_{yy}\frac{\partial Q_n}{\partial \phi_n} - M_q \right) Q_n = -\left(J_{yy} Q_n \tan\phi_n + M_q \right) Q_n, \tag{5.21}$$

where both J_{yy} and M_q are determined in advance for the entire nominal trajectory, (h_n, v_n, ϕ_n). Thus, a good estimate of the nominal gimbal angle time profile can be obtained from equation (5.3) as follows:

$$\mu_{1n} = \left(J_{yy} Q_n \tan\phi_n + M_q \right) \frac{Q_n}{T\xi}. \tag{5.22}$$

When the pitch program is based upon a fed-forward nominal pitch rate, the instantaneous flight-path angle is computed by integration of Q_n, beginning from an initial value, $\phi_n(0)$:

$$\phi_n(t) = \phi_n(0) + \int_0^t Q_n(\tau) d\tau. \tag{5.23}$$

Equations (5.22) and (5.23) completely specify the pitch program that can be used as a feedforward pitch controller with M_q, J_{yy}, T stored onboard as either average values or pre-determined functions of Q_n.

Example 5.1 Using the nominal trajectory given in Chapter 4, the pitch dynamic parameters given in Table 5.1, and the constants $T = 133\,200$ N and $\xi = 5$ m, estimate the nominal gimbal angle profile for the first 20 s of flight.

With the given data, the nominal pitch rate, Q_n, and the nominal gimbal angle, μ_{1n}, are plotted in Figures 5.3 and 5.4, respectively. Note the initially large pitch rate – and consequently a large gimbal angle – immediately after launch when the speed is quite small. As the speed increases, both Q_n and μ_{1n} rapidly decay to nearly zero in about 20 seconds.

5.5.2 *Pitch Guidance and Control System*

If there are no external disturbances and the vehicle's parameters are precisely known, the nominal trajectory would be perfectly followed with $\alpha = 0$, merely by the open-loop moment (equation (5.21)) applied by the pitch program. However, since a practical flight always involves off-nominal deviations and uncertain parameter values, a feedback control system is crucial in maintaining $\alpha = 0$. Such a feedback consists of an accurate measurement (or estimate) of both the pitch rate and the angle of attack.

The likely sensors used for pitch feedback control are a rate gyro for measuring the pitch rate, Q, an accelerometer for sensing the normal acceleration, a_z^o, at a sensor location, ξ_n, from the center of mass (Figure 5.1), and/or an angle-of-attack sensor for directly sensing α through air data. Generally, Q and a_z^o are the outputs of a single IMU unit, which may also include a rate-integrating gyro (Tewari 2006) for estimating the pitch angle, θ.

Table 5.1 Pitch moment of inertia and damping in pitch for stage I of a launch vehicle

t (s)	J_{yy} (kg.m^2)	M_q (N.m.s/rad)
0	153065.6	−34.4
6.061	150552.53	−7349.09
12.121	148039.47	−11385.68
18.182	145526.4	−11692.64
24.242	143013.33	−9619.5
30.303	140500.27	−6792.28
36.363	137987.2	−4270.42
42.424	135474.13	−2448.08
48.485	132961.07	−1301.83
54.545	130448	−650.75
60.606	127934.93	−309.08
66.667	125421.87	−140.74
72.727	122908.8	−61.91
78.788	120395.73	−26.48
84.849	117882.67	−11.07
90.909	115369.6	−4.55
96.97	112856.53	−1.85
103.03	110343.47	−0.74

For a normal acceleration at the center of mass, a_z, the angle-of-attack rate is given by equation (5.12) to be as follows:

$$\dot{\alpha} = Q + \frac{a_z}{v}, \tag{5.24}$$

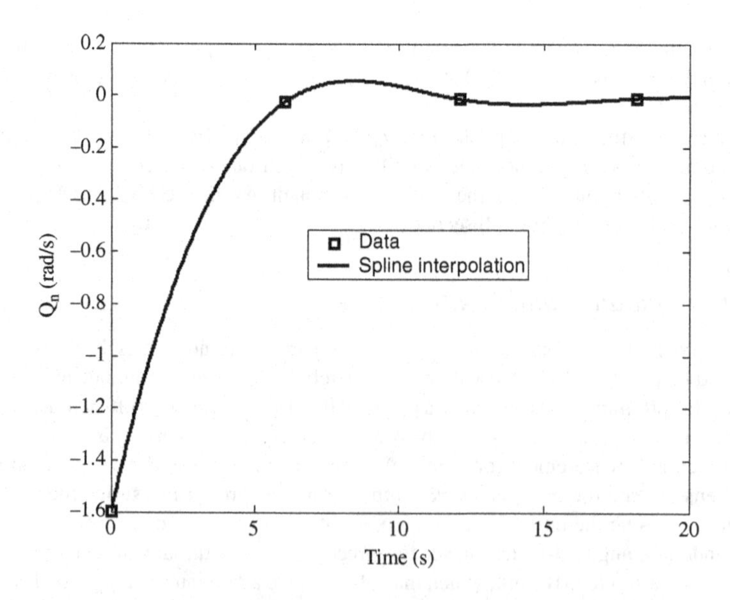

Figure 5.3 Nominal pitch rate of a gravity-turn launch vehicle

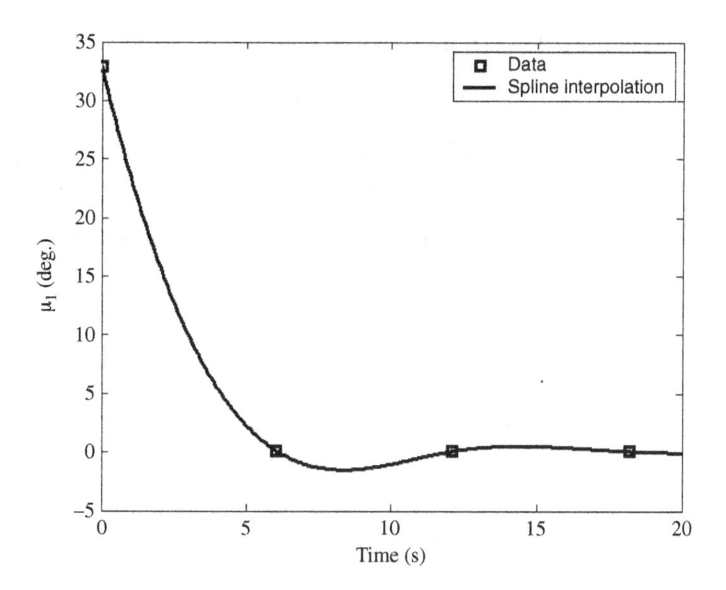

Figure 5.4 Nominal gimbal angle of a gravity-turn launch vehicle

which, after accounting for the nominal acceleration of the pitch program, $a_{zn} = -vQ_n$, can be expressed as

$$\dot{\alpha} = q + \frac{\Delta a_z}{v} = q + \Delta\dot{\phi}, \tag{5.25}$$

that is the same as the first part of equation (5.13). The normal acceleration output, a_z^o, sensed by an accelerometer located a distance, ξ_n, along ox from the center of mass is related to a_z by

$$a_z^o = a_z - \xi_n \dot{Q} \tag{5.26}$$

and

$$\Delta a_z^o = \Delta a_z - \xi_n \dot{q}. \tag{5.27}$$

We express the thrust deflection control input by

$$\mu_1(t) = \mu_{1n}(t) + \mu_{1e}(t), \tag{5.28}$$

where μ_{1n} is the feedforward gimbal angle demanded by the guidance system in order to rotate the vehicle according to the pitch program and is given by equation (5.22). Any small deviation from the nominal flight-path angle, $\Delta\phi = \phi - \phi_n$ (thus a pitch error, $q = Q - Q_n$), and a small angle of attack, α, are corrected by a linear feedback control,

$$\mu_{1e}(t) = -\mathbf{K}(t)\mathbf{e}(t), \tag{5.29}$$

such that the error vector,

$$\mathbf{e} = \mathbf{x} - (Q_n, 0)^T = (q, \alpha)^T, \tag{5.30}$$

tends to zero in the steady state. The state feedback gain matrix of the regulator is given by

$$\mathbf{K}(t) = \left[k_q(t),\ k_z(t)\right]. \tag{5.31}$$

The pitch controller is thus a linear regulator for driving the state error vector, $\mathbf{e} = (q, \alpha)^T$, to zero for a given pitch rate reference, $Q_{\mathrm{n}}(t)$, and a feedforward control input, $\mu_{1\mathrm{n}}(t)$, supplied by the guidance system.

To derive the linearized pitch dynamics plant we substitute equations (5.22) and (5.29) into the pitching moment equation,

$$J_{yy}\dot{Q} = -T\xi\mu_1 + M_\alpha\alpha + M_q Q, \tag{5.32}$$

leading to

$$J_{yy}\dot{q} = -T\xi\mu_{1\mathrm{e}} + M_\alpha\alpha + M_q q. \tag{5.33}$$

By writing the control force as

$$\Delta F_z = -T\mu_{1\mathrm{e}}, \tag{5.34}$$

and substituting into equation (5.14), we have

$$mv\dot{\alpha} = mvq + Z_\alpha\alpha - T\mu_{1\mathrm{e}}. \tag{5.35}$$

Finally, by substituting equations (5.25), (5.33), and (5.35) into equation (5.27), the accelerometer output perturbation is given by

$$\Delta a_z^o = \frac{1}{m}\left(Z_\alpha\alpha - T\mu_{1\mathrm{e}}\right) - \frac{\xi_{\mathrm{n}}}{J_{yy}}\left(-T\xi\mu_{1\mathrm{e}} + M_\alpha\alpha + M_q q\right). \tag{5.36}$$

Therefore, the single-input pitch dynamics plant has the following state-space representation:

$$\dot{\mathbf{x}} = \mathbf{A}\mathbf{x} + \mathbf{B}u, \tag{5.37}$$

$$\mathbf{y} = \mathbf{C}\mathbf{x} + \mathbf{D}u, \tag{5.38}$$

where

$$\begin{aligned} \mathbf{x} &= (q, \alpha)^T, \\ u &= \mu_{1\mathrm{e}}, \\ \mathbf{y} &= \left(q, \Delta a_z^o\right)^T, \end{aligned} \tag{5.39}$$

and

$$\mathbf{A} = \begin{pmatrix} \frac{M_q}{J_{yy}} & \frac{M_\alpha}{J_{yy}} \\ 1 & \frac{Z_\alpha}{mv} \end{pmatrix}, \quad \mathbf{B} = \begin{pmatrix} -\frac{T\xi}{J_{yy}} \\ -\frac{T}{mv} \end{pmatrix}, \tag{5.40}$$

$$\mathbf{C} = \begin{bmatrix} 1 & 0 \\ -\frac{M_q\xi_{\mathrm{n}}}{J_{yy}} & \frac{Z_\alpha}{m} - \frac{M_\alpha\xi_{\mathrm{n}}}{J_{yy}} \end{bmatrix}, \quad \mathbf{D} = \begin{bmatrix} 0 \\ T\left(\frac{\xi_{\mathrm{n}}\xi}{J_{yy}} - \frac{1}{m}\right) \end{bmatrix}. \tag{5.41}$$

Due to the variation of the coefficients, $v, m, J_{yy}, Z_\alpha, M_q, M_\alpha, T, \xi$, along the trajectory, the plant is a time-varying one and must be treated with the same caution as the guidance and roll control plants. The plant is unconditionally controllable, even in the space where the aerodynamic derivatives, Z_α, M_q, M_α, vanish. Hence, a linear feedback regulator can be designed by either the pole-placement or the LQR approach (Chapter 2).

Since the output vector, $\mathbf{y}(t)$ – rather than the state vector – is actually measured and fed back, the regulator control law is modified by virtue of the output equation (5.38) to the following:

$$\mu_{1e} = -\mathbf{K}^{o}(\mathbf{y} - \mathbf{y}_{d}) = -\mathbf{K}^{o}(\mathbf{Ce} + \mathbf{D}\mu_{1e}), \tag{5.42}$$

or

$$\mu_{1e} = -\frac{\mathbf{K}^{o}\mathbf{C}}{1 + \mathbf{K}^{o}\mathbf{D}}\mathbf{e}, \tag{5.43}$$

thereby implying

$$\mathbf{K} = \frac{\mathbf{K}^{o}\mathbf{C}}{1 + \mathbf{K}^{o}\mathbf{D}} \tag{5.44}$$

or

$$\mathbf{K}^{o} = \mathbf{K}(\mathbf{C} - \mathbf{DK})^{-1}, \tag{5.45}$$

where

$$\mathbf{K}^{o}(t) = \left[k_q^{o}(t),\ k_z^{o}(t)\right]. \tag{5.46}$$

A block diagram of a typical pitch guidance and control system is shown in Figure 5.5. Note the outer pitch guidance loop shown by dashed lines that is necessary for correcting trajectory errors, $(\Delta h, \Delta v, \Delta\phi)$.

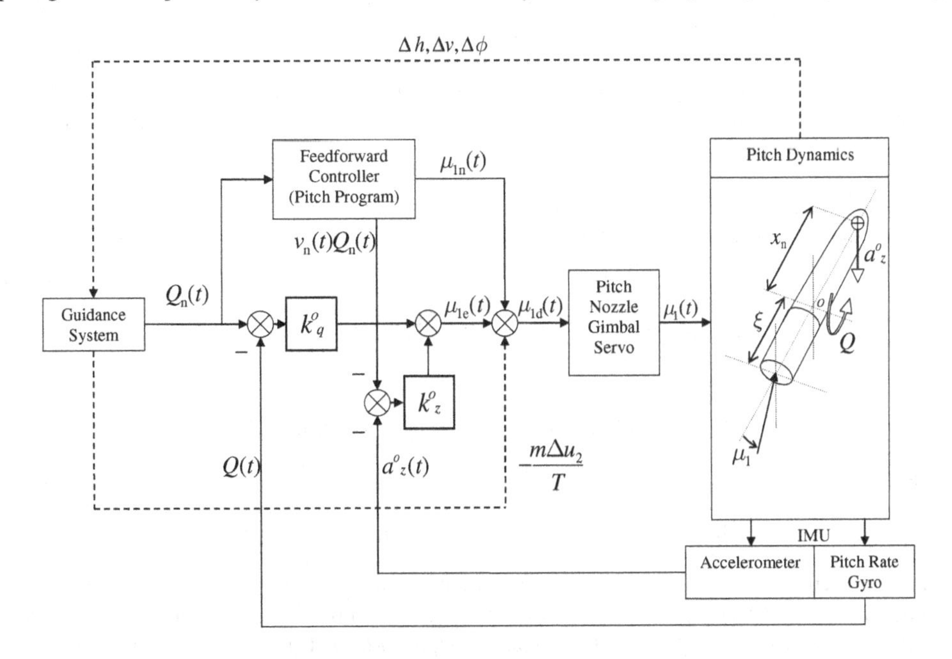

Figure 5.5 A typical pitch control system for a rocket

5.5.3 Adaptive Pitch Control System

Determination of feedback regulator gains, $k_q^o(t)$, $k_z^o(t)$, for the second-order, time-varying, pitch dynamics plant can be carried out in closed form with the pole-placement method. Using the state feedback law (equation (5.31)), we write the closed-loop dynamics matrix as

$$\mathbf{A}_c = \mathbf{A} - \mathbf{BK} = \begin{pmatrix} \frac{M_q+Tk_q\xi}{J_{yy}} & \frac{M_\alpha+Tk_\alpha\xi}{J_{yy}} \\ 1+\frac{Tk_q}{mv} & \frac{Z_\alpha+Tk_\alpha}{mv} \end{pmatrix}. \tag{5.47}$$

The closed-loop characteristic polynomial is thus

$$\begin{aligned}|s\mathbf{I} - \mathbf{A}_c| &= s^2 - \left(\frac{M_q + Tk_q\xi}{J_{yy}} + \frac{Z_\alpha + Tk_\alpha}{mv}\right)s \\ &+ \frac{k_qT(\xi Z_\alpha - M_\alpha) + k_\alpha T(M_q - \xi mv) + M_qZ_\alpha - mvM_\alpha}{mvJ_{yy}} \\ &= s^2 + 2\zeta\omega s + \omega^2. \end{aligned} \tag{5.48}$$

If we select a closed-loop damping ratio, $\zeta = 1/\sqrt{2}$, and a settling time, $t_s \simeq 4/\zeta\omega$, the state feedback gains are given by

$$k_q = \frac{\frac{32mvJ_{yy}}{t_s^2} + \left[mvM_\alpha + (M_q - mv\xi)\left(Z_\alpha + \frac{8mv}{t_s} + \frac{mvM_q}{J_{yy}}\right)\right]}{T\left(\xi Z_\alpha - M_\alpha + \frac{m^2v^2\xi^2}{J_{yy}} - \frac{mv\xi M_q}{J_{yy}}\right)} \tag{5.49}$$

and

$$k_\alpha = -\frac{1}{T}\left[Z_\alpha + \frac{8mv}{t_s} + \frac{mv}{J_{yy}}\left(M_q + T\xi k_q\right)\right]. \tag{5.50}$$

Thus, the adaptive controller gains can either be determined online in real time, t, as functions of the nominal time-varying parameters, m, v, ξ, J_{yy}, T, Z_α, M_α, and M_q, or stored onboard as pre-determined time functions, $k_q(t)$, $k_\alpha(t)$.

Example 5.2 Using the data given in Example 5.1 and the following additional data, compute the state feedback gains, $k_q(t)$, $k_\alpha(t)$, and the output feedback gains, $k_q^o(t)$, $k_z^o(t)$, for a sensor location, $\xi_n = 3$ m, and three different values of the settling time, $t_s = 5$, 10 and 20 s, for the first 100 s of flight:

$$m(t) = \begin{cases} 30\,860 - (308.6 - 225)t \text{ kg}, & 0 \le t \le 103.03 \text{ s}, \\ 22\,246.692 \text{ kg}, & t > 103.03 \text{ s}, \end{cases}$$

$$Z_\alpha = -3.13\bar{q}S \text{ N/rad}, \quad M_\alpha = 11.27\bar{q}Sb \text{ N.m/rad},$$

where

$$\bar{q} = 0.876v^2e^{-h/6700} \text{ N/m}^2, \quad b = 1.1433 \text{ m}, \quad S = 1.0262 \text{ m}^2.$$

The computation of the feedback gains is carried out by the following MATLAB® statements and the stored nominal trajectory parameters, tn, rn, vn, in seconds, kilometers, and kilometers per second, respectively, as well as (1×18) row vectors for pitch damping, Mq, and pitch inertia, J, for the first 103 s of flight:

```
ts=20;% s
qbar=0.876*exp((6378.14-rn)/6.7)*1e6;
t=tn(:,1:18); %First 103 seconds of flight
m=1000*(30.86-(30.86-22.5)*t/100); %kg
T=133200; %N
xi=5; %m
xn=3; %m
Cza=-3.13; %per rad.
Cma=11.27; %per rad.
b=1.1433; %m
S=1.0262; %m^2
Malfa=qbar(:,1:18)*S*b*Cma;
Zalfa=qbar(:,1:18)*S*Cza;
v=vn(:,1:18)*1000;
k1=(32*J.*m.*v/ts^2+(Malfa.*m.*v+(8*m.*v/ts+Zalfa+m.*v.*Mq./J).* ...
(Mq-m.*v*xi))))./(T*(Zalfa*xi-Malfa+m.*m.*v.*v*xi^2./J-Mq.*m.*v*xi./J));
k2=-(8*m.*v/ts+Zalfa+m.*v.*(Mq+T*k1.*xi)./J)/T;
E=[];Kbar=[];
for i=1:size(t,2)
    A=[Mq(:,i)/J(:,i) Malfa(:,i)/J(:,i); 1 Zalfa(:,i)/(m(:,i)*v(:,i))];
    B=-T*[xi/J(:,i); 1/(m(:,i)*v(:,i))];
    C=[1 0; -Mq(:,i)*xn/J(:,i)  Zalfa(:,i)/m(:,i)-Malfa(:,i)*xn/J(:,i)];
    D=[0; T*(xn*xi/J(:,i)-1/m(:,i))];
    K=[k1(:,i) k2(:,i)];
    Kb=K*inv(C-D*K);
    Kbar(i,:)=Kb;
end
```

The nominal aerodynamic parameters of the vehicle are plotted for the first 100 seconds of flight in Figure 5.6, while the computed state feedback gains are shown in Figure 5.7. Note that while k_q is roughly doubled in magnitude as the value of t_s is halved, there is a negligible variation of k_α with t_s. It

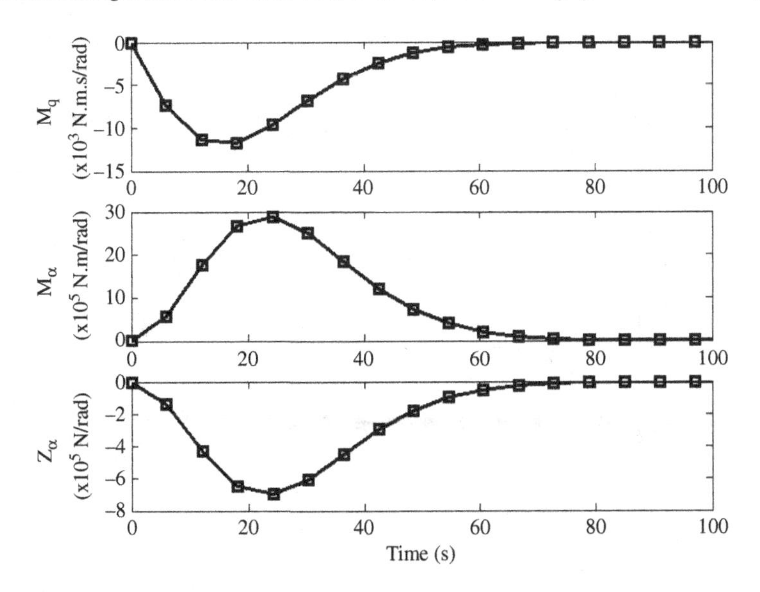

Figure 5.6 Aerodynamic parameters for gravity-turn rocket's pitch dynamics plant

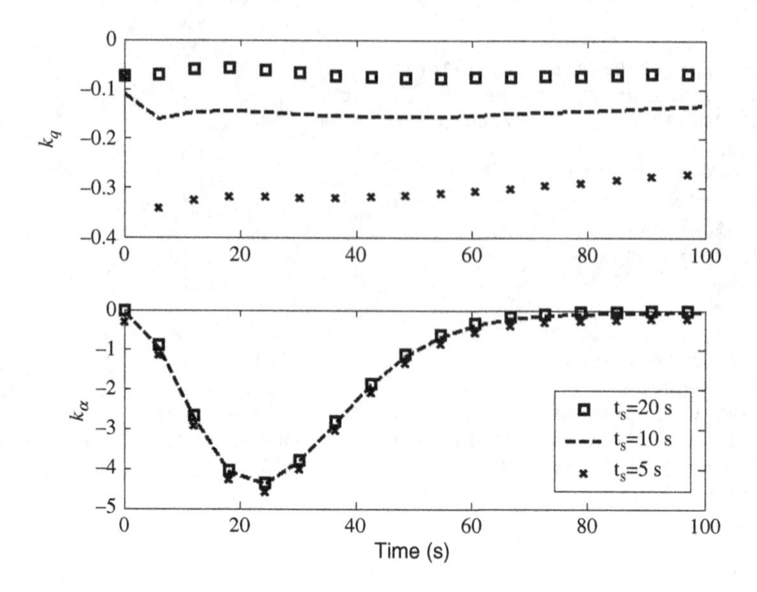

Figure 5.7 State feedback gains for gravity-turn rocket's pitch control system

is also evident that the magnitude of k_α decays to zero, while k_q becomes a constant, in about 100 s as the vehicle leaves the sensible atmosphere. The output feedback gains, k_q^o, k_z^o, plotted in Figure 5.8, show a large fluctuation in the interval $60 < t < 90$ s due to a small aerodynamic stiffness and damping (thus increased controller activity) in that region. Such a fluctuation is absent in the state feedback controller gains, k_q, k_α. Thus we expect the acceleration-based output feedback controller to be less robust to parametric variations than that based upon the angle of attack.

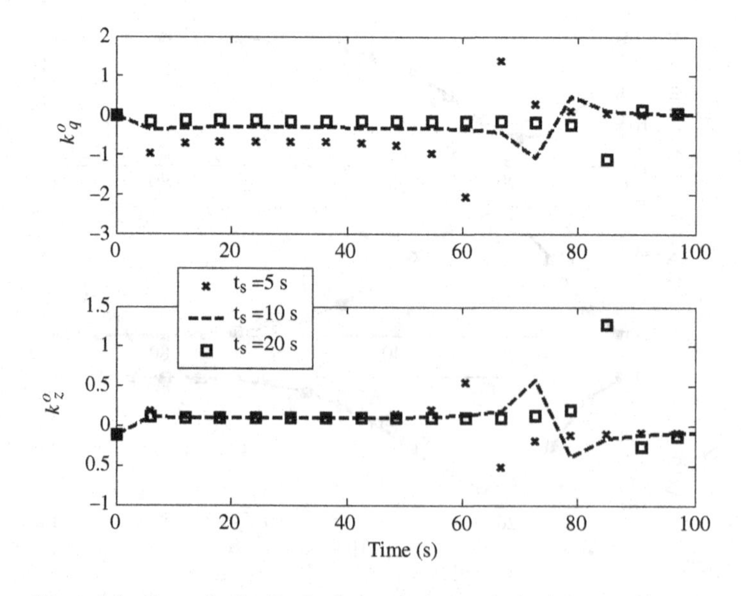

Figure 5.8 Output feedback gains for gravity-turn rocket's pitch control system

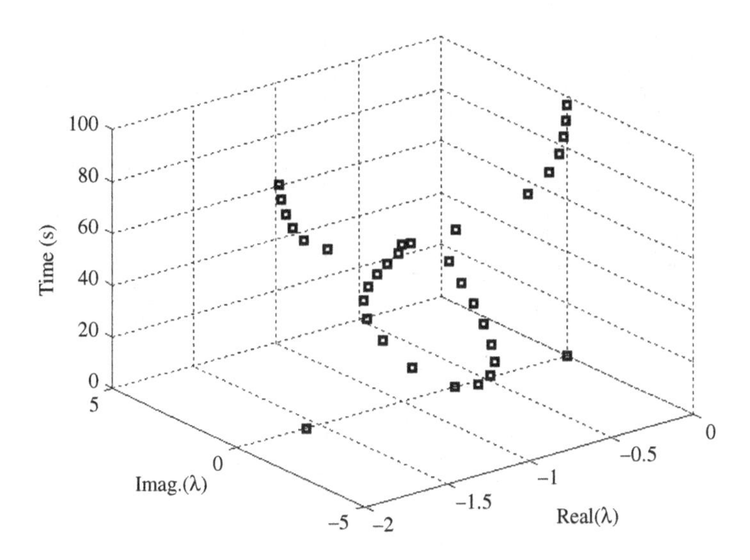

Figure 5.9 Eigenvalues of the pitch control system for a gravity-turn rocket with constant output feedback gains

Example 5.3 As a simpler alternative to the time-varying pitch controller designed in Example 5.2, consider a fixed gain design with $k_q^o = -1$ s, $k_z^o = 0.2$ s^2/m. Plot the eigenvalues of the closed-loop system for the first 100 s of flight.

A plot of the closed-loop system's eigenvalues with fixed gains is shown in Figure 5.9. Both the eigenvalues remain in the left-half s-plane throughout the flight. For $5 < t < 40$ s, the eigenvalues are complex, with real part approximately -0.8. At other times, the eigenvalues are real and negative, and, apart from $s = -0.000182$ at $t = 0$, have smallest value $s = -0.0042$ at $t = 100$ s.

Example 5.4 With the acceleration-based pitch control system designed in Examples 5.2 and 5.3, simulate the closed-loop response of the gravity-turn rocket for an initial perturbation of $q(0) = 0.05$ rad/s, $a_z^o(0) = -0.01$ m/s^2. with: (a) time-varying feedback controller designed with $t_s = 5$ s; (b) controller with constant feedback gains, $k_q^o = -1$ s, $k_z^o = 0.2$ s^2/m.

Include a first-order gimbal actuator with a time constant of 0.2 s, saturation limits ± 0.2 rad, and actuation rate limit ± 1 rad/s. Account for measurement noise in both pitch rate and normal acceleration channels. Pitch rate noise is white with an amplitude of 10^{-4} rad/s (20 deg/hr), while the accelerometer has a white noise of 10^{-4} m/s^2 (0.001 g).[2]

The Simulink block diagrams for the necessary simulations are shown in Figures 5.10 and 5.11 for cases (a) and (b), respectively. The *Pitch Coefficients* (Figure 5.12) and *Time Varying State-Space* (Figure 5.13) subsystem blocks are common to both the simulations. The simulations are carried out with noise spectral density 10^{-8} and sampling interval 0.1 s in both output channels, a fourth-order Runge–Kutta solver, and a relative tolerance of 10^{-5}. Case (a) is able to stabilize the noise-free system without actuator transfer function and nonlinearities, as shown in the results plotted in Figure 5.14. However, when either the noise or the actuator dynamics is added to the closed-loop system (Figure 5.12), the system becomes unstable. Thus, a lack of robustness of the regulator with time-varying gains is demonstrated.

[2] The specified sensors are rather crude by modern standards, where a rate sensitivity of 0.01 deg/hr and acceleration sensitivity of 10^{-6} g is the norm. Hence, the simulation can be regarded as a worst-case scenario.

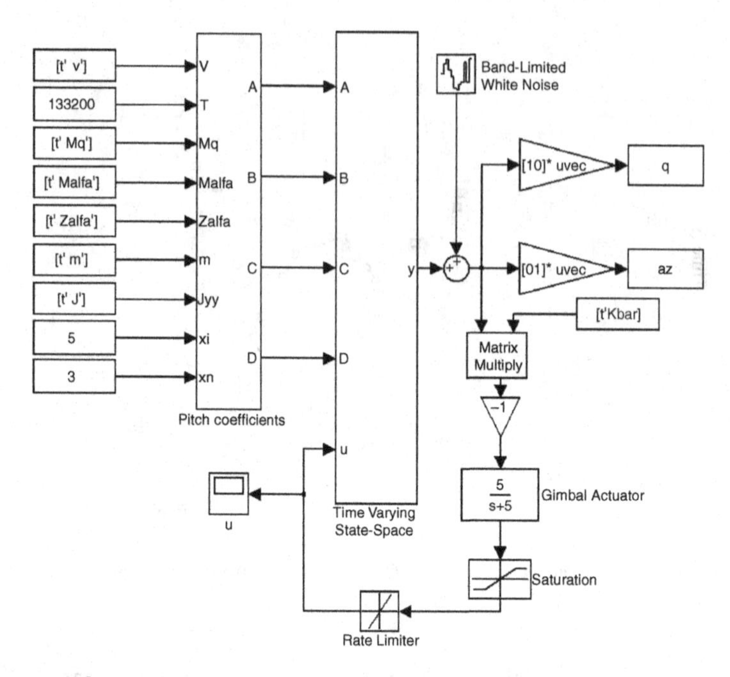

Figure 5.10 Simulink® block diagram of pitch control system for a gravity-turn rocket with time-varying output feedback gains

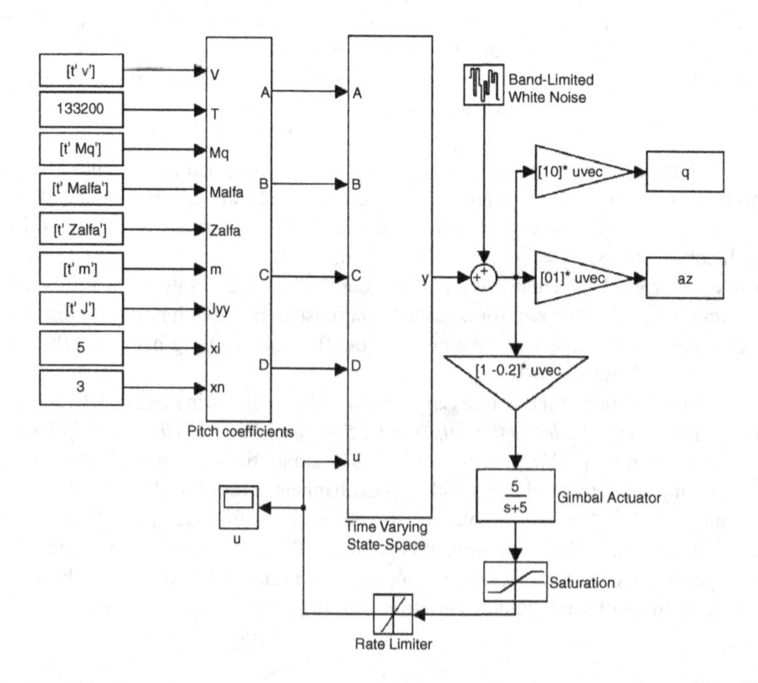

Figure 5.11 Simulink block diagram of pitch control system for a gravity-turn rocket with constant output feedback gains

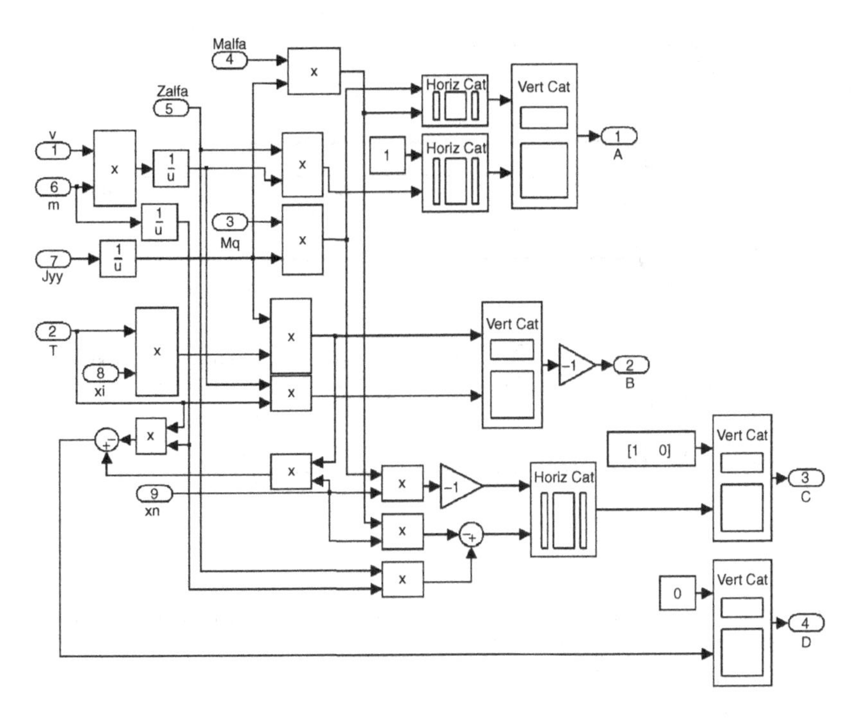

Figure 5.12 Simulink block diagram of *Pitch Coefficients* subsystem block for a gravity-turn rocket's pitch dynamics

With constant output feedback gains as specified in case (b), the simulation results with noise and actuator dynamics are plotted in Figure 5.15. Note the quick initial decay of pitch rate and normal acceleration error in about 10 s, followed by a small-amplitude oscillation during $15 < t < 40$ s due to reduced closed-loop damping in that region. After $t = 40$ s, the system error is maintained quite close to zero despite sensor noise. Therefore, it is possible to have a pitch stabilization of the time-varying plant with constant feedback gains.

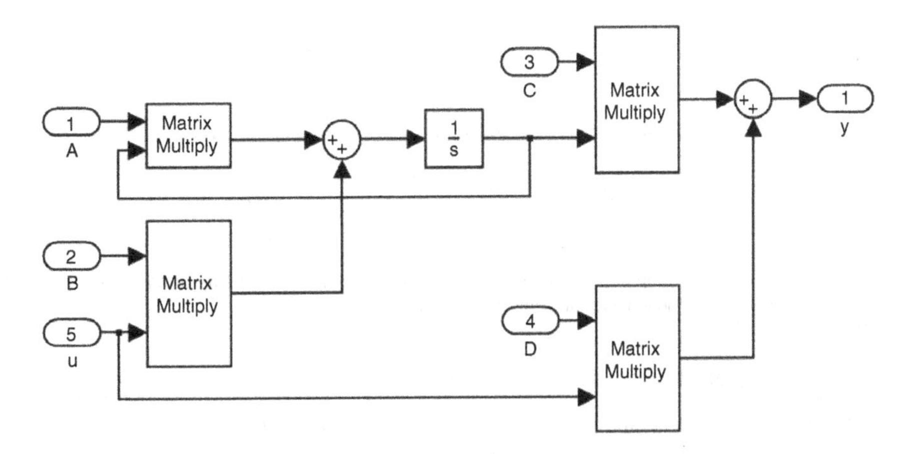

Figure 5.13 Simulink block diagram of *Time Varying State-Space* subsystem block for a gravity-turn rocket's pitch dynamics

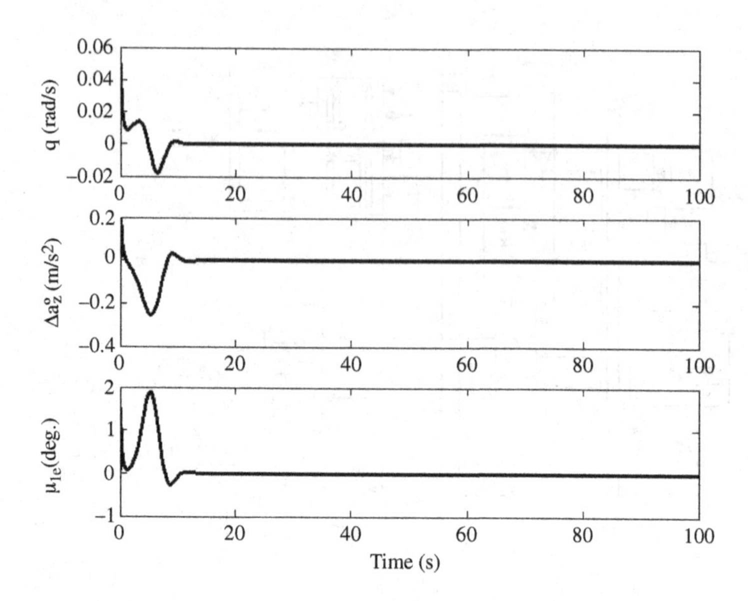

Figure 5.14 Simulated closed-loop initial response of pitch control system for a gravity-turn rocket with time-varying output feedback gains, without either measurement noise or actuator dynamics

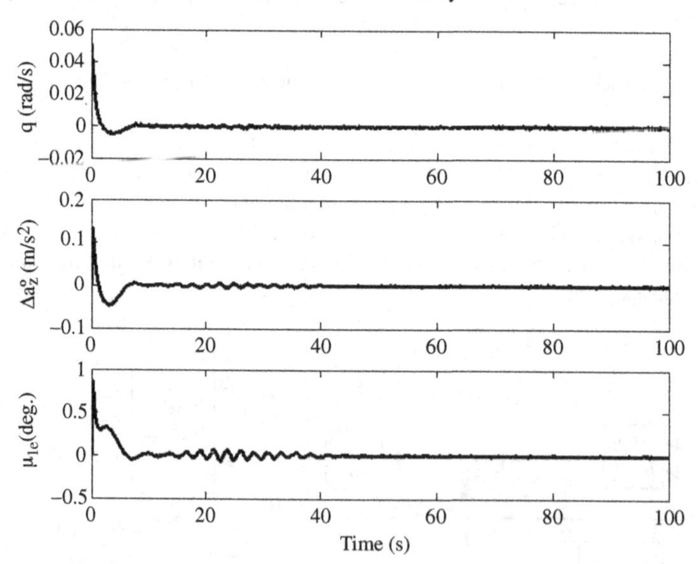

Figure 5.15 Simulated closed-loop initial response of pitch control system for a gravity-turn rocket with constant output feedback gains, measurement noise and actuator dynamics

5.6 Yaw Control of Rockets

Due to pitch–yaw aerodynamic and inertial symmetry, we have $J_{zz} = J_{yy}$, $Y_\beta = Z_\alpha$, $N_\beta = -M_\alpha$, and $N_r = M_q$. Therefore, the yaw and sideslip dynamics are described by

$$J_{yy}\dot{R} = T\xi\mu_2 - M_\alpha\beta + M_q R \tag{5.51}$$

and

$$mv(\dot{\beta} + R) = -T\mu_2 + Z_\alpha \beta, \tag{5.52}$$

which can be expressed in a state-space form as

$$\dot{\mathbf{x}} = \mathbf{A}\mathbf{x} + \mathbf{B}u, \tag{5.53}$$

where

$$\mathbf{x} = (-R, \beta)^T,$$
$$u = \mu_2, \tag{5.54}$$

and

$$\mathbf{A} = \begin{pmatrix} \frac{M_q}{J_{yy}} & \frac{M_\alpha}{J_{yy}} \\ 1 & \frac{Z_\alpha}{mv} \end{pmatrix}, \quad \mathbf{B} = \begin{pmatrix} \frac{T\xi}{J_{yy}} \\ -\frac{T}{mv} \end{pmatrix}, \tag{5.55}$$

which is identical to the pitch dynamics state equation (5.37). Similarly, the output equation with $\mathbf{y} = (-R, a_y^o)^T$, is identical to that of pitch dynamics equation (5.38). Thus one can use exactly the same feedback gains as those of the pitch controller to regulate the state vector, $\mathbf{x} = (-R, \beta)^T$, resulting in the following yaw control-law:

$$\mu_2 = k_q^o R - k_z^o a_y^o. \tag{5.56}$$

In contrast to the pitch controller, a feedforward path such as the pitch program is not required for yaw control because both R and β, as well as the yaw gimbal angle, μ_2, are zero for the nominal trajectory. However, a small control acceleration, $a_y(t)$, may be generated by the guidance system in order to correct an angular error, ψ, in the flight plane orientation caused by planetary rotation (Tewari 2006).

5.7 Summary

Attitude control of a rocket is crucial for maintaining zero angle of attack and sideslip during flight-path corrections by the outer, guidance loop. Roll control systems can be decoupled from those for pitch and yaw, and are designed by the single-variable approach. The pitch control system is a combined feedforward/feedback tracking system for maintaining a nominal pitch rate of the gravity-turn trajectory (pitch program), while regulating the pitch and angle-of-attack errors. Due to pitch–yaw symmetry, the feedback loops for pitch and yaw have identical gains.

Exercises

(1) Design a time-optimal (Chapter 2) roll control system with roll angle feedback, $\varphi(t)$, for a rocket with nominal roll dynamic parameters, $J_{xx} = 4208$ kg.m^2 and $L_p = -130.15$ N.m.s/rad. The vehicle is equipped with a gas jet actuator with a first-order time constant of 0.2 s, which can exert a maximum torque of ± 150 N.m with a deadband of ± 5 N.m (i.e., the actuator does not respond to commanded torque less than 5 N.m in magnitude). The performance should be such that a step desired roll angle of 10° is achieved in about 1 second. Study the closed-loop system's robustness with respect to roll angle measurement noise through an appropriate Simulink model.

(2) Redesign the roll controller of Exercise 1 with a roll rate feedback. What changes are observed from the previous design simulation?

(3) Redesign the pitch controller of Example 5.3 by the frozen LQR method. Select the cost coefficients, $\mathbf{Q}, \mathbf{R}$, such that the maximum gimbal angle is not exceeded at $t = 20$ s. Simulate the closed-loop response with the given actuator model.

(4) Design a reduced-order observer for the pitch dynamics of Example 5.3 using the measurement of the pitch angle, θ, by a rate-integrating gyro and normal acceleration, a_z^o, at the given sensor location. Select the observer coefficients such that the given performance objectives are met without exceeding the maximum gimbal angle. Simulate the closed-loop response with the given actuator model.

(5) Repeat Exercise 3 for a reduced-order observer for the pitch dynamics of Example 5.3 using the measurement of the pitch angle, θ, by a rate-integrating gyro and pitch rate, Q, by a rate gyro. Which of the two designs has the better performance and why?

(6) Build a closed-loop simulation model with the guidance system designed in Chapter 4 with normal acceleration input as well as the pitch attitude controller of Example 5.3, in order to simulate the complete three-degrees-of-freedom motion of the rocket in the vertical plane. Do you observe any change from the closed-loop errors plotted in Figure 4.34?

Reference

Tewari, A. (2006) *Atmospheric and Space Flight Dynamics*. Birkhäuser, Boston.

6

Spacecraft Guidance Systems

6.1 Introduction

Space flight is much more predictable than either aircraft or rocket flight, since it is not subject to substantially large disturbances at any given time. Being governed primarily by the gravitational field of a spherical body (called orbital mechanics), the equations of motion are amenable to simplified solution in several important cases. However, the small magnitude of perturbations – such as non-uniform gravity, atmospheric drag, solar radiation pressure – can appreciably modify the trajectory over a long duration. Since spacecraft flight times are orders of magnitude larger than those of either aircraft or rockets, it is necessary to account for the perturbations while guiding a vehicle in its space mission. We shall begin with an overview of orbital mechanics and then demonstrate how practical guidance systems can be designed for powered space flight in the presence of small orbital perturbations. Instead of covering the linear, single-input, single-output guidance systems that can be handled by classical control theory, our focus in this chapter will be on nonlinear, two-point boundary value problems (2PBVPs) and those having multi-input feedback control.

6.2 Orbital Mechanics

The relative motion of two spherical bodies in mutual gravitational attraction is the fundamental problem of translational space dynamics, called *orbital mechanics*, and possesses an analytical solution. The motion of a spacecraft under the influence of a celestial body is usually approximated as a two-body problem by neglecting the small gravitational effects caused by the other objects, as well as the actual, non-spherical shapes of the two bodies. With some exceptions, such as a lunar trajectory, the two-body assumption is generally valid. Consider spherical masses m_1, m_2 in mutual attraction given by Newton's law of gravitation. The equation of motion of each mass is given by Newton's second law of motion referred to an inertial frame. Subtracting the two equations of motion from one another, we have

$$\frac{\mathrm{d}^2\mathbf{r}}{\mathrm{d}t^2} + \frac{\mu}{r^3}\mathbf{r} = \mathbf{0}, \tag{6.1}$$

where $\mathbf{r}(t)$ is the position of the center of mass m_2 relative to the center of mass m_1 and $\mu = G(m_1 + m_2)$, with $G = 6.6726 \times 10^{-11}$ m^3/kg/s^2 being the *universal gravitational constant*. Usually, the spacecraft's mass, m_2, is negligible in comparison with that of the celestial body, m_1, and we can approximate $\mu \approx Gm_1$.

6.2.1 Orbit Equation

Let us begin the solution of equation (6.1) by taking the vector product of both the sides with **r**:

$$\mathbf{r} \times \frac{d^2\mathbf{r}}{dt^2} + \frac{\mu}{r^3}(\mathbf{r} \times \mathbf{r}) = \mathbf{0} \tag{6.2}$$

or

$$\frac{d}{dt}\left(\mathbf{r} \times \frac{d\mathbf{r}}{dt}\right) = \mathbf{0}, \tag{6.3}$$

which implies that the *specific angular momentum* of m_2 relative to m_1, defined by

$$\mathbf{h} \doteq \mathbf{r} \times \frac{d\mathbf{r}}{dt} = \mathbf{r} \times \mathbf{v}, \tag{6.4}$$

is conserved. Here $\mathbf{v} \doteq \frac{d\mathbf{r}}{dt}$ is the velocity. Since **h** is a constant vector, we have the following consequences:

(a) The direction of **h** is a constant. This implies that the vectors **r** and **v** are always in the same plane – called the *orbital plane* – and **h** is normal to that plane.
(b) The magnitude of **h** is constant. Writing **h** in polar coordinates, (r, θ),

$$h \doteq \mid \mathbf{h} \mid = \mid \mathbf{r} \times \mathbf{v} \mid = r^2 \frac{d\theta}{dt}, \tag{6.5}$$

which implies that the radius vector, **r**, sweeps out area at a constant rate, $\frac{1}{2}r^2\frac{d\theta}{dt}$.[1]

The trajectories in the orbital plane – called *orbits* – are classified according to the magnitude and direction of a constant **h**. The case of $\mathbf{h} = \mathbf{0}$ represents rectilinear motion along the line joining the two bodies, while $\mathbf{h} \neq \mathbf{0}$ represents the more common orbits involving rotation of m_2 about m_1.

We next take the vector product of both sides of equation (6.1) with **h**:

$$\frac{d^2\mathbf{r}}{dt^2} \times \mathbf{h} + \frac{\mu}{r^3}(\mathbf{r} \times \mathbf{h}) = \mathbf{0}. \tag{6.6}$$

Since **h** is constant, we have

$$\frac{d^2\mathbf{r}}{dt^2} \times \mathbf{h} = \frac{d}{dt}(\mathbf{v} \times \mathbf{h}). \tag{6.7}$$

Furthermore, the following interesting identity can be derived by differentiating r^2 with respect to time:

$$\mathbf{r} \cdot \mathbf{v} = r\dot{r}, \tag{6.8}$$

[1] This is the general form of *Kepler's second law of planetary motion.*

Hence, the second term on the left-hand side of equation (6.6) becomes

$$\begin{aligned}\frac{\mu}{r^3}(\mathbf{r}\times\mathbf{h}) &= \frac{\mu}{r^3}(\mathbf{r}\times(\mathbf{r}\times\mathbf{v}))\\ &= \frac{\mu}{r^3}[(\mathbf{r}\cdot\mathbf{v})\mathbf{r}-(\mathbf{r}\cdot\mathbf{r})\mathbf{v}]\\ &= \frac{\mu\dot{r}}{r^2}\mathbf{r}-\frac{\mu}{r}\mathbf{v}\\ &= -\mu\left(\frac{\mathbf{v}}{r}-\frac{\dot{r}}{r^2}\mathbf{r}\right)\\ &= -\mu\frac{\mathrm{d}}{\mathrm{d}t}\left(\frac{\mathbf{r}}{r}\right).\end{aligned} \tag{6.9}$$

Substituting equations (6.7) and (6.9) into equation (6.6), we have

$$\frac{\mathrm{d}}{\mathrm{d}t}\left(\mathbf{v}\times\mathbf{h}-\frac{\mu\mathbf{r}}{r}\right)=\mathbf{0}. \tag{6.10}$$

Thus, a constant vector $\mathbf{e}$, called the *eccentricity vector*, can be defined such that

$$\mu\mathbf{e}\doteq\mathbf{v}\times\mathbf{h}-\frac{\mu\mathbf{r}}{r}. \tag{6.11}$$

Since the eccentricity vector is normal to $\mathbf{h}$, it lies in the orbital plane and is employed as a reference to specify the direction of the relative position vector, $\mathbf{r}(t)$. The angle, $\theta(t)$, made by $\mathbf{r}(t)$ with $\mathbf{e}$ (measured along the flight direction) is called the *true anomaly*, $\theta(t)$.

Taking the scalar product of $\mu\mathbf{e}$ with $\mathbf{r}(t)$, we have

$$r+\mathbf{e}\cdot\mathbf{r}=\frac{1}{\mu}\mathbf{r}\cdot(\mathbf{v}\times\mathbf{h})=\frac{h^2}{\mu} \tag{6.12}$$

or

$$r(\theta)=\frac{h^2/\mu}{1+e\cos\theta}. \tag{6.13}$$

Equation (6.13) – called the *orbit equation* – defines the shape of the orbit in polar coordinates, (r,θ), and indicates that the orbit is symmetrical about $\mathbf{e}$. Furthermore, the minimum separation of the two bodies (called *periapsis*) occurs for $\theta=0$, implying that $\mathbf{e}$ points toward the periapsis. From the orbit equation it is clear that the general orbit is a *conic section* – the shape obtained by cutting a right-circular cone in a particular way. For $e<1$, the orbit is an *ellipse*. The circle is a special ellipse with $e=0$ and $r=h^2/\mu$. For $e=1$, the orbit is a *parabola*. The *rectilinear* trajectory is a special parabola with $h=0$. For $e>1$, the orbit is a *hyperbola*. In all cases, the *focus* of the orbit is at the center of the celestial body, and the *semi-major axis*, a, is given by

$$a=\frac{h^2/\mu}{1-e^2}. \tag{6.14}$$

6.2.2 Perifocal and Celestial Frames

A right-handed coordinate frame fixed to the orbital plane, with unit vectors $\mathbf{i_h}\doteq\mathbf{h}/h$, $\mathbf{i_e}\doteq\mathbf{e}/e$, and $\mathbf{i_p}\doteq\mathbf{i_h}\times\mathbf{i_e}$, is used to specify the relative position and velocity vectors of the spacecraft in the orbital plane. This frame, $(\mathbf{i_e},\mathbf{i_p},\mathbf{i_h})$, is called the *perifocal frame*. The position and velocity of the spacecraft

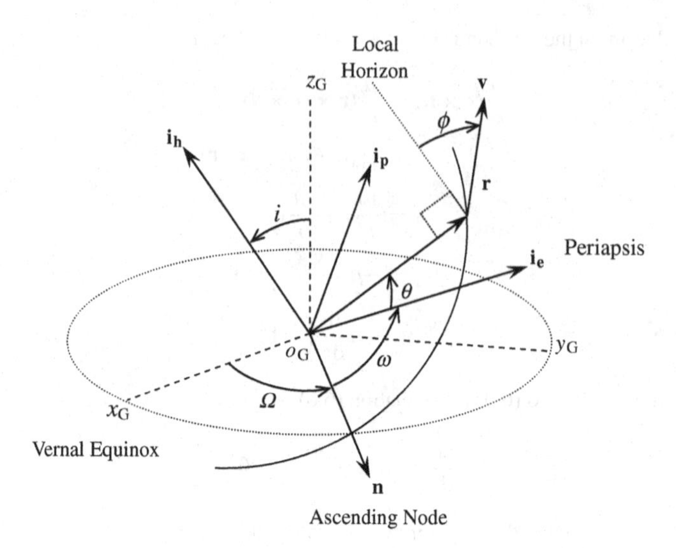

Figure 6.1 Orientation of the perifocal frame relative to the celestial plane given by Euler angles, $(\Omega)_3, (i)_1, (\omega)_3$

in the perifocal frame are given by

$$\begin{aligned}\mathbf{r} &= r\cos\theta\mathbf{i_e} + r\sin\theta\mathbf{i_p}, \\ \mathbf{v} &= v(\sin\phi\cos\theta - \cos\phi\sin\theta)\mathbf{i_e} + v(\sin\phi\sin\theta + \cos\phi\cos\theta)\mathbf{i_p},\end{aligned} \tag{6.15}$$

where $\phi(t)$ is the flight-path angle (taken to be positive above the local horizon) shown in Figure 6.1, and given by (Tewari 2006)

$$\tan\phi = \frac{e\sin\theta}{1 + e\cos\theta}. \tag{6.16}$$

On eliminating the flight-path angle from the velocity expression, we have

$$\mathbf{v} = -\frac{\mu}{h}\sin\theta\mathbf{i_e} + \frac{\mu}{h}(e + \cos\theta)\mathbf{i_p}. \tag{6.17}$$

In most practical applications, the spacecraft's position and velocity are resolved in a stationary *celestial frame* that is fixed with respect to distant stars and has its origin at the center of the celestial body. Examples of Earth-based celestial frames are the geocentric frames, $(o_G x_G y_G z_G)$, where the plane $(x_G y_G)$ is either the *equatorial plane*, or the *ecliptic plane*, and the axis $o_G x_G$ points toward the *vernal equinox*.[2] Normally, the equatorial plane is employed as the reference plane, $(x_G y_G)$, when an orbit close to a planet is of interest, while the ecliptic plane is used for interplanetary trajectories. The intersection of the orbital plane with the reference plane, $(x_G y_G)$, yields the *line of nodes*, as shown in Figure 6.1. The *ascending node* is the name given to the point on the line of nodes where the orbit crosses the plane $(x_G y_G)$ from south to north. A unit vector, $\mathbf{n}$, pointing toward the ascending node makes an angle Ω with the axis $o_G x_G$. The angle Ω is measured in the plane $(x_G y_G)$ in an anti-clockwise direction (Figure 6.1), and is

[2] The vernal equinox is the unique location of the Sun against a background of distant stars, as it crosses a planet's equatorial plane from south to north. The vernal equinox thus indicates a direction along the intersection of a planet's equatorial plane, and the plane of the orbit of the planet around the Sun (called the *ecliptic plane*). On Earth, the vernal equinox occurs at noon on the first day of spring, around March 21.

termed *right ascension of the ascending node*. The *inclination*, i, is the angle between the orbital plane and $(x_G y_G)$, and is the positive rotation about $\mathbf{n}$ required to produce $\mathbf{i_h}$ from the axis $o_G z_G$. The angle ω represents a positive rotation of $\mathbf{n}$ about $\mathbf{i_h}$ to produce $\mathbf{i_e}$ in the orbital plane, and is called the *argument of periapsis*.

A coordinate transformation between the perifocal and celestial frames is necessary. The orientation of the perifocal frame relative to the celestial frame is usually described by the $(\Omega)_3, (i)_1, (\omega)_3$ Euler angles, as shown in Figure 6.1, leading to the rotation matrix

$$\mathbf{C} = \mathbf{C}_3(\omega)\mathbf{C}_1(i)\mathbf{C}_3(\Omega). \tag{6.18}$$

In order to transform the perifocal position and velocity (given above) to the celestial frame, $(o_G x_G y_G z_G)$, we write

$$\begin{Bmatrix} \mathbf{I} \\ \mathbf{J} \\ \mathbf{K} \end{Bmatrix} = \mathbf{C}^T \begin{Bmatrix} \mathbf{i_e} \\ \mathbf{i_p} \\ \mathbf{i_h} \end{Bmatrix}, \tag{6.19}$$

where the triad $\mathbf{I}, \mathbf{J}, \mathbf{K}$, denotes the frame $(o_G x_G y_G z_G)$.

6.2.3 Time Equation

The vectors $\mathbf{h}$ and $\mathbf{e}$ completely determine the shape and orientation of a two-body trajectory, but do not provide any information about the location of the spacecraft at a given time. This missing data is usually expressed as an equation that relates the variation of true anomaly with time, $\theta(t)$, beginning with the *time of periapsis*, τ, for which $\theta = 0$. On substituting the orbit equation (6.13) into the angular momentum magnitude, equation (6.5), we have

$$\frac{\mathrm{d}\theta}{\mathrm{d}t} = \sqrt{\frac{\mu}{p^3}}(1 + e\cos\theta)^2. \tag{6.20}$$

Integration of equation (6.20) provides τ, thereby determining the function $\theta(t)$, and completing the solution to the two-body problem. However, such an integration is usually carried out by a numerical procedure, depending upon whether we have an elliptical, parabolic, or hyperbolic orbit.

Elliptical Orbit ($0 \le e < 1$)

For an elliptical orbit, equation (6.20) becomes

$$\frac{(1 - e^2)^{3/2}\mathrm{d}\theta}{(1 + e\cos\theta)^2} = n\mathrm{d}t, \tag{6.21}$$

where

$$n \doteq \sqrt{\frac{\mu}{a^3}} \tag{6.22}$$

is referred to as the orbital *mean motion*. In order to simplify equation (6.21), we introduce an *eccentric anomaly*, E, defined by

$$\cos E \doteq \frac{e + \cos\theta}{1 + e\cos\theta}, \tag{6.23}$$

which, substituted into equation (6.21), yields *Kepler's equation*:

$$E - e\sin E = M, \tag{6.24}$$

where $M \doteq n(t - \tau)$ is called the *mean anomaly*. An unambiguous relationship between the true and eccentric anomalies is the following:

$$\tan\frac{\theta}{2} = \sqrt{\frac{1+e}{1-e}}\tan\frac{E}{2}. \tag{6.25}$$

Note that $\frac{E}{2}$ and $\frac{\theta}{2}$ are always in the same quadrant. Kepler's equation can be solved by the Newton–Raphson method. Consider a linear approximation of the Taylor series expansion of the function, $f(E) = E - e\sin E - M$, as follows:

$$f(E + \Delta E) \simeq f(E) + f'(E)\Delta E, \tag{6.26}$$

where $f' \doteq \frac{\mathrm{d}f(E)}{\mathrm{d}E}$ and ΔE is the step size.

1. Given a mean anomaly, M, guess an initial value for the eccentric anomaly, E. (A good initial guess in most cases is $E = M + e\sin M$.)
2. Calculate the change required in the value of E so that $f(E + \Delta E) = 0$, using

$$\Delta E = -\frac{f(E)}{f'(E)} = \frac{-E + e\sin E + M}{1 - e\cos E}. \tag{6.27}$$

3. Update E, using $E = E + \Delta E$.
4. Recalculate $f(E) = E - e\sin E - M$.
5. If $|f(E)| \leq \delta$, where δ is the specified *tolerance*, then the new value of E is acceptable. Otherwise, go back to step 2.

Such a scheme usually converges to a very small tolerance in only a few iterations. However, an increased number of iterations may be necessary when the eccentricity, e, is close to unity, or when M is small. The number of iterations also depends upon the initial guess for E in the first step.

The position and velocity vectors in an elliptical orbit can be expressed directly in terms of the eccentric anomaly as follows:

$$\mathbf{r} = a(\cos E - e)\mathbf{i_e} + \sqrt{ap}\sin E\mathbf{i_p},$$
$$\mathbf{v} = -\frac{\sqrt{\mu a}}{r}\sin E\mathbf{i_e} + \frac{\sqrt{\mu p}}{r}\cos E\mathbf{i_p}. \tag{6.28}$$

Hyperbolic Orbit ($e > 1$)

The equivalent form of Kepler's equation can be obtained for hyperbolic orbits by introducing the *hyperbolic anomaly*, H. The parametric equation of a hyperbola centered at $x = 0,\ y = 0$, with semi-major axis a, can be written as

$$x = a\cosh H$$

or

$$r\cos\theta = a(\cosh H - e).$$

Therefore, the relationship between the hyperbolic anomaly and the true anomaly is given by

$$\cos\theta = \frac{\cosh H - e}{1 - e\cosh H} \tag{6.29}$$

or

$$\tan\frac{\theta}{2} = \sqrt{\frac{1+e}{e-1}}\tanh\frac{H}{2}. \tag{6.30}$$

Substitution of equation (6.29) into equation (6.20) yields

$$(e\cosh H - 1)\mathrm{d}H = \sqrt{-\frac{\mu}{a^3}}\mathrm{d}t, \tag{6.31}$$

which, upon integration from $H = 0, t = \tau$ yields

$$e\sinh H - H = n(t - \tau). \tag{6.32}$$

Here the *hyperbolic mean motion* is given by

$$n = \sqrt{-\frac{\mu}{a^3}}. \tag{6.33}$$

Equation (6.32) can be solved numerically using the Newton-Raphson method in a manner similar to Kepler's equation.

Parabolic Escape Trajectory ($e = 1$)

Whereas an elliptical orbit represents a closed trajectory, a hyperbolic trajectory is an open orbit corresponding to escape from (or arrival at) a planet's gravity with a positive speed at infinite radius, $r \to \infty$. The boundary between these two trajectories is that of the parabolic trajectory, which has zero speed at $r \to \infty$. A parabolic orbit is thus the minimum energy trajectory for escaping the gravitational influence of a planet. While impractical for interplanetary travel, the parabolic trajectory is sometimes a valuable mathematical aid for quickly determining the minimum fuel requirements for a given mission. Equation (6.20) for a parabolic orbit is simply

$$\frac{1}{4}\sec^4\frac{\theta}{2}\mathrm{d}\theta = \sqrt{\frac{\mu}{p^3}}\mathrm{d}t, \tag{6.34}$$

which, on integration, results in

$$\tan^3\frac{\theta}{2} + 3\tan\frac{\theta}{2} = 6\sqrt{\frac{\mu}{p^3}}(t - \tau). \tag{6.35}$$

Equation (6.35) represents the parabolic form of Kepler's equation, and is called *Barker's equation*. A real, unique, and closed-form solution to Barker's equation can be obtained by substituting

$$\tan\frac{\theta}{2} = \alpha - \frac{1}{\alpha}, \tag{6.36}$$

and solving the resulting quadratic equation for α^3, resulting in

$$\tan\frac{\theta}{2} = (C + \sqrt{1 + C^2})^{1/3} - (C + \sqrt{1 + C^2})^{-1/3}, \tag{6.37}$$

where

$$C \doteq 3\sqrt{\frac{\mu}{p^3}}(t-\tau) \tag{6.38}$$

is an equivalent mean anomaly.

6.2.4 Lagrange's Coefficients

The two-body trajectory expressed in the perifocal frame with Cartesian axes along $\mathbf{i_e}$, $\mathbf{i_p}$, and $\mathbf{i_h}$ is given by

$$\mathbf{r} = r\cos\theta\mathbf{i_e} + r\sin\theta\mathbf{i_p}, \tag{6.39}$$

$$\mathbf{v} = -\frac{\mu}{h}\sin\theta\mathbf{i_e} + \frac{\mu}{h}(e+\cos\theta)\mathbf{i_p}. \tag{6.40}$$

Given the position and velocity at a time, t_0, one would like to determine the position and velocity at some other time, t. In order to do so, we write the known position and velocity as

$$\begin{aligned}\mathbf{r}_0 &= r_0\cos\theta_0\mathbf{i_e} + r_0\sin\theta_0\mathbf{i_p},\\ \mathbf{v}_0 &= -\frac{\mu}{h}\sin\theta_0\mathbf{i_e} + \frac{\mu}{h}(e+\cos\theta_0)\mathbf{i_p},\end{aligned} \tag{6.41}$$

or, in matrix equation form,

$$\begin{Bmatrix}\mathbf{r}_0\\ \mathbf{v}_0\end{Bmatrix} = \begin{bmatrix} r_0\cos\theta_0 & r_0\sin\theta_0\\ -\frac{\mu}{h}\sin\theta_0 & \frac{\mu}{h}(e+\cos\theta_0)\end{bmatrix}\begin{Bmatrix}\mathbf{i_e}\\ \mathbf{i_p}\end{Bmatrix}. \tag{6.42}$$

Since the square matrix in equation (6.42) is non-singular (its determinant is equal to h), we can invert the matrix to obtain $\mathbf{i_e}$ and $\mathbf{i_p}$ as follows:

$$\begin{Bmatrix}\mathbf{i_e}\\ \mathbf{i_p}\end{Bmatrix} = \begin{bmatrix} \frac{1}{p}(e+\cos\theta_0) & -\frac{r_0}{h}\sin\theta_0\\ \frac{1}{p}\sin\theta_0 & \frac{r_0}{h}(e+\cos\theta_0)\end{bmatrix}\begin{Bmatrix}\mathbf{r}_0\\ \mathbf{v}_0\end{Bmatrix}. \tag{6.43}$$

On substituting equation (6.43) into equations (6.39) and (6.40), we have

$$\begin{aligned}\begin{Bmatrix}\mathbf{r}\\ \mathbf{v}\end{Bmatrix} &= \begin{bmatrix} r\cos\theta & r\sin\theta\\ -\frac{\mu}{h}\sin\theta & \frac{\mu}{h}(e+\cos\theta)\end{bmatrix}\\ &\times\begin{bmatrix} \frac{1}{p}(e+\cos\theta_0) & -\frac{r_0}{h}\sin\theta_0\\ \frac{1}{p}\sin\theta_0 & \frac{r_0}{h}(e+\cos\theta_0)\end{bmatrix}\begin{Bmatrix}\mathbf{r}_0\\ \mathbf{v}_0\end{Bmatrix}\end{aligned} \tag{6.44}$$

or

$$\begin{Bmatrix}\mathbf{r}\\ \mathbf{v}\end{Bmatrix} = \begin{pmatrix} f & g\\ \dot{f} & \dot{g}\end{pmatrix}\begin{Bmatrix}\mathbf{r}_0\\ \mathbf{v}_0\end{Bmatrix}, \tag{6.45}$$

where

$$
\begin{aligned}
f &= 1 + \frac{r}{p}[\cos(\theta - \theta_0) - 1], \\
g &= \frac{r r_0}{h} \sin(\theta - \theta_0), \\
\dot{f} &= \frac{\mathrm{d}f}{\mathrm{d}t} = -\frac{h}{p^2}[\sin(\theta - \theta_0) + e(\sin\theta - \sin\theta_0)], \\
\dot{g} &= \frac{\mathrm{d}g}{\mathrm{d}t} = 1 + \frac{r_0}{p}[\cos(\theta - \theta_0) - 1].
\end{aligned}
\tag{6.46}
$$

The functions f and g were first derived by Lagrange, and are thus called *Lagrange's coefficients*. They are very useful in determining a two-body orbit from a known position and velocity. The matrix

$$
\mathbf{\Phi}(t, t_0) \doteq \begin{pmatrix} f & g \\ \dot{f} & \dot{g} \end{pmatrix},
$$

called the *state transition matrix*, has a special significance, because it uniquely determines the *current state*, $(\mathbf{r},\ \mathbf{v})$, from the *initial state*, $(\mathbf{r}_0,\ \mathbf{v}_0)$. Such a relationship between initial and final states is rarely possible for the solution to a nonlinear differential equation, and is thus a valuable property of the two-body problem. The state transition matrix is seen to have the following properties:

1. From the conservation of angular momentum,

   $$
   \mathbf{h} = \mathbf{r} \times \mathbf{v} = (f\dot{g} - g\dot{f})\mathbf{r}_0 \times \mathbf{v}_0 = \mathbf{r}_0 \times \mathbf{v}_0,
   $$

   it follows that

   $$
   |\mathbf{\Phi}| = f\dot{g} - g\dot{f} = 1.
   $$

   A consequence of the unit determinant is that the inverse of the state transition matrix is given by

   $$
   \mathbf{\Phi}^{-1} \doteq \begin{pmatrix} \dot{g} & -g \\ -\dot{f} & f \end{pmatrix}.
   $$

   Such a matrix is said to be *symplectic*, as we saw in Chapter 2.

2. Given any three points $(t_0,\ t_1,\ t_2)$ along the trajectory, it is true that

   $$
   \mathbf{\Phi}(t_2, t_0) = \mathbf{\Phi}(t_2, t_1)\mathbf{\Phi}(t_1, t_0).
   $$

6.3 Spacecraft Terminal Guidance

A spacecraft's nominal trajectory relative to the central object (either the Sun or a planetary body) is generally the solution to the two-body problem, unless significant orbital perturbations are present in the form of atmospheric drag, non-spherical and third-body gravity (Tewari 2006). Even if orbital perturbations are present, the two-body orbit can be taken to be a nominal flight path, which should either be followed despite small perturbations by the application of control inputs, or used as an initial guess in obtaining the closed-loop solution to the perturbed problem by iterative techniques (Tewari 2006). Therefore, the problem of guiding the spacecraft from an initial to a desired final position in space can generally be treated as the problem of determining an optimal trajectory that passes through the two given positions.

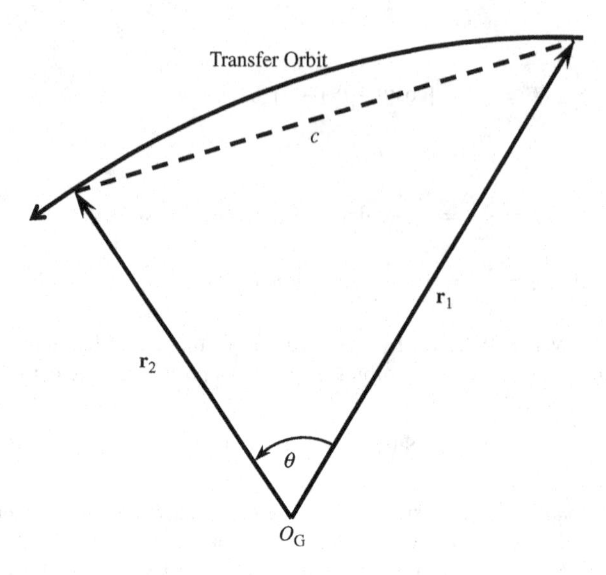

Figure 6.2 Geometry of orbital transfer for spacecraft guidance

Consider the orbital transfer from an initial position vector, $\mathbf{r}_1(t_1)$, to a final position, $\mathbf{r}_2(t_2)$, as shown in Figure 6.2. If the *time of flight*, $t_f = t_2 - t_1$, is known (or specified) *a priori*, the problem has a unique solution and is referred to as *Lambert's problem*. However – as is commonly the case – if we have a choice of varying the time of flight according to the requirement of minimizing an objective function (such as the total propellant mass) for the transfer, one arrives at the problem of *optimal orbital transfer*, which must be solved by optimal control techniques (Chapter 2). In either case, one has to solve a 2PBVP yielding the required orbit, $\mathbf{r}(t)$, that takes the spacecraft from $\mathbf{r}_1$ to $\mathbf{r}_2$, which can be used as a nominal flight path to be followed by a closed-loop guidance system despite perturbations.

For a two-body orbital transfer with an arbitrary transfer time, $t_2 - t_1$, we express the final position vector, $\mathbf{r}_2$, in terms of the initial position, $\mathbf{r}_1$, and initial velocity, $\mathbf{v}_1$, through Lagrange's coefficients (equations (6.45) and (6.46)) as follows:

$$\mathbf{r}_2 = f\mathbf{r}_1 + g\mathbf{v}_1, \tag{6.47}$$

where

$$\begin{aligned}
f &= 1 + \frac{r_2}{p}(\cos\theta - 1),\\
g &= \frac{r_1 r_2}{h}\sin\theta,\\
\dot{f} &= \frac{\mathrm{d}f}{\mathrm{d}t} = -\frac{h}{p^2}[\sin\theta + e(\sin\theta_2 - \sin\theta_1)],\\
\dot{g} &= \frac{\mathrm{d}g}{\mathrm{d}t} = 1 + \frac{r_1}{p}(\cos\theta - 1).
\end{aligned} \tag{6.48}$$

Furthermore, the initial position vector in terms of the final position and velocity is given by

$$\mathbf{r}_1 = f'\mathbf{r}_2 + g'\mathbf{v}_2, \tag{6.49}$$

where

$$f' = 1 + \frac{r_1}{p}(\cos\theta - 1) = \dot{g},$$
$$g' = -\frac{r_1 r_2}{h}\sin\theta = -g. \quad (6.50)$$

Thus, we have

$$\mathbf{r}_1 = \dot{g}\mathbf{r}_2 - g\mathbf{v}_2. \quad (6.51)$$

By virtue of equations (6.47) and (6.51), the initial and final velocity vectors can be resolved into components along the chord and the local radius[3] as follows:

$$\mathbf{v}_1 = \frac{1}{g}\left[(\mathbf{r}_2 - \mathbf{r}_1) + (1 - f)\mathbf{r}_1\right] = v_c\mathbf{i}_c + v_r\mathbf{i}_{r1},$$
$$\mathbf{v}_2 = \frac{1}{g}\left[(\mathbf{r}_2 - \mathbf{r}_1) - (1 - \dot{g})\mathbf{r}_2\right] = v_c\mathbf{i}_c - v_r\mathbf{i}_{r2}, \quad (6.52)$$

where

$$\mathbf{i}_c = \frac{\mathbf{r}_2 - \mathbf{r}_1}{c},$$
$$\mathbf{i}_{r1} = \frac{\mathbf{r}_1}{r_1},$$
$$\mathbf{i}_{r2} = \frac{\mathbf{r}_2}{r_2},$$
$$v_c = \frac{c}{g} = \frac{ch}{r_1 r_2 \sin\theta},$$
$$v_r = \frac{h(1 - \cos\theta)}{p\sin\theta}. \quad (6.53)$$

The last two expressions in equation (6.53) clearly imply that

$$p = \frac{r_1 r_2(1 - \cos\theta)}{c}\frac{v_c}{v_r}. \quad (6.54)$$

Furthermore, we have

$$v_c v_r = \frac{c\mu\sec^2(\theta/2)}{2r_1 r_2}, \quad (6.55)$$

which is an important result showing that the product of the two velocity components depends only upon the geometry of transfer specified by (r_1, r_2, θ), and not on the shape or size of the transfer orbit. Thus, we can select various transfer orbits to meet other requirements (such as time or fuel for transfer).

6.3.1 *Minimum Energy Orbital Transfer*

There is an interesting and practical solution for the orbital 2PBVP for the case of minimum energy transfer, which corresponds to the smallest value of orbital energy $\epsilon = -\mu/2a$ for the transfer orbit. Clearly, the minimum energy orbit involves the smallest positive value of the semi-major axis, a, of the

[3] Note from Figure 6.2 that the chord and local radius are not necessarily perpendicular to each other. Thus $v_r \neq \dot{r}$.

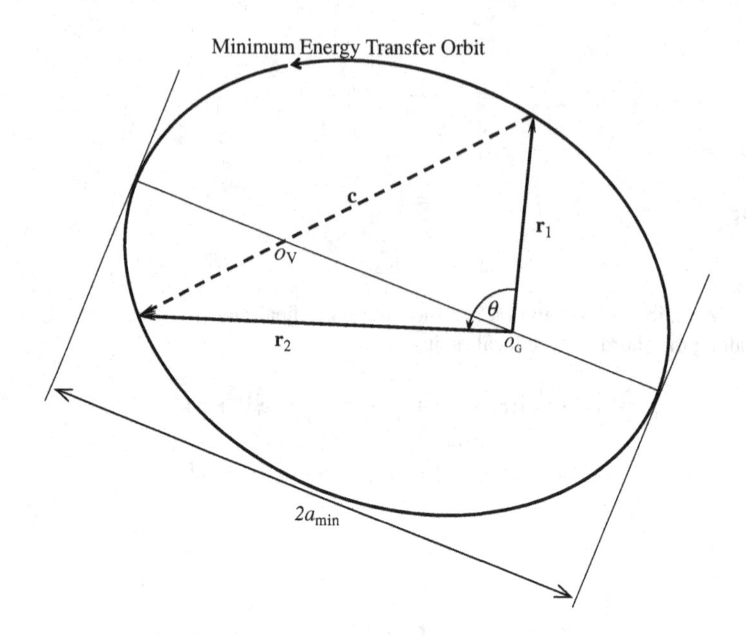

Figure 6.3 Geometry of minimum energy orbital transfer for spacecraft guidance

orbit joining the two given radii, (r_1, r_2), and would normally result in the smallest propellant expenditure. Let the minimum energy elliptical orbit have the semi-major axis $a_{\min}$ as shown in Figure 6.3 depicting the geometry of the minimum energy transfer ellipse. Let the respective distances of the initial and final positions from the *vacant focus*, o_V, be (r_1^*, r_2^*). For an ellipse of major axis a, it must be true that

$$r_1 + r_1^* = 2a, \quad r_2 + r_2^* = 2a, \tag{6.56}$$

or

$$r_1 + r_2 + r_1^* + r_2^* = 4a. \tag{6.57}$$

From geometry it follows that the smallest possible value for the distance $r_1^* + r_2^*$ is the *chord length*, $c = |\mathbf{r}_2 - \mathbf{r}_1|$, implying that o_V must lie on the chord line, as shown in Figure 6.3. Therefore, from a purely geometrical argument, we have

$$2a_{\min} = \frac{1}{2}(r_1 + r_2 + c). \tag{6.58}$$

We note from Figure 6.2 that the semi-perimeter, s, of the transfer triangle is given by

$$s = \frac{r_1 + r_2 + c}{2}. \tag{6.59}$$

Hence, the major axis of the minimum energy transfer ellipse is the same as the semi-perimeter,

$$2a_{\min} = s. \tag{6.60}$$

For a transfer with $\theta = \pi$, we have $c = r_1 + r_2$ or $s = c$. Since the ratio v_c/v_r is unity for the minimum energy transfer, we have the following expression for the corresponding parameter from equation (6.54):

$$p_m = \frac{r_1 r_2}{c}(1 - \cos\theta). \tag{6.61}$$

An algorithm for determining the minimum energy transfer orbit and time of flight between two given positions, $(\mathbf{r}_1, \mathbf{r}_2)$, can be devised as follows:

1. Compute the chord, the two radii, and semi-major axis of transfer orbit,

$$c = |\mathbf{r}_2 - \mathbf{r}_1|, \quad r_1 = |\mathbf{r}_1|, \quad r_2 = |\mathbf{r}_2|,$$

$$a = \frac{1}{4}(r_1 + r_2 + c).$$

2. Calculate the orbital speeds at the two positions,

$$v_1 = \sqrt{\frac{2\mu}{r_1} - \frac{\mu}{a}}, \quad v_2 = \sqrt{\frac{2\mu}{r_2} - \frac{\mu}{a}}.$$

3. Find the unit vectors along the chord and the two radii, and the transfer angle, θ, as follows:

$$\mathbf{i}_c = \frac{\mathbf{r}_2 - \mathbf{r}_1}{c}, \quad \mathbf{i}_{r1} = \frac{\mathbf{r}_1}{r_1}, \quad \mathbf{i}_{r2} = \frac{\mathbf{r}_2}{r_2},$$

$$\cos\theta = \mathbf{i}_{r1} \cdot \mathbf{i}_{r2}, \quad \sin\theta = |\mathbf{i}_{r1} \times \mathbf{i}_{r2}|.$$

4. Calculate the parameter and the orbital eccentricity as follows:

$$p = \frac{r_1 r_2}{c}(1 - \cos\theta),$$

$$e = \sqrt{1 - \frac{p}{a}}.$$

5. Referring to Figure 6.3, determine the initial and final eccentric anomalies, time of periapsis, and time of flight as follows:

$$E_1 = \cos^{-1}\left(\frac{1}{e} - \frac{r_1}{ae}\right),$$

$$E_2 = 2\pi - \cos^{-1}\left(\frac{1}{e} - \frac{r_2}{ae}\right),$$

$$n = \sqrt{\frac{\mu}{a^3}}, \quad \tau = t_1 - \frac{E_1 - e\sin E_1}{n},$$

$$t_2 - t_1 = \frac{E_2 - E_1 - e(\sin E_2 - \sin E_1)}{n}.$$

Example 6.1 Determine the minimum energy transfer trajectory required to reach an altitude 300 km, latitude $-45°$, and right ascension $60°$, beginning from an initial altitude 120 km, latitude $10°$, and right ascension $256°$. Also calculate the corresponding time of flight. Assume gravitational constant $\mu = 398\,600.4$ km^3/s^2 and surface radius $R_0 = 6378.14$ km for Earth.

The necessary computation is performed with the aid of the following MATLAB® statements:

```
>> mu=398600.4;R0=6378.14;
>> r1=R0+120; r2=R0+300;
>> dtr=pi/180; del1=10*dtr; lam1=256*dtr;
>> del2=-45*dtr; lam2=60*dtr;
>> R1=r1*[cos(del1)*cos(lam1) cos(del1)*sin(lam1) sin(del1)]

R1 =  -1548.15949106612          -6209.32856884053            1128.39016922459

>> R2=r2*[cos(del2)*cos(lam2) cos(del2)*sin(lam2) sin(del2)]

R2 =   2361.07903985657           4089.50885771751           -4722.15807971313

>> c=norm(R2-R1)

c =    12473.0554814557

>> a=(r1+r2+c)/4

a =    6412.33387036392

>> costheta=dot(R1,R2)/(r1*r2)

costheta =   -0.792176074609793

>> p=r1*r2*(1-costheta)/c

p =   6235.22894108535

>> e=sqrt(1-p/a)

e =   0.166190908181437

>> E1=acos((1-r1/a)/e)

E1 = 1.65140194535795

>> E2=2*pi-acos((1-r2/a)/e)

E2 = 4.46030157456608

>> n=sqrt(mu/a^3)

n =  0.00122954569947427

>> t21=(E2-E1-e*(sin(E2)-sin(E1)))/n

t21 =  2550.12009388545
```

```
%% Check time computation:
>> tau=-(E1-e*sin(E1))/n

tau =  -1208.37366094101

>> n*(t21-tau)

ans =  4.62123983274777

>> E2-e*sin(E2)

ans =  4.62123983274777
```

Thus, the required transfer orbit has elements, $a = 6412.33387$ km, $e = 0.166190908$, and $\tau = -1208.373661$ s (measured from $t_1 = 0$ at launch), and the required time of flight is 2550.12 s.

Since the vacant focus must lie on the chord for the minimum energy ellipse (Figure 6.3), the corresponding time of flight for even a small transfer angle, θ, can be quite large. In fact, it can be shown that the minimum energy orbit yields the maximum time of flight between a given pair of positions. In any case, a different transfer strategy becomes necessary when the time of flight is fixed, such as in an orbital rendezvous between two spacecraft, or interception of a target in space by a free-flight interceptor.

6.3.2 *Lambert's Theorem*

For a general orbital transfer (Figure 6.2) about a given planet, *Lambert's theorem* states that the transfer time, $t_2 - t_1$, is a function of only the semi-major axis, a, of the transfer orbit, the sum of the two radii, $r_1 + r_2$, and the chord, $c = |\mathbf{r}_2 - \mathbf{r}_1|$, joining the two positions. Thus, we have

$$t_2 - t_1 = f(a, r_1 + r_2, c). \tag{6.62}$$

Since there is no dependence of the transfer time on the orbital eccentricity, one can choose any value of e for the orbit, provided $(a, r_1 + r_2, c)$ are unchanged. This is tantamount to changing the orbital shape for given initial and final positions by moving the two focii in such a way that both $(a, r_1 + r_2)$ are unaffected. Battin (1999) discusses in detail the geometrical implications of Lambert's theorem.

For the special case of parabolic transfer, Barker's equation yields the following *Euler equation*[4] (Battin 1999):

$$6\sqrt{\mu}(t_2 - t_1) = (r_1 + r_2 + c)^{3/2} \pm (r_1 + r_2 - c)^{3/2}, \tag{6.63}$$

where the positive sign is taken for the transfer angle $\theta > \pi$, and the negative for $\theta < \pi$. Clearly, parabolic transfer is a special case of Lambert's theorem,

$$t_2 - t_1 = f(r_1 + r_2, c). \tag{6.64}$$

A corollary to Lambert's theorem is the *mean value theorem*, which states that the eccentric (or hyperbolic) anomaly of the *mean position* in an orbital transfer is the arithmetic mean of the eccentric (or hyperbolic) anomalies of the initial and final positions. The mean position, $\mathbf{r}_0$, in this case is the intermediate position where – by the mean value theorem of differential calculus (Kreyszig 1998) – the

[4] Euler derived equation (6.63) in 1743 and applied it to orbital calculations for comets.

tangent to the orbit is parallel to the chord joining $\mathbf{r}_1$ and $\mathbf{r}_2$. Thus for an elliptical transfer, we have

$$r_0 = a\left[1 - e\cos\left(\frac{E_1 + E_2}{2}\right)\right], \tag{6.65}$$

where

$$r_1 = a(1 - e\cos E_1), \quad r_2 = a(1 - e\cos E_2). \tag{6.66}$$

Similarly, for a hyperbolic transfer, we have

$$r_0 = a\left[1 - e\cosh\left(\frac{H_1 + H_2}{2}\right)\right], \tag{6.67}$$

where

$$r_1 = a(1 - e\cosh H_1), \quad r_2 = a(1 - e\cosh H_2). \tag{6.68}$$

For an elliptical transfer, Lagrange derived the following equation (Battin 1999) for the transfer time with the help of Kepler's equation:

$$\sqrt{\mu}(t_2 - t_1) = 2a^{3/2}(\psi - \cos\phi\sin\psi), \tag{6.69}$$

where

$$\begin{aligned}\cos\phi &= e\cos\frac{1}{2}(E_1 + E_2),\\ \psi &= \frac{1}{2}(E_2 - E_1),\end{aligned} \tag{6.70}$$

and (E_1, E_2) denote the respective eccentric anomalies of the initial and final positions. Noting that

$$\begin{aligned}r_1 + r_2 &= 2a(1 - \cos\phi\cos\psi),\\ c &= 2a\sin\phi\sin\psi,\end{aligned} \tag{6.71}$$

Lambert's theorem is clearly proved for the elliptical orbit by eliminating ϕ, ψ from equations (6.69) and (6.71). An interesting case is that of the *minimum eccentricity* ellipse (also called the *fundamental ellipse*) for which it can be shown (Battin 1999) that the semi-major axis, parameter, and eccentricity are given by

$$\begin{aligned}a_f &= \frac{1}{2}(r_1 + r_2),\\ p_f &= \frac{p_m}{c}(r_1 + r_2),\\ e_f &= \frac{|r_1 - r_2|}{c},\end{aligned} \tag{6.72}$$

where p_m is the parameter for the minimum energy orbit given by equation (6.61).

For a hyperbola, Lambert's theorem can be proved similarly by writing

$$\sqrt{\mu}(t_2 - t_1) = 2(-a)^{3/2}(\cosh\phi\sinh\psi - \psi), \tag{6.73}$$

where

$$\cosh\phi = e\cosh\frac{1}{2}(H_1 + H_2),$$
$$\psi = \frac{1}{2}(H_2 - H_1), \tag{6.74}$$

and (H_1, H_2) denote the hyperbolic anomalies of the initial and final positions, respectively. The expressions for the sum of radii and chord for the hyperbolic case are

$$r_1 + r_2 = 2a(1 - \cosh\phi\cosh\psi),$$
$$c = -2a\sinh\phi\sinh\psi. \tag{6.75}$$

6.3.3 Lambert's Problem

Lambert's problem is the name given to the general 2PBVP resulting from a two-body orbital transfer between two position vectors in a given time. Such a problem is commonly encountered in the guidance of spacecraft and ballistic missiles, as well as in the orbital determination of space objects from two observed positions separated by a specific time interval.[5] For the initial and final positions, $(\mathbf{r}_1, \mathbf{r}_2)$, and the transfer angle, θ (Figure 6.2), for the required time of flight, $t_{12} = t_2 - t_1$, we must determine a transfer orbit that is coplanar with $(\mathbf{r}_1, \mathbf{r}_2)$. In the orbital plane, the two positions are uniquely specified through the radii, (r_1, r_2), and the respective true anomalies, (θ_1, θ_2).

Lambert's problem given by equation (6.62) subject to the boundary conditions, $\mathbf{r}(0) = \mathbf{r}_1$ and $\mathbf{r}(t_{12}) = \mathbf{r}_2$, is to be solved for the unique transfer orbit, (a, e, τ). However, equation (6.62) yields a transfer time t_{12} that is a double-valued function of the semi-major axis, a, which implies that an iterative solution may not converge to a single value of a. Furthermore, the derivative of t_{12} with respect to a is infinite for a minimum energy orbit, therefore Newton's method cannot be used for solving the time equation with respect to a, if $a = a_m$. A practical solution to Lambert's problem – that can be used for guiding spacecraft – must avoid such difficulties. Thus we need a robust algorithm which converges to a unique, non-singular solution. Examples of Lambert algorithms are the universal variable methods by Battin (1964; 1999), which can be applied to elliptical, parabolic, and hyperbolic transfer trajectories. These methods can be formulated either by evaluation of infinite power series (Stumpff functions) or continued fractions (hypergeometric functions).

Stumpff Function Method

Let us define an auxiliary variable, x, by

$$x = E\sqrt{a}, \tag{6.76}$$

where $E \doteq E_2 - E_1$ is the difference between the eccentric anomalies of the final and initial positions, and a is the semi-major axis of the transfer orbit. Lagrange's coefficients, f, g, can be expressed as

[5] Lambert's problem attracted the attention of the greatest mathematicians, such as Euler, Gauss, and Lagrange, and is responsible for many advances in analytical and computational mechanics. A solution to Lambert's problem was first presented by Gauss in his *Theoria Motus* (1801), using three angular positions of the asteroid Ceres measured at three time instants.

functions of x as follows (Battin 1964):

$$\begin{aligned} f &= 1 - \frac{x^2}{r_1}C(z), \\ \dot{f} &= \frac{x\sqrt{\mu}}{r_1 r_2}[zS(z) - 1], \\ g &= t_{12} - \frac{1}{\sqrt{\mu}}x^3 S(z), \\ \dot{g} &= 1 - \frac{x^2}{r_2}C(z), \end{aligned} \tag{6.77}$$

where

$$z = \frac{x^2}{a}, \tag{6.78}$$

and $C(z)$, $S(z)$ are the following *Stumpff functions* (Stiefel and Scheifele 1971):

$$\begin{aligned} C(z) &= \frac{1}{2!} - \frac{z}{4!} + \frac{z^2}{6!} - \dots, \\ S(z) &= \frac{1}{3!} - \frac{z}{5!} + \frac{z^2}{7!} - \dots. \end{aligned} \tag{6.79}$$

The elliptical transfer is given by $z > 0$, while the hyperbolic transfer has $z < 0$. Clearly, the parabolic case, $z = 0$, results in $C(0) = \frac{1}{2}$, $S(0) = \frac{1}{6}$, and can be solved analytically.

In order to solve for the universal variable, x, in the general case, we define another univeral variable of transfer, y, as follows:

$$y = \frac{r_1 r_2}{p}(1 - \cos\theta), \tag{6.80}$$

where p is the parameter of the transfer orbit. A comparison of the definition of the Lagrange's coefficients, equation (6.46), written for the orbital transfer as

$$\begin{aligned} f &= 1 - \frac{r_2}{p}(1 - \cos\theta), \\ \dot{f} &= \sqrt{\frac{\mu}{p}}\left(\frac{1 - \cos\theta}{\theta}\right)\left[\frac{1 - \cos\theta}{p} - \frac{1}{r_1} - \frac{1}{r_2}\right], \\ g &= \frac{r_1 r_2}{\sqrt{\mu p}}\sin\theta, \\ \dot{g} &= 1 - \frac{r_1}{p}(1 - \cos\theta), \end{aligned} \tag{6.81}$$

with equation (6.77) yields

$$x = \sqrt{\frac{y}{C(z)}}, \tag{6.82}$$

where x is the solution to the cubic equation,

$$t_{12}\sqrt{\mu} = Ax\sqrt{C(z)} + S(z)x^3, \tag{6.83}$$

and A is the constant

$$A = \sin\theta\sqrt{\frac{r_1 r_2}{1-\cos\theta}}. \tag{6.84}$$

Thus, we have

$$f = 1 - \frac{y}{r_1},$$

$$g = A\sqrt{\frac{y}{\mu}}, \tag{6.85}$$

$$\dot{g} = 1 - \frac{y}{r_2}, \tag{6.86}$$

while the remaining Lagrange's coefficient, $\dot{f}$, is obtained from the relationship $f\dot{g} - g\dot{f} = 1$. By expressing the transfer time in terms of the universal variables, (x, y), rather than the classical elements, (a, e) (or (a, p)), we have avoided the singularity and convergence issues discussed earlier. However, the convergence of the present formulation is guaranteed only for transfer angles less than 180°.

In a terminal control application, it is usually required to find the initial and final velocities of the transfer, which can finally be derived from the Lambert solution, f, g, $\dot{g}$, as follows:

$$\mathbf{v}_1 = \frac{1}{g}(\mathbf{r}_2 - f\mathbf{r}_1),$$

$$\mathbf{v}_2 = \dot{f}\mathbf{r}_1 + \dot{g}\mathbf{v}_1 = \frac{1}{g}(\dot{g}\mathbf{r}_2 - \mathbf{r}_1). \tag{6.87}$$

Clearly, the relationship among transfer geometry, time, and Lagrange's coefficients is a transcendental one requiring an iterative solution for the universal variables, (x, y), that satisfies the transfer time equation (6.83) within a given tolerance, ϵ. The steps involved in the solution are given by the flowchart of Figure 6.4, which is coded in *lambert_stumpff.m* (Table 6.1) and employs Newton's method for x iteration. To carry out Newton's iteration, we write

$$F(x) = Ax\sqrt{C(z)} + S(z)x^3 - t_{12}\sqrt{\mu} \tag{6.88}$$

and take the derivative with respect to x,

$$F'(x) = A\sqrt{C(z)} + 3S(z)x^2 + \left(Ax\frac{C'(z)}{2\sqrt{C(z)}} + S'(z)x^3\right)\frac{\mathrm{d}z}{\mathrm{d}x}, \tag{6.89}$$

where

$$\frac{\mathrm{d}z}{\mathrm{d}x} = \frac{2x}{a} \tag{6.90}$$

and

$$C'(z) = -\frac{1}{4!} + \frac{2z}{6!} - \frac{3z^2}{8!} - \ldots,$$

$$S'(z) = -\frac{1}{5!} + \frac{2z}{7!} - \frac{3z^2}{9!} - \ldots. \tag{6.91}$$

The correction in each step is then calculated by

$$\Delta x = -\frac{F(x)}{F'(x)}. \tag{6.92}$$

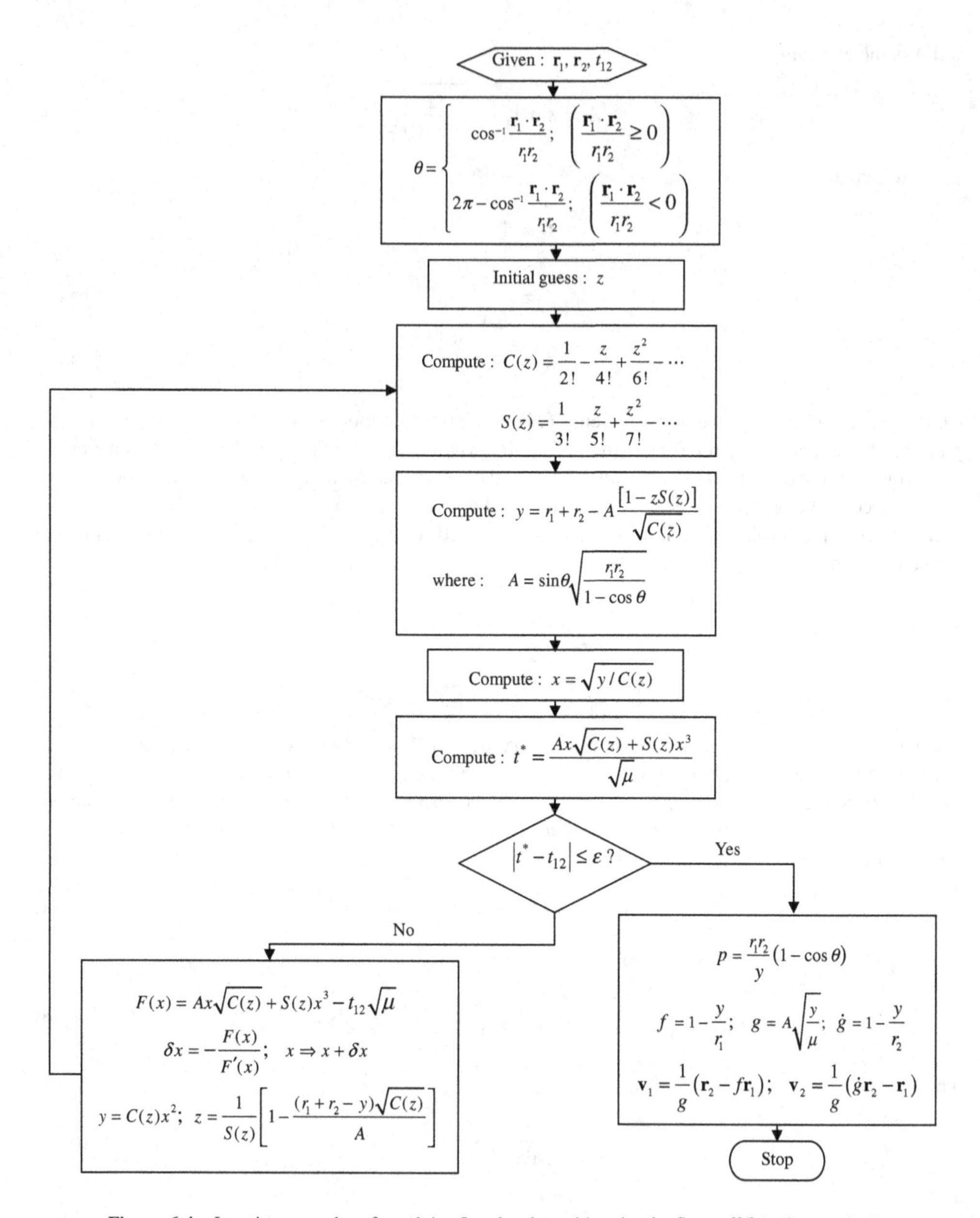

Figure 6.4 Iteration procedure for solving Lambert's problem by the Stumpff function method

The determination of the transfer angle, θ, depends upon whether the orbit is direct ($0 \leq i < \pi/2$) or retrograde ($\pi/2 \leq i \leq \pi$). Also note the small tolerance of 10^{-9} s specified in the code for transfer time convergence. Of course, increasing the tolerance results in a smaller number of iterations.

Example 6.2 Determine the transfer trajectory and the initial and final velocity vectors for a ballistic missile from a burn-out point of altitude 200 km, latitude 20°, and right ascension 80°, to a re-entry

Table 6.1

```
function [a,p,V1,V2]=lambert_stumpff(mu,R1,R2,t12,id)
% Solution to Lambert's two-point boundary value problem by Stumpff fn.
% mu: gravitational constant of central mass.
% R1,V1: initial position and velocity vectors in celestial frame.
% R2,V2: final position and velocity vectors in celestial frame.
% t12: transfer time.
% id: indicator of transfer direction (id=1:direct; -1:retrograde)
% (c) 2010 Ashish Tewari
r1=norm(R1); r2=norm(R2);
costh=dot(R1,R2)/(r1*r2) % cosine transfer angle
theta=acos(costh)
if costh<0
    theta=2*pi-theta
end
cosi=dot([0 0 1]',cross(R1,R2)/norm(cross(R1,R2)))
if id*cosi<0
    theta=2*pi-theta
end
A=sqrt(r1*r2/(1-costh))*sin(theta);
z=0.01;
n=1;
[C,S]=stumpff(z,5);
y=r1+r2-A*(1-z*S)/sqrt(C);
x=sqrt(y/C);
tc=(x^3*S+A*sqrt(y))/sqrt(mu);
% Newton's iteration for 'x' follows:
while abs(t12-tc)>1e-9
    fx=A*sqrt(C)*x+S*x^3-t12*sqrt(mu);
    [Cp,Sp]=dstumpff(z,20);
    p=r1*r2*(1-cos(theta))/y;
    f=1-y/r1;
    g=A*sqrt(y/mu);
    v1=norm((R2-f*R1)/g);
    a=-mu/(v1^2-2*mu/r1);
    fxp=A*sqrt(C)+3*S*x^2+A*x^2*Cp/(a*sqrt(C))+2*x^4*Sp/a;
    dx=-fx/fxp;
    x=x+dx;
    n=n+1;
    y=C*x*x;
    tc=(x^3*S+A*sqrt(y))/sqrt(mu);
    z=(1-sqrt(C)*(r1+r2-y)/A)/S;
    [C,S]=stumpff(z,20);
end
n
z
abs(t12-tc)
p=r1*r2*(1-cos(theta))/y;
f=1-y/r1;
g=A*sqrt(y/mu);
gd=1-y/r2;
V1=(R2-f*R1)/g;
V2=(gd*R2-R1)/g;
```

Table 6.1 (*Continued*)

```
function [C,S]=stumpff(z,n)
% Stumpff functions evaluation:
C=0.5;
S=1/6;
for i=1:n
    C=C+(-1)^i*z^i/factorial(2*(i+1));
    S=S+(-1)^i*z^i/factorial(2*i+3);
end

function [Cp,Sp]=dstumpff(z,n)
% Stumpff function derivatives:
Cp=-1/24;
Sp=-1/120;
for i=1:n
    Cp=Cp+(-1)^(i+1)*(i+1)*z^i/factorial(2*(i+2));
    Sp=Sp+(-1)^(i+1)*(i+1)*z^i/factorial(2*i+5);
end
```

point given by altitude 100 km, latitude 45°, and right ascension −5°. The transfer must take place in a retrograde direction in exactly 30 min. Assume a gravitational constant $\mu = 398\,600.4$ km^3/s^2 and a constant surface radius of $R_0 = 6378.14$ km for Earth.

The necessary computation is performed for the retrograde transfer case (with transfer angle less than 180°) with the aid of the following MATLAB statements and the code *lambert_stumpff.m*:

```
>> mu=398600.4;R0=6378.14;
>> r1=R0+200; r2=R0+100;
>> dtr=pi/180;
>> del1=20*dtr; lam1=80*dtr; del2=45*dtr; lam2=-5*dtr
>> R1=r1*[cos(del1)*cos(lam1) cos(del1)*sin(lam1) sin(del1)]'
R1 =
         1073.39398828103
         6087.51981102515
         2249.85638561631

>> R2=r2*[cos(del2)*cos(lam2) cos(del2)*sin(lam2) sin(del2)]'
R2 =
         4563.30563728077
        -399.237511466009
         4580.73672347582

>> [a,p,V1,V2]=lambert_stumpff(mu,R1,R2,30*60,-1)
cosi = -0.693840277353644

theta =   1.2663588873311

n =     16

z =     4.49267769246433

a = 5197.62671746848
```

```
p = 3772.93508012859

V1 =    4.56006018866584
        0.794010191037251
        4.80373489230704

V2 = -1.98537060969878
        -5.7227235312436
        -3.11126222296681
```

The results converge after 16 iterations to an error of 4×10^{-10} s, producing the transfer trajectory, $a = 5197.62672$ km, $p = 3772.93508$ km, $\mathbf{v}_1 = (4.56, 0.794, 4.8037)^T$ km/s, and $\mathbf{v}_2 = (-1.98537, -5.72272, -3.11126)^T$ km/s, from which e and τ are determined as follows:

$$v_1 = ||\mathbf{v}_1|| = 6.6708672617 \text{ km/s}, \quad v_2 = ||\mathbf{v}_2|| = 6.8096412309 \text{ km/s},$$

$$e = \sqrt{1 - \frac{p}{a}} = 0.52354967515,$$

$$\cos E_1 = \frac{1}{e}\left(1 - \frac{r_1}{a}\right) = -0.5073148895, \quad E_1 = 2.102862406955 \text{ rad},$$

$$n = \sqrt{\frac{\mu}{a^3}} = 0.00168485 \text{ rad/s}, \quad \tau = -\frac{E_1 - e \sin E_1}{n} = -980.3175 \text{ s}.$$

Note that E_1 must be in the second quadrant as the flight-path angle at burn-out, given by

$$\sin \phi_1 = \frac{\mathbf{V}_1 \cdot \mathbf{R}_1}{r_1 v_1} = 0.4679836,$$

is positive. Similarly, E_2 must be in the third quadrant, as

$$\sin \phi_2 = \frac{\mathbf{V}_2 \cdot \mathbf{R}_2}{r_2 v_2} = -0.4766535.$$

Therefore,

$$E_2 = \frac{\pi}{2} - \cos^{-1}\{\cos E_2\} = \frac{\pi}{2} - \cos^{-1}\{-0.4705666\} = 4.222456 \text{ rad}.$$

Of course, τ must also satisfy Kepler's equation for $E = E_2$, which we leave to the reader to verify. Plots of the desired trajectory and velocity profile are shown in Figures 6.5 and 6.6, respectively.

Hypergeometric Function Method

Battin (1999) presents an alternative procedure for the general Lambert problem through the use of *hypergeometric* functions evaluated by continued fractions, instead of the Stumpff functions. Such a formulation improves convergence, even for transfer angles greater than 180°, and renders the method insensitive to the initial choice of z. Consider a universal variable x such that the semi-major axis, a, of

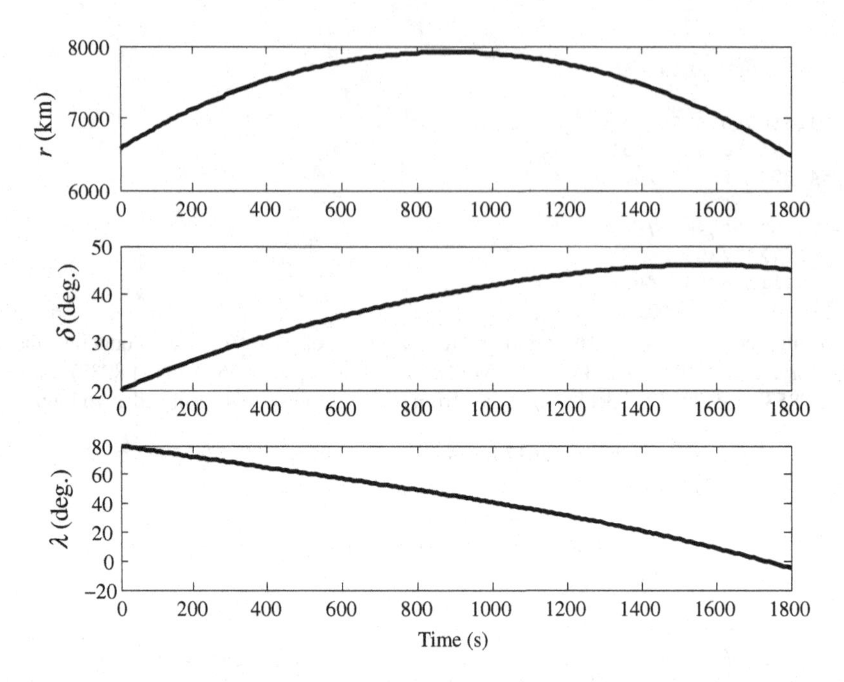

Figure 6.5 Trajectory for Lambert's transfer of a ballistic missile from burn-out to re-entry

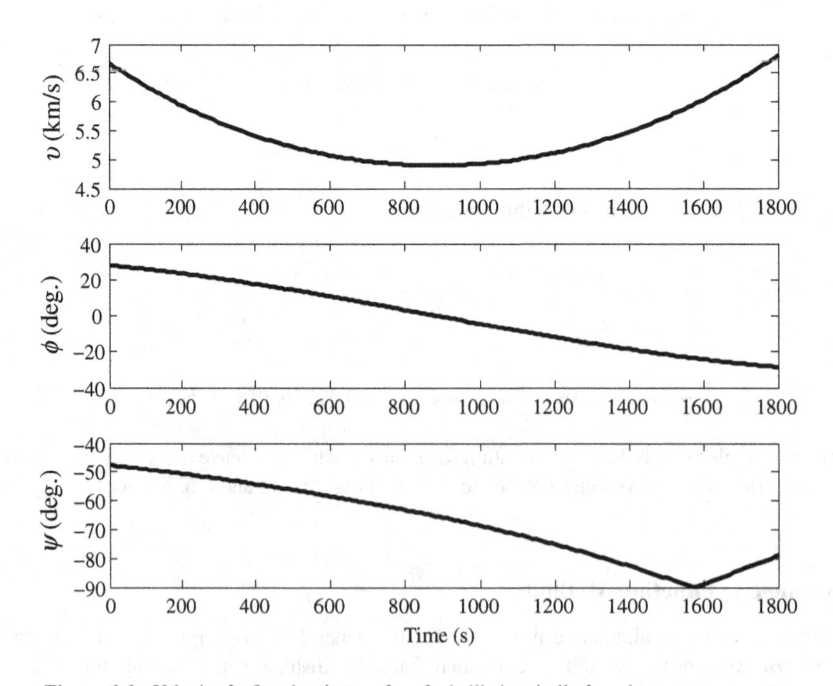

Figure 6.6 Velocity for Lambert's transfer of a ballistic missile from burn-out to re-entry

the transfer orbit is given by

$$a = \frac{a_{\rm m}}{1 - x^2}, \tag{6.93}$$

where $a_{\rm m}$ is the semi-major axis of the minimum energy ellipse connecting the initial and final radii. Clearly, for an elliptical transfer we have $-1 < x < 1$ ($x = 0$ for minimum energy transfer), while for a parabola, $x = 1$. A hyperbolic transfer requires $x > 1$. Furthermore, for a given transfer, r_1, r_2, θ, we define the parameter,

$$\lambda = \frac{r_1 r_2}{s} \cos\frac{\theta}{2}, \tag{6.94}$$

where s is the semi-perimeter,

$$s = 2a_{\rm m} = \frac{r_1 + r_2 + c}{2}. \tag{6.95}$$

Note that $-1 < \lambda < 1$. Additional variables are defined by

$$y = \sqrt{1 - \lambda^2(1 - x^2)}, \tag{6.96}$$

$$\eta = y - x\lambda, \tag{6.97}$$

and

$$z = \frac{1}{2}(1 - \lambda - x\eta), \tag{6.98}$$

which transform the time equation to

$$t_{12}\sqrt{\frac{\mu}{a_m^3}} = \frac{4}{3}\eta^3 F(z) + 4\lambda\eta. \tag{6.99}$$

Here $F(z)$ is a hypergeometric function (Battin 1999) that can be evaluated by a continued fraction as follows:

$$F(z) = \cfrac{1}{1 - \cfrac{\gamma_1 z}{1 - \cfrac{\gamma_2 z}{1 - \cfrac{\gamma_3 z}{1 - \ddots}}}} \tag{6.100}$$

and

$$\gamma_n = \begin{cases} \frac{(n+2)(n+5)}{(2n+1)(2n+3)}, & n \text{ odd}, \\ \frac{n(n-3)}{(2n+1)(2n+3)}, & n \text{ even}. \end{cases} \tag{6.101}$$

When adopting Newton's method to solve the time equation for x, we write

$$f(x) = \frac{4}{3}\eta^3 F(z) + 4\lambda\eta - t_{12}\sqrt{\frac{\mu}{a_{\rm m}^3}} \tag{6.102}$$

and

$$f'(x) = 4\left[\eta^2 F(z) + \lambda\right]\frac{\mathrm{d}\eta}{\mathrm{d}x} + \frac{4}{3}\eta^3 F'(z)\frac{\mathrm{d}z}{\mathrm{d}x}, \tag{6.103}$$

where prime denotes differentiation with respect to the argument of a function. From the definitions given above, we have

$$\frac{\mathrm{d}\eta}{\mathrm{d}x} = -\frac{\lambda\eta}{y}, \tag{6.104}$$

$$\frac{\mathrm{d}z}{\mathrm{d}x} = -\frac{\eta^2}{2y}, \tag{6.105}$$

and

$$F'(z) = \frac{6 - 3\gamma_1 G(z)}{2(1-z)[1 - z\gamma_1 G(z)]}, \tag{6.106}$$

where

$$G(z) = \cfrac{1}{1 - \cfrac{\gamma_2 z}{1 - \cfrac{\gamma_3 z}{1 - \cfrac{\gamma_4 z}{1 - \ddots}}}}. \tag{6.107}$$

A solution procedure is given by the algorithm of Figure 6.7, which is coded in the program, *lambert_hypergeom.m* (Table 6.2), with the evaluations by continued fractions coded in *hypergeom.m* and *hypergeomd.m* (Table 6.3).

6.3.4 Lambert Guidance of Rockets

An interesting example of Lambert guidance is the ascent of a rocket to a given terminal point defined by altitude, latitude, and longitude. It is assumed that the vehicle follows a nominal gravity-turn trajectory (Chapter 4) until the guidance scheme takes over, generating acceleration inputs for guiding the vehicle to its terminal point. As guidance is active until the destination is reached, the flight time is limited by the burn-out of the rocket motor. Since longitude is one of the state variables, the closed-loop simulation must be carried out including the planetary rotational effects. Furthermore, a realistic model of rocket stages, mass, drag, and thrust must be employed.

Suppose the rocket has reached a geocentric, planet fixed position, $\mathbf{r}_0 = (r_0, \delta_0, \ell_0)^T$ at $t = t_0$. To achieve the terminal point, $\mathbf{r}_\mathrm{f} = (r_\mathrm{f}, \delta_\mathrm{f}, \ell_\mathrm{f})$, on or before burn-out at $t = t_\mathrm{b}$, the required initial velocity relative to a rotating Earth, $\mathbf{v}_{0r}$, must be obtained from the solution of Lambert's problem associated with $\mathbf{r}_0, \mathbf{r}_\mathrm{f}, (t_\mathrm{b} - t_0)$. However, since the rocket is producing thrust and may also have atmospheric drag, the motion is non-Keplerian and hence the initial velocity vector calculated from Lambert's problem at $t = t_0$ – even if exactly attained – will not take the vehicle to the destination in the required time. Therefore, the velocity vector must be continuously updated by solving Lambert's problem repeatedly at various instants until the position error becomes sufficiently small. This is the crux of the Lambert guidance method. Needless to say, real-time online computation of acceleration inputs requires an efficient Lambert algorithm.

The rocket thrust is completely determined by staging; therefore, normal acceleration is the only control input and can be approximated by

$$\mathbf{u}_2 \simeq -(\mathbf{v}_{0r} - \mathbf{v}_0) \times \frac{\mathbf{v}_0}{v_0} = -\mathbf{v}_{0r} \times \frac{\mathbf{v}_0}{v_0}, \tag{6.108}$$

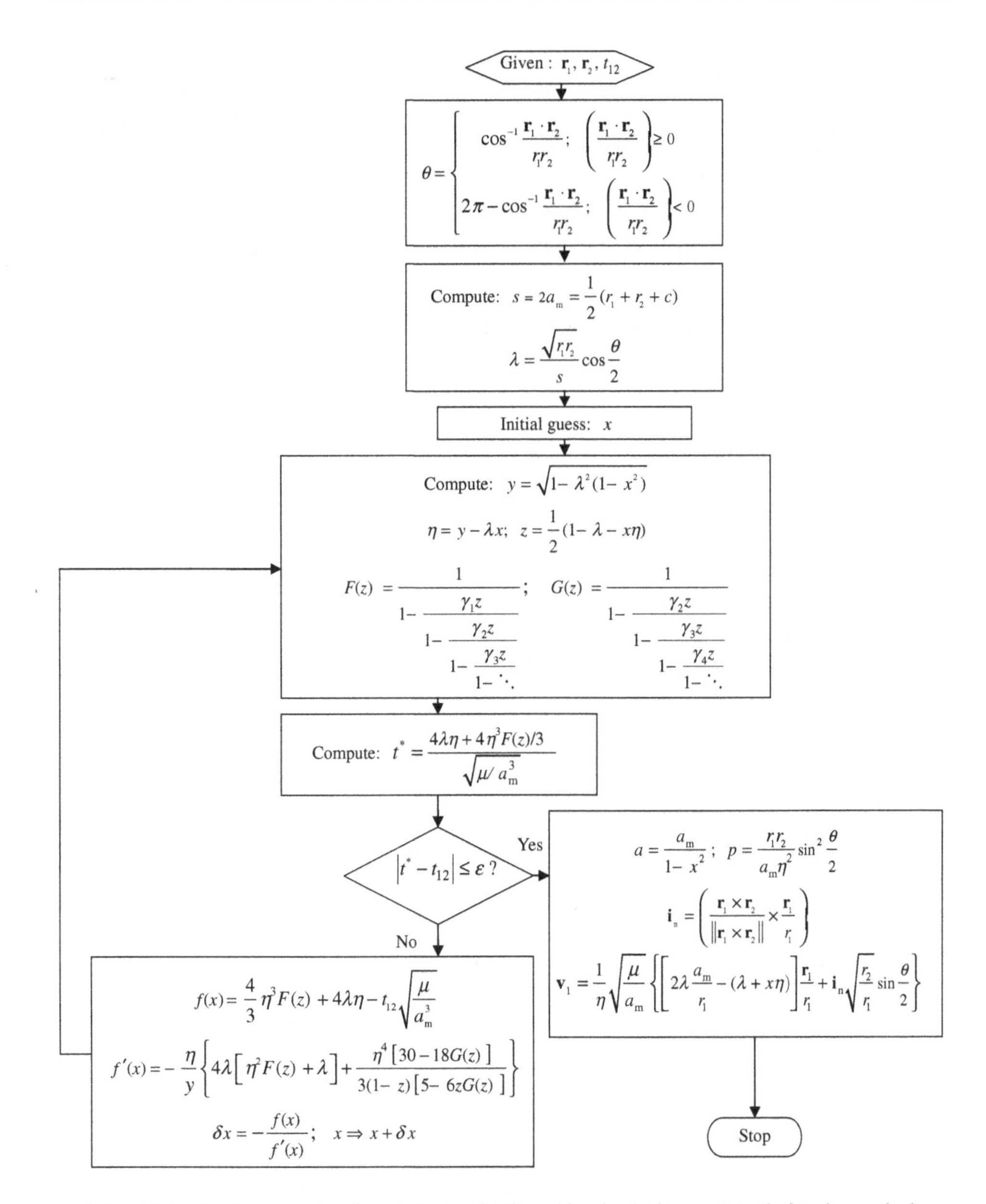

Figure 6.7 Iteration procedure for solving Lambert's problem by the hypergeometric function method

where $\mathbf{v}_{0r}$ is the required velocity at $t = t_0$ from the Lambert solution. Clearly, the required normal acceleration vector is also normal to the velocity error, $(\mathbf{v}_{0r} - \mathbf{v}_0)$ and thus the navigation law is quite similar to cross-product steering (Chapter 1).

Example 6.3 Consider Lambert guidance of a two-stage rocket for launch from Cape Canaveral $(\delta = 28.5°, \ell = -80.55°)$ with initial relative speed, $v = 0.0001$ m/s, and relative flight path,

Table 6.2

```
function [a,p,V1]=lambert_hypergeom(mu,R1,R2,t12,id)
% Solution to Lambert's two-point boundary value problem in space navigation.
% mu: gravitational constant of central mass.
% R1,V1: initial position and velocity vectors in celestial frame.
% R2,V2: final position and velocity vectors in celestial frame.
% t12: transfer time.
% id: indicator of transfer direction (id=1:direct; -1:retrograde)
% (c) 2010 Ashish Tewari
r1=norm(R1); r2=norm(R2);
costh=dot(R1,R2)/(r1*r2); % cosine transfer angle
theta=acos(costh);
if costh<0
    theta=2*pi-theta;
end
cosi=dot([0 0 1]',cross(R1,R2)/norm(cross(R1,R2)));
if id*cosi<0
    theta=2*pi-theta;
end
c=norm(R2-R1);
s=0.5*(r1+r2+c);
am=s/2;
L=sqrt(r1*r2)*cos(theta/2)/s;
x=0.1;
y=sqrt(1-L^2*(1-x^2));
eta=y-L*x;
z=0.5*(1-L-x*eta);
n=1;
F=hypergeom(z);
tc=(4*eta^3*F/3+4*L*eta)/sqrt(mu/am^3);
% Newton's iteration for 'x' follows:
while abs(t12-tc)>1e-9
    fx=4*eta^3*F/3+4*L*eta-t12*sqrt(mu/am^3);
    G=hypergeomd(z);
    Fd=(6-3*6*G/5)/(2*(1-z)*(1-z*6*G/5));
    fxp=-2*eta^5*Fd/(3*y)-L*eta*(4*eta^2*F+4*L)/y;
    dx=-fx/fxp;
    x=x+dx;
    n=n+1;
    y=sqrt(1-L^2*(1-x^2));
    eta=y-L*x;
    z=0.5*(1-L-x*eta);
    F=hypergeom(z);
    tc=(4*eta^3*F/3+4*L*eta)/sqrt(mu/am^3);
    end
a=am/(1-x^2);
p=r1*r2*(sin(0.5*theta))^2/(am*eta^2);
V1=sqrt(mu/am)*((2*L*am/r1-(L+x*eta))*R1/r1+ ...
   sqrt(r2/r1)*sin(0.5*theta)*cross(cross(R1,R2) ...
   /norm(cross(R1,R2)),R1/r1))/eta;
```

Table 6.3

```
function F=hypergeom(z)
% Hypergeometric function evaluation
% by continued fraction
% (c) 2010 Ashish Tewari
d=1;u=1;F=1;
eps=1e-9;
i=1;
while abs(u)>eps
    if rem(i,2)==0
    gamma=i*(i-3)/((2*i+1)*(2*i+3));
    else
    gamma=(i+2)*(i+5)/((2*i+1)*(2*i+3));
    end
    d=1/(1-gamma*z*d);
    u=u*(d-1);
    F=F+u;
    i=i+1;
end

function G=hypergeomd(z)
% Hypergeometric function, G(z)
% evaluated by continued fraction
d=1;u=1;G=1;
eps=1e-9;
i=1;
while abs(u)>eps
    n=i+1;
    if rem(n,2)==0
    gamma=n*(n-3)/((2*n+1)*(2*n+3));
    else
    gamma=(n+2)*(n+5)/((2*n+1)*(2*n+3));
    end
    d=1/(1-gamma*z*d);
    u=u*(d-1);
    G=G+u;
    i=i+1;
end
```

$\phi = 90^\circ$, $\psi = 170^\circ$, which translates into the best inertial launch direction (due east) for taking advantage of the Earth's rotation (Tewari 2006). The first stage employs a solid propellant with specific impulse 200 s, while the second stage has kerosene/LO_2 with a specific impulse 350 s. The first- and second-stage burn times are 100 and 175 seconds, respectively, whereas the respective initial mass, propellant mass, and thrust are as follows:

$$m_{01} = 30\,856.513 \text{ kg}, \quad m_{02} = 8262.362 \text{ kg},$$

$$m_{p1} = 21\,012.561 \text{ kg}, \quad m_{p2} = 7516.744 \text{ kg},$$

$$f_{T1} = 412\,266.44 \text{ N}, \quad f_{T2} = 147\,478.51 \text{ N}.$$

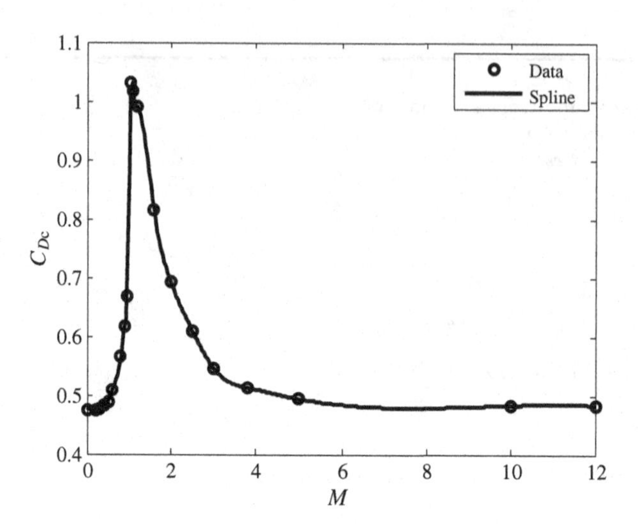

Figure 6.8 Continuum drag coefficient of payload of the two-stage rocket guided by Lambert's algorithm

The payload mass at burn-out is 350 kg. The thrust loss due to incorrect expansion of exhaust gases is negligible.

The drag coefficient of the payload, C_{Dp}, based upon base area, $S = 4$ m^2, is

$$C_{Dp} = C_{Dc}, \quad Kn < 0.0146,$$
$$C_{Dp} = C_{Dfm}, \quad Kn > 14.5,$$
$$C_{Dp} = C_{Dc} + (C_{Dfm} - C_{Dc})\left(\frac{1}{3}\log_{10}\frac{Kn}{\sin 30^\circ} + 0.5113\right), \quad 0.0146 < Kn < 14.5,$$

where C_{Dc} is the drag coefficient in the continuum limit (Tewari 2006), plotted in Figure 6.8 as a function of Mach number, C_{Dfm} is the drag coefficient in the free-molecular flow limit with cold-wall approximation (Tewari 2006), given by

$$C_{Dfm} = 1.75 + \frac{\sqrt{\pi}}{2s},$$

with $s = \frac{v}{\sqrt{2RT}}$ denoting the molecular speed ratio (Tewari 2006), and the Knudsen number, Kn, is based on a nose radius of 0.5 m. The drag coefficient of the first and second stages is $8C_{Dp}$ and $3C_{Dp}$, respectively.

The destination of $h_f = 120$ km, $\delta_f = 26^\circ$, $\ell_f = -82^\circ$ must be reached simultaneously with the second-stage burn-out, that is, at $t = t_{b1} + t_{b2} = 275$ s. The guidance system is activated as the vehicle reaches the altitude of 100 km, and ended when about 0.1 s of powered flight remains in order to avoid solving Lambert's problem for almost zero transfer angle. We carry out closed-loop simulation with Lambert guidance by fourth-order Runge–Kutta solution using the MATLAB codes given in Tables 6.4 and 6.5. The simulation uses the standard atmosphere code *atmosphere.m* and an oblate gravity model code *oblate.m* that are both listed in (Tewari 2006) and can be freely downloaded from the website http://home.iitk.ac.in/~ashtew/index_files. Another file *CLH.m* required for coordinate transformation from planet-fixed to local horizon frame is tabulated in Table 6.6.

The results of the simulation with Lambert guidance begun at $h = 100$ km are plotted in Figure 6.9, showing that the final destination is reached with 0.01 s of flight time to spare, with a radial error of 0.016

Table 6.4

```
% (c) 2010 Ashish Tewari
global mu; mu=398600.4e9;
global R0; R0=6378.14e3;
global omega; omega = 2*pi/(23*3600+56*60+4.0905);%rad/sec
global hf; hf=120e3;
global delf; delf=26*pi/180;
global lamf; lamf=-82*pi/180;
global S; S =4; %reference base area, m2
global c; c=0.5; % nose-radius, m
global tb1; tb1=100; %first-stage burn-time (s)
global tb2; tb2=175; %second-stage burn-time (s)
global fT1;fT1=824532.879800791/2; %first-stage thrust (N)
global fT2;fT2=161304.620971835*(1.6/1.75); %second-stage thrust (N)
global m01; m01=30856.5129807023; %first-stage initial mass (kg)
global m02; m02=8262.36174702614; %second-stage initial mass (kg)
global mp1; mp1=21012.5606473188; %first-stage propellant mass (kg)
global mp2; mp2=7516.74365967484; %second-stage propellant mass (kg)
global mL; mL=350; %payload mass (kg)
global Gamma; Gamma=1.41; %specific heat ratio, cp/cv
global machr; machr = [0 0.2 0.3 0.4 0.5 0.6 0.8 0.9 0.95 1.05 1.1 ...
                        1.2 1.6 2.0 2.5 3 3.8 5 10 99];
global Cdr; Cdr =[.475475 .475475 .47576 .48336 .488965 .508345 ...
                  .56563 .618165 .668135 1.031795 1.01707 .990565 ...
                  .815955 .69236 .60971 .54606 .513 .494 .48317 .48317];
dtr=pi/180;
%Initial conditions:
long = -80.55*dtr;    %initial longitude.
lat = 28.5*dtr;       %initial latitude.
rad=R0;               %initial radius (m)
vel=0.0001;           %initial velocity (m/s)
fpa=90*dtr;           %initial f.p.a
chi=170*dtr; %initial heading angle (measured from north)
orbinit = [long; lat; rad; vel; fpa; chi];
[t,x]=ode45('lambert_rocket',[0 275],orbinit);
er=(x(size(x,1),3)-6378.14e3-hf)/1000 %final radial error (km)
edel=(x(size(x,1),2)-delf)/dtr %final latitude error (deg.)
elam=(x(size(x,1),1)-lamf)/dtr %final longitude error (deg.)
```

km, latitude error of $-0.052°$ and an almost zero longitudinal error (-3.5×10^{-4} degrees). The errors are reduced further if guidance is activated slightly earlier, say at 80 km altitude.

6.3.5 *Optimal Terminal Guidance of Re-entry Vehicles*

The problem of re-entry vehicle guidance is little different from that of a rocket, in that both have their motion largely confined to the vertical plane. The only difference between a rocket and a re-entry vehicle is that the latter is always decelerating and descending, thereby losing its total energy due to atmospheric drag, while a rocket is always gaining energy by doing work through its engine. In some cases, the deceleration of re-entry may be assisted by firing a retro-rocket, because deceleration purely by atmospheric drag could cause excessive aerodynamic heating. Without going into the details of aerothermodynamic modeling, we would like to pose the re-entry guidance problem by minimizing

Table 6.5

```
% (c) 2010 Ashish Tewari
function deriv=lambert_rocket(t,o);
global hf; global delf; global lamf;
global mu; global omega; global S; global c;
global R0; global tb1; global tb2; global fT1;
global fT2; global m01; global m02;
global mL; global mp1; global mp2; global Gamma;
global machr; global Cdr;
[g,gn]=gravity(o(3),o(2)); %accln. due to gravity (oblate earth)
lo = o(1);la = o(2);
cLH=CLH(la,lo);
clo = cos(lo); slo = sin(lo); cla = cos(la); sla = sin(la);
fpa = o(5); chi = o(6); cfpa = cos(fpa); sfpa = sin(fpa);
cchi = cos(chi); schi = sin(chi);
%%%atmospheric drag determination
if o(3)<R0; o(3)=R0; end
alt = o(3) - R0;
v  = o(4);
if v<0; v=0; end
if alt<=2000e3
atmosp = atmosphere(alt,v,c);
rho = atmosp(2);
Qinf = 0.5*rho*v^2;
mach = atmosp(3);
Kn=atmosp(4);
CDC=interp1(machr, Cdr, mach);
s = mach*sqrt(Gamma/2);
CDFM=1.75+sqrt(pi)/(2*s);
iflow=atmosp(6);
if iflow==2
    CD=CDC;
elseif iflow==1
    CD=CDFM;
else
    CD = CDC + (CDFM - CDC)*(0.333*log10(Kn/sin(pi/6))+0.5113);
end
else
    rho=0;Qinf=0;CD=0;mach=0;
end
if t<=tb1
    fT=fT1; m=m01-mp1*t/tb1; CD=8*CD;
elseif t<=(tb1+tb2)
    fT=fT2; m=m02-mp2*(t-tb1)/tb2; CD=3*CD;
else
    fT=0; m=mL;
end
D=Qinf*S*CD;
Xfo = fT-D; Yfo = 0; Zfo = 0;
R=o(3)*[cla*clo cla*slo sla]';
V=v*[sfpa cfpa*schi cfpa*cchi]';
Rf=(hf+R0)*[cos(delf)*cos(lamf) cos(delf)*sin(lamf) sin(delf)]';
if alt>100e3 && t<tb1+tb2-.1
        [a,p,Vi]=lambert_hypergeom(mu,R,Rf,tb1+tb2-t,-1);
```

Table 6.5 (*Continued*)

```
        Vi=cLH*Vi; %planet-fixed to local horizon frame
        u=-(Vi-V)/v;
    else
        u=zeros(3,1);
end
%trajectory equations follow :
longidot = o(4)*cfpa*schi/(o(3)*cla); %Longitude
latidot =  o(4)*cfpa*cchi/o(3); %Latitude
raddot = o(4)*sfpa;  %Radius
veldot = -g*sfpa +gn*cchi*cfpa + Xfo/m + ...
          omega*omega*o(3)*cla*(sfpa*cla - cfpa*cchi*sla);
if t<=10; headdot=0; gammadot=0;
else
gammadot = u(1,1)+(o(4)/o(3)-g/o(4))*cfpa+gn*cchi*sfpa/o(4) ...
           + Zfo/(o(4)*m) + 2*omega*schi*cla ...
           + omega*omega*o(3)*cla*(cfpa*cla + sfpa*cchi*sla)/o(4);
if abs(cfpa)>1e-6
headdot = u(2,1)+ o(4)*schi*tan(o(2))*cfpa/o(3)-gn*schi/o(4) ...
         -Yfo/(o(4)*cfpa*m)-2*omega*(tan(o(5))*cchi*cla-sla) ...
         +omega*omega*o(3)*schi*sla*cla/(o(4)*cfpa);
else
    headdot=0;
end
end
deriv = [longidot; latidot; raddot; veldot; gammadot; headdot];
```

the total deceleration required from an entry at near-orbital speed to a terminal point at small altitude and a much smaller speed.

We revisit the equations of motion for flight in a vertical plane, referred to a non-rotating, spherical planet (Chapter 4):

$$\dot{\delta} = \frac{v\cos\phi}{r}, \tag{6.109}$$

$$\dot{r} = v\sin\phi, \tag{6.110}$$

$$\dot{v} = u_1 - \frac{\mu}{r^2}\sin\phi, \tag{6.111}$$

$$v\dot{\phi} = u_2 + \left(\frac{v^2}{r} - \frac{\mu}{r^2}\right)\cos\phi, \tag{6.112}$$

Table 6.6

```
% Rotation matrix from planet-fixed to local horizon frame
% delta: Latitude (rad.)
% lambda: Longitude (rad.)
function C=CLH(delta,lambda)
C= [cos(delta)*cos(lambda) cos(delta)*sin(lambda) sin(delta);
   -sin(lambda) cos(lambda) 0;
   -sin(delta)*cos(lambda) -sin(delta)*sin(lambda) cos(delta)];
```

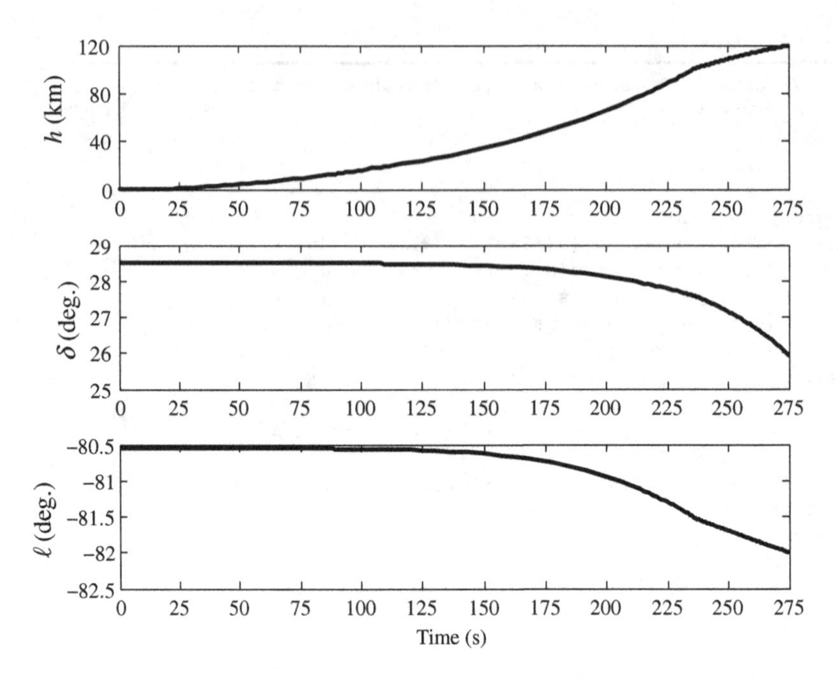

Figure 6.9 Trajectory of the two-stage rocket guided by Lambert's algorithm, initiated at 100 km altitude

where the two acceleration inputs, u_1 and u_2, are the same as the axial and normal acceleration components, respectively, shown in Figure 4.15. In a typical re-entry, u_1 is generated entirely by drag and u_2 by the lift. In a ballistic re-entry, the lift is negligible compared to the drag, hence u_2 is taken to be nearly zero. However, a lifting re-entry vehicle (of which the Space Shuttle was an example) can take advantage of a lift-assisted vertical maneuver in which the downrange (declination) can be controlled, and the aerodynamic deceleration (thus heating) is minimized compared with the ballistic case. We will derive optimal guidance laws for both ballistic and lifting re-entry examples. As in the case of a rocket, the declination, $\delta(t)$, radius, $r(t)$, speed, $v(t)$, and flight-path angle, $\phi(t)$ are state variables, and the forward acceleration input, $u_1(t)$, and normal acceleration input, $u_2(t)$, are the control inputs.

For optimization with a quadratic cost and without terminal penalty, that is, $\varphi = 0$, $L = R_1 u_1^2 + R_2 u_2^2$ (Chapter 2), the Hamiltonian is given by $H = L + \boldsymbol{\lambda}^T \mathbf{f}$, where

$$\mathbf{f} = \left\{ \begin{array}{c} \frac{v}{r}\cos\phi \\ v\sin\phi \\ u_1 - \frac{\mu}{r^2}\sin\phi \\ \frac{u_2}{v} + \left(\frac{v}{r} - \frac{\mu}{vr^2}\right)\cos\phi \end{array} \right\}. \tag{6.113}$$

The necessary conditions for optimality yield the following extremal control input:

$$H_{\mathbf{u}} = L_{\mathbf{u}} + \boldsymbol{\lambda}^T \mathbf{B}, \tag{6.114}$$

where

$$\mathbf{B} = \frac{\partial \mathbf{f}}{\partial \mathbf{u}} = \begin{pmatrix} 0 & 0 \\ 0 & 0 \\ 1 & 0 \\ 0 & \frac{1}{v} \end{pmatrix},$$

which results in

$$2\begin{Bmatrix} R_1 u_1^* \\ R_2 u_2^* \end{Bmatrix} + \begin{Bmatrix} \lambda_3 \\ \frac{\lambda_4}{v} \end{Bmatrix} = \mathbf{0} \tag{6.115}$$

or

$$\begin{Bmatrix} u_1^* \\ u_2^* \end{Bmatrix} = -\begin{Bmatrix} \frac{\lambda_3}{2R_1} \\ \frac{\lambda_4}{2R_2 v} \end{Bmatrix}. \tag{6.116}$$

Substitution of the extremal control and trajectory yields the following set of co-state equations for $\boldsymbol{\lambda}(t) = [\lambda_1(t), \lambda_2(t), \lambda_3(t), \lambda_4(t)]^T$:

$$\dot{\boldsymbol{\lambda}} = -\mathbf{A}^T \boldsymbol{\lambda}, \tag{6.117}$$

where

$$\mathbf{A} = \frac{\partial \mathbf{f}}{\partial \mathbf{x}} = \begin{pmatrix} 0 & -\frac{v\cos\phi}{r^2} & \frac{\cos\phi}{r} & -\frac{v\sin\phi}{r} \\ 0 & 0 & \sin\phi & v\cos\phi \\ 0 & \frac{2\mu\sin\phi}{r^3} & 0 & -\frac{\mu\cos\phi}{r^2} \\ 0 & \left(-\frac{v}{r^2} + \frac{2\mu}{vr^3}\right)\cos\phi & -\frac{u_2^*}{v^2} + \left(\frac{1}{r} + \frac{\mu}{v^2 r^2}\right)\cos\phi & -\left(\frac{v}{r} - \frac{\mu}{vr^2}\right)\sin\phi \end{pmatrix},$$

resulting in

$$\dot{\lambda}_1 = 0, \tag{6.118}$$

$$\dot{\lambda}_2 = \frac{v\cos\phi}{r^2}\lambda_1 - \frac{2\mu\sin\phi}{r^3}\lambda_3 - \lambda_4\left(\frac{2\mu}{vr^3} - \frac{v}{r^2}\right)\cos\phi, \tag{6.119}$$

$$\dot{\lambda}_3 = -\frac{\cos\phi}{r}\lambda_1 - (\sin\phi)\lambda_2 - \left[\frac{\lambda_4}{2R_2 v^3} + \left(\frac{1}{r} + \frac{\mu}{v^2 r^2}\right)\cos\phi\right]\lambda_4, \tag{6.120}$$

$$\dot{\lambda}_4 = \frac{v\sin\phi}{r}\lambda_1 - (v\cos\phi)\lambda_2 + \frac{\mu\cos\phi}{r^2}\lambda_3 + \lambda_4\left(\frac{v}{r} - \frac{\mu}{vr^2}\right)\sin\phi. \tag{6.121}$$

The nemesis of a re-entry vehicle is aerodynamic heating caused by deceleration due to atmospheric drag. Being roughly proportional to the cube of the speed relative to the atmosphere and directly proportional to the atmospheric density (Tewari 2006), heat flux can be minimized either by flying a trajectory that results in a higher altitude (thus smaller density) when the speed is large, or by reducing the speed by firing a retro-rocket. In a ballistic entry, there are no means of controlling the flight path; therefore the flight-path angle at entry determines the altitude and speed profile, and hence the maximum heating experienced by a vehicle. However, a lifting (or maneuvering) entry vehicle can control its flight path, not only minimizing aerodynamic heating by flying a suitable profile, but also controlling the downrange from the point of entry.

Ballistic Entry

For a ballistic entry, $u_2(t) = 0$, resulting in $\lambda_4(t) = 0$ and $L = Ru_1^2$. Furthermore, the steepness of the trajectory is defined by the conditions at entry and thus the downrange (or declination) cannot be precisely controlled in a ballistic trajectory. Hence, state and co-state equations corresponding to $\delta(t)$ are dropped. The co-state equations are thus simplified to the following:

$$\begin{aligned} \dot{\lambda}_2 &= -\frac{2\mu \sin\phi}{r^3}\lambda_3, \\ \dot{\lambda}_3 &= -\lambda_2 \sin\phi. \end{aligned} \tag{6.122}$$

Practical boundary conditions for a ballistic entry are as follows:

$$\begin{aligned} r(0) &= r_0, \\ v(0) &= v_0, \\ \phi(0) &= \phi_0, \\ r(t_f) &= r_f, \\ v(t_f) &= v_f, \end{aligned} \tag{6.123}$$

The 2PBVP resulting from the state and co-state equations as well as the boundary conditions must be solved numerically in order to yield the optimal control history,

$$u_1^*(t) = -\frac{\lambda_3(t)}{2R}. \tag{6.124}$$

Example 6.4 Consider a ballistic entry from $r(0) = 100$ km, $v(0) = 8$ km/s and $\phi(0) = -0.1°$ to $r(t_f) = 5$ km, $v(t_f) = 2$ km/s. The terminal time, t_f, is to be adjusted such that the maximum deceleration does not exceed 30 m/s^2 at any point in the trajectory.

The necessary computations are carried out by the following MATLAB calling program using the intrinsic 2PBVP solver, *bvp4c.m*, and the codes *entryode.m* and *entrybc.m* (Table 6.7) specifying the differential equations and the boundary conditions, respectively:

```
>> global mu; mu=398600.4;
>> global r0; r0=6378.14;
>> global R; R=1;
>> tf=250;
>> dtr=pi/180;
>> solinit = bvpinit(linspace(0,tf,5),[6380+100 8 -0.1*dtr 0 0]);
>> sol = bvp4c(@entryode,@entrybc,solinit);
>> x = linspace(0,tf);
>> y = deval(sol,x);
>> u=-0.5*y(5,:)/R;
```

The resulting trajectories for $t_f = 250$, 260, and 270 s are plotted in Figure 6.10. A remarkable qualitative change in the optimum deceleration profile is observed between $t_f = 250$ and 260 s, with the

Table 6.7

```
% (c) 2010 Ashish Tewari
function dydx=entryode(x,y)
global R;
global mu;
global r0;
dydx=[y(2)*sin(y(3))
      -0.5*y(5)/R-mu*sin(y(3))/y(1)^2
      (y(2)-mu/(y(1)*y(2)))*cos(y(3))/y(1)
      -2*mu*y(5)*sin(y(3))/y(1)^3
      -y(4)*sin(y(3))];

%%%%%%%%%%%%%%%%%%%%%%%%%%%%%%%%%%%%%%%%%%%

function res=entrybc(ya,yb)
global r0;
dtr=pi/180;
res=[ya(1)-r0-100
     ya(2)-8
     ya(3)+0.1*dtr
     yb(1)-r0-5
     yb(2)-2];
```

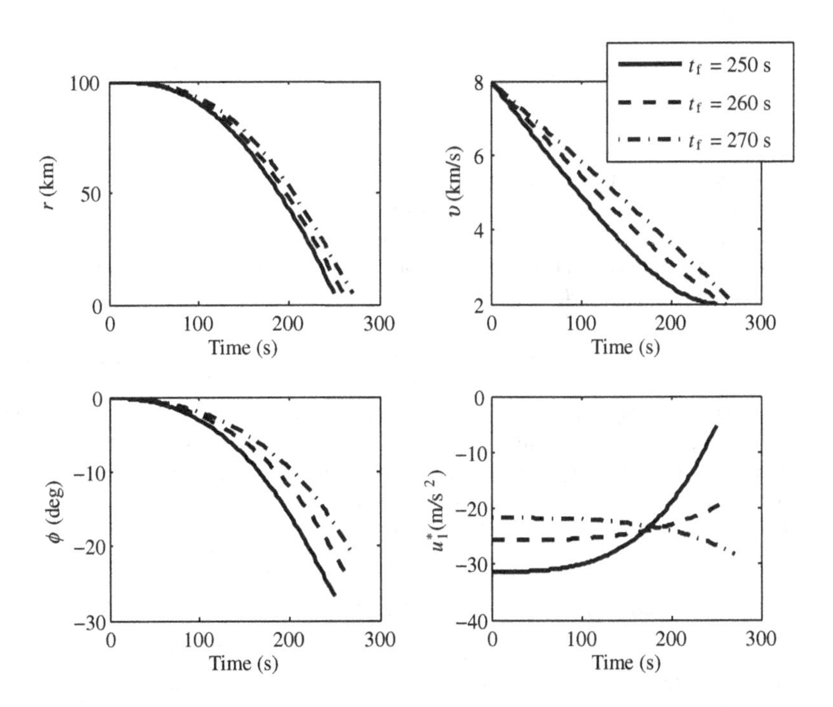

Figure 6.10 Optimal ballistic re-entry trajectories for three different terminal times

former requiring a large deceleration to be applied initially that tapers off to the end, while the latter needs a nearly constant deceleration almost right up to the terminal point. Such deceleration profiles can only be applied by a rocket engine, since the atmospheric drag increases almost exponentially with a decrease of altitude (i.e., increase of density), reaching a maximum value and then tapering off due to speed reduction (Tewari 2006). For $t_f = 270$ s, the deceleration profile shows an ever increasing nature, but remains smaller than the maximum deceleration limit of 30 m/s^2. The velocity profile for $t_f = 270$ s shows an almost linear reduction of speed with time, and also has the shallowest flight path of the three cases.

Maneuvering Entry

When atmospheric lift and/or vectored thrust are available, we have normal acceleration, $u_2(t)$, as an additional control variable, allowing a re-entry vehicle to be maneuvered to a desired destination (altitude and declination) while alleviating axial deceleration (thus aerodynamic heating) caused by atmospheric drag. However, the optimal guidance problem is now much more difficult as the complete set of equations (6.109)–(6.112) and (6.118)–(6.121) must be solved for the optimal control inputs, $u_1^*(t)$, $u_2^*(t)$, subject to the following boundary conditions:

$$\begin{aligned}
\delta(0) &= \delta_0, \\
r(0) &= r_0, \\
v(0) &= v_0, \\
\phi(0) &= \phi_0, \\
\delta(t_f) &= \delta_f, \\
r(t_f) &= r_f, \\
v(t_f) &= v_f.
\end{aligned} \tag{6.125}$$

Example 6.5 Repeat the entry problem of Example 6.4 with both axial and normal acceleration inputs and additional boundary conditions, $\delta(0) = 30°$ and $\delta(t_f) = 60°$ The terminal time, t_f, is to be adjusted such that the maximum axial deceleration does not exceed 15 m/s^2 while requiring a maximum normal acceleration of only ± 10 m/s^2 throughout the trajectory.

The necessary computations are carried out by the MATLAB calling program listed in Table 6.8, which uses the intrinsic 2PBVP solver, *bvp4c.m*, and the codes *normballisticbc.m* (Table 4.12) and *manentrybc.m* (Table 6.8) specifying the differential equations and the boundary conditions, respectively. The results are plotted in Figures 6.11 and 6.12 for $t_f = 650$ s, $R_1 = 1$, and three different values of the control cost parameter, $R_2 = 2, 5, 10$. Note the initially ascending trajectories for the cases $R_2 = 5$ and $R_2 = 10$ which delay the actual entry by about 550 s while making the vehicle traverse the requisite downrange. In contrast, $R_2 = 2$ produces most of the maneuvering within the atmosphere, in the altitude range, $75 \le h \le 100$ km, thereby requiring the largest normal acceleration (but smallest axial deceleration) among the three cases. The case of $R_2 = 10$ has the smallest normal acceleration requirement and an almost linear deceleration profile between $-14.75 \le u_1 \le -5.5$ m/s^2.

6.4 General Orbital Plant for Tracking Guidance

A spacecraft must maintain a desired (nominal) orbit in the presence of disturbing forces, called *orbital perturbations*. Such perturbations are caused by non-Keplerian effects, such as propulsive thrust, non-spherical shape of the planet, presence of a third body, atmospheric drag, and solar radiation pressure. While gravitational perturbations are conservative and can be represented by the position derivative of a potential function, atmospheric drag, thrust, and solar radiation are non-conservative effects capable of

Table 6.8

```
% (c) 2010 Ashish Tewari
global R1; R1=1;
global R2; R2=10;
global tf; tf=650;
dtr=pi/180;
solinit = bvpinit(linspace(0,tf,5),[30*dtr 6478 8 -0.1*dtr 0 0 0 0]);
sol = bvp4c(@normballisticode,@manentrybc,solinit);
x = linspace(0,tf);
y = deval(sol,x);
u1=-0.5*y(7,:)/R1;
u2=-0.5*y(8,:)./(R2*y(3,:));

%%%%%%%%%%%%%%%%%%%%%%%%%%%%%%%%%%%%%%%%%%%%%%%%%%%%%%%%%%%%%%%%%%%%%%

function res=manentrybc(ya,yb)
mu=398600.4;
r0=6378.14;
dtr=pi/180;
res=[ya(1)-30*dtr
    ya(2)-r0-100
    ya(3)-8
    ya(4)+0.1*dtr
    yb(1)-60*dtr
    yb(2)-r0-5
    yb(3)-2
    yb(8)];
```

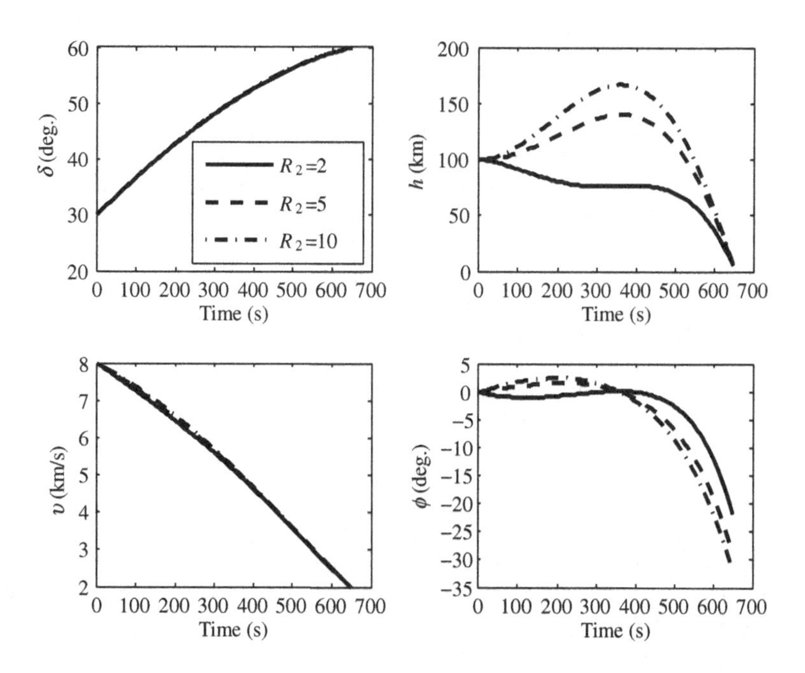

Figure 6.11 Optimal maneuvering re-entry trajectories for three different values of the control cost parameter, R_2

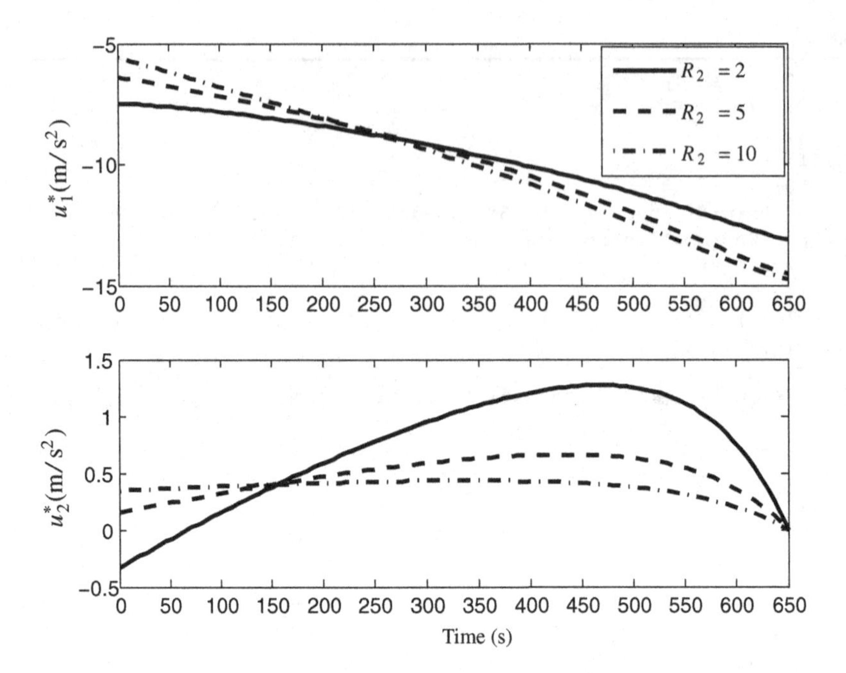

Figure 6.12 Optimal acceleration inputs of maneuvering re-entry for three different values of the control cost parameter, R_2

changing both energy and angular momentum of an orbit. For our present purposes, let us approximate the effect of orbital perturbation to be a small deviation from the desired orbit that may be modeled as either a point perturbation causing an initial deviation, or a continuous, random disturbance. Needless to say, an orbit control system is invariably required that can compensate for orbital perturbations by bringing their effect to zero in a steady state. An orbit controller requires an actuator capable of applying an external force on the spacecraft (generally a rocket thruster), a sensor capable of measuring instantaneous orbital position, and a control law that can produce an input signal for the actuator depending upon the sensor output.

Since the nominal trajectory is a two-body orbit which remains confined to an orbital plane, an orbit control system must apply corrective acceleration inputs to maintain both the plane and the shape of the orbit. The orbit shape control requires a coplanar acceleration input, which for orbit plane control must be out of the plane of the orbit.

The general equation of orbital motion can be expressed as

$$\frac{d\mathbf{v}}{dt} + \frac{\mu}{r^3}\mathbf{r} = \mathbf{u} + \mathbf{a}_\text{d}, \tag{6.126}$$

where $\mathbf{a}_\text{d}(t)$ represents a disturbing acceleration, and $\mathbf{u}(t)$ is the required acceleration control input. Unless modeled by appropriate physical processes, $\mathbf{a}_\text{d}(t)$ is commonly treated as a stochastic disturbance applied to the control system in the form of a process noise, $\boldsymbol{\nu}(t)$, and a measurement noise, $\mathbf{w}(t)$ (Figure 6.13). The nominal orbit, described by

$$\frac{d\mathbf{v}_\text{n}}{dt} + \frac{\mu}{{r_n}^3}\mathbf{r}_\text{n} = \mathbf{0}, \tag{6.127}$$

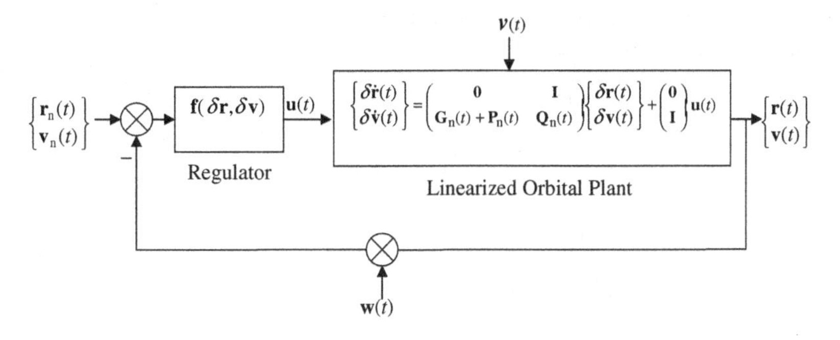

Figure 6.13 General orbit control system

results in a first-order Taylor series expansion of the position and velocity vectors about their nominal values as follows:

$$\mathbf{r}(t) \simeq \mathbf{r}_n(t) + \delta\mathbf{r}(t),$$
$$\mathbf{v}(t) \simeq \mathbf{v}_n(t) + \delta\mathbf{v}(t), \tag{6.128}$$

which is a valid approximation as long as

$$|\mathbf{a}_d| \ll \frac{\mu}{r^2}, \quad |\mathbf{u}| \ll \frac{\mu}{r^2}.$$

Writing the spherical gravitational acceleration as

$$\mathbf{a}_g = -\frac{\mu}{r^3}\mathbf{r} = -\frac{\mu}{r_n^3}\mathbf{r}_n + \left(\frac{\partial \mathbf{a}_g}{\partial \mathbf{r}}\right)_{\mathbf{r}=\mathbf{r}_n} \delta\mathbf{r} = -\frac{\mu}{r_n^3}\mathbf{r}_n + \delta\mathbf{a}_g \tag{6.129}$$

or

$$\delta\mathbf{a}_g = \mathbf{G}_n\delta\mathbf{r}, \tag{6.130}$$

where $\mathbf{G}_n$ is the *gravity-gradient* matrix evaluated on the nominal trajectory,

$$\mathbf{G}_n = \left(\frac{\partial \mathbf{a}_g}{\partial \mathbf{r}}\right)_{\mathbf{r}=\mathbf{r}_n}.$$

Similarly, the perturbed acceleration can be linearized about the nominal trajectory as follows:

$$\mathbf{a}_d = \mathbf{P}_n\delta\mathbf{r} + \mathbf{Q}_n\delta\mathbf{v}, \tag{6.131}$$

where

$$\mathbf{P}_n = \left(\frac{\partial \mathbf{a}_d}{\partial \mathbf{r}}\right)_{\mathbf{r}=\mathbf{r}_n}$$

and

$$\mathbf{Q}_n = \left(\frac{\partial \mathbf{a}_d}{\partial \mathbf{v}}\right)_{\mathbf{v}=\mathbf{v}_n}.$$

The design of an orbit control system is thus based upon the following linearized plant derived above with the small perturbation approximation:

$$\frac{\mathrm{d}\delta\mathbf{r}}{\mathrm{d}t} = \delta\mathbf{v},$$
$$\frac{\mathrm{d}\delta\mathbf{v}}{\mathrm{d}t} = (\mathbf{G}_\mathrm{n} + \mathbf{P}_\mathrm{n})\,\delta\mathbf{r} + \mathbf{Q}_\mathrm{n}\delta\mathbf{v} + \mathbf{u}, \tag{6.132}$$

which can be expressed in state-space form as

$$\dot{\mathbf{x}}(t) = \mathbf{A}(t)\mathbf{x}(t) + \mathbf{B}(t)\mathbf{u}(t), \quad \mathbf{x}(0) = \mathbf{x}_0, \tag{6.133}$$

where

$$\mathbf{x} = \begin{Bmatrix} \delta\mathbf{r} \\ \delta\mathbf{v} \end{Bmatrix},$$

$$\mathbf{A} = \begin{pmatrix} \mathbf{0} & \mathbf{I} \\ \mathbf{G}_\mathrm{n} + \mathbf{P}_\mathrm{n} & \mathbf{Q}_\mathrm{n} \end{pmatrix}, \quad \mathbf{B} = \begin{pmatrix} \mathbf{0} \\ \mathbf{I} \end{pmatrix}.$$

As the nominal control input is zero for a given orbit, we have no need for a feedforward controller and the orbit control problem can be posed as that of a time-varying, linear feedback regulator as depicted in Figure 6.13. If the disturbance coefficient matrices, $\mathbf{P}_\mathrm{n}$, $\mathbf{Q}_\mathrm{n}$, cannot be determined through a physical model, one has to include their effects by means of the stochastic process and measurement noise vectors, $\boldsymbol{\nu}(t)$ and $\mathbf{w}(t)$, respectively. In such a case, the state vector is not determined directly, but instead estimated from the measurement of (noisy) output variables by an observer (Kalman filter). Therefore, the stochastic tracking plant is expressed by

$$\dot{\mathbf{x}}(t) = \mathbf{A}(t)\mathbf{x}(t) + \mathbf{B}(t)\mathbf{u}(t) + \mathbf{F}(t)\boldsymbol{\nu}(t), \quad \mathbf{x}(0) = \mathbf{x}_0, \tag{6.134}$$

where

$$\mathbf{A} = \begin{pmatrix} \mathbf{0} & \mathbf{I} \\ \mathbf{G}_\mathrm{n} & \mathbf{0} \end{pmatrix}, \quad \mathbf{B} = \begin{pmatrix} \mathbf{0} \\ \mathbf{I} \end{pmatrix},$$

along with the output equation,

$$\mathbf{y}(t) = \mathbf{C}(t)\mathbf{x}(t) + \mathbf{D}(t)\mathbf{u}(t) + \mathbf{w}(t). \tag{6.135}$$

We note that the state transition matrix, $\boldsymbol{\Phi}(t, t_0)$, corresponding to the homogeneous part of the state equation (6.134), is a symplectic matrix (Chapter 2), because the gravity-gradient matrix, $\mathbf{G}_\mathrm{n}(t)$, is always symmetric. This gives us a distinct advantage, because we can express the state transition matrix (Chapter 1) of the nominal plant $[\boldsymbol{\nu}(t) = \mathbf{0}, \mathbf{w}(t) = \mathbf{0}]$ as follows (Battin 1999):

$$\boldsymbol{\Phi}(t, t_0) = \begin{pmatrix} \tilde{\mathbf{R}}(t) & \mathbf{R}(t) \\ \tilde{\mathbf{V}}(t) & \mathbf{V}(t) \end{pmatrix}, \tag{6.136}$$

with the property

$$\boldsymbol{\Phi}^{-1}(t, t_0) = \boldsymbol{\Phi}(t_0, t) = \begin{pmatrix} \mathbf{V}^T(t) & -\mathbf{R}^T(t) \\ -\tilde{\mathbf{V}}^T(t) & \tilde{\mathbf{R}}^T(t) \end{pmatrix}. \tag{6.137}$$

Here the partition matrices, $\mathbf{R}$, $\tilde{\mathbf{R}}$, $\mathbf{V}$, $\tilde{\mathbf{V}}$, must obey the basic identities indicated by equations (6.136) and (6.137), which reveal that the following matrices are symmetric (Battin 1999): $\mathbf{R}\tilde{\mathbf{R}}^T$, $\mathbf{V}\tilde{\mathbf{V}}^T$, $\tilde{\mathbf{R}}^T\tilde{\mathbf{V}}$, $\mathbf{R}^T\mathbf{V}$, $\mathbf{V}\mathbf{R}^{-1}$, $\mathbf{R}^{-1}\tilde{\mathbf{R}}$, $\mathbf{V}^{-1}\tilde{\mathbf{V}}$, $\tilde{\mathbf{V}}\tilde{\mathbf{R}}^{-1}$, $\mathbf{R}\mathbf{V}^{-1}$, $\tilde{\mathbf{R}}^{-1}\mathbf{R}$, $\tilde{\mathbf{V}}^{-1}\mathbf{V}$, $\tilde{\mathbf{R}}\tilde{\mathbf{V}}^{-1}$. Furthermore, the following identity must be satisfied:

$$\mathbf{V}\tilde{\mathbf{R}}^T - \tilde{\mathbf{V}}\mathbf{R}^T = \mathbf{I}. \tag{6.138}$$

Of course, the partition matrices should satisfy the governing homogeneous equations and initial conditions,

$$\begin{aligned} \frac{\mathrm{d}\tilde{\mathbf{R}}}{\mathrm{d}t} &= \tilde{\mathbf{V}}, \\ \frac{\mathrm{d}\tilde{\mathbf{V}}}{\mathrm{d}t} &= \mathbf{G}_\mathrm{n}\tilde{\mathbf{R}}, \\ \tilde{\mathbf{R}}(t_0) &= \mathbf{I}, \\ \tilde{\mathbf{V}}(t_0) &= \mathbf{0}, \end{aligned} \tag{6.139}$$

and

$$\begin{aligned} \dot{\mathbf{R}} &= \mathbf{V}, \\ \dot{\mathbf{V}} &= \mathbf{G}_\mathrm{n}\mathbf{R}, \\ \mathbf{R}(t_0) &= \mathbf{0}, \\ \mathbf{V}(t_0) &= \mathbf{I}. \end{aligned} \tag{6.140}$$

However, solving even these uncoupled sets of linear equations in real time is no easy task. Battin (1999) presents an algorithm for the linearized state transition matrix based on Lagrange's coefficients and universal functions that require series (or continued fraction) evaluations. For practical guidance problems, it appears much simpler to solve the nonlinear 2PBVP by a Lambert algorithm, rather than determine a linearized solution in this manner.

6.5 Planar Orbital Regulation

When both radial and circumferential thrust inputs of independent magnitudes are applied in order to control the general orbital shape, we have a multi-input, linear, time-varying plant for which a multi-variable planar orbit controller can be designed. The orbital plant can be expressed in state-space form as follows:

$$\dot{\mathbf{x}}(t) = \mathbf{A}(t)\mathbf{x}(t) + \mathbf{B}(t)\mathbf{u}(t), \quad \mathbf{x}(0) = \mathbf{x}_0, \tag{6.141}$$

where

$$\mathbf{x}(t) = \left\{ \begin{array}{c} \delta r(t) \\ \delta \dot{r}(t) \\ \delta h(t) \end{array} \right\},$$

$$\mathbf{u}(t) = \left\{ \begin{array}{c} \frac{f_\mathrm{T}(t)}{m(t)} \sin\alpha(t) \\ \frac{f_\mathrm{T}(t)}{m(t)} \cos\alpha(t) \end{array} \right\},$$

$$\mathbf{A}(t) = \begin{pmatrix} 0 & 1 & 0 \\ -\left(\frac{3h_n^2}{r_n^4(t)} - \frac{2\mu}{r_n^3(t)}\right) & 0 & \frac{2h_n}{r_n^3(t)} \\ 0 & 0 & 0 \end{pmatrix},$$

and

$$\mathbf{B}(t) = \begin{pmatrix} 0 & 0 \\ 1 & 0 \\ 0 & r_n(t) \end{pmatrix}.$$

The stability of the linear planar orbital plant is analyzed by finding the eigenvalues of $\mathbf{A}$ as follows:

$$|\lambda\mathbf{I} - \mathbf{A}| = \lambda\left(\lambda^2 + \frac{3h_n^2}{r_n^4(t)} - \frac{2\mu}{r_n^3(t)}\right). \tag{6.142}$$

Clearly, for stability we require

$$\frac{3h_n^2}{r_n^4(t)} > \frac{2\mu}{r_n^3(t)} \tag{6.143}$$

or

$$h_n^2 > \frac{2}{3}\mu r_n(t).$$

Since there will be points on any given orbit where the stability condition will not be satisfied, we need a feedback controller for stability augmentation at such points. Even for those points where the stability condition is satisfied, the plant is not asymptotically stable, thereby requiring a feedback controller to provide active damping in order to bring the orbital deviations to zero in the steady state.

The general planar orbit control task with both radial and circumferential thrust (Figure 6.14) requires the solution of a linear time-varying regulator problem. Linear optimal control (Chapter 2) is best suited to this multi-input plant. Such a controller can be later modified to include an observer and to account for stochastic disturbances, $\boldsymbol{\nu}(t)$, $\mathbf{w}(t)$, by a robust multi-variable scheme such as the LQG or H_∞ method (Chapter 2). In order to derive the feedback control input vector, $\mathbf{u}(t)$, such that the state vector, $\mathbf{x}(t)$, asymptotically approaches zero in the steady state, we consider the infinite-time linear quadratic regulator (LQR) with cost function

$$\mathcal{J}_\infty(t) = \frac{1}{2}\int_t^\infty \left[\mathbf{x}^T(\tau)\mathbf{Q}(\tau)\mathbf{x}(\tau) + \mathbf{u}^T(\tau)\mathbf{R}(\tau)\mathbf{u}(\tau)\right] d\tau, \tag{6.144}$$

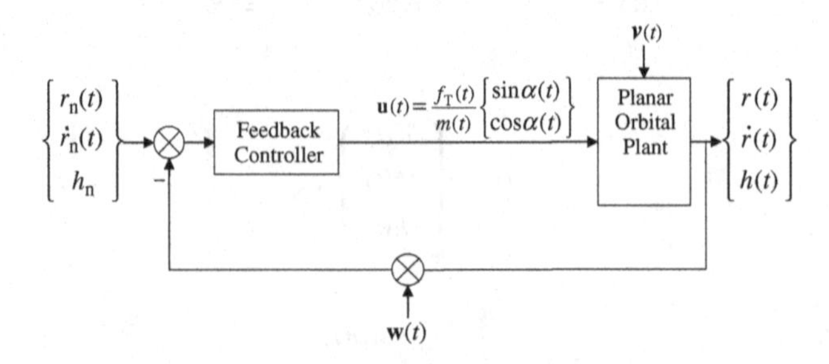

Figure 6.14 Multi-input, planar orbit control system

subject to the state constraint equation

$$\dot{\mathbf{x}}(t) = \mathbf{A}(t)\mathbf{x}(t) + \mathbf{B}(t)\mathbf{u}(t), \quad \mathbf{x}(0) = \mathbf{x}_0, \tag{6.145}$$

and the linear full-state feedback control law,

$$\mathbf{u}(t) = -\mathbf{K}_\infty(t)\mathbf{x}(t). \tag{6.146}$$

Since the plant is slowly time-varying, the stabilizing steady-state solution for the feedback gain matrix at each time instant, $\mathbf{K}_\infty(t)$, is given by

$$\mathbf{K}_\infty(t) = \mathbf{R}^{-1}(t)\mathbf{B}^T(t)\mathbf{P}(t), \tag{6.147}$$

where $\mathbf{P}_\infty(t)$ is the symmetric positive semi-definite quasi-steady solution of the following algebraic Riccati equation (Chapter 2):

$$\mathbf{P}_\infty(t)\mathbf{A}(t) + \mathbf{A}^T(t)\mathbf{P}_\infty(t) - \mathbf{P}_\infty(t)\mathbf{B}(t)\mathbf{R}^{-1}(t)\mathbf{B}^T(t)\mathbf{P}_\infty(t) + \mathbf{Q}(t) = \mathbf{0}. \tag{6.148}$$

The quasi-steady approximation is tantamount to finding a constant steady-state solution to equation (6.148) at each time instant, provided the variation of $\mathbf{A}, \mathbf{B}, \mathbf{Q}, \mathbf{R}$ with time is much slower than the settling time of all the transients of the closed-loop dynamics matrix, $\mathbf{A} - \mathbf{BK}$.

Since the plant is unconditionally controllable, that is, the controllability test matrix

$$\begin{bmatrix} 0 & 0 & 1 & \frac{2h_n}{r_n^3(t)} & 0 & 0 \\ 1 & 0 & 0 & 0 & \frac{2\mu}{r_n^3(t)} - \frac{3h_n^2}{r_n^4(t)} & \frac{4\mu h_n}{r_n^6(t)} - \frac{6h_n^3}{r_n^7(t)} \\ 0 & r_n(t) & 0 & 0 & 0 & 0 \end{bmatrix}$$

is of rank 3 (the order of the plant) at all times, we can satisfy the sufficient conditions for the existence of a unique positive semi-definite solution to the algebraic Riccati equation by selecting suitable state and control cost coefficient matrices, $\mathbf{Q}(t), \mathbf{R}(t)$. A possible choice is the following constant pair:

$$\mathbf{Q} = \begin{pmatrix} Q_1 & 0 & 0 \\ 0 & Q_2 & 0 \\ 0 & 0 & Q_3 \end{pmatrix}, \quad \mathbf{R} = \begin{pmatrix} R_1 & 0 \\ 0 & R_2 \end{pmatrix},$$

where $Q_i \geq 0, R_i > 0$ are selected based upon the relative importance of the state and control variables. For the orbit control plant, generally a much lower cost is specified for the angular momentum than either the radial displacement or velocity, that is, $Q_3 \ll Q_1, Q_3 \ll Q_2$.

In the interests of simplicity, we choose to avoid solving the algebraic Riccati equation (6.148) in real time, as such a controller is rarely implementable. Instead, we select a constant feedback gain matrix based on a specific time instant, and apply it throughout the control interval. A linear regulator based upon the initial plant dynamics, $\mathbf{A}(0), \mathbf{B}(0)$, can be designed to be stabilizing if the control interval is sufficiently small compared to the time scale of a slowly varying plant. In the present application, a control interval of a few hundred seconds is quite small compared to the orbital period of a satellite. Therefore, we modify the control law by taking $t = 0$ for the constant gain matrix,

$$\mathbf{K}_\infty(0) = \mathbf{R}^{-1}(0)\mathbf{B}^T(0)\mathbf{P}_\infty(0), \tag{6.149}$$

where $\mathbf{P}_\infty(0)$ is the constant solution to

$$\mathbf{P}_\infty(0)\mathbf{A}(0) + \mathbf{A}^T(0)\mathbf{P}_\infty(0) - \mathbf{P}_\infty(0)\mathbf{B}(0)\mathbf{R}^{-1}(0)\mathbf{B}^T(0)\mathbf{P}_\infty(0) + \mathbf{Q}(0) = \mathbf{0}. \tag{6.150}$$

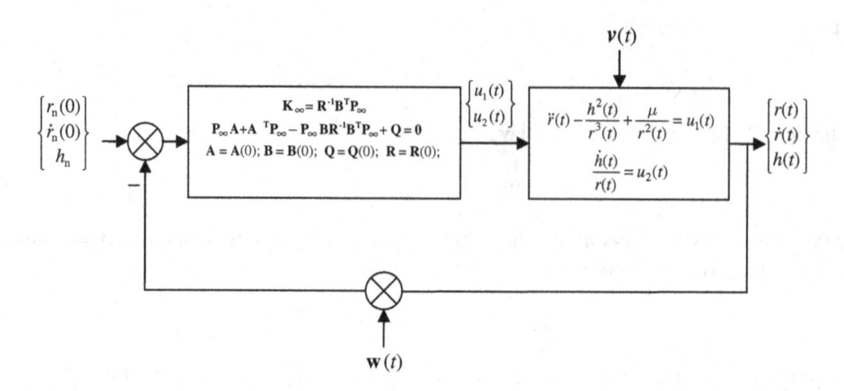

Figure 6.15 Constant gain linear quadratic regulator for time-varying planar orbit control system

Referring to the orbital plant dynamics, the constant gain approximation consists of neglecting the variation in the nominal orbital radius with time, $r_n(t) \approx r_n(0)$. The schematic diagram of the constant gain LQR regulator applied to orbit control is shown in Figure 6.15.

Example 6.6 A spacecraft has the following initial condition relative to the Earth:

$$r(0) = 8500 \text{ km}, \quad \dot{r}(0) = 9 \text{ km/s}, \quad h(0) = 80000 \text{ km}^2/\text{s}.$$

It is desired to change the orbit to that given by the following orbital elements:

$$a = 7000 \text{ km}, \quad e = 0.5, \quad \tau = -2000 \text{ s}.$$

Using both radial acceleration, $u_1 = (f_T/m)\sin\alpha$, and the circumferential acceleration, $u_2 = (f_T/m)\cos\alpha$, as control inputs, design an orbital regulator for this task.

The orbital elements at $t = 0$ of the perturbed orbit can be calculated from the initial condition as follows:

$$v(0) = \sqrt{\dot{r}^2 + r^2\dot{\theta}^2} = \sqrt{\dot{r}^2(0) + \frac{h^2(0)}{r^2(0)}} = 13.0223 \text{ km/s},$$

$$a(0) = \frac{\mu}{2\mu/r(0) - v^2(0)} = -5259.06715 \text{ km},$$

$$e(0) = \sqrt{1 - \frac{h^2(0)}{\mu a(0)}} = 2.01321814,$$

$$\cosh H(0) = \frac{1}{e}\left(1 - \frac{r(0)}{a(0)}\right) = 1.29953937,$$

$$\sin\phi(0) = \frac{\dot{r}(0)}{v(0)} = 0.69112008,$$

$$\sinh H(0) = \frac{(e\cosh H(0) - 1)\, h(0) v(0) \sin\phi(0)}{\mu e\sqrt{e^2 - 1}} = 0.82994130,$$

Table 6.9

```
% Program for solving Kepler's equation by Newton--Raphson iteration.
function E=kepler(e,M);
eps=1e-10; % Tolerance
% Initial guess:
B=cos(e)-(pi/2-e)*sin(e);
E=M+e*sin(M)/(B+M*sin(e));
fE=E-e*sin(E)-M;
% Iteration for 'E' follows:
while abs(fE)>eps
fE=E-e*sin(E)-M;
fpE=1-e*cos(E);
dE=-fE/fpE;
E=E+dE;
end
```

$$H(0) = 0.75587813 \text{ rad},$$

$$n(0) = \sqrt{-\frac{\mu}{a^3}} = 0.00165541 \text{ rad/s},$$

$$\tau(0) = \frac{H(0) - e \sinh H(0)}{n} = -552.71763 \text{ s}.$$

Thus, the orbital elements of the initial orbit are

$$a(0) = -5259.06715 \text{ km}, \quad e(0) = 2.01321814, \quad \tau = -552.71763 \text{ s}.$$

Hence, a significant change is required from a hyperbolic orbit to an elliptical orbit, along with a rotation of the line of apsides (i.e., a change in the time of perigee).

We begin the design process by choosing the cost coefficients matrices as follows:

$$\mathbf{Q} = \begin{pmatrix} 10^{-8} & 0 & 0 \\ 0 & 10^{-8} & 0 \\ 0 & 0 & 10^{-12} \end{pmatrix}, \quad \mathbf{R} = \mathbf{I}.$$

At $t = 0$, the nominal plant is determined and the constant gain LQR regulator is designed by the following MATLAB statements, including the code *kepler.m* for iteratively solving Kepler's equation by the Newton–Raphson method (Table 6.9):

```
>> mu=398600.4; %Gravitational constant of planet
>> a=7000; % Semi-major axis of nominal orbit
>> e=0.5; % Eccentricity of nominal orbit (Elliptical)
>> t0=-2000; % Time of periapsis of nominal orbit
>> p=a*(1-e^2); % Parameter of nominal orbit
>> h=sqrt(mu*p); % Angular momentum of nominal orbit
>>      n=sqrt(mu/a^3);
>>      M=-n*t0;
>>      E=kepler(e,M);
>>      r=a*(1-e*cos(E));
>>      rdot=e*sqrt(mu*a)*sin(E)/r;
```

```
>> A=[0  1 0; -3*h^2/r^4+2*mu/r^3  0  2*h/r^3; 0 0 0];
>> B=[0 0;1 0;0 r];
>> [k,S,v]=lqr(A,B,[1e-8 0 0;0 1e-8 0;0 0 1e-12],eye(2));
```

which produces the following output:

```
n =     1.078007556348836e-003

M =     2.156015112697673e+000

E =     2.467938235705962e+000

r =     9.735414831667225e+003

rdot =     1.692431086032995e+000

A =

                        0    1.000000000000000e+000                        0
   1.651047895043663e-007                         0   9.915511603292160e-008
                        0                         0                        0

B =

                        0                         0
   1.000000000000000e+000                         0
                        0    9.735414831667225e+003

k =

   1.001249565511754e-004    1.413452700199927e-002   7.079776644590912e-008
   2.838185231637881e-006    6.892456255064159e-004   1.004503657414909e-006

S =

   1.414841430594429e-006    1.001249565511754e-004   2.915320282404270e-010
   1.001249565511754e-004    1.413452700199927e-002   7.079776644590912e-008
   2.915320282404270e-010    7.079776644590912e-008   1.031803651702106e-010

v =

   -7.076637928420101e-003 +7.081901976920645e-003i
   -7.076637928420101e-003 -7.081901976920645e-003i
   -9.760510949989759e-003
```

Therefore, the constant regulator gain matrix is

$$\mathbf{Q} = \begin{pmatrix} 1.0012 \times 10^{-4} & 1.4135 \times 10^{-2} & 7.0798 \times 10^{-8} \\ 2.8382 \times 10^{-6} & 6.8925 \times 10^{-4} & 1.0045 \times 10^{-6} \end{pmatrix}.$$

Next, a code called *orbitplantfixk.m* (Table 6.10) is written in order to generate the equations of motion for the perturbed trajectory with the constant gain regulator, to be integrated using the intrinsic fourth-order, fifth-stage, Runge–Kutta solver of MATLAB, *ode45.m*. The code *orbitplantfixk.m* requires

Table 6.10

```
% Program 'orbitplantfixk.m' for generating the closed-loop equations
% with a constant gain, planar orbit regulator for an
% elliptical nominal orbit.
% (c) 2009 Ashish Tewari
% Requires the codes 'kepler.m' and 'nominalorbit.m'
function xdot=orbitplantfixk(t,x)
global f9;
mu=398600.4; %Gravitational constant of planet (km^3/s^2)
[r, rdot, h]=nominalorbit(t); % Nominal state variables
% Following is the constant regulator gain matrix computed
% at t=0 for the nominal plant:
k =[1.001249565511754e-4 1.413452700199927e-2 7.079776644590912e-8
   2.838185231637881e-6 6.892456255064159e-4 1.004503657414909e-6];
u=-k*[x(1,1)-r; x(2,1)-rdot; x(3,1)-h]; %Control input (km/s^2)
U=norm(u); % magnitude of control input (f_\mathrmT/m) (km/s^2)
alpha=180*atan(u(1,1)/u(2,1))/pi; %Direction of control input (deg.)
fprintf(f9, '\t%1.5e\t%1.5e\t%1.5e\n', t, U, alpha );
% Equations of motion for the perturbed trajectory:
xdot(1,1)=x(2,1); % Radial rate (km/s)
xdot(2,1)=u(1,1)-mu/x(1,1)^2+x(3,1)^2/x(1,1)^3;%Radial accln(km/s^2)
xdot(3,1)=x(1,1)*u(2,1); % Rate of angular momentum (km^2/s^2)
```

kepler.m (Table 6.9) in order to create nominal state variables at each time instant, necessary for the linear feedback control law. The time step integration is carried out and the results plotted using the calling program given in Table 6.11, including the comparison with the nominal state variables computed by the code *nominalorbit.m* (Table 6.12).

The resulting plots of the state variables are shown in Figure 6.16, while the radial and angular velocity components are compared with the nominal orbit in Figure 6.17. The control input magnitude and thrust direction are plotted in Figures 6.18 and 6.19, respectively. Note that all transients settle down to the required nominal values in about 600 seconds, which is only about a ninth of the nominal orbital period. For this rather large initial velocity perturbation (Figure 6.17), the maximum required control input magnitude is $u = 55$ m/s^2 and the maximum deflection angle is $\alpha = 85°$.

6.6 Optimal Non-planar Orbital Regulation

We finally consider tracking a nominal orbit with non-planar (three-dimensional) acceleration inputs. Taking advantage of the symplectic character of the state transition matrix, we can apply a time-varying linear quadratic regulator by solving the algebraic Riccati equation for an infinite horizon at different time instants. Extending the formulation of the previous section, we have

$$\mathbf{u}(t) = -\mathbf{R}^{-1}(t)\mathbf{B}^T(t)\mathbf{P}_\infty(t)\mathbf{x}(t), \tag{6.151}$$

where $\mathbf{P}_\infty(t)$ is the symmetric, positive semi-definite, quasi-steady solution of the algebraic Riccati equation (6.148) and the time-varying coefficient matrices, $\mathbf{A}$, $\mathbf{B}$, are the same as those in equation (6.134). It can easily be shown that the pair $\mathbf{A}$, $\mathbf{B}$ is unconditionally controllable at all times, thus satisfying the sufficient conditions for the existence of a unique positive semi-definite algebraic Riccati solution (Chapter 2). Hence, we can solve the infinite-horizon LQR problem at various instants to determine the regulator gains. A suitable Kalman filter (Chapter 2) can be designed for a practical LQG implementation, with loop-transfer recovery (LTR) at either the plant's input or its output.

Table 6.11

```
% Program for simulating perturbed trajectory with a constant gain,
% planar orbit regulator for an elliptical nominal orbit,
% using Runge-Kutta 4(5) scheme, 'ode45.m'.
% (c) 2009 Ashish Tewari
global f9;f9 = fopen('u.txt', 'w');
OPTIONS = odeset('RelTol', 1e-8);
[t,x]=ode45(@orbitplantfixk,[0 5829],[8500,9,80000]','OPTIONS');
n=size(t,1);
rn=[];rndot=[];hn=[];
for i=1:n
[rn(i,1), rndot(i,1), hn(i,1)]=nominalorbit(t(i,1));
end
N=round(n/6);
subplot(311),plot(t(1:N,1),x(1:N,1),t(1:N,1),rn(1:N,1),':'),
ylabel('r (km)'),hold on,
subplot(312),plot(t(1:N,1),x(1:N,2),t(1:N,1),rndot(1:N,1),':'),
ylabel('dr/dt (km/s)'),hold on,
subplot(313),plot(t(1:N,1),x(1:N,3),t(1:N,1),hn(1:N,1),':'),
xlabel('Time (s)'),ylabel('h (km^2/s)')
r=x(:,1);rdot=x(:,2);h=x(:,3);
v=sqrt(rdot.^2+(h./r).^2);
vn=sqrt(rndot.^2+(hn./rn).^2);
tanphi=rdot.*r./h;
tanphin=rndot.*rn./hn;
phi=atan(tanphi);
phin=atan(tanphin);
figure
plot(v.*cos(phi),v.*sin(phi),vn.*cos(phin),vn.*sin(phin),':'),
xlabel('v cos\phi'),ylabel('v sin\phi'),;
fclose('all');
load u.txt -ascii
figure
n=round(size(u,1)/6);
plot(u(1:10:n,1),1000*u(1:10:n,2)),xlabel('Time (s)'),
ylabel('f_\mathrmT/m (m/s^2)')
figure
plot(u(1:10:n,1),u(1:10:n,3)),xlabel('Time (s)'),
ylabel('\alpha (deg.)')
```

Example 6.7 A spacecraft is desired to be in an elliptical Earth orbit with the following classical orbital elements at $t = 0$:

$$a = 14000 \text{ km}, \quad e = 0.5, \quad \tau = -12000 \text{ s},$$

$$i = 40^\circ, \quad \Omega = 240^\circ, \quad \omega = 100^\circ.$$

However, the spacecraft is initially displaced from the nominal orbit by the following distances (in kilometers) measured in the geocentric, celestial frame:

$$\mathbf{r}(0) - \mathbf{r}_n(0) = -100\mathbf{i} + 50\mathbf{j} + 200\mathbf{k}.$$

Table 6.12

```
% Program 'nominalorbit.m' for calculating radius,
% radial speed, and angular momentum of spacecraft in a
% nominal elliptical orbit.
% Requires 'kepler.m'
% (c) 2009 Ashish Tewari
function [r,rdot,h]=nominalorbit(t);
mu=398600.4; %Gravitational constant of planet (km^3/s^2)
a=7000; % Semi-major axis of nominal orbit (km)
e=0.5; % Eccentricity of nominal orbit (Elliptical)
t0=-2000; % Time of periapsis of nominal orbit (s)
p=a*(1-e^2); % Parameter of nominal orbit (km)
h=sqrt(mu*p); % Angular momentum of nominal orbit (km^2/s)
n=sqrt(mu/a^3); % Mean motion of nominal orbit (rad/s)
M=n*(t-t0); % Mean anomaly in the nominal orbit (rad.)
E=kepler(e,M); % Eccentric anomaly in the nominal orbit (rad.)
r=a*(1-e*cos(E)); % Nominal radius (km)
rdot=e*sqrt(mu*a)*sin(E)/r; % Nominal radial speed (km/s)
```

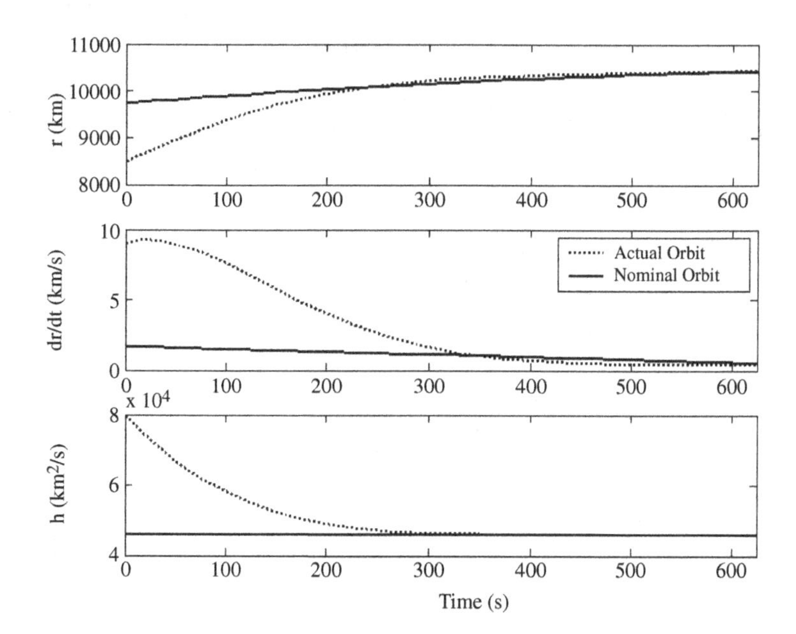

Figure 6.16 Simulated closed-loop response of radius, radial speed, and angular momentum with constant feedback gain matrix

It is desired to bring the spacecraft back to the nominal orbit in about 10 minutes with the acceleration inputs, $\mathbf{u} = (u_x, u_y, u_z)^T$, not exceeding ± 20 m/s^2 in any direction.

We carry out a full-state feedback LQR design by dividing the total control interval into blocks of 100 s, in each of which a Riccati solution with constant gains computed with the cost parameters

$$\mathbf{Q} = 10^{-8}\mathbf{I}, \quad \mathbf{R} = \mathbf{I}$$

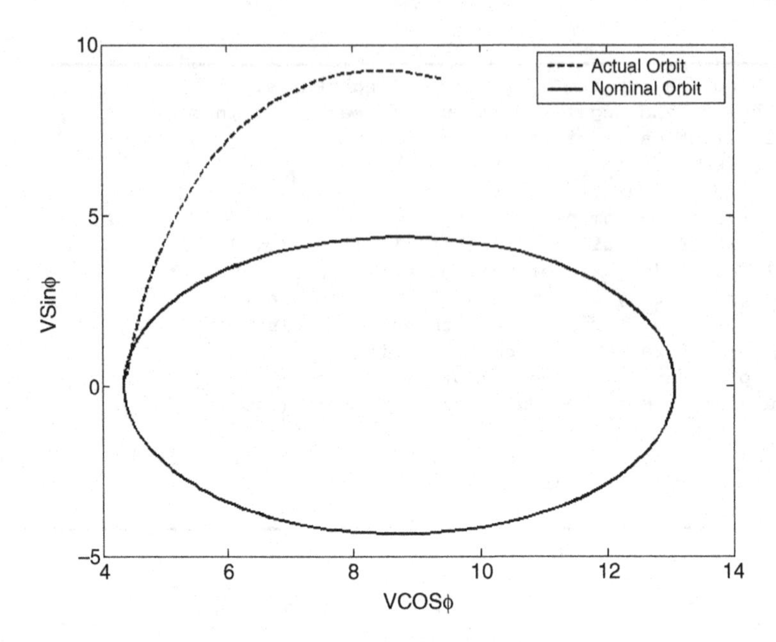

Figure 6.17 Simulated closed-loop response of velocity vector with constant feedback gain matrix, compared with the nominal values

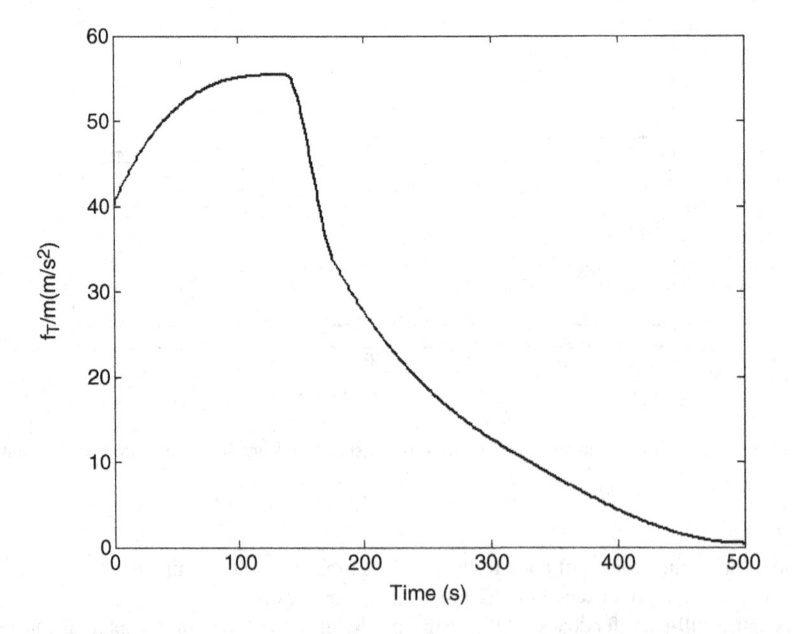

Figure 6.18 Feedback control input magnitude (thrust acceleration) of simulated closed-loop system with constant feedback gain matrix

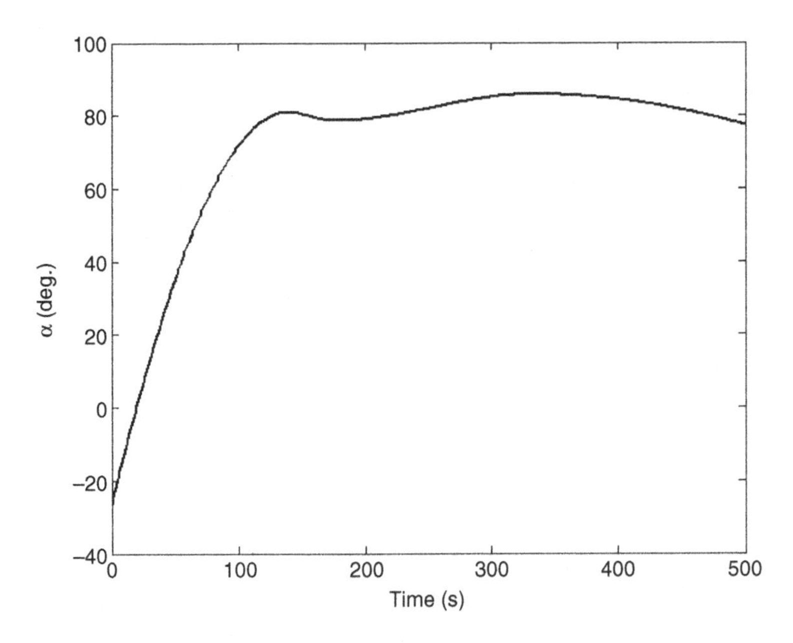

Figure 6.19 Feedback control deflection (thrust direction) of simulated closed-loop system with constant feedback gain matrix

Table 6.13

```
% Program 'gravitygradient.m' for calculating
% the gravity-gradient matrix in a spherical
% gravity field.
% (c) 2009 Ashish Tewari
function G=gravitygradient(mu,r)
x=r(1,1);
y=r(2,1);
z=r(3,1);
rad=norm(r);
G=mu*[3*x^2-rad^2 3*x*y 3*x*z;
      3*x*y 3*y^2-rad^2 3*y*z;
      3*x*z 3*y*z 3*z^2-rad^2]/rad^5;
```

is employed. However, since the plant is a time-varying one, the gains are allowed to change from one block to the next (in contrast with Example 6.6, where the gains were constant for the entire control interval). The necessary closed-loop computations are carried out by the following MATLAB Control System Toolbox statements with the gravity-gradient matrix, $\mathbf{G}_n$, computed by the code *gravitygradient.m* that is tabulated in Table 6.13, and the nominal state updated with the help of perifocal position and velocity computed by the trajectory program *trajellip.m* (Table 6.14):

```
dtr=pi/180;
a=14000;
e=0.5;
i=40*dtr;
Om=290*dtr;
w=100*dtr;
tau=-12000;
mu=398600.4;
n=sqrt(mu/a^3);
C=rotation(i,Om,w);
E0=kepler(e,-n*tau)
r0=a*(1-e*cos(E0))
v0=sqrt(2*mu/r0-mu/a)
R0=a*[cos(E0)-e sqrt(1-e^2)*sin(E0) 0]'
V0=sqrt(mu*a)*[-sin(E0);sqrt(1-e^2)*cos(E0);0]/r0;
T=0;
x0=[-100 50 200 0 0 0]';
for k=1:10;
    [R,V]=trajellip(mu,T,R0,V0,T+100);
    R0=R;V0=V;
    Rn=C*R;
    G=gravitygradient(mu,R);
    A=[zeros(3,3) eye(3);G zeros(3,3)];B=[zeros(3,3);eye(3)];
    [k,s,E]=lqr(A,B,1e-8*eye(6),eye(3));
    sys=ss(A-B*k,B,[eye(3) zeros(3,3)],zeros(3,3));
    [y,t,x]=initial(sys,x0,100);
    u=-x*k';
    plot(t+T,y),hold on
    T=T+100;
    x0=x(size(t,1),:);
end
```

Table 6.14

```
% Program for updating the perifocal position
% and velocity in an elliptical orbit
% (c) 2009 Ashish Tewari
function [R,V]=trajellip(mu,t0,R0,V0,t)
eps=1e-10;
r0=norm(R0);
v0=norm(V0);
alpha=dot(R0,V0);
H=cross(R0,V0);
h=norm(H);
p=h^2/mu;
ecv=cross(V0,H)/mu-R0/r0;
e=norm(ecv);
ecth0=p/r0-1;
esth0=norm(cross(ecv,R0))/r0;
if abs(ecth0)>=eps;
th0=atan(esth0/ecth0);
if ecth0<0
    if esth0>=0;
    th0=th0+pi;
    end
```

```
elseif esth0<0
    th0=th0+2*pi;
end
elseif esth0>=0
  th0=pi/2;
else
    th0=3*pi/2;
end
    ainv=-(v0^2)/mu+2/r0;
    a=1/ainv;
    n=sqrt(mu/a^3);
    E0=2*atan(sqrt((1-e)/(1+e))*tan(0.5*th0));
    tau=t0+(-E0+e*sin(E0))/n;
    M=n*(t-tau);
    E=kepler(e,M);
    r=a*(1-e*cos(E));
end
f=1+a*(cos(E-E0)-1)/r0;
g=a*alpha*(1-cos(E-E0))/mu+r0*sqrt(a/mu)*sin(E-E0);
fd=-sqrt(mu*a)*(sin(E-E0))/(r*r0);
gd=1+a*(cos(E-E0)-1)/r;
R=f*R0+g*V0;
V=fd*R0+gd*V0;
```

The resulting plots of the displacement errors and the control inputs are shown in Figures 6.20 and 6.21, respectively. The excellent performance of LQR control is evident with the errors dissipating to nearly zero in about 600 s, without exceeding the desired control input magnitudes.

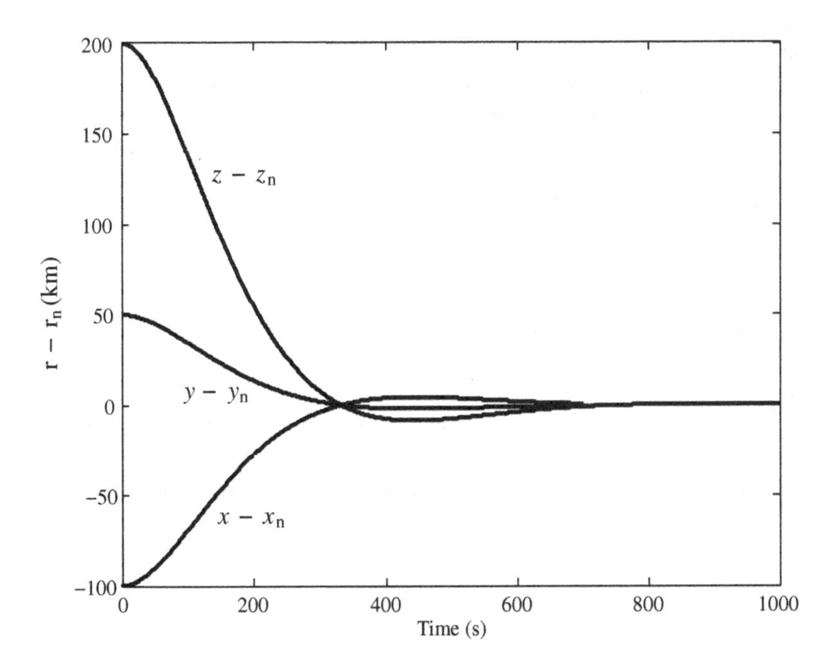

Figure 6.20 Closed-loop displacement from nominal orbit

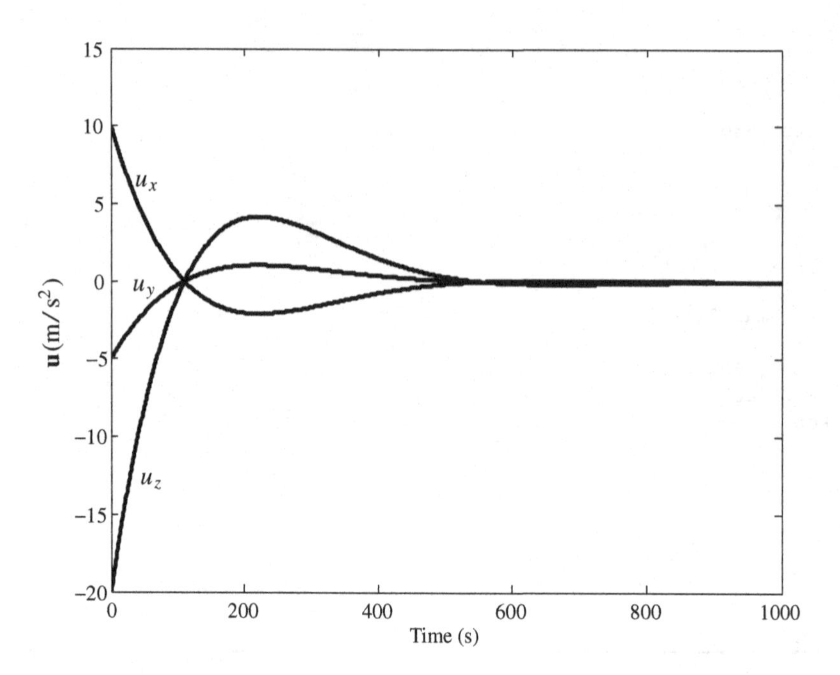

Figure 6.21 Closed-loop acceleration inputs with linear optimal state feedback

6.7 Summary

Spacecraft guidance and navigation consist of controlling both the shape and the plane of its orbit around a central body. The terminal orbit controller is based upon a two-point boundary value problem that is solved by either optimal control or Lambert's algorithm. The orbit tracking plant can be linearized with respect to a nominal orbit, about which the vehicle experiences small disturbances caused by orbital perturbations. Simultaneous control of the radius as well as the orbital angular momentum (orbital plane) requires a multi-variable controller design that can be carried out with either the fixed or time-varying feedback gains by the infinite-horizon LQR (or LQG/LTR) method.

Exercises

(1) With the help of Lagrange's equation for elliptical transfer between two given positions, show that the minimum energy orbit leads to the maximum time of flight.

(2) Since the transfer time is invariant with eccentricity according to Lambert's theorem, we can vary e for a given time of flight. Show geometrically that it is possible to change the eccentricity for an elliptical transfer orbit while keeping the values of a, c, $r_1 + r_2$ (thus $t_2 - t_1$) constant.

(3) Repeat Example 6.1 for the case of minimum eccentricity (rather than minimum energy) transfer orbit. (*Ans.* $t_2 - t_1 = 2104.948$ *s*.)

(4) For Lambert's problem involving an elliptical transfer orbit ($z > 0$), show that Stumpff functions converge to

$$C(z) = \frac{1 - \cos\sqrt{z}}{2},$$
$$S(z) = \frac{\sqrt{z} - \sin\sqrt{z}}{z^{3/2}}. \tag{6.152}$$

(5) Show that the Stumpff functions for a hyperbolic orbital transfer ($z < 0$) are given by

$$C(z) = \frac{\cosh\sqrt{-z} - 1}{-z},$$
$$S(z) = \frac{-\sqrt{-z} + \sinh\sqrt{-z}}{(-z)^{3/2}}. \tag{6.153}$$

(6) For a spherical gravitational field, show that the gravity-gradient matrix,

$$\mathbf{G}_\mathrm{n} = \left(\frac{\partial \mathbf{a}_g}{\partial \mathbf{r}}\right)_{\mathbf{r}=\mathbf{r}_\mathrm{n}}.$$

in terms of planet-centered Cartesian coordinates, $\mathbf{r} = x\mathbf{i} + y\mathbf{j} + z\mathbf{k}$, is given by

$$\mathbf{G}_\mathrm{n} = \frac{\mu}{r_n^5}\begin{pmatrix} 3x_n^2 - r_n^2 & 3x_n y_n & 3x_n z_n \\ 3x_n y_n & 3y_n^2 - r_n^2 & 3y_n z_n \\ 3x_n z_n & 3y_n z_n & 3z_n^2 - r_n^2 \end{pmatrix}.$$

(7) Consider the planar flight of a low-thrust orbiter that produces a thrust acceleration, $\mathbf{a}_\mathrm{T}$, very small in magnitude compared to the spherical gravitational acceleration, $\mathbf{a}_g$. For such a spacecraft:

(a) Express the equations of motion in polar coordinates, (r, θ), relative to an inertial reference frame at planetary center.

(b) Solve the equations of motion when a constant magnitude thrust acceleration is applied tangentially (i.e., in the direction of increasing θ) for a flight beginning from a circular orbit of radius r_0.

(c) Repeat part (b) for the case of a radial (rather than tangential) thrust acceleration of constant magnitude.

(d) Now, let a thrust acceleration of variable magnitude, $a_\mathrm{T}(t)$, be applied at a variable angle, $\alpha(t)$, measured from the instantaneous flight direction. Formulate a 2PBVP (Chapter 2) to minimize the total fuel expenditure expressed by the performance index

$$\mathcal{J} = \int_0^{t_\mathrm{f}} Q\, a_\mathrm{T}^2(t)\mathrm{d}t,$$

where $Q > 0$, by suitably varying the control input, $\mathbf{u}(t) = \mathbf{a}_\mathrm{T}(t)$, for a flight from an initial radius, r_0, to a final radius, r_f.

(e) Write a MATLAB code to solve the optimal control problem formulated in part (d) with $0 \leq a_\mathrm{T} \leq 0.01$ m/s^2 for a flight around Earth, from an initial radius of 7000 km to a final radius of 43000 km. Plot the extremal trajectory, $r(t)$, $\theta(t)$, and the optimal control variable history, $a_\mathrm{T}(t)$, $\alpha(t)$.

(8) Show that the translational equations of planar motion of a vectored thrust spacecraft (Figure 6.22) can be written as

$$\dot{r} = v\sin\phi, \tag{6.154}$$

$$\dot{v} = \frac{f_\mathrm{T}}{m}\cos\beta - \frac{\mu}{r^2}\sin\phi, \tag{6.155}$$

and

$$v\dot{\phi} = \frac{f_\mathrm{T}}{m}\sin\beta + \left(\frac{v^2}{r} - \frac{\mu}{r^2}\right)\cos\phi, \tag{6.156}$$

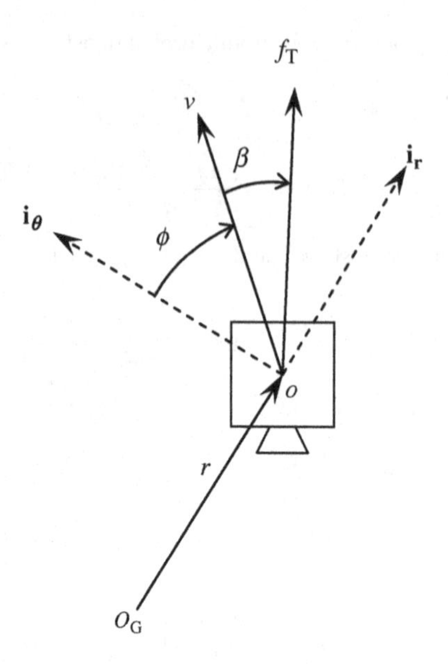

Figure 6.22 Planar motion of a spacecraft with vectored rocket thrust

where ϕ is the flight-path angle (taken to be positive above the horizon) and β is the angle made by the thrust vector with the velocity vector, as shown in Figure 6.22.

(9) For the spacecraft of Exercise 8, design a control system to decrease the radius and flight-path angle of a satellite from an initial circular orbit of radius r_0 with a variable thrust acceleration, $u(t) = f_T(t)/m(t)$. The controller must maintain the initial speed during the control interval, $0 \leq t \leq t_f$, and achieve a given final flight-path angle, $\phi(t_f) = \phi_f < 0$ (where $|\phi_0|$ is small), simultaneously with $r(t_f) = r_f = 0.95r_0$.

(10) Write a program to integrate the equations of motion of Exercise 8 with the MATLAB's intrinsic Runge–Kutta 4(5) solver, *ode45.m*, for a given set of initial conditions, $r(0) = r_0$, $v(0) = v_0$, $\phi(0) = \phi_0$, and a general thrust acceleration, $(f_T/m, \beta)$. Using the new code, generate a nominal trajectory for de-orbiting a satellite from a circular Earth orbit of radius $r_0 = 6600$ km with a constant tangential thrust acceleration, $(u = 0.01\mu/r_0^2, \beta = 180°)$, acting for $t_f = 100$ s. What are the radius, speed, and flight-path angle at $t = t_f$?

(11) Consider small deviations, $\delta r(t)$, $\delta v(t)$, and $\delta\phi(t)$, from the nominal trajectory derived in Exercise 10. Using first-order Taylor series expansions about the respective nominal values, derive the linear state equations for perturbed motion around the nominal trajectory, with tangential thrust acceleration, $\delta u(t)$, and thrust direction, $\delta\beta(t)$, as control inputs over and above their nominal values. Analyze:
(a) the stability of the plant at $t = 0$ and $t = t_f$;
(b) the controllability of the plant with both the inputs;
(c) the controllability of the plant with $\delta u(t)$ as the only input;
(d) the controllability of the plant with $\delta\beta(t)$ as the only input;
(e) the observability of the plant with $\delta r(t)$ as the only output;
(f) the observability of the plant with $\delta v(t)$ as the only output;
(g) the observability of the plant with $\delta\phi(t)$ as the only output.

(12) With the linear orbital plant derived in Exercise 11, and with either one (or both) of the control inputs, design a full-state feedback, regulated control system to maintain the spacecraft on the

nominal trajectory obtained in Exercise 10 despite small initial deviations, $\delta r(0) = \delta r_0, \delta v(0) = \delta v_0$, and $\delta\phi(0) = \delta\phi_0$.

(13) With the regulated control system designed in Exercise 12, modify the code developed in Exercise 10 to carry out a Runge–Kutta simulation for the closed-loop nonlinear state equations with $\delta r_0 = 20$ km and $\delta\phi_0 = -2°$. Plot the simulation results, $\delta r(t)$, $\delta v(t)$, and $\delta\phi(t)$ for $t_f = 100$ s. and compute and plot the required feedback control input(s).

(14) Repeat Exercises 12 and 13 by including a linear observer based upon the measurement of only one state variable. The observer should be suitably designed so that the necessary control input magnitude(s) do not increase by more than 10%, compared to the case of full-state feedback (Exercise 12).

(15) Using the definition of a symplectic matrix given in Chapter 2, show that the state transition matrix, $\mathbf{\Phi}(t, t_0)$, corresponding to the homogeneous part of the state equation (6.134), is symplectic.

(16) Using the LQG/LTR method (Chapter 2), add a fixed-gain Kalman filter to the controller designed in Example 6.6 with measurement and feedback of the radial position, $r(t)$, and speed, $v(t)$. Try to recover the return ratio at the plant's output in a frequency range of $10^{-3} \leq \omega \leq 0.02$ rad/s. What (if any) degradation of closed-loop performance is observed compared with the state feedback results of Example 6.6?

(17) Using the LQG/LTR (Chapter 2), add a suitable Kalman filter with time-varying gains to the controller designed in Example 6.7, based upon the measurement and feedback of the displacement vector, $\mathbf{r}(t) - \mathbf{r}_n(t)$. Try to recover the return ratio at the plant's output in a bandwidth of 0.01 rad/s. What (if any) degradation of closed-loop performance is observed compared with the state feedback results of Example 6.7?

References

Battin, R.H. (1964) *Astronautical Guidance*. McGraw-Hill, New York.

Battin, R.H. (1999) *An Introduction to the Mathematics and Methods of Astrodynamics*. American Institute of Aeronautics and Astronautics, Reston, VA.

Kreyszig, E. (1998) *Advanced Engineering Mathematics*. John Wiley & Sons, Inc., New York.

Stiefel, E.L. and Scheifele, G. (1971) *Linear and Regular Celestial Mechanics*. Springer-Verlag, New York.

Tewari, A. (2006) *Atmospheric and Space Flight Dynamics*. Birkhäuser, Boston.

7

Optimal Spacecraft Attitude Control

7.1 Introduction

A spacecraft's attitude is generally described by the orientation of principal body axes fixed to a rigid body relative to a reference frame that may either be stationary (inertial) or rotating at a small, constant rate (orbital). In order to carry out its mission successfully, a spacecraft must maintain the desired orientation, thus attitude stability and control are essential for all spacecraft. Since the reference orientation is generally an equilibrium state, the attitude dynamics plant can be linearized about the equilibrium and the orientation given by a small deviation from equilibrium through an appropriate Euler angle representation – either 3-1-3 or 3-2-1 (Tewari 2006). The spacecraft is always under the influence of both environmental disturbances and control torques needed to stabilize the attitude in the presence of the disturbances. If an attitude control system is absent, the spacecraft may depart from equilibrium under the influence of environmental torques and become useless in its mission. In the case of large, low-orbiting spacecraft (such as the International Space Station) it is also necessary to maintain an orientation that produces the smallest aerodynamic drag, which can otherwise degrade the orbit or require additional fuel expenditure in orbit-raising maneuvers.

While most of a spacecraft's mission is spent in tracking a desired nominal orientation with small deviations, sometimes large and rapid rotational maneuvers may also be required for various reasons. Examples of these include docking and undocking of two spacecraft, initial deployment of satellites from launch vehicles, and change of equilibrium rotational states during a mission redeployment. Such maneuvers are essentially governed by nonlinear differential equations, and thus require a solution of optimal, two-point boundary value problems by the terminal controller.

7.2 Terminal Control of Spacecraft Attitude

We shall begin by studying both single-axis and multi-axis, nonlinear, optimal, terminal attitude control of spacecraft modeled as a rigid body. Since the rotations involved may be large and rapid, we will not consider the small gravity-gradient, geomagnetic, and solar radiation torques during the terminal control interval. These are, however, important in the tracking control of satellite attitude dynamics where small displacements from an equilibrium state are to be minimized, which will be discussed later. Furthermore, the optimal torques required for terminal control can be generated in a variety of ways (such as rocket

Advanced Control of Aircraft, Spacecraft and Rockets, Ashish Tewari.

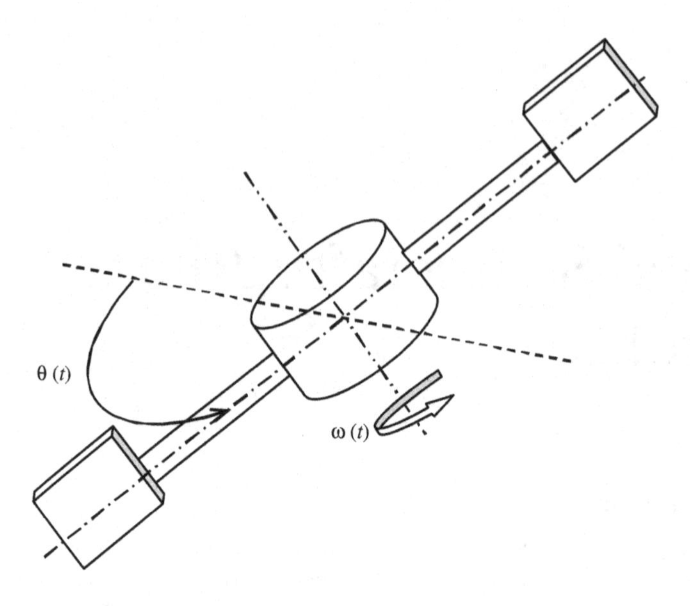

Figure 7.1 Rigid spacecraft rotating about a single axis

thrusters, reaction wheels, control moment gyroscope, and variable-speed control moment gyroscope) and thus a specific actuator model is not considered.

7.2.1 Optimal Single-Axis Rotation of Spacecraft

Consider the rotational maneuvers of a rigid spacecraft of moment of inertia, J, about a given axis (Figure 7.1). It is desired to carry out a rotation by angle θ_f, and to spin up the spacecraft (initially at rest) to an angular speed, ω_f, both at a given time t_f, by applying a control torque $u(t)$. The governing equations of motion are linear, and can be written as

$$\begin{aligned} J\dot{\omega} &= u, \\ \dot{\theta} &= \omega, \end{aligned} \tag{7.1}$$

which can be expressed in the following time-invariant state-space form with state vector $\mathbf{x}(t) = [\theta(t), \omega(t)]^T$:

$$\dot{\mathbf{x}} = \mathbf{f}(\mathbf{x}, u) = \mathbf{A}\mathbf{x} + \mathbf{B}u, \quad \mathbf{x}(0) = \mathbf{0}, \tag{7.2}$$

where

$$\mathbf{A} = \frac{\partial \mathbf{f}}{\partial \mathbf{x}} = \begin{pmatrix} 0 & 1 \\ 0 & 0 \end{pmatrix}$$

and

$$\mathbf{B} = \frac{\partial \mathbf{f}}{\partial u} = \begin{pmatrix} 0 \\ \frac{1}{J} \end{pmatrix}.$$

The following quadratic objective function is chosen in order to minimize the control input with the final state $\mathbf{x}_f = \mathbf{x}(t_f) = [\theta_f, \omega_f]^T$:

$$\mathcal{J} = [\mathbf{x}(t_f) - \mathbf{x}]^T \begin{pmatrix} Q_1 & 0 \\ 0 & Q_2 \end{pmatrix} [\mathbf{x}(t_f) - \mathbf{x}] + \int_{t_0}^{t_f} \left\{ Ru^2(t) + \boldsymbol{\lambda}^T(t) \left(\mathbf{f}[\mathbf{x}(t), \mathbf{u}(t)] - \dot{\mathbf{x}} \right) \right\} \mathrm{d}t. \tag{7.3}$$

Thus, we have selected the terminal penalty function, $\varphi = Q_1(\theta_f - \theta)^2 + Q_2(\omega_f - \omega)^2$, $L = Ru^2$, and the Hamiltonian, $H = Ru^2 + \boldsymbol{\lambda}^T\mathbf{f}$.

The necessary conditions for optimality (Chapter 2) yield the following set of governing equations for $\boldsymbol{\lambda}(t) = [\lambda_1(t), \lambda_2(t)]^T$, and hence the optimal control, $u^*(t)$:

$$\dot{\boldsymbol{\lambda}} = -H_{\mathbf{x}}^T = -\mathbf{A}^T\boldsymbol{\lambda} = \begin{pmatrix} 0 \\ -\lambda_1 \end{pmatrix}, \tag{7.4}$$

$$\boldsymbol{\lambda}(t_f) = \left(\frac{\partial \varphi}{\partial \mathbf{x}} \right)^T_{t=t_f} = \begin{Bmatrix} -2Q_1(\theta_f - \theta) \\ -2Q_2(\omega_f - \omega) \end{Bmatrix}, \tag{7.5}$$

and

$$H_u = \frac{\partial H}{\partial u} = 2Ru + \boldsymbol{\lambda}^T \begin{pmatrix} 0 \\ \frac{1}{J} \end{pmatrix} = 2Ru^* + \frac{\lambda_2(t_f)}{J} = 0. \tag{7.6}$$

The closed-form solution to equations (7.4)–(7.6) for $0 \le t \le t_f$ is obtained as follows:

$$\begin{aligned} \lambda_1(t) &= \lambda_1(t_f) = -2Q_1[\theta_f - \theta(t_f)], \\ \lambda_2(t) &= 2\left\{ Q_1[\theta_f - \theta(t_f)](t - t_f) - Q_2[\omega_f - \omega(t_f)] \right\}, \end{aligned} \tag{7.7}$$

$$u^*(t) = -\frac{\lambda_2(t)}{2RJ} = \frac{Q_2[\omega_f - \omega(t_f)] - Q_1[\theta_f - \theta(t_f)](t - t_f)}{RJ}, \tag{7.8}$$

where

$$\begin{aligned} \mathbf{x}(t_f) = \begin{Bmatrix} \theta(t_f) \\ \omega(t_f) \end{Bmatrix} &= \int_0^{t_f} e^{\mathbf{A}(t_f - t)} \mathbf{B} u^*(t) \mathrm{d}t \\ &= \frac{1}{RJ^2} \begin{Bmatrix} Q_1[\theta_f - \theta(t_f)]\frac{t_f^3}{3} + Q_2[\omega_f - \omega(t_f)]\frac{t_f^2}{2} \\ Q_1[\theta_f - \theta(t_f)]\frac{t_f^2}{2} + Q_2[\omega_f - \omega(t_f)]t_f \end{Bmatrix}. \end{aligned} \tag{7.9}$$

Solving equation (7.9), we have

$$\begin{aligned} \theta(t_f) &= \frac{t_f^2}{\Delta} \left[(Q_2 t_f + RJ^2) \left(\frac{Q_1 t_f \theta_f}{3} + \frac{Q_2 \omega_f}{2} \right) - \frac{Q_2}{2} \left(\frac{Q_1 t_f \theta_f}{2} + Q_2 \omega_f \right) t_f \right], \\ \omega(t_f) &= \frac{t_f}{\Delta} \left[-\frac{Q_1}{2} \left(\frac{Q_1 t_f \theta_f}{3} + \frac{Q_2 \omega_f}{2} \right) t_f^3 + \left(\frac{Q_2 t_f^3}{3} + RJ^2 \right) \left(\frac{Q_1 t_f \theta_f}{2} + Q_2 \omega_f \right) \right], \end{aligned} \tag{7.10}$$

where

$$\Delta = \left(\frac{Q_1 t_f^3}{3} + RJ^2 \right) (Q_2 t_f + RJ^2) - \frac{Q_1 Q_2 t_f^4}{4}. \tag{7.11}$$

We can substitute equation (7.11) into equation (7.8) to obtain the optimal control input, and hence the optimal trajectory, in the control interval $0 \le t \le t_f$, by the linear state solution,

$$\begin{aligned} \mathbf{x}(t) &= \begin{Bmatrix} \theta(t) \\ \omega(t) \end{Bmatrix} = \int_0^t e^{\mathbf{A}(t-\tau)} \mathbf{B} u^*(\tau) d\tau \\ &= \frac{1}{RJ^2} \begin{Bmatrix} \frac{Q_1}{3}[\theta_f - \theta(t_f)] \left[t_f^3 - (t_f - t)^3 \right] + \frac{Q_2}{2}[\omega_f - \omega(t_f)] \left[t_f^2 - (t_f - t)^2 \right] \\ Q_1[\theta_f - \theta(t_f)] \left(t_f t - \frac{t}{2} \right) + Q_2[\omega_f - \omega(t_f)] t \end{Bmatrix}. \end{aligned} \tag{7.12}$$

Since terminal conditions are satisfied at $t = t_f$, we have $u^*(t) = 0, t > t_f$, for which

$$\begin{Bmatrix} \theta(t) \\ \omega(t) \end{Bmatrix} = \begin{Bmatrix} \theta(t_f) + \omega(t_f) t \\ \omega(t_f) \end{Bmatrix}, \quad t > t_f. \tag{7.13}$$

The weighting factors, R, Q_1, and Q_2, must be carefully selected according to the required performance (i.e., precision in meeting terminal conditions and the torque input magnitudes). It is important to note that only the relative magnitudes of these factors are important for calculating the optimal trajectory. A common choice that makes the optimal trajectory independent of J is $R = 1/J^2$, which leaves Q_1 and Q_2 to be determined from the performance requirements. We will next consider two different classes of rotational maneuvers.

Spin Maneuvers

Spin maneuvers are solely concerned with increasing (or decreasing) the angular speed to a desired terminal value, ω_f, without regard to the final orientation of the spacecraft. At the outset, it appears that the required optimal solution can be obtained merely by removing the terminal penalty on the angle ($Q_1 = 0$). We shall pursue this solution first, noting that equation (7.8) produces in this case (for zero initial condition)

$$u^*(t) = \begin{cases} \dfrac{Q_2[\omega_f - \omega(t_f)]}{RJ}, & 0 \le t \le t_f, \\ 0, & t > t_f, \end{cases} \tag{7.14}$$

where

$$\omega(t_f) = \frac{Q_2 \omega_f t_f}{RJ^2 + Q_2 t_f}. \tag{7.15}$$

Clearly, the optimal torque remains constant in the control interval, but abruptly increases (or decreases) from zero at $t = 0$, and abruptly returns to zero at $t = t_f$, thereby requiring a discontinuous input. Such a control is termed *on–off control* and can be achieved by switching on a pair of rocket thrusters for the duration $0 \le t \le t_f$. For a good precision in the final speed, we require $Q_2 t_f \gg RJ^2$.

In order to improve the performance by having a continuous control input, we force the angular speed to meet the following constraint at the fixed final time, t_f:

$$\omega(t_f) = \omega_f. \tag{7.16}$$

By doing so, the terminal penalty on angular speed is removed ($Q_2 = 0$), but that on the rotation angle is kept ($Q_1 = Q \neq 0$), that is, we have $\varphi = Q(\theta_f - \theta)^2$, $L = Ru^2$, and $H = Ru^2 + \boldsymbol{\lambda}^T \mathbf{f}$.

The necessary conditions for optimality yield the following set of governing equations for $\boldsymbol{\lambda}(t) = [\lambda_1(t), \lambda_2(t)]^T$, and hence the optimal control, $u^*(t)$:

$$\dot{\boldsymbol{\lambda}} = -H_{\mathbf{x}}^T = \begin{pmatrix} 0 \\ -\lambda_1 \end{pmatrix}, \tag{7.17}$$

$$\boldsymbol{\lambda}(t_f) = \left(\frac{\partial \varphi}{\partial \mathbf{x}}\right)^T_{t=t_f} + \mathbf{G}^T \mu = \begin{Bmatrix} -2Q(\theta_f - \theta) \\ \mu \end{Bmatrix}, \tag{7.18}$$

and

$$H_u = 2Ru^* + \frac{\lambda_2(t_f)}{J} = 0. \tag{7.19}$$

The closed-form solution to equations (7.17)–(7.19) for $0 \le t \le t_f$ is found to be

$$\begin{aligned} \lambda_1(t) &= -2Q[\theta_f - \theta(t_f)],, \\ \lambda_2(t) &= 2Q[\theta_f - \theta(t_f)](t - t_f) + \mu, \end{aligned} \tag{7.20}$$

$$u^*(t) = -\frac{\lambda_2(t)}{2RJ} = \frac{2Q[\theta_f - \theta(t_f)](t_f - t) + \mu}{2RJ}, \tag{7.21}$$

where

$$\begin{aligned} \mathbf{x}(t_f) = \begin{Bmatrix} \theta(t_f) \\ \omega(t_f) \end{Bmatrix} &= \int_0^{t_f} e^{\mathbf{A}(t_f - t)} \mathbf{B} u^*(t) \mathrm{d}t \\ &= \frac{1}{RJ^2} \begin{Bmatrix} \frac{\left(Q\theta_f t_f + \frac{3}{4}\mu\right) t_f^2}{Q t_f^3 + 3RJ^2} \\ Q[\theta_f - \theta(t_f)]\frac{t_f^2}{2} + \frac{\mu t_f}{2} \end{Bmatrix}. \end{aligned} \tag{7.22}$$

We are free to select μ according to the desired terminal conditions, which may produce conflicting requirements. For example, it is impossible to achieve both $\omega(t_f) = \omega_f$ and $\theta(t_f) = \theta_f$ for an arbitrary pair, (θ_f, ω_f), by any value of μ. Since the former is the end-point constraint, we choose to satisfy it and have $\theta(t_f) \neq \theta_f$ by selecting $\mu = 0$ so as to produce a continuous control input (i.e., $u^*(t_f) = 0$) at $t = t_f$,

$$u^*(t) = \frac{Q[\theta_f - \theta(t_f)](t_f - t)}{RJ}, \tag{7.23}$$

and the following rotational response:

$$\mathbf{x}(t) = \begin{Bmatrix} \theta(t) \\ \omega(t) \end{Bmatrix} = \int_0^t e^{\mathbf{A}(t-\tau)}\mathbf{B}u^*(\tau)\mathrm{d}\tau$$
$$= \frac{1}{RJ^2}\begin{Bmatrix} Q[\theta_f - \theta(t_f)]\left(t_f\frac{t^2}{2} - \frac{t^3}{6}\right) \\ Q[\theta_f - \theta(t_f)]\left(tt_f - \frac{t^2}{2}\right) \end{Bmatrix}, \quad 0 \leq t \leq t_f, \tag{7.24}$$

and

$$\begin{Bmatrix} \theta(t) \\ \omega(t) \end{Bmatrix} = \begin{Bmatrix} \theta(t_f) + \omega_f(t - t_f) \\ \omega_f \end{Bmatrix}, \quad t > t_f, \tag{7.25}$$

where

$$\theta(t_f) = \frac{Q\theta_f t_f^3}{Qt_f^3 + 3RJ^2}. \tag{7.26}$$

From equation (7.24) it is evident that the cost coefficients, Q, R, are related as follows in order to achieve the end-point constraint, $\omega(t_f) = \omega_f$:

$$Qt_f^2 = \frac{RJ^2\omega_f}{\frac{\theta_f}{2} - \frac{\omega_f t_f}{3}}, \tag{7.27}$$

which, substituted into equation (7.26), yields

$$\theta(t_f) = \frac{2}{3}\omega_f t_f. \tag{7.28}$$

Thus, the terminal angle is independent of the cost coefficients. It is interesting to note that for a given R, the optimal control power, $\int u^*(t)\mathrm{d}t$, is the same for both unconstrained (equation (7.14)) and constrained (equation (7.23)) problems (why?). Thus, no extra expenditure of control energy is required in order to enforce the terminal constraint.

Example 7.1 Consider the optimal rotation of a spacecraft (initially at rest) about an axis with $J = 1000$ kg.m^2 in order to achieve a final angular speed of $\omega_f = 0.01$ rad/s after $t_f = 100$ s. Such a maneuver may be necessary to synchronize and align a payload with a target body whose angular rate with the spacecraft is constant. While a small error is acceptable in the final orientation, the final speed must be matched with a greater precision. Therefore, we select $Q_1 = 3RJ^2/t_f^3$, $Q_2 = RJ^2/t_f$, and $R = 1/J^2$. The unrestricted optimal control torque and the optimal trajectory are plotted in Figures 7.2 and 7.3, respectively.

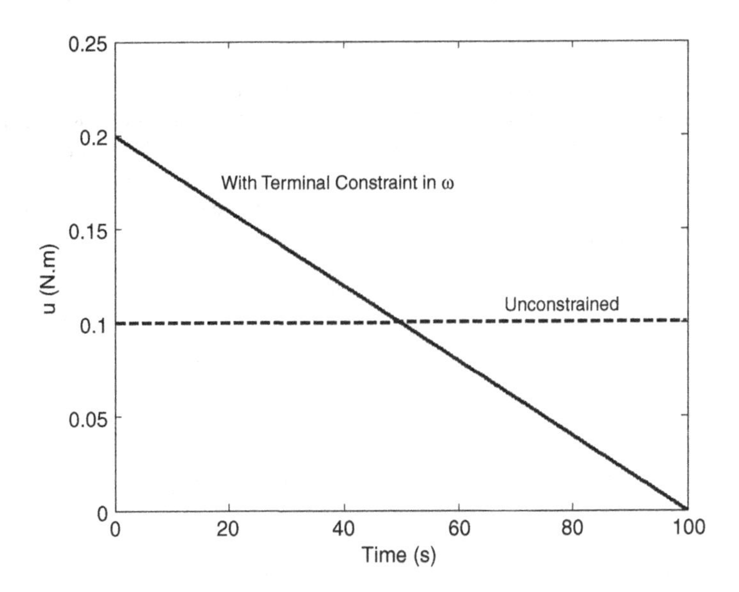

Figure 7.2 Optimal spin maneuver of a rigid spacecraft: control input for terminal constraint and unconstrained cases

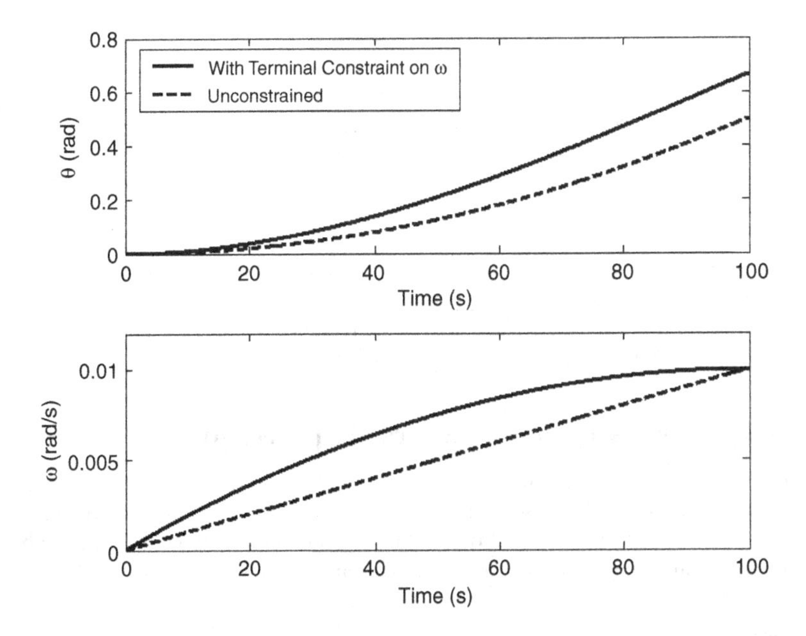

Figure 7.3 Optimal spin maneuver of a rigid spacecraft: trajectory for terminal constraint and unconstrained cases

Rest Maneuvers

In contrast to spin maneuvers where the terminal angular speed is controlled, rest maneuvers involve achieving a target angular orientation in a fixed time with zero terminal speed.[1] Rest maneuvers are important in the attitude control of most aerospace vehicles. We follow the formulation of the previous subsection with $\varphi = Q[\theta_f - \theta(t_f)]^2$, $L = Ru^2$, and terminal angular speed constraint, $\omega(t_f) = 0$. The optimality conditions and optimal control input are the same as those given by equations (7.17)–(7.22). Putting the terminal constraint, $\omega(t_f) = 0$, in equation (7.22), we have

$$\mu = -Q[\theta_f - \theta(t_f)]t_f, \tag{7.29}$$

which, substituted into equation (7.21), yields

$$u^*(t) = \frac{Q[\theta_f - \theta(t_f)](t_f - 2t)}{2RJ}. \tag{7.30}$$

Here we note that the optimal control torque is linear in time, with $u^*(0) = -u^*(t_f)$. Solving for the optimal trajectory, we have

$$\theta(t) = \int_0^t \omega(\tau)\mathrm{d}\tau = \frac{Q}{2RJ^2}[\theta_f - \theta(t_f)]\left(t_f\frac{t^2}{2} - \frac{t^3}{3}\right), \tag{7.31}$$

and the terminal angle is given by

$$\theta(t_f) = \frac{Q}{12RJ^2}[\theta_f - \theta(t_f)]t_f^3 \tag{7.32}$$

or

$$\theta(t_f) = \frac{\theta_f}{1 + \dfrac{12RJ^2}{Qt_f^3}}. \tag{7.33}$$

Therefore, by choosing cost parameters such that $Qt_f^3 \gg 12RJ^2$ one can achieve good accuracy in the terminal angle (θ_f). However, increasing the ratio Q/RJ causes an increase in the optimal control magnitude (equation (7.30)). Thus, a trade-off between the desired accuracy and control effort may be necessary.

Example 7.2 Reconsider the spacecraft of Example 7.1. Instead of a spin maneuver, it is now desired to reorient the spacecraft (initially at rest) such that a final angle, $\theta = \pi$ rad, is achieved after $t_f = 100$ s. The optimal control torque and the optimal trajectory are plotted in Figures 7.4 and 7.5, respectively, for $Q = RJ^2$. Note the smooth trajectory and linear control torque with $u^*(0) = -u^*(t_f)$, as expected.

7.3 Multi-axis Rotational Maneuvers of Spacecraft

Optimal control with end-point constraints (Chapter 2) can be applied to multi-axis rotations of rigid spacecraft. However, as opposed to single-axis rotation, a closed-form solution to this nonlinear problem is unavailable, thus requiring numerical 2PBVP solution. Consider *oxyz* to be the principal body-fixed frame of the spacecraft (Figure 7.6) for which the rotational dynamics equations can be written in an

[1] While we are only considering the case $\mathbf{x}(0) = \mathbf{0}$, the treatment can be easily extended for non-zero initial conditions.

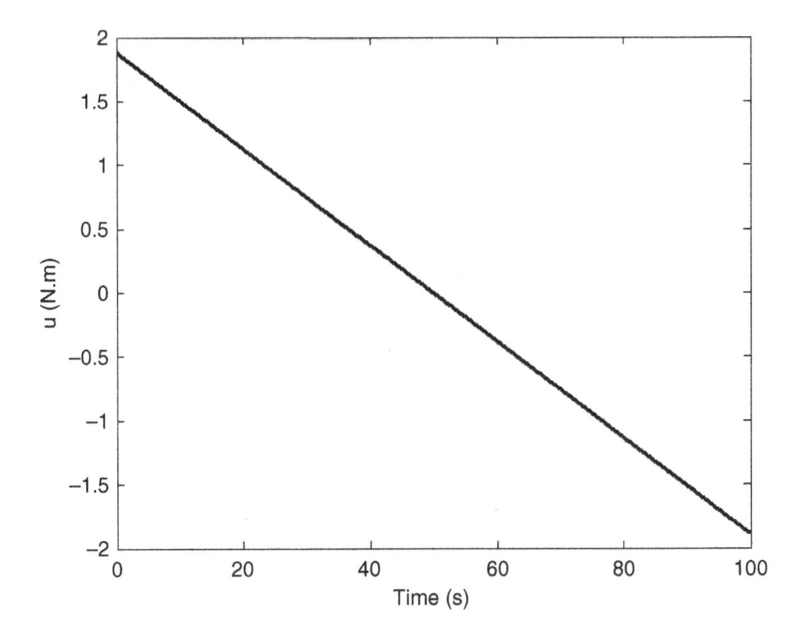

Figure 7.4 Optimal control input for the rest maneuver of a rigid spacecraft

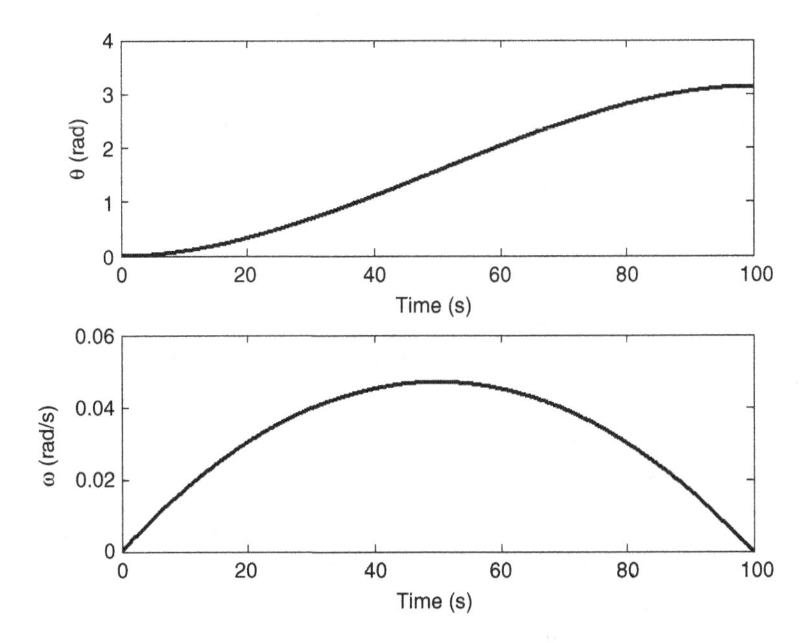

Figure 7.5 Optimal trajectory for the rest maneuver of a rigid spacecraft

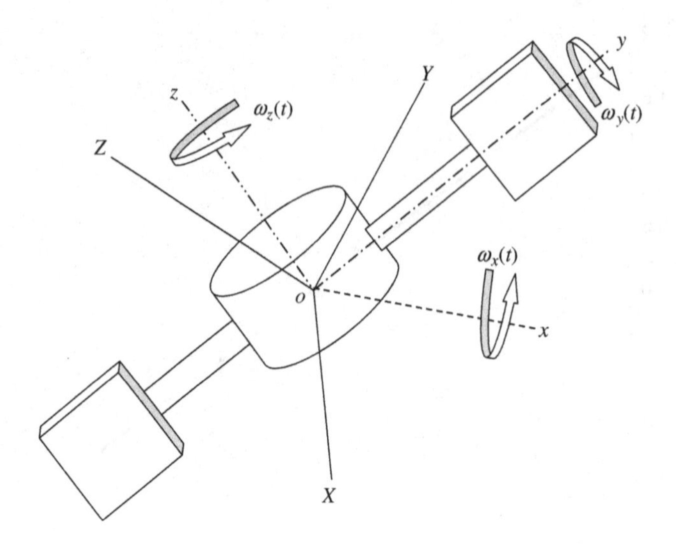

Figure 7.6 Rigid spacecraft with principal body-fixed frame, *oxyz*, rotating about multiple axes relative to an inertial frame, *oXYZ*

inertial reference frame, *oXYZ*, as follows:

$$\boldsymbol{\tau} = \mathbf{J}\dot{\boldsymbol{\omega}} + \mathbf{S}(\boldsymbol{\omega})\mathbf{J}\boldsymbol{\omega}. \tag{7.34}$$

Here the inertia tensor is the following diagonal matrix of principal moments of inertia:

$$\mathbf{J} = \begin{pmatrix} J_{xx} & 0 & 0 \\ 0 & J_{yy} & 0 \\ 0 & 0 & J_{zz} \end{pmatrix}, \tag{7.35}$$

and $\mathbf{S}(\mathbf{a})$ is the following skew-symmetric matrix function formed out of the elements of a vector, $\mathbf{a} = (a_1, a_2, a_3)^T$:

$$\mathbf{S}(\mathbf{a}) = \begin{pmatrix} 0 & -a_3 & a_2 \\ a_3 & 0 & -a_1 \\ -a_2 & a_1 & 0 \end{pmatrix}. \tag{7.36}$$

The orientation of the principal body-fixed frame, *oxyz*, relative to the inertial reference frame, *oXYZ*, is given by the following general rotational kinematics equation in terms of the rotation matrix, $\mathbf{C}(t)$:

$$\frac{\mathrm{d}\mathbf{C}}{\mathrm{d}t} = -\mathbf{S}(\boldsymbol{\omega})\mathbf{C}. \tag{7.37}$$

However, instead of working with nine dependent elements of $\mathbf{C}(t)$ as state variables, or adopting the Euler angles (since they may encounter singularities during a large rotation), we choose to express the orientation (attitude) of the frame *oxyz* by the quaternion representation, q_1, q_2, q_3, q_4. The attitude kinematics equation is expressed in terms of the *quaternion* (see Chapter 3; see also Tewari 2006) as follows:

$$\frac{\mathrm{d}\{\mathbf{q}, q_4\}^T}{\mathrm{d}t} = \frac{1}{2}\boldsymbol{\Omega}\{\mathbf{q}^T(t), q_4(t)\}^T, \tag{7.38}$$

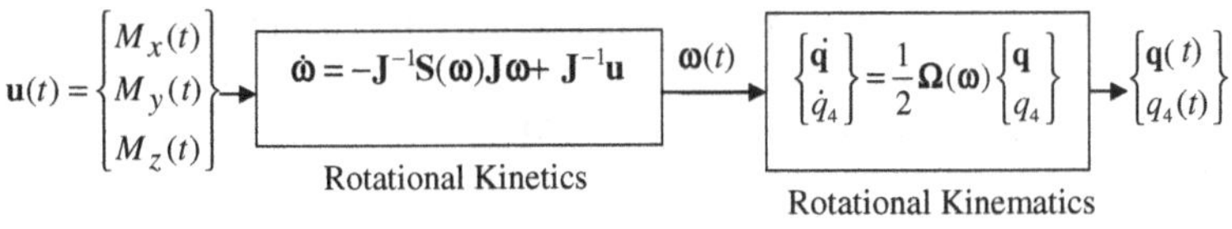

$$\mathbf{u}(t)=\begin{Bmatrix} M_x(t) \\ M_y(t) \\ M_z(t) \end{Bmatrix} \rightarrow \boxed{\dot{\boldsymbol{\omega}} = -\mathbf{J}^{-1}\mathbf{S}(\boldsymbol{\omega})\mathbf{J}\boldsymbol{\omega} + \mathbf{J}^{-1}\mathbf{u}} \xrightarrow{\boldsymbol{\omega}(t)} \boxed{\begin{Bmatrix} \dot{\mathbf{q}} \\ \dot{q}_4 \end{Bmatrix} = \frac{1}{2}\boldsymbol{\Omega}(\boldsymbol{\omega})\begin{Bmatrix} \mathbf{q} \\ q_4 \end{Bmatrix}} \rightarrow \begin{Bmatrix} \mathbf{q}(t) \\ q_4(t) \end{Bmatrix}$$

Rotational Kinetics

Rotational Kinematics

Figure 7.7 Plant for multi-axis rotations of a rigid spacecraft

where

$$\boldsymbol{\Omega} = \begin{pmatrix} 0 & \omega_z & -\omega_y & \omega_x \\ -\omega_z & 0 & \omega_x & \omega_y \\ \omega_y & -\omega_x & 0 & \omega_z \\ -\omega_x & -\omega_y & -\omega_z & 0 \end{pmatrix}. \tag{7.39}$$

Considering the fact that the angular velocity does not depend upon the attitude, $\mathbf{q}$, q_4, the differential equation (7.39) governing the attitude is essentially a linear one for a given angular velocity, $\boldsymbol{\omega}(t)$. Such a linear algebraic form of the attitude kinematics equation is an obvious advantage of the quaternion representation over the nonlinear transcendental kinematics equation of the Euler angle representation. The overall plant for multi-axis rotations is depicted in Figure 7.7, clearly showing the cascade relationship between rotational kinetics and kinematics.

Multi-axis Spin Maneuvers

A multi-axis spin maneuver involves changing the angular velocity vector from an initial value, $\boldsymbol{\omega}(0)$, to a final one, $\boldsymbol{\omega}(t_f)$, in a given time, t_f, irrespective of the initial and final attitudes. The state equation for a spin maneuver is thus written as

$$\dot{\mathbf{x}} = \mathbf{f}(\mathbf{x}, \mathbf{u}), \tag{7.40}$$

where

$$\mathbf{x}(t) = \boldsymbol{\omega}(t_f) = \begin{Bmatrix} \omega_x \\ \omega_y \\ \omega_z \end{Bmatrix}, \quad \mathbf{u}(t) = \boldsymbol{\tau}(t) = \begin{Bmatrix} M_x \\ M_y \\ M_z \end{Bmatrix},$$

$$\mathbf{f} = \begin{Bmatrix} \frac{J_{yy}-J_{zz}}{J_{xx}}\omega_y\omega_z + \frac{M_x}{J_{xx}} \\ \frac{J_{zz}-J_{xx}}{J_{yy}}\omega_x\omega_z + \frac{M_y}{J_{yy}} \\ \frac{J_{xx}-J_{yy}}{J_{zz}}\omega_x\omega_y + \frac{M_z}{J_{zz}} \end{Bmatrix}. \tag{7.41}$$

One can pose the problem for an extremal trajectory with end-point constraints, a zero terminal cost, $\varphi = 0$, and a quadratic control cost, $L = \mathbf{u}^T\mathbf{R}\mathbf{u}$, using the Hamiltonian

$$H = \mathbf{u}^T\mathbf{R}\mathbf{u} + \boldsymbol{\lambda}^T(t)\mathbf{f}. \tag{7.42}$$

The necessary conditions for optimality yield the following extremal control input:

$$H_{\mathbf{u}} = L_{\mathbf{u}} + \boldsymbol{\lambda}^T \mathbf{B}, \tag{7.43}$$

where

$$\mathbf{B} = \frac{\partial \mathbf{f}}{\partial \mathbf{u}} = \begin{pmatrix} \frac{1}{J_{xx}} & 0 & 0 \\ 0 & \frac{1}{J_{yy}} & 0 \\ 0 & 0 & \frac{1}{J_{zz}} \end{pmatrix}.$$

A symmetric, positive definite matrix is chosen for the control cost coefficients such as the diagonal matrix, $\mathbf{R} = \text{diag}(R_1, R_2, R_3)$, which results in

$$\mathbf{u}^* = \begin{Bmatrix} M_x^* \\ M_y^* \\ M_y^* \end{Bmatrix} = -\frac{1}{2}\mathbf{R}^{-1}\mathbf{B}^T\boldsymbol{\lambda} = -\begin{Bmatrix} \frac{\lambda_1}{2R_1 J_{xx}} \\ \frac{\lambda_2}{2R_2 J_{yy}} \\ \frac{\lambda_3}{2R_3 J_{zz}} \end{Bmatrix}. \tag{7.44}$$

Substitution of the extremal control and trajectory yields the following set of governing equations for $\boldsymbol{\lambda}(t) = [\lambda_1(t), \lambda_2(t), \lambda_3(t)]^T$:

$$\dot{\boldsymbol{\lambda}} = -\mathbf{A}^T\boldsymbol{\lambda} - \left(\frac{\partial L}{\partial \mathbf{x}}\right)^T,$$

where

$$\mathbf{A} = \frac{\partial \mathbf{f}}{\partial \mathbf{x}} = \begin{pmatrix} 0 & j_x\omega_z & j_x\omega_y \\ j_y\omega_z & 0 & j_y\omega_x \\ j_z\omega_y & j_z\omega_x & 0 \end{pmatrix}, \tag{7.45}$$

and

$$\begin{aligned} j_x &= \frac{J_{yy} - J_{zz}}{J_{xx}}, \\ j_y &= \frac{J_{zz} - J_{xx}}{J_{yy}}, \\ j_z &= \frac{J_{xx} - J_{yy}}{J_{zz}}, \end{aligned} \tag{7.46}$$

resulting in

$$\begin{aligned} \dot{\lambda}_1 &= -j_y\omega_z\lambda_2 - j_z\omega_y\lambda_3, \\ \dot{\lambda}_2 &= -j_x\omega_z\lambda_1 - j_z\omega_x\lambda_3, \\ \dot{\lambda}_3 &= -j_x\omega_y\lambda_1 - j_y\omega_x\lambda_2. \end{aligned} \tag{7.47}$$

The state equations (7.40) and co-state equations (7.47) are to be solved subject to the following boundary conditions applied at the ends of the control interval, $0 \leq t \leq t_f$:

$$\mathbf{x}(0) = \boldsymbol{\omega}_0, \quad \mathbf{x}(t_f) = \boldsymbol{\omega}_f. \tag{7.48}$$

Due to the nonlinear nature of both state and co-state equations, a numerical solution to the boundary value problem is required.

Example 7.3 Consider a spacecraft with the principal moments of inertia $J_{xx} = 1000$ kg.m^2, $J_{yy} = 1500$ kg.m^2, and $J_{zz} = 2000$ kg.m^2. It is desired to spin the spacecraft from rest to a final angular velocity, $\omega = (0.1, -0.1, 0.3)^T$ rad/s in 50 seconds while minimizing the required control effort.

In order to solve the 2PBVP comprising the state equations (7.40), the co-state equations (7.47), and the boundary conditions (7.48), we utilize the inbuilt MATLAB® code *bvp4c.m* that is called by the program listed in Table 7.1. The differential equations and the boundary conditions required by *bvp4c.m* are specified by the respective codes *spinode.m* and *spinbc.m*, also listed in Table 7.1. The control cost coefficient matrix is chosen as

$$\mathbf{R} = \mathbf{J}^{-1} = \begin{pmatrix} \frac{1}{J_{xx}} & 0 & 0 \\ 0 & \frac{1}{J_{yy}} & 0 \\ 0 & 0 & \frac{1}{J_{zz}} \end{pmatrix}.$$

With this choice of $\mathbf{R}$, the optimal control input is given simply by

$$\mathbf{u}^* = -\frac{1}{2}\boldsymbol{\lambda}.$$

The 2PBVP solution is obtained using 50 collocation points and the resulting extremal trajectory and optimal control torques are plotted in Figures 7.8 and 7.9, respectively.

General Multi-axis Maneuvers

In a general maneuver, it is necessary to control not only the angular velocity, $\boldsymbol{\omega}$, but also the attitude given by a set of kinematical parameters, $\boldsymbol{\zeta}$ (e.g., the quaternion, $\boldsymbol{\zeta} = (\mathbf{q}^T, q_4)^T$). Thus, the full state vector,

$$\mathbf{x} = (\boldsymbol{\omega}, \boldsymbol{\zeta})^T,$$

must be accounted for in a general maneuver. Writing the state equations in the vector form

$$\dot{\mathbf{x}} = \mathbf{h}(\mathbf{x}, \mathbf{u}), \tag{7.49}$$

where

$$\mathbf{h} = \begin{Bmatrix} \mathbf{f}(\boldsymbol{\omega}, \mathbf{u}) \\ \mathbf{g}(\boldsymbol{\omega}, \boldsymbol{\zeta}) \end{Bmatrix}, \tag{7.50}$$

the rotational dynamics functional, $\mathbf{f}(\boldsymbol{\omega}, \mathbf{u})$, is given by equation (7.41), and the functional, $\mathbf{g}(\boldsymbol{\omega}, \boldsymbol{\zeta})$, satisfies the following rotational kinematics equation:

$$\dot{\boldsymbol{\zeta}} = \mathbf{g}(\boldsymbol{\omega}, \boldsymbol{\zeta}). \tag{7.51}$$

Table 7.1

```
% Calling program for solving the 2PBVP for optimal spin maneuver
% by collocation method by MATLAB's intrinsic code 'bvp4c.m'.
% Requires 'spinode.m' and 'spinbc.m'.
% (c) 2009 Ashish Tewari
%  dy/dx=f(y,x); a<=x<=b
%  y(x=a), y(x=b): Boundary conditions
%  y(1:3,1)=omega (rad/s) (Angular velocity vector)
%  y(4:6,1)=lambda (Lagrange multipliers vector)

global tf; tf=50; % Terminal time (s)
global J; J=[1000 1500 2000]; % Moms. of inertia (kg-m^2)
global wi; wi=[0 0 0]; % Initial ang. velocity (rad/s)
global wf; wf=[0.1 -0.1 0.3]; % Final ang. velocity (rad/s)
global R; R=[1/J(1) 1/J(2) 1/J(3)];% Control cost coefficients
% Collocation points & initial guess follow:
solinit = bvpinit(linspace(0,tf,50),[wf 0 0 0]);
% 2PBVP Solution by collocation method:
sol = bvp4c(@spinode,@spinbc,solinit);
x = linspace(0,tf); % Time vector (s)
y = deval(sol,x); % Solution state vector
u=-inv(diag(R))*inv(diag(J))*y(4:6,:)/2; % Control vector (N-m)

% Program for specifying governing ODEs expressed as
% state equations for the 2PBVP (to be called by 'bvp4c.m')
function dydx=spinode(x,y)
global tf;
global J;
global R;
if x<tf
u1=-y(4)/(2*R(1)*J(1));
u2=-y(5)/(2*R(2)*J(2));
u3=-y(6)/(2*R(3)*J(3));
else
    u1=0; u2=0; u3=0;
end

jx=(J(2)-J(3))/J(1);
jy=(J(3)-J(1))/J(2);
jz=(J(1)-J(2))/J(3);
dydx=[jx*y(2)*y(3)+u1/J(1)
      jy*y(1)*y(3)+u2/J(2)
      jz*y(1)*y(2)+u3/J(3)
      -jy*y(3)*y(5)-jz*y(2)*y(6)
      -jx*y(3)*y(4)-jz*y(1)*y(6)
      -jx*y(2)*y(4)-jy*y(1)*y(5)];

% Program for specifying boundary conditions for the 2PBVP.
% (To be called by 'bvp4c.m')
function res=spinbc(ya,yb)
global wi;
global wf;
res=[ya(1)-wi(1)
     ya(2)-wi(2)
     ya(3)-wi(3)
     yb(1)-wf(1)
     yb(2)-wf(2)
     yb(3)-wf(3)];
```

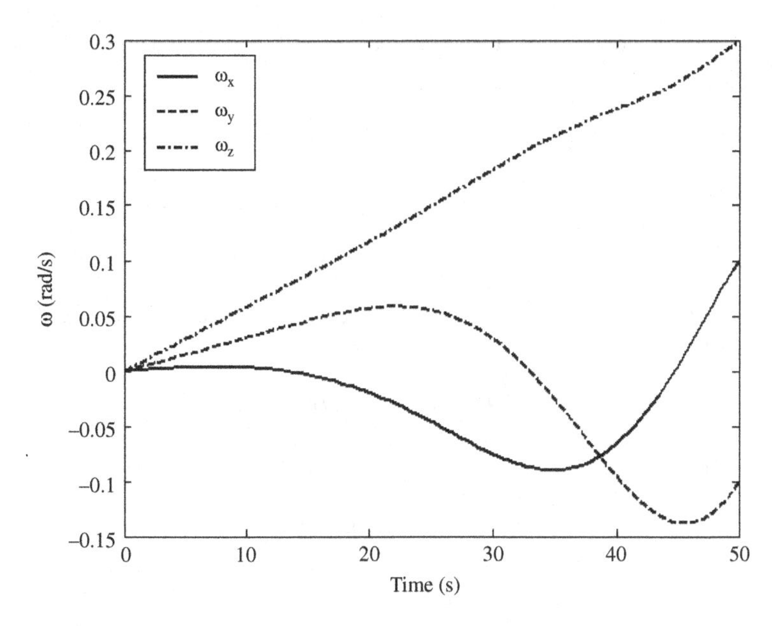

Figure 7.8 Extremal angular velocity for the multi-axis spin maneuver of a spacecraft

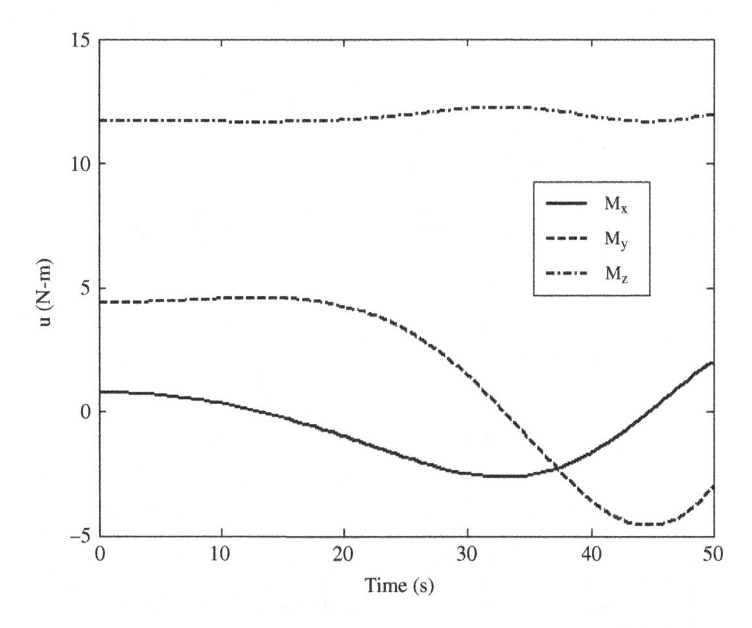

Figure 7.9 Optimal control torques for the multi-axis spin maneuver of a spacecraft

The optimal control problem for a general rotation with end-point constraints consisting of initial conditions,

$$\boldsymbol{\omega}(0) = \boldsymbol{\omega}_0, \quad \boldsymbol{\zeta}(0) = \boldsymbol{\zeta}_0, \tag{7.52}$$

and final conditions,

$$\boldsymbol{\omega}(t_f) = \boldsymbol{\omega}_f, \quad \boldsymbol{\zeta}(t_f) = \boldsymbol{\zeta}_f, \tag{7.53}$$

for a fixed time, t_f, can be posed with a zero terminal cost, $\varphi = 0$, and a quadratic control cost, $L = \mathbf{u}^T\mathbf{R}\mathbf{u}$, resulting in the Hamiltonian

$$H = \mathbf{u}^T\mathbf{R}\mathbf{u} + \boldsymbol{\lambda}^T(t)\mathbf{h}. \tag{7.54}$$

The necessary conditions for optimality yield the following extremal control input:

$$H_{\mathbf{u}} = L_{\mathbf{u}} + \boldsymbol{\lambda}^T\mathbf{B} = \mathbf{0}, \tag{7.55}$$

where

$$\mathbf{B} = \frac{\partial \mathbf{h}}{\partial \mathbf{u}} = \begin{Bmatrix} \frac{\partial \mathbf{f}}{\partial \mathbf{u}} \\ \frac{\partial \mathbf{g}}{\partial \mathbf{u}} \end{Bmatrix} = \begin{Bmatrix} \mathbf{f}_{\mathbf{u}} \\ \mathbf{0} \end{Bmatrix} \tag{7.56}$$

and

$$\mathbf{f}_{\mathbf{u}} = \begin{pmatrix} \frac{1}{J_{xx}} & 0 & 0 \\ 0 & \frac{1}{J_{yy}} & 0 \\ 0 & 0 & \frac{1}{J_{zz}} \end{pmatrix}.$$

Substituting $\mathbf{R} = \text{diag}(R_1, R_2, R_3)$ into equation (7.55), we have the following optimal control torque:

$$\mathbf{u}^* = \begin{Bmatrix} M_x^* \\ M_y^* \\ M_y^* \end{Bmatrix} = -\frac{1}{2}\mathbf{R}^{-1}\mathbf{B}^T\boldsymbol{\lambda} = -\begin{Bmatrix} \frac{\lambda_1}{2R_1 J_{xx}} \\ \frac{\lambda_2}{2R_2 J_{yy}} \\ \frac{\lambda_3}{2R_3 J_{zz}} \end{Bmatrix}, \tag{7.57}$$

which is the same as equation (7.44). Adopting the quaternion as the kinematical set, $\boldsymbol{\zeta} = (\mathbf{q}^T, q_4)^T$, we have

$$\mathbf{g}(\boldsymbol{\omega}, \boldsymbol{\zeta}) = \frac{1}{2}\boldsymbol{\Omega}(\boldsymbol{\omega})\boldsymbol{\zeta}, \tag{7.58}$$

where $\boldsymbol{\Omega}(\boldsymbol{\omega})$ is given by equation (7.39). Hence, the co-state equations (to be solved for Lagrange multipliers) can be written as

$$\dot{\boldsymbol{\lambda}} = -\mathbf{A}^T\boldsymbol{\lambda}, \tag{7.59}$$

where

$$\mathbf{A} = \frac{\partial \mathbf{h}}{\partial \mathbf{x}} = \begin{pmatrix} \mathbf{f}_\omega & \mathbf{f}_\zeta \\ \mathbf{g}_\omega & \mathbf{g}_\zeta \end{pmatrix}, \tag{7.60}$$

$$\mathbf{f}_\omega \doteq \frac{\partial \mathbf{f}}{\partial \boldsymbol{\omega}} = \begin{pmatrix} 0 & j_x\omega_z & j_x\omega_y \\ j_y\omega_z & 0 & j_y\omega_x \\ j_z\omega_y & j_z\omega_x & 0 \end{pmatrix}, \tag{7.61}$$

$$\mathbf{f}_\zeta \doteq \frac{\partial \mathbf{f}}{\partial \boldsymbol{\zeta}} = \mathbf{0}, \tag{7.62}$$

$$\mathbf{g}_\omega \doteq \frac{\partial \mathbf{g}}{\partial \boldsymbol{\omega}} = \frac{1}{2}\begin{pmatrix} q_4 & -q_3 & q_2 \\ q_3 & q_4 & -q_1 \\ -q_2 & q_1 & q_4 \\ -q_1 & -q_2 & -q_3 \end{pmatrix}, \tag{7.63}$$

$$\mathbf{g}_\zeta \doteq \frac{\partial \mathbf{g}}{\partial \boldsymbol{\zeta}} = \frac{1}{2}\boldsymbol{\Omega}(\boldsymbol{\omega}). \tag{7.64}$$

Thus, we have

$$\dot{\boldsymbol{\lambda}} = \begin{Bmatrix} \dot{\boldsymbol{\lambda}}_\omega \\ \dot{\boldsymbol{\lambda}}_\zeta \end{Bmatrix} = -\begin{Bmatrix} \mathbf{f}_\omega^T \boldsymbol{\lambda}_\omega + \mathbf{g}_\omega^T \boldsymbol{\lambda}_\zeta \\ \mathbf{g}_\zeta^T \boldsymbol{\lambda}_\zeta \end{Bmatrix}. \tag{7.65}$$

The set of ordinary differential equations (7.49) and (7.65), along with the boundary conditions given by equations (7.52) and (7.53), constitute the 2PBVP for an optimal general maneuver with end-point constraints.

Example 7.4 It is sometimes required to synchronize the rotation of a spacecraft such that both angular velocity and angular orientation are equal to specified values at a given time. Tracking another orbiting body with an antenna always pointing toward the body could be such an application. Consider the spacecraft with the principal moments of inertia given in Example 7.3. It is desired to spin the spacecraft from rest and initial orientation, $\mathbf{q} = \mathbf{0}$, to a final angular velocity, $\omega = (0.03, -0.02, 0.01)^T$ rad/s and a final orientation, $\mathbf{q} = (0.1, -0.2, 0.3)^T$, $q_4 = \sqrt{1 - q_1^2 - q_2^2 - q_3^2} = 0.9274$, in exactly 70 seconds, while minimizing the required control effort.

Using the optimal control formulation with end-point constraints described in the present section, we solve the 2PBVP consisting of the state equations (7.49), the co-state equations (7.65), and the boundary conditions (7.52)–(7.53), using MATLAB's inbuilt code *bvp4c.m* that is called by a program listed in Table 7.2. The differential equations and the boundary conditions required by *bvp4c.m* are specified by the respective codes *genrotode.m* and *genrotbc.m*, listed in Table 7.3. The control cost coefficient matrix is selected as $\mathbf{R} = \mathbf{J}^{-1}$. The resulting extremal trajectory is plotted in Figures 7.10 and 7.11, and the required optimal control torques in Figure 7.12.

Table 7.2

```
% Calling program for solving the 2PBVP for optimal general
% rotational maneuver by collocation method using MATLAB's
% intrinsic code 'bvp4c.m'.
% Requires 'genrotode.m' and 'genrotbc.m'.
% (c) 2009 Ashish Tewari
%  dy/dx=f(y,x); a<=x<=b
%  y(x=a), y(x=b): Boundary conditions
%  y(1:3,1)=omega (rad/s) (Angular velocity vector)
%  y(4:7,1)=q_1,q_2,q_3,q_4 (Quaternion)
%  y(8:14,1)=lambda (Lagrange multipliers vector)

global tf; tf=70; % Terminal time (s)
global J; J=diag([1000 1500 2000]); % Moms. of inertia (kg-m^2)
global wi; wi=[0 0 0]'; % Initial ang. velocity (rad/s)
global wf; wf=[0.03 -0.02 0.01]'; % Final ang. velocity (rad/s)
global Qi; Qi=[0 0 0 1]'; % Initial quaternion
global Qf; Qf=[0.1 -0.2 0.3 sqrt(1-.1^2-.2^2-.3^2)]';% Final quaternion
global R; R=inv(J); % Control cost coefficients
% Collocation points & initial guess follow:
solinit = bvpinit(linspace(0,tf,5),[wf; Qf; zeros(7,1)]);
% 2PBVP Solution by collocation method:
options=bvpset('Nmax',100);
sol = bvp4c(@genrotode,@genrotbc,solinit,options);
x = linspace(0,tf); % Time vector (s)
y = deval(sol,x); % Solution state vector
plot(x,y(1:3,:)),xlabel('Time (s)'),ylabel('\omega (rad/s)')
figure
plot(x,y(4:7,:)),xlabel('Time (s)'),ylabel('q')
figure
u=-0.5*inv(R)*inv(J)*y(8:10,:);
plot(x,u),xlabel('Time (s)'),ylabel('u (N-m)')
```

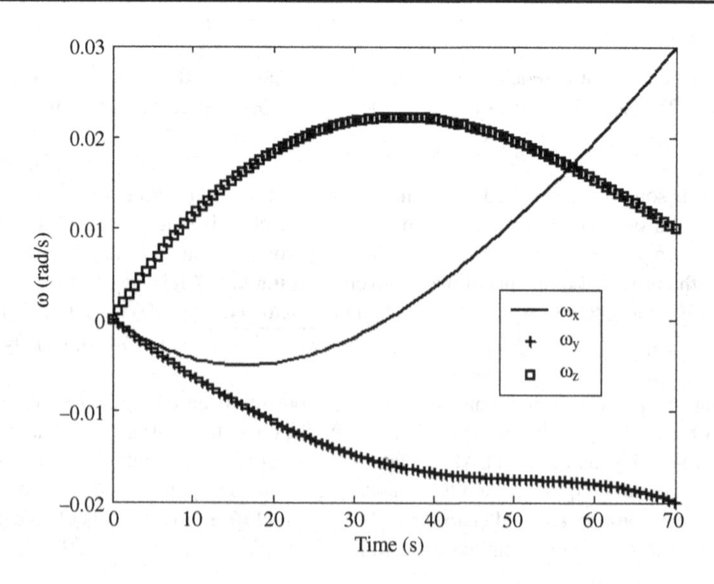

Figure 7.10 Extremal angular velocity for the multi-axis, general rotational maneuver of a spacecraft

Table 7.3

```
% Program for specifying governing ODEs expressed as
% state equations for the 2PBVP (to be called by 'bvp4c.m')
function dydx=genrotode(x,y)
global tf;
global J;
global R;
wx=y(1);wy=y(2);wz=y(3);
q1=y(4);q2=y(5);q3=y(6);q4=y(7);
jx=(J(2,2)-J(3,3))/J(1,1);
jy=(J(3,3)-J(1,1))/J(2,2);
jz=(J(1,1)-J(2,2))/J(3,3);
Fw=[0 jx*wz jx*wy; jy*wz 0 jy*wx;
    jz*wy jz*wx 0];
Gw=0.5*[q4 -q3 q2; q3 q4 -q1;
    -q2 q1 q4; -q1 -q2 -q3];
Gq=0.5*[0 wz -wy wx; -wz 0 wx wy;
    wy -wx 0 wz; -wx -wy -wz 0];
if x<tf
    u=-0.5*inv(R)*inv(J)*y(8:10,1);
else
    u=zeros(3,1);
end
dydx(1:3,1)=[jx*y(2)*y(3)+u(1,1)/J(1,1);
             jy*y(1)*y(3)+u(2,1)/J(2,2);
             jz*y(1)*y(2)+u(3,1)/J(3,3)];
 dydx(4:7,1)=0.5*Gq*y(4:7,1);
 dydx(8:14,1)=-[Fw' Gw'; zeros(4,3) Gq']*y(8:14,1);

% Program for specifying boundary conditions for the 2PBVP.
% (To be called by 'bvp4c.m')
function res=genrotbc(ya,yb)
global wi;
global wf;
global Qi;
global Qf;
res=[ya(1:3,1)-wi
     ya(4:7,1)-Qi
     yb(1:3,1)-wf
     yb(4:7,1)-Qf];
```

7.4 Spacecraft Control Torques

Until now we have not considered the mechanisms by which a control torque can be applied to a spacecraft. The attitude control of spacecraft is dependent upon the ability of a set of actuators to produce a sufficiently large control torque. Such a torque can be either external (e.g., that applied by rocket thrusters and magnetic coils) or internal (e.g., that produced by rotors mounted on the spacecraft).

7.4.1 *Rocket Thrusters*

Most spacecraft typically have a *reaction control system* (RCS) that employs a pair of rocket thrusters – called *attitude thrusters* – about each principal axis for performing attitude maneuvers. When a torque

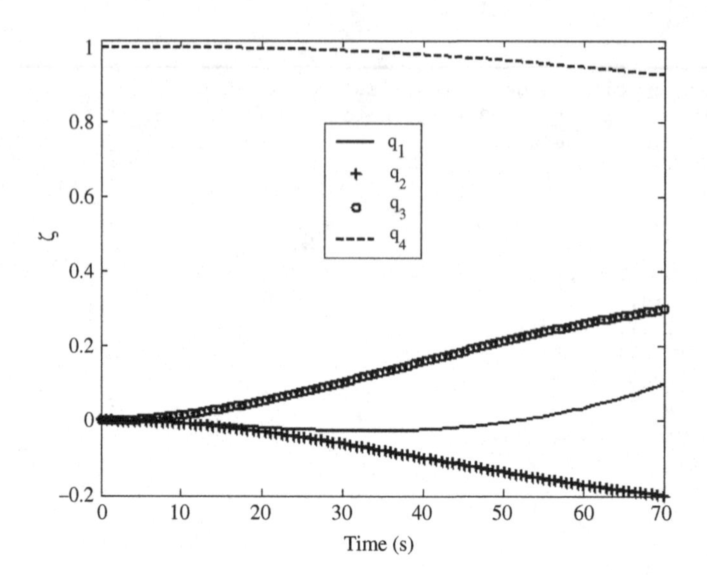

Figure 7.11 Extremal quaternion for the multi-axis, general rotational maneuver of a spacecraft

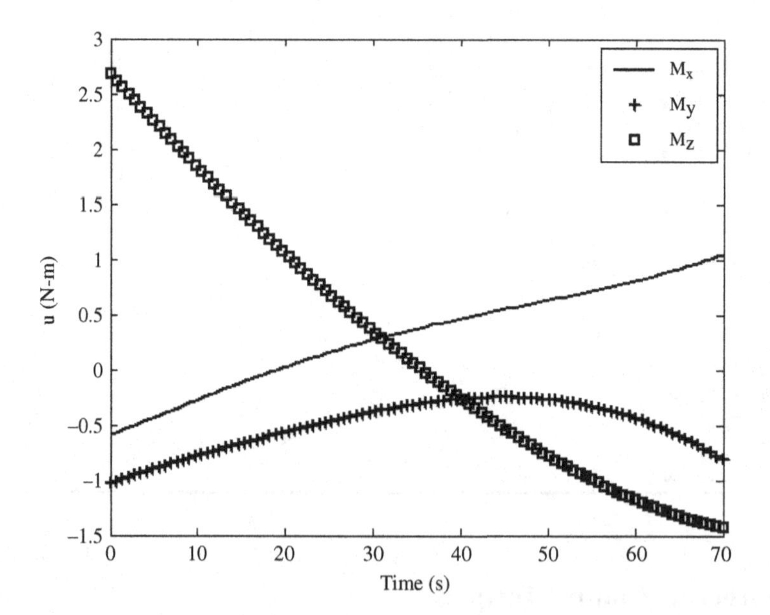

Figure 7.12 Optimal control torques for the multi-axis, general rotational maneuver of a spacecraft

about each principal axis is applied for stability and control, the spacecraft is said to be *three-axis stabilized*, as opposed to a torque-free, spin stabilized spacecraft. The attitude thrusters of an RCS are operated in pairs with an equal and opposite thrust, such that the net external force is zero. The firing of thrusters is limited to short bursts, which can be approximated by *torque impulses*. A torque impulse is mathematically defined as a torque of infinite magnitude acting for an infinitesimal duration, thereby causing an instantaneous change in the angular momentum of the spacecraft about the axis of application.

The concept of the torque impulse is very useful in analyzing the single-axis rotation of spacecraft, as it allows us to utilize the well-known linear systems theory (Tewari 2002), wherein the governing linear differential equation is solved in a closed form with the use of the *unit impulse function*, $\delta(t)$. The change in angular momentum caused by an impulsive torque, $\boldsymbol{\tau}(t) = \boldsymbol{\tau}(0)\delta(t)$, can be obtained as the total area under the torque vs. time graph, given by

$$\Delta \mathbf{H} \doteq \int_{-\infty}^{\infty} \boldsymbol{\tau}(t)\mathrm{d}t = \int_{-\infty}^{\infty} \boldsymbol{\tau}(0)\delta(t)\mathrm{d}t = \boldsymbol{\tau}(0). \tag{7.66}$$

Thus, the torque impulse causes an instantaneous change in the angular momentum, equal to the value of the torque at the instant of impulse application, $t = 0$. A complex multi-axis maneuver can be designed as a sequence of single-axis rotations carried out by the open-loop firing of rocket thrusters. While neither the thruster torque is infinite, nor the time interval over which the torque acts tends to zero, it is a good approximation to assume an impulsive thruster torque because its duration is much smaller than the period of the maneuver. Furthermore, a spacecraft can generally be assumed to have negligible environmental torques when undergoing an impulsive maneuver. As we briefly discussed in Chapter 4 for the translation of a mass (with negligible drag and gravitational acceleration), the response to an impulse input is a linearly increasing displacement and a step change in the speed. Therefore, if the maneuvering requirement is for a step change in angular velocity, a single impulse is sufficient. However, if a given rotational displacement is desired, one has to apply another impulse in the opposite direction in order to stop the rotation when the desired displacement has been reached. Since the governing differential equation of sequential single-axis rotation is linear, its solution obeys the principle of linear superposition, which allows a weighted addition of the responses to individual impulses, to yield the total displacement caused by multiple impulses. The application of two equal and opposite impulses of the maximum magnitude in order to achieve a desired displacement in the shortest possible time is called *bang-bang* control, which is an open-loop strategy requiring only the desired displacement. When used in a closed loop, the near impulsive torque results in a nonlinear control system that should be analyzed by suitable techniques (Slotine and Li 1991).

7.4.2 Reaction Wheels, Momentum Wheels and Control Moment Gyros

Rotors used for applying internal control torques on a spacecraft by changing their angular momenta can be categorized into *reaction wheels*, *momentum wheels*, and *control moment gyroscopes*. A reaction wheel is a rotor with its axis fixed relative to the spacecraft's body frame, and thus produces a torque by changing its angular velocity according to

$$\frac{\mathrm{d}\mathbf{H}}{\mathrm{d}t} = \mathbf{J}\dot{\boldsymbol{\omega}} + \boldsymbol{\omega} \times \mathbf{J}\boldsymbol{\omega} + J_{\mathrm{R}}\dot{\omega}_{\mathrm{R}}\mathbf{i}_{\mathrm{R}}^{\mathrm{e}} = \mathbf{0}, \tag{7.67}$$

where J_{R}, ω_{R}, and $\mathbf{i}_{\mathrm{R}}^{\mathrm{e}}$ are the moment of inertia, angular speed, and constant spin axis of the wheel, respectively, relative to the spacecraft. The torque applied by the reaction wheel on the spacecraft is thus given by

$$\boldsymbol{\tau} = -J_{\mathrm{R}}\dot{\omega}_{\mathrm{R}}\mathbf{i}_{\mathrm{R}}^{\mathrm{e}}. \tag{7.68}$$

Clearly, in order to control the general attitude we must either have a single reaction wheel with an inclination with respect to all three principal axes, or one reaction wheel about each principal axis. For redundancy and better control, there are usually four reaction wheels mounted in a tetrahedral configuration such that any two wheels are capable of applying a general torque.

A momentum wheel has a fixed axis and a constant angular speed relative to the spacecraft, and thus can produce a control torque only when there is a change in the spacecraft's attitude. The wheel's constant angular momentum relative to the spacecraft is given by

$$\mathbf{H}_{\mathrm{R}}^{\mathrm{e}} = J_{\mathrm{R}} \omega_{\mathrm{R}}^{\mathrm{e}} \mathbf{i}_{\mathrm{R}}, \tag{7.69}$$

and the control torque it applies to the spacecraft rotating with angular velocity, $\boldsymbol{\omega}$, is

$$\boldsymbol{\tau} = -\boldsymbol{\omega} \times J_{\mathrm{R}} \omega_{\mathrm{R}}^{\mathrm{e}} \mathbf{i}_{\mathrm{R}}. \tag{7.70}$$

A control moment gyroscope (CMG) is a wheel spinning at a constant rate, $\omega_{\mathrm{R}}^{\mathrm{e}}$, that generates its control torque by changing its spin axis relative to the spacecraft:

$$\boldsymbol{\tau} = -J_{\mathrm{R}} \omega_{\mathrm{R}}^{\mathrm{e}} \frac{\mathrm{d}\mathbf{i}_{\mathrm{R}}}{\mathrm{d}t}. \tag{7.71}$$

A variable-speed control moment gyroscope (VSCMG) can modify both speed and axis of spin relative to the spacecraft, and thus offers a better control compared with either a CMG or a reaction wheel. The control torque generated by a VSCMG is given by

$$\boldsymbol{\tau} = -J_{\mathrm{R}} \omega_{\mathrm{R}} \frac{\mathrm{d}\mathbf{i}_{\mathrm{R}}}{\mathrm{d}t} - J_{\mathrm{R}} \dot{\omega}_{\mathrm{R}} \mathbf{i}_{\mathrm{R}}. \tag{7.72}$$

A single VSCMG with a general spin axis inclination can control all the axes of the spacecraft. However, designing control systems for CMGs and VSCMGs is difficult due to the nonlinear nature of the control torque produced by them. Nonlinear dynamical response to VSCMG actuation is presented in Chapter 14 of Tewari (2006).

It is often practical to include the actuator model for a rotor while designing or simulating a closed-loop attitude control system. Such an actuator is usually a servomotor that can quickly change either the reaction wheel speed (by using a tachometer or rev counter) or the CMG spin axis (by using an angle encoder) for feedback. The servo gains must be designed such that the servo dynamics is several times faster than that of the wheel.

7.4.3 Magnetic Field Torque

A significant control torque can be generated by a low-orbiting spacecraft from a planet's magnetic field through magnetic coils and eddy conductors that have a flowing electric current. By adjusting the magnitude and direction of the current loops, a control torque can be created about a desired axis. Furthermore, since the control torque depends upon the spacecraft's orientation, a natural closed-loop attitude regulation is provided by the magnetic field torque. An example of magnetic torquers is the magnetic eddy current damper, which consists of a conductor plate moving between the poles of an electromagnet, and provides a damping force proportional to the relative velocity between the plate and the magnet (Haines and Leondes 1973). Due to the dependence of the control torque on orbital declination, the magnetic torque controller is essentially a linear time-varying system and can be used to increase closed-loop damping and natural frequency in pitch. However, a magnetic torquer provides only a weak roll–yaw coupling and therefore must be used in conjunction with other torque-generating devices (such as reaction and momentum wheels). Furthermore, the magnetically generated control torques are seldom large enough to overcome atmospheric drag and gravity-gradient perturbations.

7.5 Satellite Dynamics Plant for Tracking Control

A satellite in a circular orbit is the most common space flight application, and is utilized for communication, planetary observation, and mapping missions. A certain spacecraft attitude relative to the planet must be maintained for a successful mission. Generally, either a camera or a communications payload is mounted on a platform that always points toward the planet's center. If the platform is fixed on the satellite, the latter itself has a principal axis pointing in the radial direction, while the other axes are along the flight direction and normal to the orbital plane, respectively, as shown in Figure 7.13. Such an orientation is called *nadir pointing* and constitutes a common reference attitude for linearizing the equations of motion. Needless to say, an attitude control system is invariably required to maintain the equilibrium orientation in the presence of disturbing torques.

If we take the equilibrium axis ox^e of a nadir pointing satellite along the flight direction, the axis oz^e pointing toward planetary center, and oy^e opposite to orbital angular momentum, we have the following angular veclocity components, $\boldsymbol{\omega} = P\mathbf{i} + Q\mathbf{j} + R\mathbf{k}$:

$$\begin{aligned} P &= p, \\ Q &= -n + q, \\ R &= r, \end{aligned} \tag{7.73}$$

where p, q, r are small disturbances from the equilibrium angular velocity, $\boldsymbol{\omega}^e = -n\mathbf{j}^e$. Furthermore, a small displacement from the initial equilibrium attitude is represented by the Euler angles $(\psi)_3, (\theta)_2, (\phi)_1$ (Tewari 2006), resulting in

$$\boldsymbol{\omega} \simeq -n\mathbf{j}^e + \dot{\phi}\mathbf{i} + \dot{\theta}\mathbf{j} + \dot{\psi}\mathbf{k}, \tag{7.74}$$

where

$$\mathbf{j}^e \simeq \psi\mathbf{i} + \mathbf{j} - \phi\mathbf{k}. \tag{7.75}$$

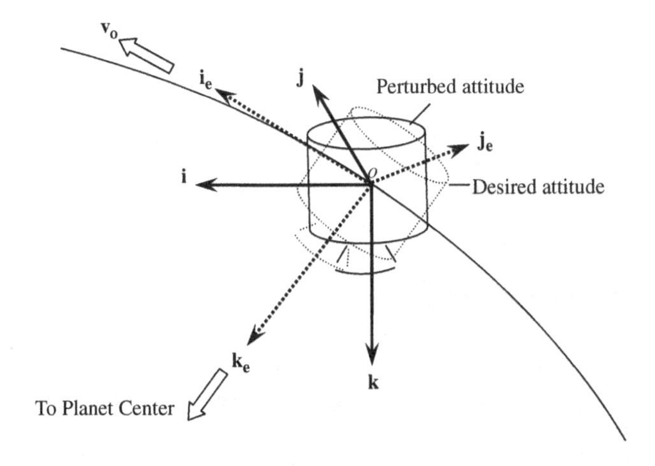

Figure 7.13 A nadir pointing satellite in a circular orbit

Substitution of equation (7.75) into equation (7.74) and comparison with equation (7.73) yields the following linearized kinematical equations for small angular displacements, ψ, θ, ϕ:

$$\dot{\phi} = p + n\psi, \tag{7.76}$$

$$\dot{\theta} = q, \tag{7.77}$$

$$\dot{\psi} = r - n\phi. \tag{7.78}$$

Hence, there is a kinematical coupling between roll and yaw.

In order to derive the rotational kinetic equations, we consider the following equilibrium angular momentum of the spacecraft with momentum wheels of constant angular momentum relative to the spacecraft, $\mathbf{H}_{\mathrm{R}}^{\mathrm{e}} = H_{\mathrm{R}x}^{\mathrm{e}}\mathbf{i} + H_{\mathrm{R}y}^{\mathrm{e}}\mathbf{j} + H_{\mathrm{R}z}^{\mathrm{e}}\mathbf{k}$:

$$\mathbf{H}^{\mathrm{e}} = -J_{yy}n\mathbf{j} + \mathbf{H}_{\mathrm{R}}^{\mathrm{e}}. \tag{7.79}$$

For the spacecraft to be in equilibrium, we must have a zero net torque,

$$-n\mathbf{j} \times \mathbf{H}^{\mathrm{e}} = -n\mathbf{j} \times \mathbf{H}_{\mathrm{R}}^{\mathrm{e}} = n(H_{\mathrm{R}x}^{\mathrm{e}}\mathbf{k} - H_{\mathrm{R}z}^{\mathrm{e}}\mathbf{i}) = \mathbf{0} \tag{7.80}$$

or

$$\begin{aligned} H_{\mathrm{R}x}^{\mathrm{e}} &= 0, \\ H_{\mathrm{R}z}^{\mathrm{e}} &= 0. \end{aligned} \tag{7.81}$$

Thus, a satellite in equilibrium cannot have momentum wheels about roll and yaw axes. However, a pitch momentum wheel, $\mathbf{H}_{\mathrm{R}}^{\mathrm{e}} = H_{\mathrm{R}y}^{\mathrm{e}}\mathbf{j}$ can be provided for biasing purposes (i.e., additional roll–yaw coupling) while keeping the vehicle in equilibrium. The linearized rotational kinetics plant for a small angular velocity deviation, p, q, r, from the equilibrium state thus becomes

$$\begin{aligned} \boldsymbol{\tau}_C + \boldsymbol{\tau}_D &= \mathbf{J}\frac{\partial \boldsymbol{\omega}}{\partial t} + \mathbf{S}(\boldsymbol{\omega})(\mathbf{J}\boldsymbol{\omega} + H_{\mathrm{R}y}^{\mathrm{e}}\mathbf{j}) \\ &= \mathbf{J}\begin{Bmatrix} \dot{p} \\ \dot{q} \\ \dot{r} \end{Bmatrix} \mathbf{S}(-n\mathbf{j}) \left[\mathbf{J}\begin{Bmatrix} p \\ (-n+q) \\ r \end{Bmatrix} + H_{\mathrm{R}y}^{\mathrm{e}}\mathbf{j} \right] \\ &= \begin{Bmatrix} J_{xx}\dot{p} + [(J_{yy} - J_{zz})n - H_{\mathrm{R}y}^{\mathrm{e}}]r \\ J_{yy}\dot{q} \\ J_{zz}\dot{r} + [(J_{xx} - J_{yy})n + H_{\mathrm{R}y}^{\mathrm{e}}]p \end{Bmatrix}, \end{aligned} \tag{7.82}$$

where $\boldsymbol{\tau}_C$ is the control torque (due to reaction wheels, control moment gyro, jet thrusters and magnetic coils) and $\boldsymbol{\tau}_D$ the disturbance torque (gravity-gradient, solar radiation, structural/propellant dynamics, etc.).

7.6 Environmental Torques

For a spacecraft in a low orbit ($h < 300$ km for the Earth), the aerodynamic torque is often the predominant disturbance that is nearly proportional to the angular deviation. As the altitude increases, the aerodynamic torque decreases exponentially to a negligible magnitude. The *gravity-gradient* and *solar radiation* torque disturbances are negligible in comparison with the aerodynamic torque within the atmosphere, but become significant in the quiet environment of space. If a spacecraft has a relatively large difference in its principal moments of inertia, a substantial gravity-gradient torque is experienced in a low orbit. Away from the immediate vicinity of a planet where aerodynamic and gravity-gradient torques are negligible, the spacecraft comes under the influence of solar radiation torque. All environmental torques depend upon the spacecraft's orientation and hence require careful modeling.

Both solar radiation and aerodynamic torques arise due to the pressure exerted by sunlight and the rarefied upper atmosphere, respectively, and can be modeled by the following common expression:

$$\boldsymbol{\tau}_D = \int \boldsymbol{\xi} \times \mathbf{df}, \tag{7.83}$$

where $\boldsymbol{\xi}$ locates an elemental surface area, dA, relative to the spacecraft's center of mass, $\mathbf{df}$ is the net differential force applied on the elemental surface by aerodynamics and solar radiation, and the integral is carried out over the entire external surface of the spacecraft. The solar radiation component of $\mathbf{df}$ is denoted by $\mathbf{df}_s$ and depends upon the angular orientation of the elemental surface with respect to the sun line (i.e., line joining spacecraft's center of mass with that of the Sun) and has different analytical models depending upon whether the radiation is completely absorbed, specularly reflected, or diffusely reflected (Wertz 1978). Most spacecraft have large solar panels arrayed nearly symmetrically about a principal axis, and are often deflected such that they are always normal to the sun line for the best electrical efficiency. Therefore, the solar radiation torque is unaffected by a small displacement from the equilibrium orientation, and can be considered nearly constant.

The aerodynamic component of $\mathbf{df}$ is denoted by $\mathbf{df}_a$ and depends upon the angular orientation of the elemental surface area, dA, with respect to the spacecraft's velocity vector, $\mathbf{v}$. It can be modeled as follows by assuming that the incident air molecules lose their entire kinetic energy upon surface impact:

$$\mathbf{df}_a = -\frac{1}{2}\rho C_d(\mathbf{n}\cdot\mathbf{v})\mathbf{v}dA, \tag{7.84}$$

where $\mathbf{n}$ is the outward unit normal of the elemental surface area, C_d is its drag coefficient, and ρ is the atmospheric density. While detailed models for C_d are available depending upon the smoothness of the surface and the local angle of attack (Schaaf and Chambre 1958), a reasonable approximation is obtained by taking $C_d = 2.0$.

A spacecraft with a small pitch and yaw angular perturbation, θ, ψ, relative to the velocity vector, has the aerodynamic disturbance torque, $\boldsymbol{\tau}_{\mathrm{Da}}$, approximately proportional to the angular perturbation:

$$\boldsymbol{\tau}_{\mathrm{Da}} = \mathbf{D}(\theta, \psi)^T, \tag{7.85}$$

where $\mathbf{D}$ is a matrix of scalar constants called *aerodynamic stiffness derivatives*. Note that for a spacecraft of characteristic length, ℓ, linear velocity, $\mathbf{v}_0$, and angular velocity, $\boldsymbol{\omega}$, we have $\ell|\boldsymbol{\omega}| \ll |\mathbf{v}_0|$. Therefore, the aerodynamic damping terms (i.e., those that depend upon the angular velocity, $\boldsymbol{\omega}$) are several orders of magnitude smaller than the stiffness terms, and we can easily neglect the former in favor of the latter. Using the aircraft nomenclature of Chapter 3 for rolling moment, L, pitching moment, M, and yawing moment, N, we write

$$\mathbf{D} = \begin{pmatrix} L_\theta & L_\psi \\ M_\theta & M_\psi \\ N_\theta & N_\psi \end{pmatrix}. \tag{7.86}$$

In aircraft nomenclature (Chapter 3), pitch and yaw deviations from the velocity vector are called angle-of-attack and angle-of-sideslip disturbances, α, β, respectively. Thus for a spacecraft in the equilibrium reference frame of Figure 7.13, which is the same as the stability axes for an aircraft, we have $\alpha = \theta$ and $\beta = \psi$. Note that whereas for an aircraft we have $L_\theta = M_\psi = N_\theta = 0$ as a result of decoupling of longitudinal and lateral-directional dynamics due to *oxz* being the plane of symmetry (Chapter 3), the plane of symmetry may either be absent, or different from *oxz*, leading to an aerodynamic cross-coupling between roll, pitch, and yaw. A nadir pointing satellite equipped with a pitch momentum wheel thus has the following equations of small perturbation:

$$\tau_C = \mathbf{J}\frac{\partial \boldsymbol{\omega}}{\partial t} + \mathbf{S}(\boldsymbol{\omega})[(\mathbf{J}-\mathbf{D})\boldsymbol{\omega} + H^{\text{e}}_{\text{Ry}}\mathbf{j}]$$
$$= \begin{Bmatrix} J_{xx}\dot{p} + [(J_{yy}-J_{zz})n - H^{\text{e}}_{\text{Ry}}]r - L_\theta\theta - L_\psi\psi \\ J_{yy}\dot{q} - M_\theta\theta - M_\psi\psi \\ J_{zz}\dot{r} + [(J_{xx}-J_{yy})n + H^{\text{e}}_{\text{Ry}}]p - N_\theta\theta - N_\psi\psi \end{Bmatrix}. \tag{7.87}$$

With the state vector of the linearized spacecraft plant, $\mathbf{x} = (\phi, \theta, \psi, p, q, r)^T$, the dynamic state coefficient matrix is given by

$$\mathbf{A} = \begin{Bmatrix} 0 & 0 & 0 & 1 & 0 & 0 \\ 0 & 0 & 0 & 0 & 1 & 0 \\ 0 & 0 & 0 & 0 & 0 & 1 \\ 0 & \frac{L_\theta}{J_{xx}} & \frac{L_\psi}{J_{xx}} & 0 & 0 & \frac{(J_{zz}-J_{yy})n + H^{\text{e}}_{\text{Ry}}}{J_{xx}} \\ 0 & \frac{M_\theta}{J_{yy}} & \frac{M_\psi}{J_{yy}} & 0 & 0 & 0 \\ 0 & \frac{N_\theta}{J_{zz}} & \frac{N_\psi}{J_{zz}} & \frac{(J_{yy}-J_{xx})n - H^{\text{e}}_{\text{Ry}}}{J_{zz}} & 0 & 0 \end{Bmatrix} \tag{7.88}$$

The aerodynamic stiffness terms can have either a stabilizing or destabilizing influence, depending on whether the effective aerodynamic center is aft or forward of the center of mass, respectively (Chapter 3). For example, $M_\theta > 0$ would cause pitch instability while $M_\psi > 0$ would lead to an unstable roll–yaw dynamics.

The solar radiation torque effects can be modeled in a manner similar to the aerodynamic torque, with the difference that the displacement angles (α, β) are now measured from the *Sun direction*, which is a unit vector from the center of mass of the spacecraft to the Sun.

7.6.1 Gravity-Gradient Torque

A spacecraft in a low-altitude orbit can generate an appreciable torque due to the variation of the gravity force along its dimensions, called the *gravity-gradient* torque. The magnitude of gravity gradient can be increased by employing a long boom in the desired direction. For a large spacecraft (such as the International Space-Station) in a low orbit, the gravity-gradient torque is capable of overwhelming the attitude control system over time, if not properly compensated for.

Consider a spacecraft in a low, circular orbit of radius R. The gravity-gradient torque experienced by the craft can be expressed as

$$\boldsymbol{\tau}_{Dg} = \int \boldsymbol{\rho} \times \mathbf{a}_g \text{d}m, \tag{7.89}$$

where $\boldsymbol{\rho}$ locates an elemental mass, dm, relative to the spacecraft's center of mass. The acceleration due to gravity, $\mathbf{a}_g$, is approximated by Newton's law of gravitation for a spherical planet, and can be expanded using the binomial theorem as follows:

$$\begin{aligned}\mathbf{a}_g &= -\mu\frac{\mathbf{R}+\boldsymbol{\rho}}{\|\ \mathbf{R}+\boldsymbol{\rho}\ \|^3} \\ &= \frac{\mu(\mathbf{R}+\boldsymbol{\rho})}{R^3}\left(1-3\frac{\mathbf{R}\cdot\boldsymbol{\rho}}{R^2}+\ldots\right).\end{aligned} \tag{7.90}$$

Ignoring the second- and higher-order terms in equation (7.90), and carrying out the integral of equation (7.89) in terms of the body referenced components of $\mathbf{R} = X\mathbf{i} + Y\mathbf{j} + Z\mathbf{k}$ and $\boldsymbol{\rho} = x\mathbf{i} + y\mathbf{j} + z\mathbf{k}$ (where $\mathbf{i}, \mathbf{j}, \mathbf{k}$ are the spacecraft's principal body axes), we have

$$\boldsymbol{\tau}_{Dg} = \tau_{gx}\mathbf{i} + \tau_{gy}\mathbf{j} + \tau_{gz}\mathbf{k}, \tag{7.91}$$

where

$$\begin{aligned}\tau_{gx} &= \frac{3\mu}{R^5}YZ(J_{zz} - J_{yy}), \\ \tau_{gy} &= \frac{3\mu}{R^5}XZ(J_{xx} - J_{zz}), \\ \tau_{gz} &= \frac{3\mu}{R^5}XY(J_{yy} - J_{xx}).\end{aligned} \tag{7.92}$$

Let the equilibrium attitude of the spacecraft be given by the undisturbed body axes, $\mathbf{i}^e, \mathbf{j}^e, \mathbf{k}^e$, with $\mathbf{i}^e$ along the direction of flight (velocity vector), $\mathbf{k}^e$ pointing toward the planetary center, and $\mathbf{j}^e$ completing the right-hand triad being opposite to the orbital angular momentum, as shown in Figure 7.13. The position vector resolved in the body axes is then given in terms of the 3-2-1 Euler angles by

$$\mathbf{R} = R(\sin\theta\mathbf{i} - \sin\phi\cos\theta\mathbf{j} - \cos\phi\cos\theta\mathbf{k}), \tag{7.93}$$

which, substituted into equation (7.92), leads to the following gravity-gradient torque components:

$$\begin{aligned}\tau_{gx} &= \frac{3}{2}n^2(J_{zz} - J_{yy})\cos^2\theta\sin(2\phi), \\ \tau_{gy} &= \frac{3}{2}n^2(J_{zz} - J_{xx})\cos\phi\sin(2\theta), \\ \tau_{gz} &= \frac{3}{2}n^2(J_{xx} - J_{yy})\sin\phi\sin(2\theta),\end{aligned} \tag{7.94}$$

where $n = \sqrt{\mu/R^3}$ is the mean motion (Chapter 6). If the attitude dynamics is stable, the small deviations, ϕ, θ, ψ, will remain small, thereby producing the following linearized approximation for the gravity-gradient torque components:

$$\begin{aligned}\tau_{gx} &\simeq 3n^2(J_{zz} - J_{yy})\phi, \\ \tau_{gy} &\simeq 3n^2(J_{zz} - J_{xx})\theta, \\ \tau_{gz} &\simeq 0.\end{aligned} \tag{7.95}$$

7.7 Multi-variable Tracking Control of Spacecraft Attitude

The linearized attitude control plant for a nadir pointing satellite with a pitch momentum wheel, roll, pitch and yaw reaction wheels in a low circular orbit can be expressed as

$$\dot{\mathbf{x}} = \mathbf{A}\mathbf{x} + \mathbf{B}\boldsymbol{\tau}_C, \tag{7.96}$$

where $\mathbf{x} = (\phi, \theta, \psi, p, q, r)^T$ is the state vector, $\boldsymbol{\tau}_C$ the control input vector generated by the three reaction wheels as well as any magnetic torquers, and the coefficient matrices are given by

$$\mathbf{A} = \begin{Bmatrix} 0 & 0 & 0 & 1 & 0 & 0 \\ 0 & 0 & 0 & 0 & 1 & 0 \\ 0 & 0 & 0 & 0 & 0 & 1 \\ a_{41} & a_{42} & a_{43} & 0 & 0 & a_{46} \\ 0 & a_{52} & a_{53} & 0 & 0 & 0 \\ 0 & a_{62} & a_{63} & a_{64} & 0 & 0 \end{Bmatrix}, \tag{7.97}$$

$$\mathbf{B} = \begin{pmatrix} 0 & 0 & 0 \\ 0 & 0 & 0 \\ 0 & 0 & 0 \\ -\frac{1}{J_{xx}} & 0 & 0 \\ 0 & -\frac{1}{J_{yy}} & 0 \\ 0 & 0 & -\frac{1}{J_{zz}} \end{pmatrix}, \tag{7.98}$$

where

$$\begin{aligned} a_{41} &= \frac{L_\phi + 3n^2(J_{zz} - J_{yy})}{J_{xx}}, \\ a_{42} &= \frac{L_\theta}{J_{xx}}, \\ a_{43} &= \frac{L_\psi}{J_{xx}}, \\ a_{46} &= \frac{(J_{zz} - J_{yy})n + H^{\text{e}}_{\text{R}y}}{J_{xx}}, \\ a_{52} &= \frac{M_\theta + 3n^2(J_{zz} - J_{xx})}{J_{yy}}, \\ a_{53} &= \frac{M_\psi}{J_{yy}}, \\ a_{62} &= \frac{N_\theta}{J_{zz}}, \\ a_{63} &= \frac{N_\psi}{J_{zz}}, \\ a_{64} &= \frac{(J_{yy} - J_{xx})n - H^{\text{e}}_{\text{R}y}}{J_{zz}}. \end{aligned} \tag{7.99}$$

Here, the aerodynamic and solar radiation effects are clubbed together in the constant coefficients $L_\phi, L_\theta, L_\psi, M_\theta, M_\psi, N_\theta, N_\psi$. We can augment the plant by adding the reaction wheel actuators (servomotors) that control the wheel speeds, $\omega_x, \omega_y, \omega_z$, by wheel torque inputs, $\mathbf{u} = (u_x, u_y, u_z)^T$, as follows:

$$\begin{aligned} j_x\dot{\omega}_x &= u_x, \\ j_y\dot{\omega}_y &= u_y, \\ j_z\dot{\omega}_z &= u_z, \end{aligned} \tag{7.100}$$

where j_x, j_y, j_z are the respective moments of inertia of the roll, pitch, and yaw reaction wheels about their own spin axes.

Since the attitude dynamics plant is controllable, one can design a suitable feedback control system using an observer-based linear optimal regulator (Chapter 2) given by

$$\mathbf{u} = -\mathbf{K}\mathbf{x}_o, \tag{7.101}$$

where $\mathbf{x}_o$ is the estimated state vector.

7.7.1 Active Attitude Control of Spacecraft by Reaction Wheels

Since the magnetic torquers can have only a small impact on the closed-loop damping of a low-orbiting satellite in the presence of gravity-gradient, solar radiation, and aerodynamic torques, one requires the greater controllability offered by either rocket thrusters or reaction wheels in order actively to bring the attitude errors to zero. However, frequent use of rocket thrusters for attitude stabilization will quickly deplete the onboard propellant, thereby reducing the satellite's life in orbit. Therefore, reaction wheels powered by solar panels are a much better proposition for active attitude regulation.

Example 7.5 Consider a small satellite equipped with roll, pitch, and yaw reaction wheels and a pitch momentum wheel in a circular Earth orbit of altitude 274.42 km, having the following parameters (Kim *et al.* 2004):

$$\begin{aligned} J_{xx} &= 150 \text{ kg.m}^2, \\ J_{yy} &= 200 \text{ kg.m}^2, \\ J_{zz} &= 100 \text{ kg.m}^2, \\ n &= 0.0011636 \text{ rad/s}, \\ H^{\text{e}}_{\text{R}y} &= 10 \text{ N.m.s}, \\ L_\psi &= -5.1354 \times 10^{-11} \text{ N.m/rad}, \\ M_\theta &= -8.559 \times 10^{-10} \text{ N.m/rad}, \\ M_\psi &= -1.7118 \times 10^{-11} \text{ N.m/rad}, \\ N_\psi &= -1.7118 \times 10^{-10} \text{ N.m/rad..} \end{aligned}$$

Using horizon scanners for measuring roll, pitch, and yaw angles, design a reaction wheel based control system to bring an initial attitude error of $\phi(0) = 1^\circ$, $\theta(0) = 0.5^\circ$, and $\psi(0) = -0.5^\circ$ to zero in about 100 s, without exceeding the maximum reaction wheel torque magnitude of 0.05 N.m.

To design the multi-variable, Kalman filter based controller, we adopt the linear quadratic Gaussian (LQG) method as follows:

```
>> A =       [0              0     0.0011636          1         0          0
              0              0             0          0         1          0
     -0.0011636              0             0          0         0          1
```

```
 -2.7077e-006             0 -3.4236e-013           0          0     0.065891
            0 -1.0154e-006   -8.559e-014           0          0            0
            0            0 -1.7118e-012   -0.099418          0            0];

>> B =      [0            0            0
            0            0            0
            0            0            0
   -0.0066667            0            0
            0       -0.005            0
            0            0        -0.01];

>> C = [eye(3) zeros(3,3)]; D=zeros(3,3);

>> damp(A) % Plant's eigenvalues

        Eigenvalue                Damping      Freq. (rad/s)
 -2.63e-012 + 8.10e-002i      3.25e-011       8.10e-002
 -2.63e-012 - 8.10e-002i      3.25e-011       8.10e-002
  2.63e-012 + 1.14e-003i     -2.30e-009       1.14e-003
  2.63e-012 - 1.14e-003i     -2.30e-009       1.14e-003
  6.76e-020 + 1.01e-003i     -6.70e-017       1.01e-003
  6.76e-020 - 1.01e-003i     -6.70e-017       1.01e-003

% open-loop plant's initial response:
>> sys=ss(A,B,[1 0 0 0 0 0],0);
>> dtr=pi/180;
>> T=0:.1:100; [y,t,x]=initial(sys,[dtr 0.5*dtr -0.5*dtr 0 0 0]',T);
>> [K,S,E]=lqr(A,B,eye(6),eye(3)) % Regulator design by LQR method

K -
     -0.82593  3.6867e-013      0.56144      -15.774 -4.9821e-011   0.00086752
-1.8666e-012      -0.9998  1.5183e-011 -3.7366e-011      -20.023  6.1164e-011
     -0.56479  2.5786e-012     -0.82641    0.0013013  1.2233e-010      -12.894

S =
       18.551  5.9169e-012     -0.78518       123.89  3.7332e-010       56.479
  5.9169e-012       20.023 -9.0771e-011 -5.5301e-011       199.96 -2.5786e-010
     -0.78518 -9.0771e-011        16.14      -84.217 -3.0367e-009       82.641
       123.89 -5.5301e-011      -84.217         2366  7.4731e-009     -0.13013
  3.7332e-010       199.96 -3.0367e-009  7.4731e-009       4004.6 -1.2233e-008
       56.479 -2.5786e-010       82.641     -0.13013 -1.2233e-008       1289.4

E =
    -0.059716 +    0.11164i
    -0.059716 -    0.11164i
    -0.057334 +   0.029535i
    -0.057334 -   0.029535i
    -0.050057 +   0.049943i
    -0.050057 -   0.049943i

>> Lp=lqr(A',C',10*eye(6),eye(3));L=Lp' % Kalman-filter design

L =
       4.0389  2.0882e-016    -0.016642
```

```
    2.0882e-016        4.0404 -1.0743e-014
      -0.016642 -1.0743e-014         4.0418
         3.1566 -6.1116e-016     -0.0034826
   -5.9061e-016        3.1623 -8.5962e-014
         -0.131 -1.7156e-015         3.1681

>> damp(A-L*C) % observer eigenvalues

        Eigenvalue                Damping      Freq. (rad/s)
 -1.06e+000 + 9.45e-002i        9.96e-001        1.07e+000
 -1.06e+000 - 9.45e-002i        9.96e-001        1.07e+000
 -2.98e+000 + 1.07e-002i        1.00e+000        2.98e+000
 -2.98e+000 - 1.07e-002i        1.00e+000        2.98e+000
 -1.06e+000                     1.00e+000        1.06e+000
 -2.98e+000                     1.00e+000        2.98e+000

>> Ac=[A  -B*K;L*C A-L*C-B*K]; % closed-loop dynamics matrix

% Closed-loop initial response:
>> sys=ss(Ac,zeros(12,3),zeros(1, 12),zeros(1,3));
>> [y,t,x]=initial(sys,[dtr 0.5*dtr -0.5*dtr zeros(1,9)]',T);
>> u=-x*[K zeros(3,6)]'; % closed-loop reaction wheel torques
```

The open-loop plant's initial response is plotted in Figure 7.14, showing an unstable system. The closed-loop response is depicted in Figures 7.15 and 7.16, indicating an asymptotically stable closed-loop system, with a settling time less than 100 s and maximum reaction wheel torques within ±0.02 N.m. The observer dynamics is about 20 times faster than the regulator, hence there is virtually no delay in the closed-loop response due to Kalman filter dynamics.

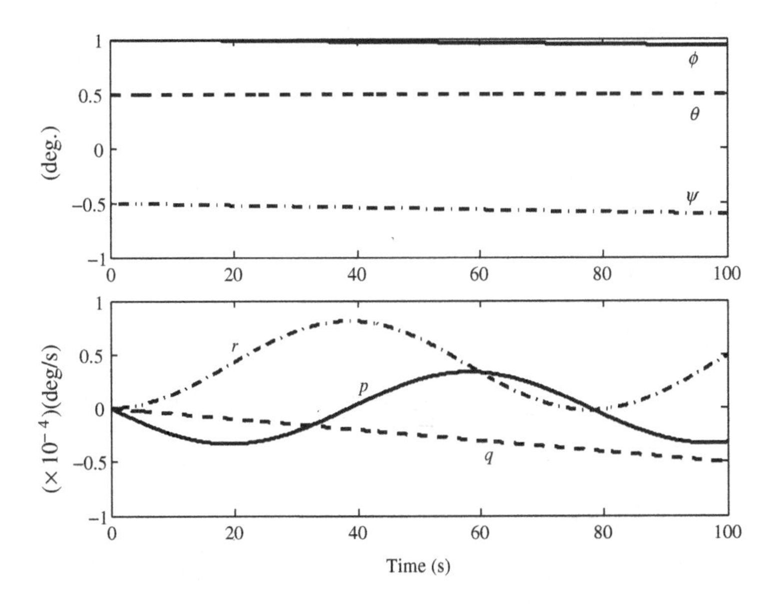

Figure 7.14 Open-loop attitude response of a spacecraft with aerodynamic and gravity-gradient torques in a low circular Earth orbit and perturbed by initial attitude error

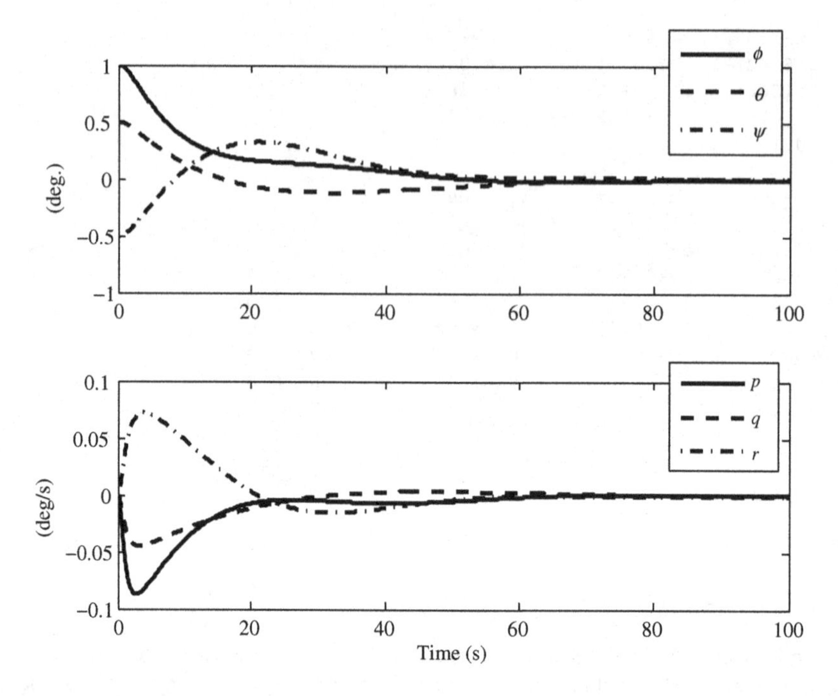

Figure 7.15 Closed-loop attitude response of a spacecraft with observer-based reaction wheel control system in a low circular Earth orbit and perturbed by initial attitude error

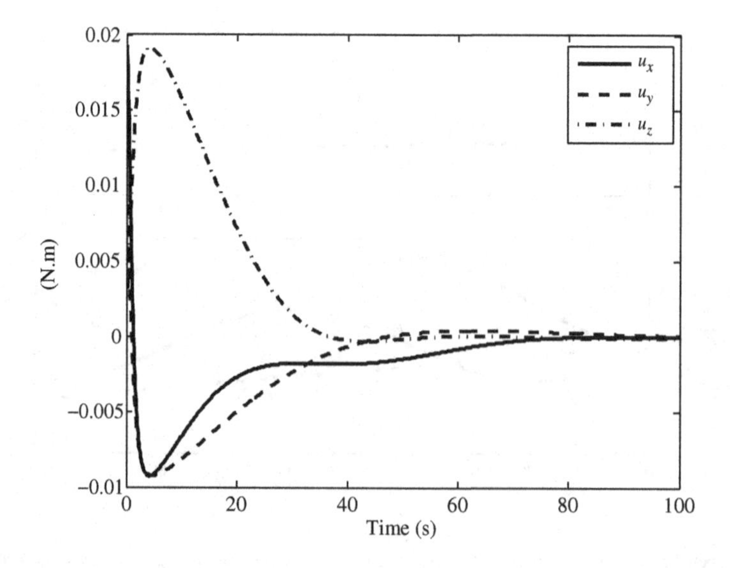

Figure 7.16 Closed-loop reaction wheel torques for the spacecraft response plotted in Figure 7.15

7.8 Summary

Large and rapid attitude maneuvers of spacecraft are controlled using terminal 2PBVP controllers designed by optimal control theory. Attitude tracking control of a spacecraft is crucial to its mission and can be carried out by rocket thrusters, rotors (reaction wheels, momentum wheels, control moment gyros, variable-speed control moment gyros), and magnetic torquers. If uncontrolled, the spacecraft attitude is quickly destabilized by aerodynamic, solar radiation, and gravity-gradient torque disturbances that are functions of the spacecraft's orientation relative to an orbital reference frame. The aerodynamic, gravity-gradient, and solar radiation disturbances can be appropriately modeled in a linearized orbital plant, which can be controlled by multi-variable design techniques, such as LQG and H_∞.

Exercises

(1) For the torque-free motion of a rigid, asymmetric spacecraft initially spinning about the principal axis, oy:
   (a) Derive the state equations of the system if the motion is a small angular velocity perturbation over the initial equilibrium state of $\boldsymbol{\omega} = (0, -n, 0)^T$, where n is a constant rate.
   (b) What are the conditions on the principal moments of inertia, J_{xx}, J_{yy}, J_{zz}, for the system's stability?
   (c) Under what conditions (if any) is the system controllable by torques applied about the principal axes, ox and oy?

(2) A rigid, axisymmetric spacecraft with $J_{xx} = J_{yy}$ and $J_{zz} = 2J_{yy}$ is initially spinning about the principal axis, oy, at a constant rate, $-n$. If the system is controllable using separate reaction wheels about the pitch and yaw principal axes, oy and oz, respectively, design separate single-input, proportional-derivative (PD) regulators (Chapter 1) for pitch and yaw such that the closed-loop system's response to an initial disturbance has a settling time of $0.1/n$ and a damping ratio of 0.707. What is the desired ratio of moments of inertia of pitch and yaw reaction wheels?

(3) Repeat Exercise 2 using multi-input LQR control (Chapter 2) instead of PD control with the two reaction wheels.

(4) Replace the full-state feedback LQR regulator of Exercise 3 by an LQG compensator (Chapter 2), using two suitable output variables for feedback.

(5) Can a rigid, symmetric ($J_{xx} = J_{yy} = J_{zz}$) spacecraft, initially spinning about the principal axis, oy, at a constant rate, $-n$, be controlled with separate reaction wheels about the pitch and yaw principal axes, oy and oz, respectively?

(6) A rigid, axisymmetric spacecraft with $J_{xx} = J_{yy}$ and $J_{zz} = 2J_{yy}$, initially spinning about the pitch principal axis, oy, at a constant rate, $-n$, has a pitch momentum wheel with $H^e_{Ry} = 0.1J_{yy}n$, as well as roll and pitch reaction wheels. Design separate single-input PD regulators for pitch and roll such that the closed-loop system's response to an initial disturbance has a settling time of $0.1/n$, and a damping ratio of 0.707. What is the ratio of the moments of inertia of the pitch and roll reaction wheels?

(7) For a rigid spacecraft under the influence of gravity-gradient torque alone:
   (a) Find the necessary condition for the pitching motion to be stable.
   (b) Under what conditions can we have a stable roll–yaw dynamics?
   (c) Is the spacecraft observable with θ as the single output?
   (d) Is the spacecraft observable with ϕ as the single output?

(8) For the spacecraft of Example 7.5, what happens when the pitch momentum wheel stops functioning? Repeat the closed-loop simulation with $H^e_{Ry} = 0$.

(9) For the spacecraft of Example 7.5, add first-order actuators for the reaction wheels with identical moments of inertia, $j = 2$ kg.m^2, and repeat the closed-loop simulation. Plot the closed-loop angular speed of each reaction wheel as a function of time.

(10) Redesign the Kalman filter of Example 7.5 for loop-transfer recovery at the plant's input in the closed-loop system's bandwidth. Plot the singular values of the return ratio concerned. Is there any change in the closed-loop performance?

(11) A satellite with $J_{xx} = 1000$ kg.m^2, $J_{yy} = 1500$ kg.m^2, and $J_{zz} = 2000$ kg.m^2 in circular Earth orbit of 90 minutes has a roll momentum wheel of angular momentum 100 N.m.s and a yaw reaction wheel with moment of inertia 50 kg.m^2. The spacecraft is initially in a nadir pointing attitude (Figure 7.13).
(a) Is the spacecraft in equilibrium?
(b) Analyze the stability and controllability of the plant.
(c) If possible, design a control system with all the closed-loop poles having real part $-3n$, where n is the orbital frequency.

References

Haines, G.A. and Leondes, C.T. (1973) Eddy current nutation dampers for dual-spin satellites. *Journal of the Astronautical Sciences*, **21**, 1.

Kim, J.W., Crassidis, J.L., Vadali, S.R., and Dershowitz, A.L. (2004) International Space Station leak localization using vent torque estimation. *Proceedings of the 55th International Astronautical Congress*, Paper IAC-04-A.4.10.

Schaaf, S.A. and Chambre, P.L. (1958) *Flow of Rarefied Gases*. Princeton University Press, Princeton, NJ.

Slotine, J.-J.E. and Li, W. (1991) *Applied Nonlinear Control*. Prentice Hall, Englewood Cliffs, NJ.

Tewari, A. (2002) *Modern Control Design with MATLAB and Simulink*. John Wiley & Sons, Ltd, Chichester.

Tewari, A. (2006) *Atmospheric and Space Flight Dynamics*. Birkhäuser, Boston.

Wertz, J.R. (ed.) (1978) *Spacecraft Attitude Determination and Control*. Kluwer Academic Publishers, Dordrecht.

Appendix A

Linear Systems

The theory of linear systems refers to the mathematical framework of a discretized (finite-dimensional) system linearized about a particular solution. For details we refer the reader to Kailath (1980).

A.1 Definition

A system with state vector, $\boldsymbol{\xi}(t)$, and input vector, $\boldsymbol{\eta}(t)$, governed by vector state equation,

$$\dot{\boldsymbol{\xi}} = \mathbf{f}(\boldsymbol{\xi}, \boldsymbol{\eta}, t), \tag{A.1}$$

initial condition,

$$\boldsymbol{\xi}(0) = \boldsymbol{\xi}_0, \tag{A.2}$$

and output equation,

$$\mathbf{y} = \mathbf{g}(\boldsymbol{\xi}, \boldsymbol{\eta}, t), \tag{A.3}$$

is said to be linear if its output vector resulting from the applied input vector,

$$\boldsymbol{\eta}(t) = c_1\boldsymbol{\eta}_1(t) + c_2\boldsymbol{\eta}_2(t), \tag{A.4}$$

is given by

$$\mathbf{y}(t) = c_1\mathbf{y}_1(t) + c_2\mathbf{y}_2(t), \tag{A.5}$$

where $\mathbf{y}_1(t)$ and $\mathbf{y}_2(t)$ are the output vectors of the system to the inputs $\boldsymbol{\eta}_1(t)$ and $\boldsymbol{\eta}_2(t)$, respectively, and c_1, c_2 are arbitrary scalar constants.

By inspecting the governing equations of a system, equations (A.1) and (A.3), it is possible to determine whether it is linear. If the functionals $\mathbf{f}(.)$, $\mathbf{g}(.)$ are continuous and do not contain nonlinear functions of the state and input variables, then the system is linear.

Advanced Control of Aircraft, Spacecraft and Rockets, Ashish Tewari.

A.2 Linearization

Let a reference state vector, $\boldsymbol{\xi}_r(t)$, and a corresponding reference input vector, $\boldsymbol{\eta}_r(t)$, satisfy the system's governing vector state equation (A.1),

$$\dot{\boldsymbol{\xi}}_r = \mathbf{f}(\boldsymbol{\xi}_r, \boldsymbol{\eta}_r, t), \tag{A.6}$$

subject to initial condition,

$$\boldsymbol{\xi}_r(0) = \boldsymbol{\xi}_{r0}. \tag{A.7}$$

Then, let $\mathbf{x}(t)$ and $\mathbf{u}(t)$ be state and control deviations, respectively, from the reference solution, $(\boldsymbol{\xi}_r, \boldsymbol{\eta}_r)$, such that the perturbed solution is given by

$$\begin{aligned}\boldsymbol{\xi}(t) &= \boldsymbol{\xi}_r(t) + \mathbf{x}(t)\\ \boldsymbol{\eta}(t) &= \boldsymbol{\eta}_r(t) + \mathbf{u}(t),\end{aligned} \tag{A.8}$$

subject to initial conditions (A.2) and (A.7), implying

$$\mathbf{x}(0) = \boldsymbol{\xi}(0) - \boldsymbol{\xi}_r(0) = \boldsymbol{\xi}_0 - \boldsymbol{\xi}_{r0} = \mathbf{x}_0. \tag{A.9}$$

Assuming that the vector functional, $\mathbf{f}(.)$, possesses continuous derivatives with respect to state and control variables up to an infinite order at the reference solution, $(\boldsymbol{\xi}_r, \boldsymbol{\eta}_r)$, we expand the state equation about the reference solution by neglecting the quadratic and higher-order terms as follows:

$$\dot{\boldsymbol{\xi}} - \dot{\boldsymbol{\xi}}_r = \dot{\mathbf{x}} = \mathbf{f}(\boldsymbol{\xi}_r + \mathbf{x}, \boldsymbol{\eta}_r + \mathbf{u}, t) - \mathbf{f}(\boldsymbol{\xi}_r, \boldsymbol{\eta}_r, t), \tag{A.10}$$

where

$$\begin{aligned}\mathbf{f}(\boldsymbol{\xi}_r + \mathbf{x}, \boldsymbol{\eta}_r + \mathbf{u}, t) \simeq{}& \mathbf{f}(\boldsymbol{\xi}_r, \boldsymbol{\eta}_r, t) + \frac{\partial \mathbf{f}}{\partial \boldsymbol{\xi}}(\boldsymbol{\xi}_r, \boldsymbol{\eta}_r, t)\mathbf{x}\\ &+ \frac{\partial \mathbf{f}}{\partial \boldsymbol{\eta}}(\boldsymbol{\xi}_r, \boldsymbol{\eta}_r, t)\mathbf{u}.\end{aligned}$$

Substitution of equation (A.11) into equation (A.10) yields the following linearized state equation about the reference solution:

$$\dot{\mathbf{x}} = \mathbf{A}(t)\mathbf{x} + \mathbf{B}(t)\mathbf{u}, \tag{A.11}$$

subject to initial condition (A.9), where $\mathbf{A}(t)$ and $\mathbf{B}(t)$ are the following Jacobian matrices:

$$\begin{aligned}\mathbf{A}(t) &= \frac{\partial \mathbf{f}}{\partial \boldsymbol{\xi}}(\boldsymbol{\xi}_r, \boldsymbol{\eta}_r, t)\\ \mathbf{B}(t) &= \frac{\partial \mathbf{f}}{\partial \boldsymbol{\eta}}(\boldsymbol{\xi}_r, \boldsymbol{\eta}_r, t).\end{aligned} \tag{A.12}$$

A.3 Solution to Linear State Equations

The solution to the linear state equation,

$$\dot{\mathbf{x}} = \mathbf{A}(t)\mathbf{x} + \mathbf{B}(t)\mathbf{u}, \tag{A.13}$$

subject to initial condition,

$$\mathbf{x}(0) = \mathbf{x}_0, \tag{A.14}$$

is expressed as the sum of homogeneous and particular solutions.

A.3.1 Homogeneous Solution

To derive the homogeneous solution, we write $\mathbf{u}(t) = \mathbf{0}$,

$$\dot{\mathbf{x}}(t) = \mathbf{A}(t)\mathbf{x}(t), \tag{A.15}$$

and

$$\mathbf{x}(t) = \mathbf{\Phi}(t, 0)\mathbf{x}_0, \quad \text{for all } t, \tag{A.16}$$

where $\mathbf{\Phi}(t, t_i)$ is the *state transition matrix* for state evolution from $t = t_i$ to the time t, with the properties of inversion,

$$\mathbf{\Phi}(t, t_i) = -\mathbf{\Phi}(t_i, t); \tag{A.17}$$

association,

$$\mathbf{\Phi}(t, t_i) = \mathbf{\Phi}(t, t_0)\mathbf{\Phi}(t_0, t_i); \tag{A.18}$$

and differentiation,

$$\frac{\mathrm{d}\mathbf{\Phi}(t, t_i)}{\mathrm{d}t} = \mathbf{A}(t)\mathbf{\Phi}(t, t_i). \tag{A.19}$$

A.3.2 General Solution

The general solution to the non-homogeneous state equation (A.13), subject to initial condition $\mathbf{x}_i = \mathbf{x}(t_i)$, is expressed as

$$\mathbf{x}(t) = \mathbf{\Phi}(t, t_i)\mathbf{x}_i + \int_{t_i}^{t} \mathbf{\Phi}(t, \tau)\mathbf{B}(\tau)\mathbf{u}(\tau)\mathrm{d}\tau, \quad \text{for all } t, \tag{A.20}$$

which can be verified by substituting into equation (A.13), along with the properties of $\mathbf{\Phi}(t, t_i)$.

The output (or *response*) of a linear system can be expressed as

$$\mathbf{y}(t) = \mathbf{C}(t)\mathbf{x}(t) + \mathbf{D}(t)\mathbf{u}(t), \tag{A.21}$$

where $\mathbf{C}(t)$ is called the *output coefficient matrix*, and $\mathbf{D}(t)$ is called the *direct transmission matrix*. If $\mathbf{D}(t) = \mathbf{0}$, the system is said to be *strictly proper*.

Substituting equation (A.20) into equation (A.21), we have the following expression for the system's response:

$$\mathbf{y}(t) = \mathbf{C}(t)\mathbf{\Phi}(t, t_i)\mathbf{x}_i + \int_{t_i}^{t} [\mathbf{C}(t)\mathbf{\Phi}(t, \tau)\mathbf{B}(\tau) + \mathbf{D}(t)\delta(t-\tau)]\mathbf{u}(\tau)\mathrm{d}\tau, \quad t \geq t_i, \tag{A.22}$$

where $\delta(t - \tau)$ is *Dirac delta* function representing a unit impulse applied at $t = \tau$. The first term on the right-hand side of equation (A.22) is called the *initial response*, while the integral term is called the *convolution integral* of the linear system. The convolution integral gives the system's response when the initial condition is zero ($\mathbf{x}_i = \mathbf{0}$), and is denoted by

$$\mathbf{y}(t) = \int_{t_i}^{t} \mathbf{G}(t, \tau)\mathbf{u}(\tau)\mathrm{d}\tau, \quad t \geq t_i, \tag{A.23}$$

where

$$\mathbf{G}(t, \tau) = \mathbf{C}(t)\mathbf{\Phi}(t, \tau)\mathbf{B}(\tau) + \mathbf{D}(t)\delta(t - \tau), \quad t \geq \tau, \tag{A.24}$$

is called the *impulse response* matrix. The (i, j)th element of $\mathbf{G}(t, \tau)$ is the value of the ith output variable at time t when the jth input variable is a unit impulse function applied at time τ, subject to zero initial condition, $\mathbf{x}_i = \mathbf{0}$. Similarly, we define the *step response* matrix as the integral of $\mathbf{G}(t, \tau)$:

$$\mathbf{S}(t, t_i) = \int_{t_i}^{t} \mathbf{G}(t, \tau)\mathrm{d}\tau, \quad t \geq t_i. \tag{A.25}$$

The (i, j)th element of $\mathbf{S}(t, t_i)$, is the value of the ith output variable at time t when the jth input variable is a unit step applied at time t_i, subject to zero initial condition, $\mathbf{x}_i = \mathbf{0}$.

A.4 Linear Time-Invariant System

When the linear system's properties are invariant with time, the system is said to be a *linear time-invariant* (LTI) system, with **A**, **B**, **C**, **D** treated as constant matrices. The LTI system's state transition matrix is expressed as

$$\mathbf{\Phi}(t, t_i) = e^{\mathbf{A}(t - t_i)}, \tag{A.26}$$

where $e^{\mathbf{M}}$ is called the *matrix exponential* of a square matrix, $\mathbf{M}$, and is defined by the following infinite series in a manner similar to the scalar exponential:

$$e^{\mathbf{M}} \doteq \mathbf{I} + \mathbf{M} + \frac{1}{2}\mathbf{M}^2 + \ldots + \frac{1}{n!}\mathbf{M}^n + \ldots. \tag{A.27}$$

Taking the Laplace transform of equation (A.13) for an LTI, subject to the initial condition, $\mathbf{x}_0 = \mathbf{x}(0)$, we have

$$s\mathbf{X}(s) - \mathbf{x}_0 = \mathbf{A}\mathbf{X}(s) + \mathbf{B}\mathbf{U}(s), \tag{A.28}$$

where $\mathbf{X}(s)$ and $\mathbf{U}(s)$ are the Laplace transforms of $\mathbf{x}(t)$ and $\mathbf{u}(t)$, respectively. The state transition matrix is then derived for the homogeneous system by the inverse Laplace transform as follows:

$$\mathbf{x}(t) = \mathcal{L}^{-1}(s\mathbf{I} - \mathbf{A})^{-1}\mathbf{x}_0 \tag{A.29}$$

or

$$e^{\mathbf{A}t} = \mathcal{L}^{-1}(s\mathbf{I} - \mathbf{A})^{-1}. \tag{A.30}$$

The general solution of an LTI system to an arbitrary, Laplace transformable input, which begins to act at time $t = 0$ when the system's state is $\mathbf{x}(0) = \mathbf{x}_0$ is thus given by

$$\mathbf{x}(t) = e^{\mathbf{A}t}\mathbf{x}_0 + \int_0^t e^{\mathbf{A}(t-\tau)}\mathbf{B}(\tau)\mathbf{u}(\tau)\mathrm{d}\tau. \tag{A.31}$$

The first term on the right-hand side (initial response) decays to zero for an asymptotically stable system (see below) in the limit $t \to \infty$. However, in the same limit, the integral term either converges to a finite value (called the *steady state*) or assumes the same functional form as that of the input (called *forced response*).

A.5 Linear Time-Invariant Stability Criteria

Equation (A.28) for $\mathbf{U}(s) = \mathbf{0}$ represents an *eigenvalue problem*, whose solution yields the *eigenvalues*, s, and *eigenvectors*, $\mathbf{X}(s)$. The eigenvalues of the linear system are obtained by the following *characteristic equation*:

$$|\, s\mathbf{I} - \mathbf{A} \,| = 0. \tag{A.32}$$

The n generally complex roots of the characteristic equation (eigenvalues of $\mathbf{A}$) signify an important system property, called *stability* (Appendix B).

Considering that an eigenvalue is generally complex, its imaginary part denotes the frequency of oscillation of the characteristic vector about the equilibrium point, and the real part signifies the growth (or decay) of its amplitude with time. We can now state the following criteria for the stability of an LTI system:

(a) If all eigenvalues have negative real parts, the system is *asymptotically stable* and regains its equilibrium in the steady state.
(b) A system having complex eigenvalues with zero real parts (and all other eigenvalues with negative real parts) displays oscillatory behavior of a constant amplitude, and is said to be stable (but not asymptotically stable).
(c) If at least one eigenvalue has a positive real part, its contribution to the system's state is an exponentially growing amplitude, and the system is said to be *unstable*.
(d) If a multiple eigenvalue of multiplicity p is at the origin (i.e., has both real and imaginary parts zero), its contribution to the system's state has terms containing the factors t^i , $i = 0, \ldots, p - 1$, which signify an unbounded behavior with time. Hence, the system is unstable.

A.6 Controllability of Linear Time-Invariant Systems

Controllability is the property of a system where it is possible to move it from *any* initial state, $\mathbf{x}(t_i)$, to *any* final state, $\mathbf{x}(t_f)$, solely by the application of the input vector, $\mathbf{u}(t)$, acting in a *finite* control interval, $t_i \leq t \leq t_f$. The words "any" and "finite" are emphasized because it may be possible to move an uncontrollable system only between some specific states by applying the control input, or to take an infinite time to change an uncontrallable system between arbitrary states. For a system to be controllable, all its state variables must be influenced, either directly or indirectly, by control inputs. If there is a subsystem that is unaffected by the control inputs, then the entire system is uncontrollable.

The necessary and sufficient condition for controllability of an LTI system ($\mathbf{A}$, $\mathbf{B}$) is that the following test matrix must have rank n, the order of the system:

$$\mathbf{P} = \left(\mathbf{B},\ \mathbf{AB},\ \mathbf{A}^2\mathbf{B}, \ldots,\ \mathbf{A}^{n-1}\mathbf{B}\right). \tag{A.33}$$

If a system is unstable but controllable, it can be stabilized by a feedback control system. It is often possible to decompose an uncontrollable LTI system into controllable and uncontrollable subsystems. A system that is both unstable and uncontrollable could also be stabilized, provided its uncontrollable subsystem is stable. In such a case, the system is said to be *stabilizable*.

A.7 Observability of Linear Time-Invariant Systems

Observability is the property of an unforced (homogeneous) system where it is possible to estimate *any* initial state, $\mathbf{x}(t_i)$, of the system solely by a *finite* record, $t \geq t_i$, of the output vector, $\mathbf{y}(t)$. For a system to be observable, all of its state variables must contribute, either directly or indirectly, to the output vector. If there is a subsystem that leaves the output vector unaffected, then the entire system is unobservable.

An unforced LTI system, $\dot{\mathbf{x}} = \mathbf{A}\mathbf{x}$, whose output is related to the state vector by

$$\mathbf{y} = \mathbf{C}\mathbf{x}, \tag{A.34}$$

is observable[1] if and only if the following test matrix has rank n, the order of the system:

$$\mathbf{N} = \left(\mathbf{C}^T,\ \mathbf{A}^T\mathbf{C}^T,\ (\mathbf{A}^T)^2\mathbf{C}^T, \ldots,\ (\mathbf{A}^T)^{n-1}\mathbf{C}^T\right). \tag{A.35}$$

It is often possible to decompose an unobservable LTI system into observable and unobservable subsystems. A system whose unobservability is caused by a stable subsystem is said to be *detectable*.

A.8 Transfer Matrix

An LTI system's response to a specified input is most efficiently analyzed and designed with the the concept of the *transfer matrix*. The transfer matrix, $\mathbf{G}(s)$, of such a system is defined as the following relationship between the output's Laplace transform, $\mathbf{Y}(s)$, and that of the input vector, $\mathbf{U}(s)$, subject to *zero initial conditions*, $\mathbf{y}(0) = \dot{\mathbf{y}}(0) = \ddot{\mathbf{y}}(0) = \ldots = 0$:

$$\mathbf{Y}(s) = \mathbf{G}(s)\mathbf{U}(s). \tag{A.36}$$

Clearly, the transfer matrix is the Laplace transform of the impulse response matrix subject to zero initial conditions. The roots of the common denominator polynomial of the transfer matrix are called the *poles* of the system and are same as the eigenvalues of the system's state dynamics matrix, $\mathbf{A}$. The roots of the numerator polynomials of the transfer matrix are called the *zeros* of the system. The transfer matrix of an LTI system can be expressed in terms of its state-space coefficients as

$$\mathbf{G}(s) = \mathbf{C}\,(s\mathbf{I} - \mathbf{A})^{-1}\,\mathbf{B} + \mathbf{D}. \tag{A.37}$$

When $s = i\omega$ the transfer matrix becomes the *frequency response* matrix, $\mathbf{G}(i\omega)$, whose elements denote the steady-state response of an output variable to a simple harmonic input variable, subject to zero initial conditions and all other inputs being zero.

A.9 Singular Value Decomposition

Singular values of a transfer matrix, $\mathbf{G}(i\omega)$, at a given frequency, ω, are the positive square roots of the eigenvalues of the square matrix

$$\mathbf{G}^T(-i\omega)\mathbf{G}(i\omega).$$

The computation of singular values is carried out by the *singular value decomposition*

$$\mathbf{G}(i\omega) = \mathbf{U}(i\omega)\mathbf{\Sigma}\mathbf{V}^T(-i\omega), \tag{A.38}$$

where $\mathbf{U}, \mathbf{V}$ are unitary complex matrices satisfying

$$\mathbf{U}(i\omega)\mathbf{U}^T(-i\omega) = \mathbf{I}, \quad \mathbf{V}(i\omega)\mathbf{V}^T(-i\omega) = \mathbf{I}, \tag{A.39}$$

and $\mathbf{\Sigma}$ contains the singular values of $\mathbf{G}(i\omega)$, denoted by

$$\sigma_k\left\{\mathbf{G}(i\omega)\right\}, \quad k = 1, 2, \ldots, n,$$

[1] This definition of observability can be extended to a forced linear system by requiring in addition that the applied inputs, $\mathbf{u}(t)$ should be known in the period of observation, $t \geq t_{\mathrm{i}}$.

as its diagonal elements. The largest of all the singular values is denoted by $\bar{\sigma}$ and the smallest by $\underline{\sigma}$.

The frequency spectrum of the singular values indicates the variation of the "magnitude" (or gain) of the frequency response matrix, which is supposed to lie between $\underline{\sigma}(\omega)$ and $\bar{\sigma}(\omega)$. The range of frequencies ($0 \leq \omega \leq \omega_b$) over which $\bar{\sigma}(\omega)$ stays above $0.707\bar{\sigma}(0)$ is called the system's *bandwidth* and is denoted by the frequency, ω_b. The bandwidth indicates the highest-frequency signal to which the system has an appreciable response. For higher frequencies ($\omega > \omega_b$), the singular values of a strictly proper system generally decline rapidly with increasing frequency, called *roll-off*. A system with a steep roll-off of the largest singular value, $\bar{\sigma}$, has a reduced sensitivity to high-frequency noise, which is a desirable property.

The highest magnitude achieved by the largest singular value over the system bandwidth is the H_∞ norm of the transfer matrix, given by

$$|| \mathbf{G} ||_\infty = \sup_\omega \left[\bar{\sigma}\left\{\mathbf{G}(i\omega)\right\}\right], \tag{A.40}$$

where $\sup_\omega(.)$ is the supremum (the maximum value) with respect to the frequency.

A.10 Linear Time-Invariant Control Design

Consider the plant dynamics expressed in a linear state-space form as follows:

$$\dot{\mathbf{x}} = \mathbf{A}\mathbf{x} + \mathbf{B}\mathbf{u}, \tag{A.41}$$

where $\mathbf{x}(t)$ is the *state vector*, $\mathbf{u}(t)$ the control input vector, and $\mathbf{A}(t)$, $\mathbf{B}(t)$ are the state space coefficient matrices, which could be time-varying. However, for the time being, we will confine our attention to a linear plant with constant coefficient matrices, $\mathbf{A}$, $\mathbf{B}$ (i.e., an LTI system).

A.10.1 Regulator Design by Eigenstructure Assignment

It is possible to design a linear state feedback regulator for the LTI plant of equation (A.41) with the control law

$$\mathbf{u} = -\mathbf{K}\mathbf{x} \tag{A.42}$$

by assigning a structure for the eigenvalues and eigenvectors of the closed-loop dynamics matrix, $\mathbf{A} - \mathbf{B}\mathbf{K}$. For single-input plants, this merely involves selecting the locations for the closed-loop poles (*pole placement*) by the following *Ackermann formula* that yields the desired closed-loop characteristics (Tewari 2002):

$$\mathbf{K} = (\mathbf{a}_d - \mathbf{a})(\mathbf{P}\mathbf{P}')^{-1}. \tag{A.43}$$

Here $\mathbf{a}$ is the row vector formed by the coefficients, a_i, of the plant's characteristic polynomial in *descending order*,

$$|s\mathbf{I} - \mathbf{A}| = s^n + a_n s^{n-1} + a_{n-1}s^{n-2} + \ldots + a_2 s + a_1; \tag{A.44}$$

$\mathbf{a}_d$ is the row vector formed by the characteristic coefficients of the closed-loop system in descending order,

$$|s\mathbf{I} - \mathbf{A} + \mathbf{B}\mathbf{K}| = s^n + a_{dn}s^{n-1} + a_{d(n-1)}s^{n-2} + \ldots + a_{d2}s + a_{d1}; \tag{A.45}$$

$\mathbf{P}$ is the controllability test matrix of the plant; and $\mathbf{P}'$ is the following upper triangular matrix:

$$\mathbf{P}' = \begin{pmatrix} 1 & a_{n-1} & a_{n-2} & \cdots & a_2 & a_1 \\ 0 & 1 & a_{n-1} & \cdots & a_3 & a_2 \\ 0 & 0 & 1 & \cdots & a_4 & a_3 \\ \cdots & \cdots & \cdots & \cdots & \cdots & \cdots \\ 0 & 0 & 0 & \cdots & 1 & a_{n-1} \\ 0 & 0 & 0 & \cdots & 0 & 1 \end{pmatrix}. \tag{A.46}$$

Of course, this requires that the plant must be controllable, $|\mathbf{P}| \neq 0$. A popular choice of closed-loop poles is the *Butterworth pattern* (Tewari 2002) where all the poles are equidistant from the origin, $s = 0$. Such a pattern generally requires the least control effort for a given bandwidth, and is thus considered to be the optimal placement of poles.

The pole-placement method is inapplicable to multi-input plants, which have many more controller gains to be found than the number of equations available from the pole locations. In such a case, we need additional equations that can be derived from the shape of the eigenvectors using the method of *eigenstructure assignment*. A popular method in this regard is the *robust pole assignment method* of (Kautsky *et al.* 1985) where the eigenvectors, $\mathbf{v}_i$, $i = 1, 2, \ldots, n$, corresponding to the eigenvalues, λ_i, respectively, and satisfying the eigenvalue problem,

$$(\mathbf{A} - \mathbf{BK})\mathbf{v}_i = \lambda_i \mathbf{v}_i, \tag{A.47}$$

are chosen such that the modal matrix,

$$\mathbf{V} = (\mathbf{v}_1, \mathbf{v}_2, \ldots, \mathbf{v}_n), \tag{A.48}$$

is as well-conditioned as possible.[2] The eigenstructure assignment algorithm of (Kautsky *et al.* 1985) is coded in the MATLAB® Control System Toolbox command called *place*.

A.10.2 Regulator Design by Linear Optimal Control

An alternative design approach to eigenstructure assignment commonly applied to multi-variable LTI systems is the linear quadratic regulator (LQR) method of linear optimal state feedback control with a quadratic objective function (Chapter 2). In this book, the LQR is the standard feedback design method due to its excellent properties and systematic procedure.

A.10.3 Linear Observers and Output Feedback Compensators

Control systems with output (rather than state) feedback require observers that can reconstruct the missing information about the system's states from the input applied to the plant, and the output fed back from the plant. An observer mimics the plant by generating an estimated state vector, $\mathbf{x}_o$, instead of the actual plant state vector, $\mathbf{x}$, and supplies it to the regulator. A control system that contains both an observer and a regulator is called a *compensator*. Due to a decoupling of the observer and plant states in the control system, it is possible to design the regulator and observer separately from each other by what is known as the *separation principle*.

[2] Conditioning of a square matrix refers to how close the matrix is to being singular (i.e., to having a zero determinant). A scalar measure called the *condition number* is assigned to the matrix that reflects its conditioning. A large condition number implies the matrix is close to being singular.

Full-Order Observer

The output equation,

$$\mathbf{y} = \mathbf{C}\mathbf{x} + \mathbf{D}\mathbf{u}, \tag{A.49}$$

is used in the design of a full-order observer with the following state equation:

$$\dot{\mathbf{x}}_o = (\mathbf{A} - \mathbf{L}\mathbf{C})\,\mathbf{x}_o + (\mathbf{B} - \mathbf{L}\mathbf{D})\,\mathbf{u} + \mathbf{L}\mathbf{y}, \tag{A.50}$$

where $\mathbf{x}_o(t)$ is the estimated state and $\mathbf{L}$ the observer gain matrix, provided the plant ($\mathbf{A}$, $\mathbf{C}$) is observable. The observer gain matrix, $\mathbf{L}$, can be selected in a manner similar to the regulator gain, $\mathbf{K}$, by either eigenstructure assignment for the observer dynamics matrix, $\mathbf{A} - \mathbf{L}\mathbf{C}$, or linear quadratic optimal control where $\mathbf{A}$ is replaced by $\mathbf{A}^T$, and $\mathbf{B}$ by $\mathbf{C}^T$.

The closed-loop control system dynamics with a desired state, $\mathbf{x}_d(t)$, and linear feedforward/feedback control with output feedback,

$$\mathbf{u} = \mathbf{K}_d\mathbf{x}_d + \mathbf{K}\,(\mathbf{x}_d - \mathbf{x}_o)\,, \tag{A.51}$$

is given by the state equation

$$\begin{Bmatrix} \dot{\mathbf{x}} \\ \dot{\mathbf{x}}_o \end{Bmatrix} = \begin{pmatrix} \mathbf{A} & -\mathbf{B}\mathbf{K} \\ \mathbf{L}\mathbf{C} & \mathbf{A} - \mathbf{B}\mathbf{K} - \mathbf{L}\mathbf{C} \end{pmatrix} \begin{Bmatrix} \mathbf{x} \\ \mathbf{x}_o \end{Bmatrix} + \begin{pmatrix} \mathbf{B}(\mathbf{K} + \mathbf{K}_d) \\ \mathbf{B}(\mathbf{K} + \mathbf{K}_d) \end{pmatrix} \mathbf{x}_d, \tag{A.52}$$

where $\mathbf{K}, \mathbf{L}$ are separately designed and the feedforward gain matrix, $\mathbf{K}_d$, is selected to ensure that the closed-loop error dynamics is independent of the desired state, $\mathbf{x}_d(t)$, with a given state equation

$$\dot{\mathbf{x}}_d = \mathbf{f}_d(\mathbf{x}_d), \tag{A.53}$$

such that

$$(\mathbf{A} + \mathbf{B}\mathbf{K}_d\mathbf{x}_d) - \mathbf{f}_d(\mathbf{x}_d) = \mathbf{0}. \tag{A.54}$$

Thus, one can design a tracking system for a plant that is both controllable and observable with the available inputs and outputs, respectively, as well as satisfying equation (A.54) with its desired state vector.

Reduced-Order Observer

When a part of the plant's state vector can be directly obtained from the output vector, it is unnecessary to estimate the entire state vector by a full-order observer. Consider a plant whose state vector is partitioned as follows:

$$\mathbf{x} = (\mathbf{x}_1^T, \mathbf{x}_2^T)^T, \tag{A.55}$$

such that

$$\begin{aligned} \dot{\mathbf{x}}_1 &= \mathbf{A}_{11}\mathbf{x}_1 + \mathbf{A}_{12}\mathbf{x}_2 + \mathbf{B}_1\mathbf{u}, \\ \dot{\mathbf{x}}_2 &= \mathbf{A}_{21}\mathbf{x}_1 + \mathbf{A}_{22}\mathbf{x}_2 + \mathbf{B}_2\mathbf{u}. \end{aligned} \tag{A.56}$$

The measurable part of the state vector, $\mathbf{x}_1$, can be directly obtained by inversion of the output equation with a square coefficient matrix, $\mathbf{C}$:

$$\mathbf{y} = \mathbf{C}\mathbf{x}_1. \tag{A.57}$$

The unmeasurable part, $\mathbf{x}_2$, needs estimation by a *reduced-order observer*, and can be expressed as follows:

$$\mathbf{x}_{o2} = \mathbf{L}\mathbf{y} + \mathbf{z}, \tag{A.58}$$

where $\mathbf{z}$ is the state vector of the reduced-order observer with state equation

$$\dot{\mathbf{z}} = \mathbf{F}\mathbf{z} + \mathbf{H}\mathbf{u} + \mathbf{G}\mathbf{y}, \tag{A.59}$$

whose coefficient matrices are determined from the requirement that the estimation error, $\mathbf{e}_o = \mathbf{x}_2 - \mathbf{x}_{o2}$, should go to zero in the steady state, irrespective of the control input and the output (Tewari 2002):

$$\begin{aligned} \mathbf{F} &= \mathbf{A}_{22} - \mathbf{L}\mathbf{C}\mathbf{A}_{12}, \\ \mathbf{G} &= \mathbf{F}\mathbf{L} + (\mathbf{A}_{21} - \mathbf{L}\mathbf{C}\mathbf{A}_{11})\,\mathbf{C}^{-1}, \\ \mathbf{H} &= \mathbf{B}_2 - \mathbf{L}\mathbf{C}\mathbf{B}_1, \end{aligned} \tag{A.60}$$

with the observer gain $\mathbf{L}$ selected by eigenstructure assignment or Kalman filter (Chapter 2) approach, such that all the eigenvalues of the oberver dynamics matrix, $\mathbf{F}$, are in the left-half s-plane.

References

Kailath, T. (1980) *Linear Systems*. Prentice Hall, Englewood Cliffs, NJ.

Kautsky, J., Nichols, N.K., and van Dooren, P. (1985) Robust pole assignment in linear state feedback. *International Journal of Control* **41**, 1129–1155.

Tewari, A. (2002) *Modern Control Design with MATLAB and Simulink*. John Wiley & Sons, Ltd, Chichester.

Appendix B

Stability

B.1 Preliminaries

Stability is broadly defined as the tendency of an unforced system to remain close to a given reference state despite small initial disturbances. Consider an unforced system with state vector, $\boldsymbol{\xi}(t)$, governed by the state equation

$$\dot{\boldsymbol{\xi}} = \mathbf{f}(\boldsymbol{\xi}, t), \tag{B.1}$$

with initial condition

$$\boldsymbol{\xi}(0) = \boldsymbol{\xi}_0. \tag{B.2}$$

Let the system have a *reference solution*, $\boldsymbol{\xi}_r(t)$, $t \geq 0$, to the state equation (B.1), subject to the given initial condition (B.2):

$$\dot{\boldsymbol{\xi}}_r = \mathbf{f}(\boldsymbol{\xi}_r, t) \tag{B.3}$$

and

$$\boldsymbol{\xi}_r(0) = \boldsymbol{\xi}_0. \tag{B.4}$$

If there is a change in the initial condition such that

$$\boldsymbol{\xi}(0) = \boldsymbol{\xi}_0 + \boldsymbol{\delta}, \tag{B.5}$$

then the perturbed solution, $\boldsymbol{\xi}(t)$, $t \geq 0$, to the state equation (B.1) may depart from the reference solution such that

$$\boldsymbol{\xi}(t) = \boldsymbol{\xi}_r(t) + \boldsymbol{\epsilon}(t), \quad t \geq 0. \tag{B.6}$$

Then stability is concerned with the boundedness of $\boldsymbol{\epsilon}(t)$, $t \geq 0$, if $\mid \boldsymbol{\delta} \mid$ is small.

The unforced system may have a constant reference solution, $\boldsymbol{\xi}_e$, called an *equilibrium point*, or a state of rest, that is,

$$\dot{\boldsymbol{\xi}}_e = \mathbf{f}(\boldsymbol{\xi}_e, t) = \mathbf{0}. \tag{B.7}$$

For an equilibrium solution it must be true that

$$\boldsymbol{\xi}(0) = \boldsymbol{\xi}_e. \tag{B.8}$$

Advanced Control of Aircraft, Spacecraft and Rockets, Ashish Tewari.

Equations (B.7) and (B.8) indicate that by redefining the system's state as the *deviation* from an equilibrium solution,

$$\mathbf{x}(t) = \boldsymbol{\xi}(t) - \boldsymbol{\xi}_e, \quad t \geq 0, \tag{B.9}$$

we have the transformed state equation

$$\dot{\mathbf{x}} = \mathbf{f}(\mathbf{x}, t), \tag{B.10}$$

with the initial condition

$$\mathbf{x}(0) = \mathbf{0}. \tag{B.11}$$

For a mechanical system, stability is usually defined with reference to an equilibrium point. Furthermore, the functional, $\mathbf{f}(\mathbf{x}, t)$, of a mechanical system is continuous and possesses continuous derivatives with respect to the state and time. There two broad mathematical definitions of stability applied to mechanical systems, namely *stability in the sense of Lagrange* (de la Grange 1788) and *stability in the sense of Lyapunov* (Lyapunov 1992). Both address the issue of the size of the solution, that is, the *Euclidean norm* of the state deviation from the origin,

$$\mid \mathbf{x} \mid = \sqrt{\sum_{i}^{n} x_i^2}, \tag{B.12}$$

where $x_i, i = 1, \ldots, n$ are the state variables of the nth-order system.

B.2 Stability in the Sense of Lagrange

A system described by equation (B.10) is said to be stable about the equilibrium point, $\mathbf{x}_e = \mathbf{0}$ in the sense of Lagrange if, for each real and positive number, δ, there exists another real and positive number, ϵ, such that

$$\mid \mathbf{x}(0) \mid \leq \delta \tag{B.13}$$

implies that

$$\mid \mathbf{x}(t) \mid \leq \epsilon, \quad t \geq 0. \tag{B.14}$$

In other words, stability in the sense of Lagrange guarantees a bounded solution about the equilibrium point at the origin, resulting from every bounded perturbation from the original initial condition.

Example B.1 Consider a simple pendulum consisting of a bob suspended by an inextensible string of length, ℓ. The unforced equation of motion describing angular displacement, $\theta(t)$, from the vertically downward position of the bob is

$$\ddot{\theta} + \frac{g}{\ell} \sin \theta = 0, \tag{B.15}$$

where g is the acceleration due to gravity (constant). The state equation with the choice of state variables, $x_1(t) = \theta(t)$ and $x_2(t) = \dot{\theta}(t)$, is expressed as

$$\dot{\mathbf{x}} = \begin{pmatrix} x_2 \\ -\frac{g}{\ell} \sin x_1 \end{pmatrix}. \tag{B.16}$$

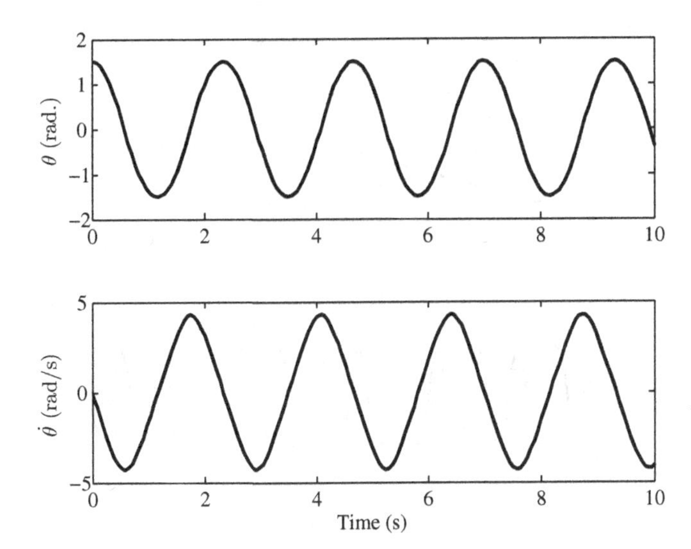

Figure B.1 Initial response of an unforced, simple pendulum to a large initial displacement ($\theta(0) = 1.5$ rad, $\dot{\theta}(0) = 0$) from the origin ($\theta(0) = 0$, $\dot{\theta}(0) = 0$)

A bounded (but large) angular *displacement*, $|\,\theta(0)\,| \leq \delta$, from the initial equilibrium state,

$$\mathbf{x}(0) = \left\{ \begin{matrix} \theta(0) \\ \dot{\theta}(0) \end{matrix} \right\} = \mathbf{0}, \tag{B.17}$$

is seen to result in a bounded solution, $|\,\mathbf{x}(t)\,| \leq \epsilon$, $t \geq 0$, as shown in Figure B.1 for the case of $g = 9.81$ m/s^2, $\ell = 1$ m, $\theta(0) = 1.5$ rad, and $\dot{\theta}(0) = 0$. The solution is obtained by the Runge–Kutta 4(5) solver in MATLAB® as follows:

```
>> [t,x]=ode45(@pend,[0 10],[1.5 0]);
>> subplot(211),plot(t,x(:,1)),ylabel('\theta (rad.)'),hold on,
>> subplot(212),plot(t,x(:,2)),xlabel('Time (s)'),
>> ylabel('d\theta/dt (rad/s)')
```

where *pend.m* is the following file specifying the state equation of the unforced pendulum:

```
function xdot=pend(t,x)
L=1;% m
g=9.81;% m/s^2
theta=x(1,1);
xdot=[x(2,1);-g*sin(theta)/L];
```

However, if a bounded (but large) initial angular *velocity*, $|\,\dot{\theta}(0)\,| \leq \delta$, is applied, the resulting solution, $\mathbf{x}(t)$, $t \geq 0$, can become unbounded. This is demonstrated in Figure B.2 for the case of $g = 9.81$ m/s^2, $\ell = 1$ m, $\theta(0) = 0$, and $\dot{\theta}(0) = 10$ rad/s:

```
>> [t,x]=ode45(@pend,[0 10],[0 10]);
```

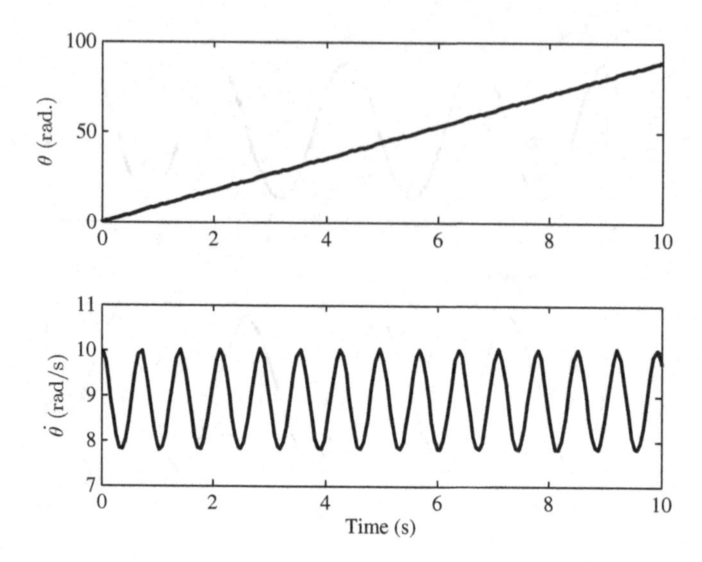

Figure B.2 Initial response of an unforced, simple pendulum to a large initial velocity ($\theta(0) = 0$, $\dot{\theta}(0) = 10$ rad/s) from the origin ($\theta(0) = 0$, $\dot{\theta}(0) = 0$)

Since *every* bounded initial perturbation may not produce a bounded response, the origin of the simple pendulum is unstable in the sense of Lagrange.

B.3 Stability in the Sense of Lyapunov

A system described by equation (B.10) is said to be stable about the equilibrium point, $\mathbf{x}_e = \mathbf{0}$, in the sense of Lyapunov if, for each real and positive number, ϵ, however small, there exists another real and positive number, δ, such that

$$\mid \mathbf{x}(0) \mid < \delta \tag{B.18}$$

implies that

$$\mid \mathbf{x}(t) \mid < \epsilon, \quad t \geq 0. \tag{B.19}$$

In other words, stability in the sense of Lyapunov requires that a solution starting in a small neighborhood of the equilibrium point at the origin should always remain in a small neighborhood of the origin. It is easily appreciated that stability in the sense of Lagrange and stability in the sense of Lyapunov do not imply each other. It is quite possible to have an unbounded solution to a large (but bounded) initial perturbation, which renders the equilibrium unstable in the sense of Lagrange, but the same system may remain in the neighborhood of the origin for small perturbations and thus be stable in the sense of Lyapunov, as is the case of the simple pendulum of Example B.1. Furthermore, it is possible for a system to have a departure from the equilibrium for even small perturbations from the origin (instability in the sense of Lyapunov), but the solution could still be bounded and thus be stable in the sense of Lagrange, as the following example demonstrates.

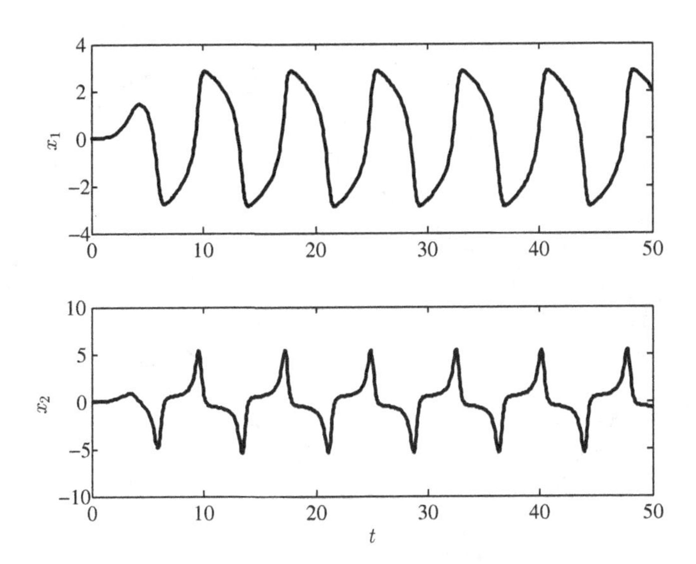

Figure B.3 Initial response of a van der Pol oscillator with $a = 2$ and initial condition $x_1(0) = 0$, $x_2(0) = 0.01$

Example B.2 Consider a *van der Pol oscillator* with state equations

$$\begin{aligned} \dot{x}_1 &= x_2, \\ \dot{x}_2 &= -x_1 + \left(a - x_1^2\right) x_2, \end{aligned} \tag{B.20}$$

where a is a real, positive constant. It is clear that the origin, (0, 0), is an equilibrium point of the system. With any small, initial perturbation from the equilibrium point the system departs from the equilibrium, but exhibits a bounded oscillation. An example of this behavior is seen in Figures B.3 and B.4 for $a = 2$ and initial condition $x_1(0) = 0$, $x_2(0) = 0.01$, computed with the following MATLAB statements:

```
>> [t,x]=ode45(@vanderpol,[0 50],[0 0.01]);
>> subplot(211),plot(t,x(:,1)),ylabel('x_1'),hold on,
>> subplot(212),plot(t,x(:,2)),xlabel('t'),ylabel('x_2')
>> figure
>> plot(x(:,1),x(:,2)),xlabel('x_1'),ylabel('x_2')
```

where *vanderpol.m* is the following file specifying the state equation of the unforced pendulum:

```
function xdot=vanderpol(t,x)
a=2;
xdot=[x(2,1); -x(1,1)+(a-x(1,1)^2)*x(2,1)];
```

Figure B.3 shows that the van der Pol oscillator eventually reaches a constant (but large) amplitude oscillation called a *limit cycle*. The plot of $x_2 = \dot{x}_1$ vs. x_1 in Figure B.4 (called the *phase portrait*) illustrates the limit cycle behavior at the outer boundary of the closed curve (called the *attracting boundary*).

Since even a small perturbation makes the system depart from the small neighborhood of the equilibrium point, the origin of the van der Pol oscillator is unstable in the sense of Lyapunov. However, since the initial response to a bounded perturbation is always bounded, the origin of the van der Pol oscillator is said to be stable in the sense of Lagrange.

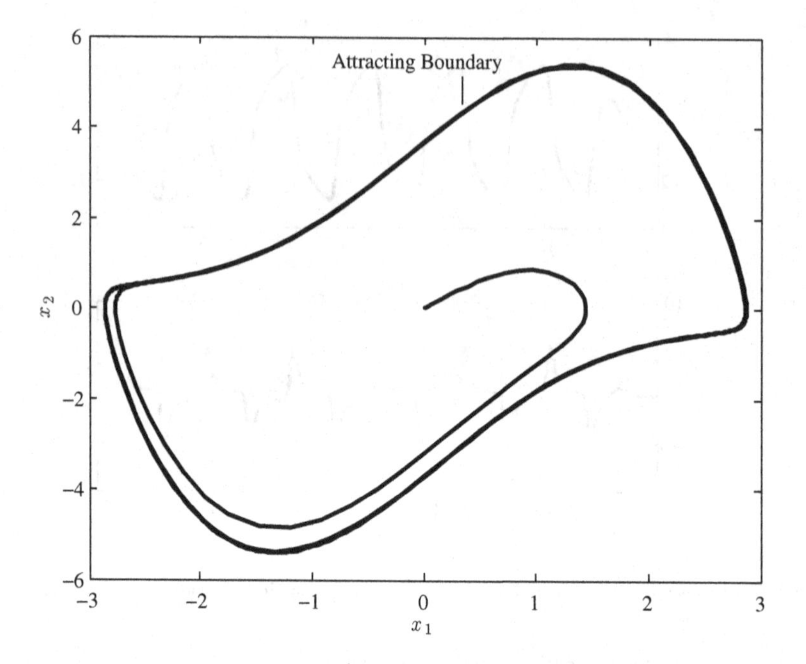

Figure B.4 Phase portrait of a van der Pol oscillator with $a = 2$ and initial condition $x_1(0) = 0$, $x_2(0) = 0.01$

In all our applications in this book, we will regard a stable equilibrium point to be the one that satisfies the stability criterion in the sense of Lyapunov.

B.3.1 Asymptotic Stability

A system described by equation (B.10) is said to be *asymptotically stable* about the origin, $\mathbf{x}_e = \mathbf{0}$, if, it is stable in the sense of Lyapunov and if for each real and positive number, ϵ, however small, there exist real and positive numbers, δ and τ, such that

$$| \mathbf{x}(0) | < \delta \tag{B.21}$$

implies that

$$| \mathbf{x}(t) | < \epsilon, \quad t > \tau. \tag{B.22}$$

Asymptotic stability is thus possessed by a special class of Lyapunov stable systems whose slightly perturbed solutions approach the origin asymptotically for large times. Thus, the equilibrium is eventually restored. Clearly, asymptotic stability guarantees stability in the sense of Lyapunov (but not vice versa).

B.3.2 Global Asymptotic Stability

A system described by equation (B.10) is said to be *globally asymptotically stable* about the origin, $\mathbf{x}_e = \mathbf{0}$, if it is stable in the sense of Lyapunov and if, for each real and positive pair, (δ, ϵ), there exists a real and positive number, τ, such that

$$| \mathbf{x}(0) | < \delta \tag{B.23}$$

implies that

$$| \mathbf{x}(t) | < \epsilon, \quad t > \tau. \tag{B.24}$$

Global asymptotic stability thus refers to asymptotic stability with respect to *all* possible initial conditions, $\mathbf{x}(0)$, such that $| \mathbf{x}(0) | < \delta$, and not just a few specific ones. There are further classifications of asymptotic stability, such as uniform asymptotic stability, and asymptotic stability in the large and whole (Loría and Panteley 2006), which we shall not consider here.

B.3.3 Lyapunov's Theorem

Let $V(\mathbf{x}, t)$ be a continuously differentiable scalar function of the time as well as of the state variables of a system described by equation (B.10), whose equilibrium point is $\mathbf{x}_e = \mathbf{0}$. If the following conditions are satisfied:

$$V(\mathbf{0}, t) = 0, \quad V(\mathbf{x}, t) > 0, \quad \frac{dV}{dt}(\mathbf{x}) < 0, \quad \text{for all } \mathbf{x} \neq \mathbf{0}, \tag{B.25}$$

$$|| \mathbf{x} || \to \infty \text{ implies } V(\mathbf{x}, t) \to \infty, \tag{B.26}$$

then the origin, $\mathbf{x}_e = \mathbf{0}$, is globally asymptotically stable.

Proof of Lyapunov's theorem (Slotine and Li 1991) is obtained from the unbounded, positive definite nature of $V(\mathbf{x}, t)$, and negative definite nature of $\dot{V}(\mathbf{x}, t)$, implying that for any initial perturbation from the origin, $\mathbf{x}(0) \neq \mathbf{0}$, the resulting solution satisfies $V(\mathbf{x}(t), t) \leq V(\mathbf{x}(0), t), t > 0$ (i.e., remains in a bounded neighborhood of the origin). Furthermore, the same also implies that $V(\mathbf{x}(t_2), t_2) \leq V(\mathbf{x}(t_1), t_1), t_2 > t_1$, which means a convergence of every solution to the origin.

Lyapunov's theorem gives a test of global asymptotic stability without the necessity of solving the system's state equations for every possible initial condition, and is thus a powerful tool in nonlinear control design. It merely requires finding a suitable Lyapunov function of the state variables.

Example B.3 Consider a system described by

$$\begin{aligned} \dot{x}_1 &= -x_1 - x_2 e^{-t}, \\ \dot{x}_2 &= x_1 - x_2. \end{aligned} \tag{B.27}$$

Clearly, the system has an equilibrium point at the origin, $x_1 = x_2 = 0$. Let us select the Lyapunov function

$$V(\mathbf{x}, t) = x_1^2 + x_2^2(1 + e^{-t}). \tag{B.28}$$

We have

$$V(\mathbf{x}, t) > 0, \quad \text{for all } \mathbf{x} \neq \mathbf{0}, \tag{B.29}$$

and

$$\begin{aligned} \frac{dV}{dt} &= \dot{V}(\mathbf{x}, t) = \frac{\partial V}{\partial t} + \frac{\partial V}{\partial \mathbf{x}} \mathbf{f}(\mathbf{x}, t) \\ &= -x_2^2 e^{-t} + 2x_1\dot{x}_1 + 2x_2\dot{x}_2(1 + e^{-t}) \\ &= -2(x_1^2 + x_2^2 - x_1x_2) - 3x_2^2 e^{-t} \qquad \text{(B.30)} \\ &\leq -2(x_1^2 + x_2^2 - x_1x_2) = -x_1^2 - x_2^2 - (x_1 - x_2)^2 \\ &< 0, \quad \text{for all } \mathbf{x} \neq \mathbf{0}. \qquad \text{(B.31)} \end{aligned}$$

Furthermore,

$$\| \mathbf{x} \| = \sqrt{x_1^2 + x_2^2} \rightarrow \infty \text{ implies } V(\mathbf{x}, t) = x_1^2 + x_2^2(1 + e^{-t}) \rightarrow \infty. \tag{B.32}$$

Therefore, all the sufficient conditions of Lyapunov's theorem are satisfied by $V(\mathbf{x}, t)$ and thus the origin is a globally asymptotically stable equilibrium point of the system.

A direct method of constructing a valid Lyapunov function for *autonomous systems* (i.e., systems that do not depend explicitly on time), which are commonly encountered in flight control applications, is offered by the following theorem (Slotine and Li 1991).

B.3.4 Krasovski's Theorem

Let $\mathbf{A}(\mathbf{x})$ be the Jacobian matrix,

$$\mathbf{A} = \frac{\partial \mathbf{f}}{\partial \mathbf{x}}, \tag{B.33}$$

of an autonomous system described by

$$\dot{\mathbf{x}} = \mathbf{f}(\mathbf{x}), \tag{B.34}$$

whose equilibrium point is $\mathbf{x}_e = \mathbf{0}$. If there exist two symmetric and positive definite matrices, $(\mathbf{P}, \mathbf{Q})$, such that, for all $\mathbf{x} \neq \mathbf{0}$, the matrix

$$\mathbf{PA} + \mathbf{A}^T\mathbf{P} + \mathbf{Q} \tag{B.35}$$

is globally negative semi-definite, and that the Lyapunov function of the system,

$$V(\mathbf{x}) = \mathbf{f}^T\mathbf{P}\mathbf{f}, \tag{B.36}$$

satisfies the condition

$$| \mathbf{x} | \rightarrow \infty \text{ implies } V(\mathbf{x}) \rightarrow \infty, \tag{B.37}$$

then the origin is globally asymptotically stable.

B.3.5 Lyapunov Stability of Linear Systems

Lyapunov's stability theorem is easily applied to linear systems. For linear time-invariant systems, Lyapunov stability requirements lead to the well-known stability criteria in terms of the eigenvalues (characteristic roots) of the state dynamics matrix (Appendix A).

References

de la Grange, J.-L. (1788) *Méchanique Analytique*. Desaint, Paris.

Loría, A., and Panteley, E. (2006) Stability, told by its developers. In A. Loría, F. Lamnabhi-Lagarrigue, and E. Panteley (eds), *Advanced Topics in Control Systems Theory*, Lecture Notes in Control and Information Sciences 328, pp. 197–258. Springer-Verlag, London.

Lyapunov, A.M. (1992) *The General Problem of Stability of Motion* (trans. and ed. A.T. Fuller). Taylor & Francis, London.

Slotine, J.-J.E., and Li, W. (1991) *Applied Nonlinear Control*. Prentice Hall, Englewood Cliffs, NJ.

Appendix C

Control of Underactuated Flight Systems

In several flight tracking system applications, the linearized plant is only weakly controllable with respect to the applied control inputs. While a stable tracking system could be designed in such cases, the design is neither easy to carry out nor very robust to disturbances. Single-variable examples of such a control system are the control of an aircraft's airspeed solely by throttle input, and the control of flight direction by differential throttle actuation of two (or more) engines. We shall consider a more interesting illustration of underactuated systems by adaptive control of a rocket's speed, altitude, and flight-path angle by the forward acceleration input generated by throttling the rocket engine.

C.1 Adaptive Rocket Guidance with Forward Acceleration Input

Consider the challenging task of guiding a rocket by regulating the altitude, speed, and flight-path angle using only the forward acceleration control input. A separate attitude regulator maintains a zero normal acceleration by keeping the vehicle's longitudinal axis always aligned with the flight direction. As discussed above, modulating the thrust with precision is not feasible for the existing rocket engines. Therefore, the achieved thrust magnitude at any given time may be quite far away from what is actually required by the control system. Furthermore, the plant is both unstable and uncontrollable with only the forward acceleration control input in the initial launch phase. Even in the later phases of the flight, the plant is only weakly controllable, as demonstrated by the large condition number of the controllability test matrix in Chapter 4. Such a plant is said to be *underactuated*, and trying to stabilize such a plant with varying parameters is a formidable task requiring a fine adaptation of the controller parameters.

The linearized plant for planar guidance with only the forward acceleration input has the following state-space coefficient matrices:

$$\mathbf{A} = \begin{pmatrix} 0 & a_{13} & a_{14} \\ a_{31} & 0 & a_{34} \\ a_{41} & a_{43} & a_{44} \end{pmatrix}, \quad \mathbf{B} = \begin{pmatrix} 0 \\ 1 \\ 0 \end{pmatrix}, \tag{C.1}$$

where the coefficients a_{ij} are the same as those given in Chapter 4. With a full-state feedback law, $\Delta u_1(t) = \mathbf{K}(t)[\Delta h, \Delta v, \Delta\phi]$, where

$$\mathbf{K}(t) = \left[k_{\rm h}(t),\ k_{\rm v}(t),\ k_{\phi}(t)\right] \tag{C.2}$$

Advanced Control of Aircraft, Spacecraft and Rockets, Ashish Tewari.

one can use the LQR design approach with variable gains, as demonstrated in Chapter 4. However, an adaptive scheduling of the gains is now necessary in order to have a stable closed-loop system for the underactuated plant. Furthermore, the linear feedback law requires a precise thrust modulation that may only be possible in the future. In the following example, we assume that a required thrust magnitude is instantly available to guide the rocket.

Example C.1 Repeat the gravity-turn trajectory guidance of Example 4.9 with a closed-loop tracking system for regulating the altitude, speed, and flight-path angle using full-state feedback and forward acceleration control input.

A regulator with the time-varying, state feedback gain matrix $\mathbf{K}(t)$ is designed by the infinite-time LQR method and the closed-loop eigenvalues computed with constant cost parameter matrices, $\mathbf{Q} = \mathbf{I}$, $\mathbf{R} = 1$, using the following MATLAB® statements:

```
n=size(t,2);
mu=398600.4;
r0=6378.14;
g0=mu/r0^2;
Eg=[];K=[];
for i=1:n;
    h=x(1,i)-r0;
    r=r0+h;
    v=x(2,i);
    phi=x(3,i);
    A=[0 sin(phi) v*cos(phi);
    2*g0*r0^2*sin(phi)/r^3    0   -g0*r0^2*cos(phi)/r^2;
    (-v/r^2+2*g0*r0^2/(v*r^3))*cos(phi) ...
    (1/r+g0*r0^2/(v^2*r^2))*cos(phi) ...
    -(v/r-g0*r0^2/(v*r^2))*sin(phi)];
    B=[0 1 0]';
    [k,S,E]=lqr(A,B,eye(3),1);
    E=sort(E);
    K(i,:)=k;
    Eg(:,i)=E;
end
```

The resulting closed-loop eigenvalues are plotted in Figure C.1, indicating an asymptotically stable closed-loop system with coefficients frozen at all times along the nominal trajectory. The variation of the LQR regulator gains, $k_h(t)$, $k_v(t)$, $k_\phi(t)$, is plotted in Figure C.2. While the assumption $k_h = 1\ \mathrm{s}^{-2}$ can be readily made as a possible adaptation law, the other two gains require a scheduling with the increasing flight speed according to the following simple adaptation laws (Figure C.2) in the given control interval:

$$k_v(t) = 1.5v^{-1/6}, \quad k_\phi(t) = 0.5v^{3.01}.$$

In order to verify how accurately the adaptation laws capture the LQR designed dynamics, we compare the closed-loop eigenvalues of the two systems in Figure C.3. Clearly, the adapted system poles have marginally decreased real-part magnitudes during the initial phase of flight, when compared to the system with LQR gains. However, as the time progresses, the difference between the two systems decreases, and near the end of the trajectory, the adapted system eigenvalues have a slightly larger real-part magnitude. In order to check the closed-loop system's response with adapted gains, we consider an initial response to a speed disturbance, $\Delta v = 0.1$ km/s, applied at $t = 54.5$ s (Figure C.4), with the coefficients frozen at their respective values also at $t = 54.5$ s. Both the control systems are seen to mitigate the disturbance

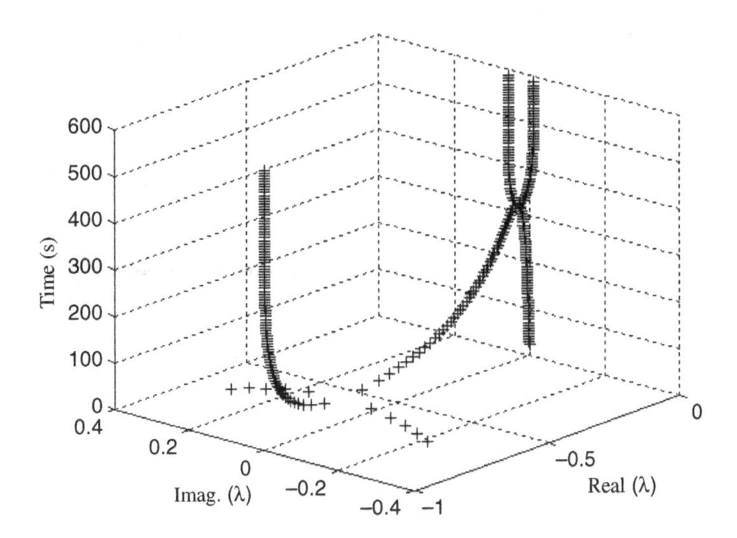

Figure C.1 Closed-loop guidance system eigenvalues along the nominal, gravity-turn trajectory of a launch vehicle with forward acceleration input

completely in about 10 s. While Figure C.4 is not an accurate picture of how the time-varying system would behave, it indicates a close agreement between the adapted and the LQR design systems, thereby validating the choice of the adaptation laws.

To simulate the actual nonlinear time-varying system with adaptive feedback gains we construct a Simulink model as shown in Figure C.5, with the previously programmed nonlinear subsystem block, *Rocket Translational Dynamics* (Chapter 4). Note the adaptation loops in Figure C.5 specified with

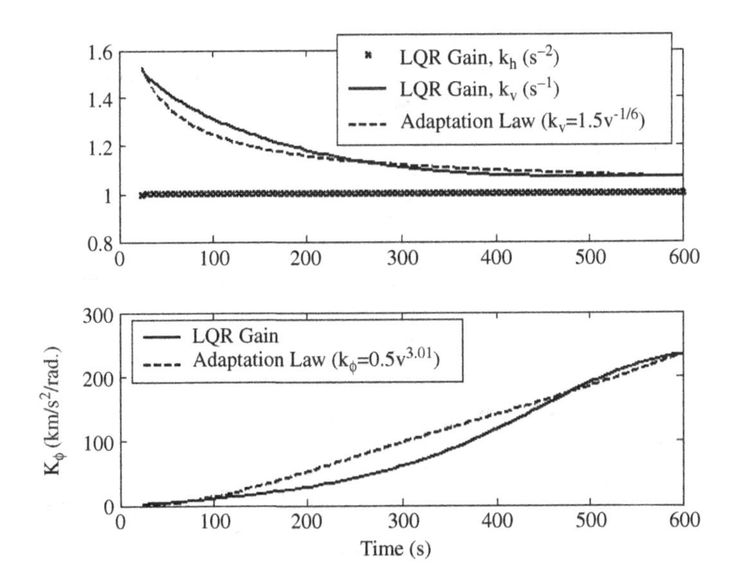

Figure C.2 Feedback gains for LQR and adaptive guidance systems along the nominal, gravity-turn trajectory of a launch vehicle with forward acceleration input

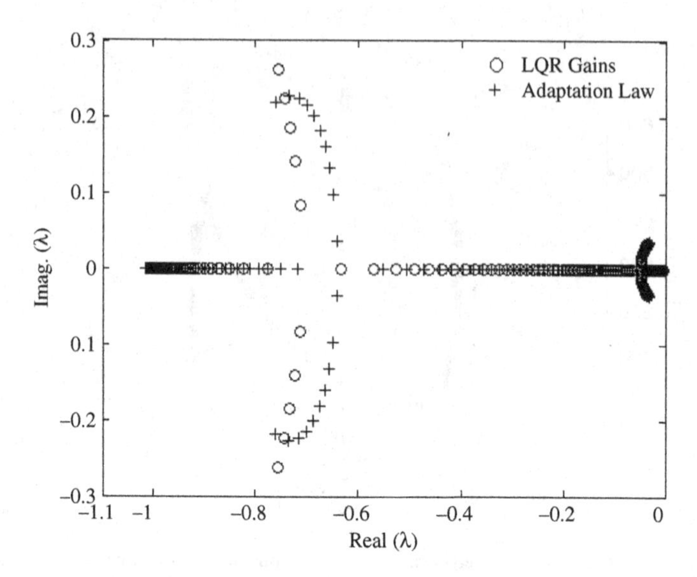

Figure C.3 Closed-loop eigenvalues comparison of LQR and adapted guidance systems along the nominal, gravity-turn trajectory of a launch vehicle with forward acceleration input

user-defined mathematical function blocks. The results of the simulation – carried out with a fourth-order Runge–Kutta scheme – are plotted in Figures C.6–C.9. The initial error of 0.1 rad (5.7°) in the flight-path angle – and the attendant undershoot in flight speed – is seen to be attenuated in about 400 s (Figure C.6), compared to about 50 s for the normal acceleration guidance case of Chapter 4, without affecting the

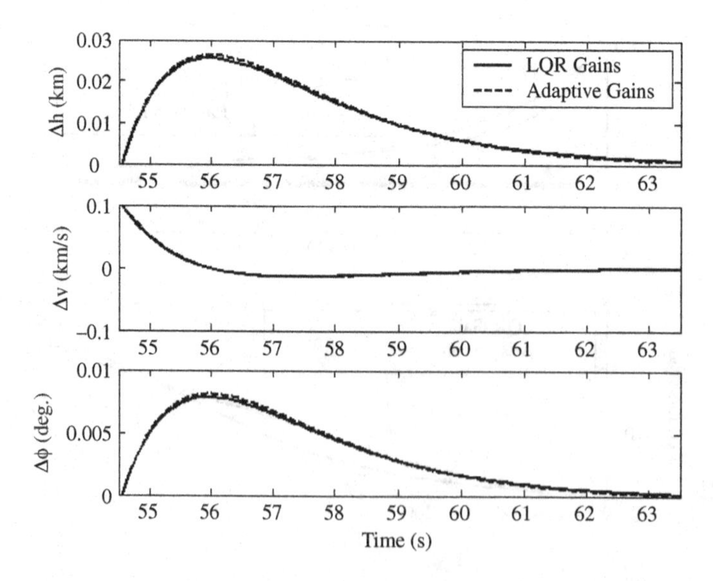

Figure C.4 Closed-loop initial response comparison of LQR and adapted guidance systems with frozen parameters and forward acceleration input

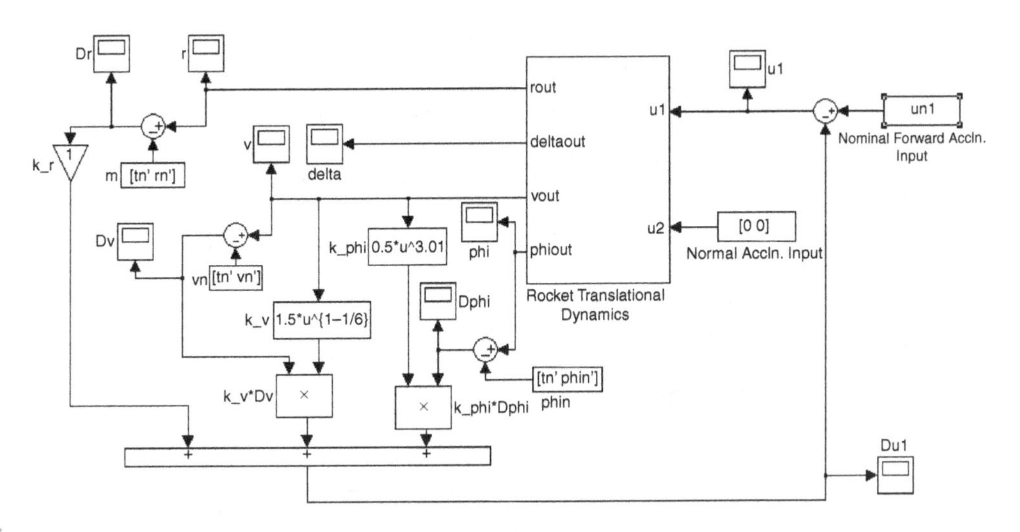

Figure C.5 Simulink block diagram for simulating closed-loop initial response of nonlinear adaptive rocket guidance system with forward acceleration input

desired terminal conditions in any way. Figure C.7 shows that the maximum altitude and speed errors of about 3 km and −3 km/s, respectively, occur nearly 225 s after launch, while the maximum flight-path angle error of 34° takes place much earlier at $t = 58$ s. The required off-nominal forward acceleration control input, Δu_1, is within ±40 m/s^2 (Figure C.8), which appears to be acceptable given the rather large initial flight-path error. Figure C.9 focuses on the simulated closed-loop flight speed in the first minute of flight, showing an oscillatory behavior due to the rapid push-pull action of the forward acceleration input, Δu_1, observed in Figure C.8. Such an oscillation in speed is capable of exciting an unwanted structural

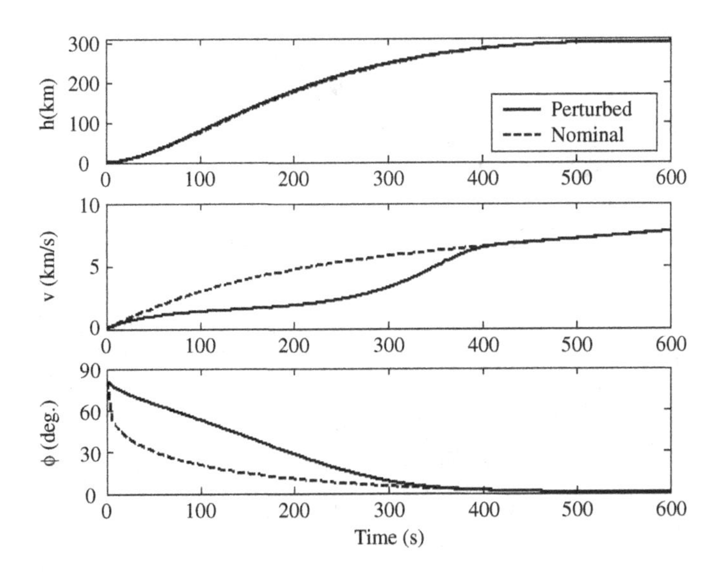

Figure C.6 Nominal (open-loop) and initially perturbed (closed-loop) trajectories with nonlinear adaptive rocket guidance system with forward acceleration input

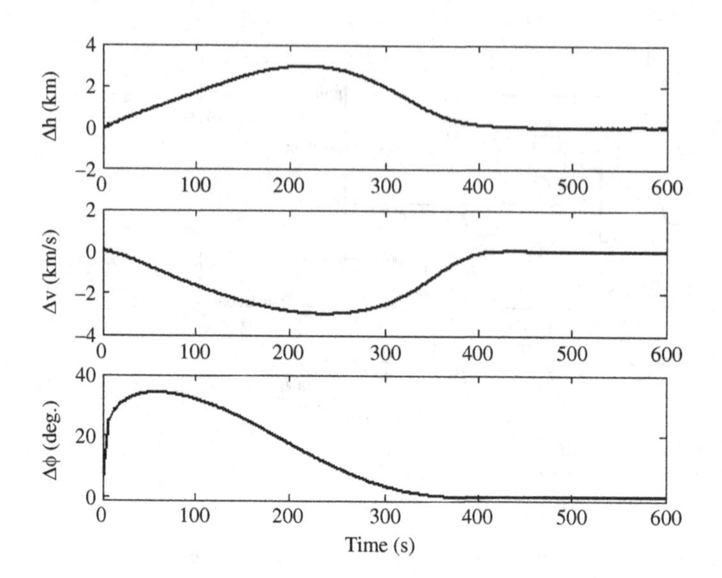

Figure C.7 Closed-loop simulated trajectory error with initial flight-path perturbation with nonlinear adaptive rocket guidance with forward acceleration input

dynamic instability of liquid propellant sloshing inside the rocket called *pogo*, which, if unattended, can destroy a rocket. The traditional (passive) way of addressing the pogo instability issue is by putting baffles in the propellant tanks that move the primary sloshing mode natural frequency away from the guidance and control system bandwidth.

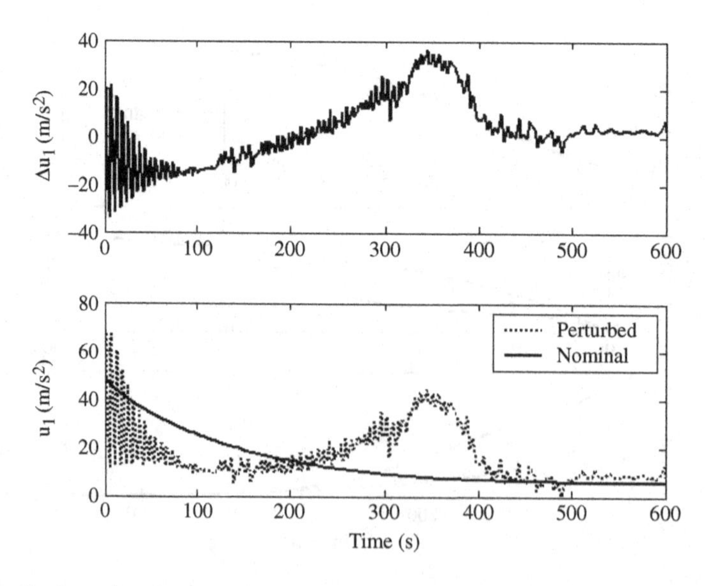

Figure C.8 Feedback, nominal, and total forward acceleration inputs with nonlinear adaptive rocket guidance with forward acceleration input

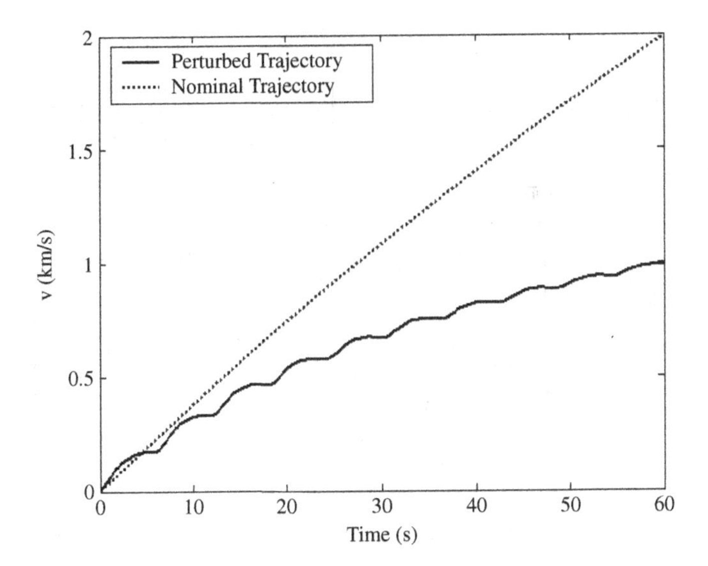

Figure C.9 Closed-loop simulated speed with initial flight-path perturbation with nonlinear adaptive rocket guidance with forward acceleration input

C.2 Thrust Saturation and Rate Limits (Increased Underactuation)

Modern liquid propellant engines can be throttled up to some extent by regulating the supply of propellants through valves. However, a propellant flow rate below a certain limit cannot sustain combustion, resulting in flame-out (zero thrust). Even when throttling is possible, the engine response is highly nonlinear due to limits on the rate of change of thrust. Thus the rocket engine can be regarded as a nonlinear actuator with output saturation and rate limits, and care must be taken to account for the engine behavior when simulating a closed-loop guidance and control system involving thrust modulation.

Due to difficulty in throttling a large rocket engine, most launch vehicles employ small liquid propellant rockets called *vernier engines* dedicated to the task of guidance and attitude control. The vernier rockets – having a faster throttle response – produce axial and normal acceleration inputs with some alacrity required for controlling the flight path and attitude, while the main engines operate at nearly full power until burn-out. However, this does not allow precise control of the flight speed.

Having magnitude and rate limits on the thrust results in a thrust-controlled vehicle becoming further underactuated due to the actuator restrictions not allowing the full level of control to be exercised, as demanded by a state feedback regulator. Furthermore, now the system's underactuated behavior is nonlinear due to the presence of actuation limits. While there are certain procedures for designing nonlinear, underactuated control systems (such as the *describing function method* (Gibson 1963), which applies a frequency domain approach through elaborate quasi-linear gains for each class of nonlinearity), we will refrain from such a design. Instead, we will test our time-varying adaptive control design for robustness in the presence of magnitude and rate limits.

Example C.2 Taking into account the following saturation and rate limits of the forward acceleration provided by the rocket engine, simulate the closed-loop response of the gravity-turn guidance system designed in Example C.1:

$$6 \leq u_1 \leq 40 \text{ m/s}^2, \qquad \left| \frac{\mathrm{d}u_1}{\mathrm{d}t} \right| \leq 30 \text{ m/s}^3.$$

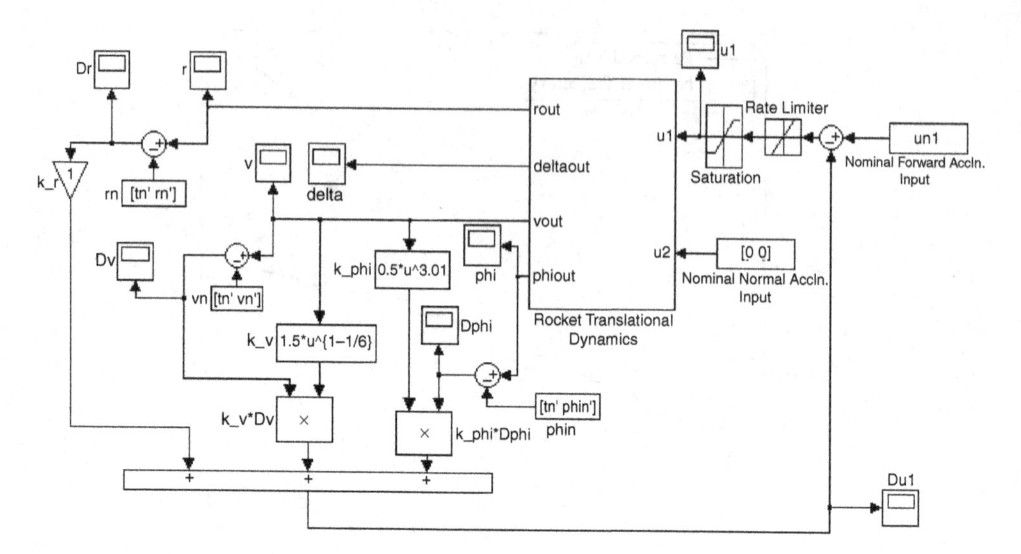

Figure C.10 Simulink block diagram for simulating the nonlinear, closed-loop, adaptive rocket guidance system with thrust saturation and rate limits in a gravity-turn launch with initial perturbation

We insert the nonlinear saturation and rate limit blocks as constituting the engine actuator into the Simulink block diagram of Figure C.5, as shown in Figure C.10, and repeat the simulation of Example C.1. The results of the simulation are plotted in Figures C.11 and C.12. As a consequence of the full level of control not being exercised, there is a terminal error (excess speed of 0.01km/s and excess altitude of 0.057km) due to the inability of the engine to throttle below $u_1 = 6$ m/s^2 as required by the guidance system designed in Example C.1. As a result, the vehicle is sent into a slightly higher orbit than necessary. However, the design can be considered conservative as the mission requirements are slightly exceeded,

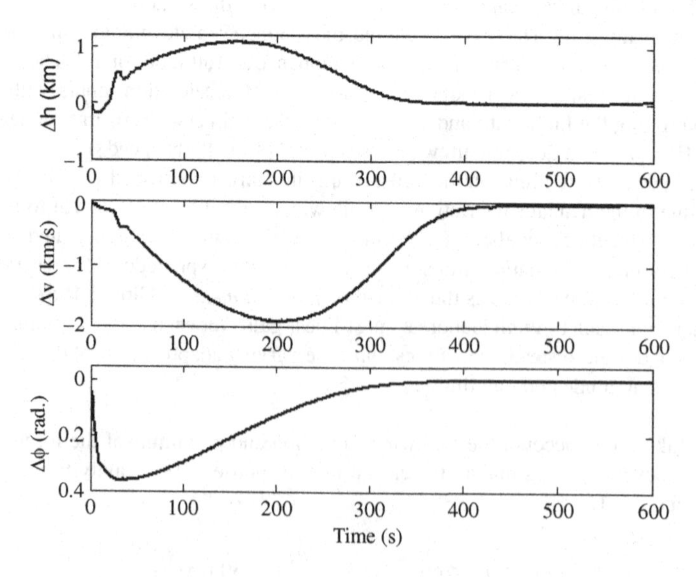

Figure C.11 Closed-loop trajectory error with an adaptive rocket guidance system simulated with thrust saturation and rate limits

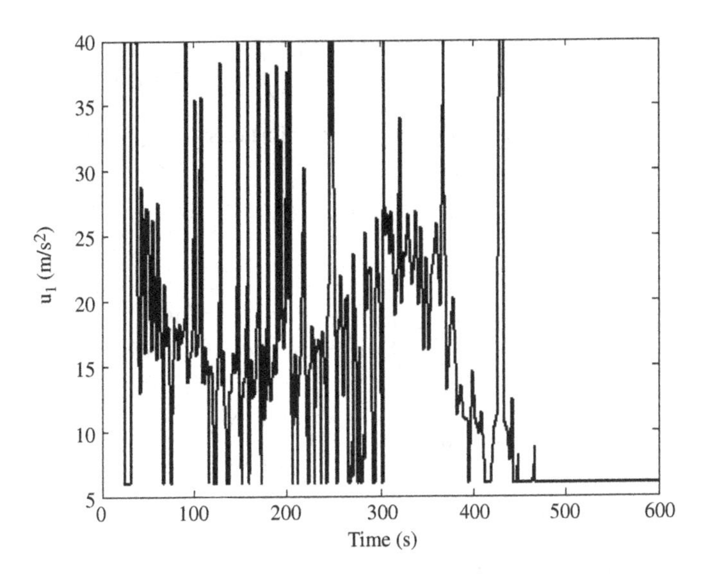

Figure C.12 Closed-loop forward acceleration input with an adaptive rocket guidance system simulated with thrust saturation and rate limits

and the desired orbit can be subsequently attained by minor orbital corrections (Chapter 6). When the atmospheric drag is taken into account, the final error can be seen to be much smaller than that observed in Figure C.11.

It is interesting to compare the closed-loop forward acceleration inputs plotted in Figures C.12 and C.8. Clearly, the curtailment of both the magnitude and the rate of actuation results in the engine switching between its upper and lower limits in order to drive the system toward a smaller error. Such a behavior of nonlinear underactuated, asymptotically stable closed-loop systems is called *switching*. It is also as if the control system were *sliding* along the nominal trajectory by alternately increasing and decreasing the input magnitude between the actuator limits. Such a nonlinear control strategy is called *sliding mode control* (Slotine and Li 1991). Although we have not designed a sliding mode controller deliberately, it arises automatically as a consequence of putting the thrust limitations in the closed-loop regulator. The ability of the nonlinear, time-varying regulator to exercise control even in the presence of substantial magnitude and rate limitations attests to its robustness.

C.3 Single- and Bi-output Observers with Forward Acceleration Input

We now extend the state feedback controller to an observer-based guidance of the underactuated rocket. Let us first consider the single output to be the speed error, $y = \Delta v$, with the rearranged state vector, $\mathbf{x} = (\mathbf{x}_1, \mathbf{x}_2)$, where $\mathbf{x}_1 = \Delta v$, $\mathbf{x}_2 = (\Delta h, \Delta\phi)$, and the following state-space coefficient matrices:

$$\mathbf{A} = \left(\begin{array}{c|cc} 0 & a_{31} & a_{34} \\ \hline a_{13} & 0 & a_{14} \\ a_{43} & a_{41} & a_{44} \end{array}\right), \quad \mathbf{B} = \left(\begin{array}{c} 1 \\ \hline 0 \\ 0 \end{array}\right). \tag{C.3}$$

The output equation, $\mathbf{y} = \mathbf{x}_1$, and the partitioning of $\mathbf{A}, \mathbf{B}$ produce the following sub-matrices required for the reduced-order observer:

$$\mathbf{C} = 1, \quad \mathbf{A}_{11} = 0, \quad \mathbf{A}_{12} = (a_{31} \;\; a_{34}), \quad \mathbf{B}_1 = 1, \tag{C.4}$$

$$\mathbf{A}_{21} = \begin{pmatrix} a_{13} \\ a_{43} \end{pmatrix}, \quad \mathbf{A}_{22} = \begin{pmatrix} 0 & a_{14} \\ a_{41} & a_{44} \end{pmatrix}, \quad \mathbf{B}_2 = \begin{pmatrix} 0 \\ 0 \end{pmatrix}. \tag{C.5}$$

The state equation of the reduced-order observer with estimated state, $\mathbf{x}_{o2} = \mathbf{L}\mathbf{y} + \mathbf{z}$, and gain matrix, $\mathbf{L} = (L_1, L_2)^T$, is

$$\dot{\mathbf{z}} = \mathbf{F}\mathbf{z} + \mathbf{H}\mathbf{u} + \mathbf{G}\mathbf{y}, \tag{C.6}$$

where

$$\mathbf{F} = \mathbf{A}_{22} - \mathbf{L}\mathbf{C}\mathbf{A}_{12} = \begin{pmatrix} -L_1 a_{31} & a_{14} - L_1 a_{34} \\ a_{41} - L_2 a_{31} & a_{44} - L_2 a_{34} \end{pmatrix}, \tag{C.7}$$

$$\mathbf{H} = \mathbf{B}_2 - \mathbf{L}\mathbf{C}\mathbf{B}_1 = -\begin{pmatrix} L_1 \\ L_2 \end{pmatrix}, \tag{C.8}$$

and

$$\mathbf{G} = \mathbf{F}\mathbf{L} + (\mathbf{A}_{21} - \mathbf{L}\mathbf{C}\mathbf{A}_{11})\,\mathbf{C}^{-1} = \begin{pmatrix} a_{13} - L_1^2 a_{31} + L_2 a_{14} - L_1 L_2 a_{34} \\ a_{43} + L_1 a_{41} - L_1 L_2 a_{31} + L_2 a_{44} - L_2^2 a_{34} \end{pmatrix}. \tag{C.9}$$

The poles of the second-order observer can be readily determined in a closed form from the desired observer characteristic equation,

$$\begin{aligned} |\lambda \mathbf{I} - \mathbf{F}| = {} & \lambda^2 + (L_1 a_{31} + L_2 a_{34} - a_{44})\,\lambda \\ & + L_1(a_{34}a_{41} - a_{31}a_{44}) + L_2 a_{14} a_{31} - a_{14}a_{41}. \end{aligned} \tag{C.10}$$

For selected observer natural frequency, ω, and damping ratio, ζ, the observer gains are thus

$$\begin{aligned} L_1 &= \frac{\omega^2 + 4v\zeta\omega \sin\phi/r + 2\mu/r^3 - v^2(\cos^2\phi + 2\sin^2\phi)/r^2}{\mu[rv(6\sin^2\phi + 1) - 2\mu/v]/r^5}, \\ L_2 &= \frac{2r^2\zeta\omega - (rv - \mu/v)\sin\phi - 2\mu L_1 \sin\phi/r}{\mu\cos\phi}. \end{aligned} \tag{C.11}$$

Selection of the second-order observer's characteristics, (ω, ζ), forms the crux of observer design. Care must be exercised in selecting the observer poles for the rocket guidance problem with thrust acceleration input, because too small a real-part magnitude may cause the observer to interact with the dominant closed-loop dynamics, thereby amplifying the transient response as well as increasing the settling time. On the other hand, observer poles too deep into the left-half plane would require a larger controller activity (Tewari 2002) that may either exceed the thrust limitations of the rocket or destabilize the time-varying system. Thus, a balance must be struck between the speed of response and observer gain magnitudes.

Example C.3 Design a reduced-order observer for the gravity-turn trajectory guidance system of Example C.1 using the speed error, Δv, as the only output. Replace the full-state feedback controller of Example C.1 with the observer-based output feedback compensator and simulate the closed-loop response to the given initial trajectory perturbation.

Ideally, a likely observer performance criterion would be a well-damped step response with settling time a fraction of that of the state feedback system. However, when applied to rocket guidance problems, one must also consider the transient control input limits that may be exceeded and the time-varying system destabilized if the observer gains are either too large or too small. Given the adaptive regulator poles of Example C.1 whose real parts lie in the range $(-1.016, -0.0035)$, we select the second-order observer's natural frequency, $\omega = 0.001\sqrt{2}$ rad/s and a critical damping ratio value of $\zeta = 1/\sqrt{2}$. While this selection produces a very large observer settling time of about 4000 s, there is little danger of destabilizing the closed-loop system by interaction with the regulator poles. However, the long settling time of the observer leads to unsuppressed transient errors and thus an increased control effort is expected, compared to the state feedback design. The observer gains computed by equation (C.11) result in the observer poles computed along the nominal trajectory being constant at $-0.001 \pm 0.001i$.

While both the observer and regulator are asymptotically stable linear systems at every time instant, an attempt to simulate the response of the closed-loop guidance system with the adaptation gains of Example C.1 quickly reveals an unstable time-varying system. This is due to the very large and initially rapidly increasing regulator gain, k_{h}, that causes large transient fluctuations (overshoots) from the nominal, which quickly destabilizes the system. Such an instability highlights our earlier caution of applying the results of a frozen time analysis to a time-varying system, where in extreme cases we have an unstable system even though the poles always remain in the left-half Laplace domain.

In order to have a stable compensator, we must fine-tune the adaptation law for the regulator gains. Unfortunately, there is no systematic procedure – like the separation principle for linear systems – for designing a nonlinear time-varying compensator; hence, one must rely upon trial and error with the given plant. After a few test simulations with the selected observer gains, we arrive at the regulator gains

$$k_{\mathrm{h}}(t) = 0.0139959, \quad k_{\mathrm{v}}(t) = 1.627v^{-1/6}, \quad k_{\phi}(t) = 0.65367v^{3.01},$$

which are substituted into the following control-law for the compensated system:

$$\Delta u_1 = k_{\mathrm{v}}\Delta v + \begin{pmatrix} k_{\mathrm{h}}, & k_{\phi} \end{pmatrix} \mathbf{x}_{o2}.$$

Table C.1 lists the MATLAB code, *rocket_obs2.m*, used for simulating the closed-loop compensated guidance system with a fourth-order Runge–Kutta method. This code is called by a separate program that lists the nominal trajectory and input (Chapter 4) as global variables tn, rn, vn, phin, un, and has the following statement:

```
[t,x]=ode45('rocket_obs2',[0 8000],[6378.14 0.001 1.5068 0 0]');
```

A separate code called *rocket_obs_u2.m* (Table C.2) computes the closed-loop forward acceleration input, $u_1(t)$, by post-processing. The results of the simulation for initial trajectory error $\Delta\phi(0) = 0.1$ rad are plotted in Figures C.13–C.15. While Figure C.13 indicates an altitude deficit in the initial launch phase, the longer-term simulation plotted in Figure C.14 confirms that the launch vehicle has placed the spacecraft in a (262 km × 338 km) elliptical orbit of mean altitude 300 km ($e = 0.0057$, $a = 6678.14$ km), even in the presence of a large initial error (5.7°) in the trajectory. The small eccentricity can be easily corrected by a final circularization maneuver in space (Chapter 6). However, the acceleration control input (Figure C.15) – requiring a peak magnitude several times that of the nominal as well as negative thrust at some points – is hardly possible with an actual rocket engine. Furthermore, the extreme sensitivity of the closed-loop trajectory to feedback gains results in a poor stability margin. If the thrust saturation and rate limits are taken into account there is no hope of obtaining a stable response. Such a lack of robustness renders the entire design practically unworthy for flight. The present example illustrates the limitations of a guidance system based upon thrust input and speed output.

Example C.3 reveals the limitation of a single-output observer for guiding a rocket with only the forward acceleration input. Let us now examine an observer that employs both altitude and speed measurements,

Table C.1

```
% Program 'rocket_obs2.m' for generating the closed-loop
% state equations for a launch vehicle to circular orbit
% while guided by a forward acceleration controlled,
% speed output based reduced-order compensator.
% (c) 2010 Ashish Tewari
% To be used by Runge-Kutta solver 'ode45.m'
% t: current time (s); x(1,1): current radius;
% x(2,1): current speed
% x(3,1): current flight-path angle;
% x(4:5,1): reduced-order observer state vector.
function xdot=rocket_obs2(t,x)
global tn;  %Nominal time listed in the calling program.
global rn;  %Nominal radius listed in the calling program.
global vn;  %Nominal speed listed in the calling program.
global phin;%Nominal flt-path angle listed in the calling program.
global un;  %Nominal control input listed in the calling program.
mu=398600.4;%Earth's gravitational constant (km^3/s^2)
zeta=1/sqrt(2);w=.001/zeta; %Observer damp.-ratio & nat. frequency.
r=interp1(tn,rn,t); %Interpolation for current nominal radius.
v=interp1(tn,vn,t); %Interpolation for current nominal speed.
phi=interp1(tn,phin,t); %Interpolation current nominal f.p.a.
u=interp1(tn,un,t); %Interpolation for current nominal control input.

% Linearized plant's state coefficient matrix:
A=[0 sin(phi) v*cos(phi);
2*mu*sin(phi)/r^3   0   -mu*cos(phi)/r^2;
(-v/r^2+2*mu/(v*r^3))*cos(phi) ...
(1/r+mu/(v^2*r^2))*cos(phi)  -(v/r-mu/(v*r^2))*sin(phi)];

% Regulator gains by adaptation law:
k=[1.39959e-2 1.627*v^(-1/6) .65367*v^3.01];

% Observer coefficients follow:
a1=A(1,2);a2=A(1,3);a6=A(2,1);a7=A(2,3);a8=A(3,1);
a9=A(3,2);a10=A(3,3);
L1=(a2*a8+w^2-a2*a6*(a10+2*zeta*w)/a7)/(a7*a8-a6*a10-a2*a6^2/a7);
L2=(2*zeta*w+a10-L1*a6)/a7;
L=[L1;L2];
F=[0 a2;a8 a10]-L*[a6 a7];
H=-L;
G=F*L+[a1;a9];

% Trajectory error variables:
dr=x(1,1)-r;
dv=x(2,1)-v;
dphi=x(3,1)-phi;
% Estimated trajectory error variables:
z=x(4:5,1);
xo2=L*dv+z;
% Feedback control input:
du=-k(1,2)*dv-k(1,1)*xo2(1,1)-k(1,3)*xo2(2,1);
if t>600
    u=0;
```

Table C.1 *(Continued)*

```
else
u=u+du;
end

% Trajectory state equations:
xdot(1,1)=x(2,1)*sin(x(3,1));
xdot(2,1)=u-mu*sin(x(3,1))/x(1,1)^2;
xdot(3,1)=(x(2,1)/x(1,1)-mu/(x(1,1)^2*x(2,1)))*cos(x(3,1));

% Observer's state equation:
xdot(4:5,1)=F*z+H*du+G*dv;
```

Table C.2

```
% Program 'rocket_obs_u2.m' for generating the closed-loop
% control input for a launch vehicle to circular orbit guided by
% forward acceleration controlled, speed output based reduced-order
% compensator, using post-processing of simulated trajectory (T,x).
% T: (nx1) vector containing simulation 'n' time instants.
% x: (nx5) vector containing trajectory and observer state.
function u=rocket_obs_u2(T,x)
global tn;
global rn;
global vn;
global phin;
global un;
mu=398600.4;
zeta=1/sqrt(2);w=.001/zeta;
n=size(T,1);
for i=1:n
t=T(i,1);
r=interp1(tn,rn,t);
v=interp1(tn,vn,t);
phi=interp1(tn,phin,t);
U=interp1(tn,un,t);
A=[0 sin(phi) v*cos(phi);
2*mu*sin(phi)/r^3   0  -mu*cos(phi)/r^2;
(-v/r^2+2*mu/(v*r^3))*cos(phi)  ...
(1/r+mu/(v^2*r^2))*cos(phi)  -(v/r-mu/(v*r^2))*sin(phi)];
a1=A(1,2);a2=A(1,3);a6=A(2,1);a7=A(2,3);a8=A(3,1);
a9=A(3,2);a10=A(3,3);
L1=(a2*a8+w^2-a2*a6*(a10+2*zeta*w)/a7)/(a7*a8-a6*a10-a2*a6^2/a7);
L2=(2*zeta*w+a10-L1*a6)/a7;
L=[L1;L2];
F=[0 a2;a8 a10]-L*[a6 a7];
H=-L;
G=F*L+[a1;a9];
dr=x(i,1)-r;
dv=x(i,2)-v;
dphi=x(i,3)-phi;
z=x(i,4:5);
xo2=L'*dv+z;
```

(Continued)

Table C.2 *(Continued)*

```
if t>600
    u(i,1)=0;
else
    k=[1.39959e-2 1.627*v^(-1/6) .65367*v^3.01];
    du=-k(1,2)*dv-k(1,1)*xo2(1,1)-k(1,3)*xo2(1,2);
    u(i,1)=U+du;
end
end
```

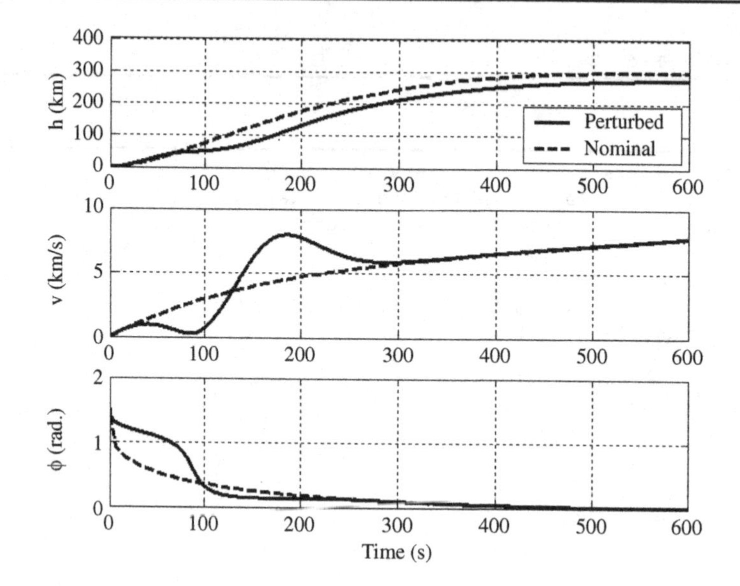

Figure C.13 Closed-loop compensated trajectory during launch phase of a guided launch vehicle with a speed output based reduced-order observer and adaptive regulator

$\mathbf{y} = \mathbf{x}_1 = (\Delta h, \Delta v)^T$, $\mathbf{x}_2 = \Delta\phi$, and the following partitioned coefficient matrices:

$$\mathbf{A} = \begin{pmatrix} 0 & a_{13} & | & a_{14} \\ a_{31} & 0 & | & a_{34} \\ \hline a_{41} & a_{43} & | & a_{44} \end{pmatrix}, \quad \mathbf{B} = \begin{pmatrix} 0 \\ 1 \\ \hline 0 \end{pmatrix}. \tag{C.12}$$

The sub-matrices required for the reduced-order observer are thus the following:

$$\mathbf{C} = \mathbf{I}, \quad \mathbf{A}_{11} = \begin{pmatrix} 0 & a_{13} \\ a_{31} & 0 \end{pmatrix}, \quad \mathbf{A}_{12} = \begin{pmatrix} a_{14} \\ a_{34} \end{pmatrix}, \quad \mathbf{B}_1 = \begin{pmatrix} 0 \\ 1 \end{pmatrix}, \tag{C.13}$$

$$\mathbf{A}_{21} = (a_{41} \quad a_{43}), \quad \mathbf{A}_{22} = a_{44}, \quad \mathbf{B}_2 = 0. \tag{C.14}$$

The state equation of the single-order observer with estimated state $x_{o2} = \mathbf{L}\mathbf{y} + z$ and gain matrix $\mathbf{L} = (L_1, L_2)$ is

$$\dot{z} = Fz + Hu + \mathbf{G}\mathbf{y}, \tag{C.15}$$

where

$$F = \mathbf{A}_{22} - \mathbf{L}\mathbf{C}\mathbf{A}_{12} = a_{44} - L_1 a_{14} - L_2 a_{34}, \quad H = \mathbf{B}_2 - \mathbf{L}\mathbf{C}\mathbf{B}_1 = -L_2, \tag{C.16}$$

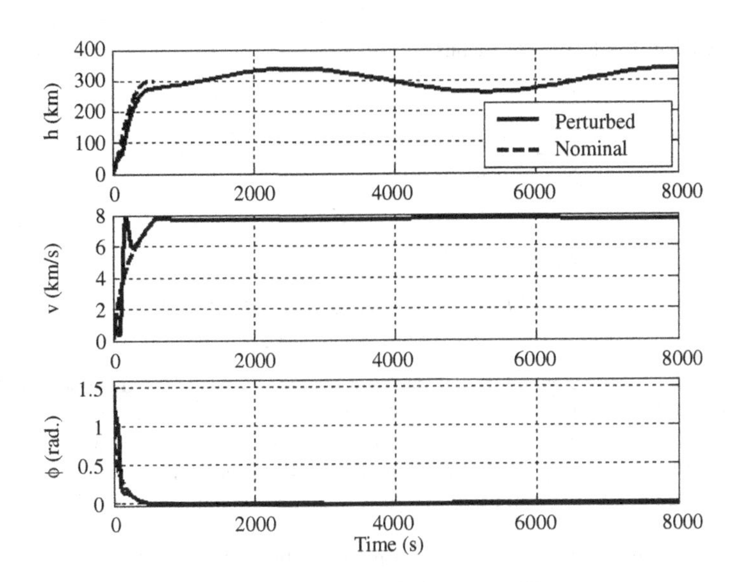

Figure C.14 Closed-loop compensated trajectory including initial orbital phase of a guided launch vehicle with a speed output based reduced-order observer and adaptive regulator

and

$$\begin{aligned}\mathbf{G} &= \mathbf{FL} + (\mathbf{A}_{21} - \mathbf{LCA}_{11})\,\mathbf{C}^{-1} \\ &= (FL_1 + a_{41} - L_2 a_{31} \quad FL_2 + a_{43} - L_1 a_{13})\,.\end{aligned} \tag{C.17}$$

The pole of the observer, $\lambda = F$, is determined from the desired settling time, t_s, by

$$F = -\frac{4}{t_s}, \tag{C.18}$$

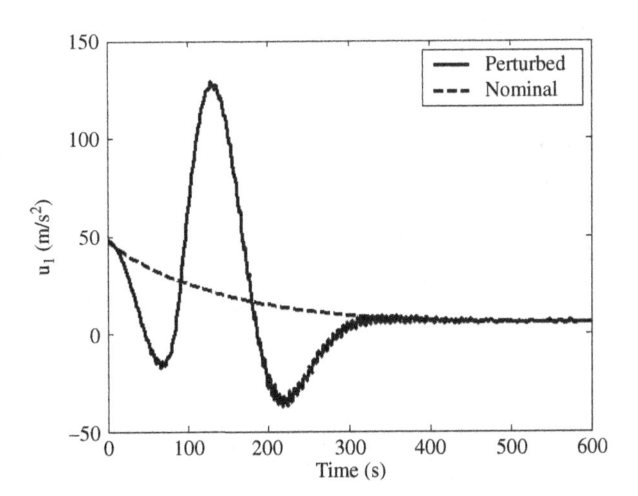

Figure C.15 Closed-loop forward acceleration input during launch phase of a guided launch vehicle with a speed output based reduced-order observer and adaptive regulator and alternative choices for the observer gains are either

$$L_1 = 0,$$
$$L_2 = \frac{a_{44} + 4/t_s}{a_{34}} = -\frac{\left(\frac{\mu}{vr^2} - \frac{v}{r}\right)\sin\phi + \frac{4}{t_s}}{\frac{\mu}{r^2}\cos\phi}, \tag{C.19}$$

or

$$L_2 = 0,$$
$$L_1 = \frac{a_{44} + 4/t_s}{a_{14}} = -\frac{\left(\frac{\mu}{vr^2} - \frac{v}{r}\right)\sin\phi + \frac{4}{t_s}}{v\cos\phi}. \tag{C.20}$$

Whichever of the above equivalent choices is made for the gains, it makes no difference to the observer design, and both produce an identical closed-loop response.

The main advantage of the bi-output observer over the single-output one is that the former is much more robust than the latter due to a better conditioning of the observability test matrix. Therefore, we expect that the bi-output observer can be successfully implemented in the actual guidance system.

Example C.4 Redesign the observer-based compensator for the gravity-turn trajectory guidance system of Example C.3 using the altitude and speed errors, Δh, Δv, as the two outputs. Simulate the closed-loop response to the initial trajectory perturbation given in Example C.1.

Selecting the observer pole as $F = -0.01$, we have

$$L_1 = 0, \quad L_2 = \frac{a_{44} + 0.01}{a_{34}},$$

$$G = (a_{41} - L_2 a_{31} \quad a_{43} - 4L_2), \quad H = -L_2.$$

The observer gain, L_2, plotted along the nominal trajectory for the first second of flight is shown in Figure C.16. Note the large initial magnitude of L_2 denoting the singularity at $v = 0$, which rapidly declines to a small value after $t = 1$ s.

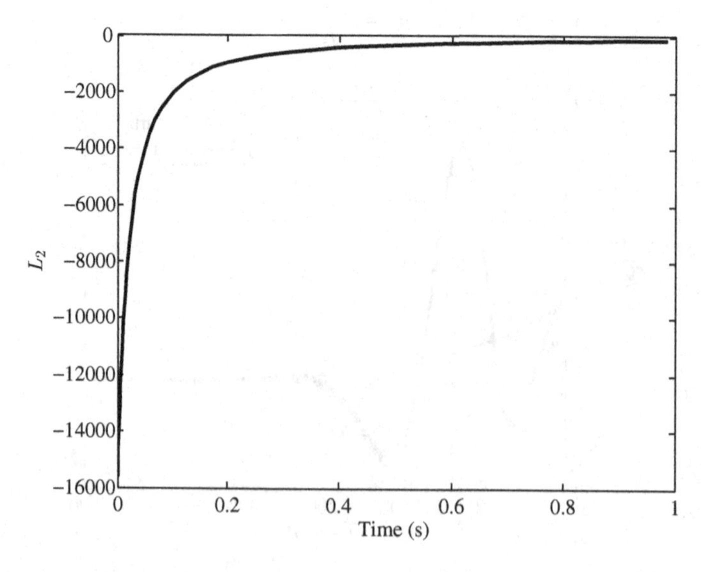

Figure C.16 Observer gain for a bi-output, speed- and altitude-based, reduced-order observer for a rocket guidance system

While we can directly utilize the adaptive regulator gains of Example C.1, it is better to fine-tune the regulator in order to avoid the usual degradation of performance with observer-based feedback. Hence, a better choice of the regulator gains for use in the reduced-order, bi-output compensator is

$$k_{\rm h}(t) = 1, \quad k_{\rm v}(t) = v^{-1/2}, \quad k_{\phi}(t) = 1.0 \times 10^{-5}.$$

For closing the loop of the compensated system, we have

$$\Delta u_1 = k_{\rm h}\Delta h + k_{\rm v}\Delta v + k_{\rm h} + k_{\phi}x_{o2}.$$

The simulation of the gravity-turn guidance system with the bi-output observer is carried out by a fourth-order Runge–Kutta method using a modification of the MATLAB code, *rocket_obs2.m* (Table C.1), listed in Table C.3. A separate code (not listed here for brevity) similar to *rocket_obs_u2.m* (Table C.2) computes the closed-loop control input for the bi-output case. The results of the simulation are plotted in Figures C.17 and C.18, showing a successful launch even in the presence of a large initial angular error while requiring a control input slightly smaller in magnitude compared to the state feedback regulator of Example C.1. The best feature of the bi-output compensator is its greater robustness with respect to variation in controller gains, $k_{\rm h}$, $k_{\rm v}$, k_{ϕ}, L_2, enabling it to have a practical application.

Table C.3

```
% Program 'rocket_obs.m' for generating the closed-loop
% state equations for a launch vehicle to circular orbit while
% guided by a forward acceleration controlled, bi-output
% reduced-order compensator based upon measurement and feedback of
% speed and altitude
% (c) 2010 Ashish Tewari
% To be used by Runge-Kutta solver 'ode45.m'
% t: current time (s); x(1,1): current radius;
% x(2,1): current speed
% x(3,1): current flight-path angle;
% x(4,1): reduced-order observer state vector.
%
function xdot=rocket_obs(t,x)
global tn;    %Nominal time listed in the calling program.
global rn;    %Nominal radius listed in the calling program.
global vn;    %Nominal speed listed in the calling program.
global phin; %Nominal f.p.a. listed in the calling program.
global un;    %Nominal control listed in the calling program.
mu=398600.4; %Earth's gravitational constant (km^3/s^2)
r=interp1(tn,rn,t); %Interpolation for current nominal radius.
v=interp1(tn,vn,t); %Interpolation for current nominal speed.
phi=interp1(tn,phin,t); %Interpolation for current nominal f.p.a.
u=interp1(tn,un,t); %Interpolation for current nominal control.

% Linearized plant's state coefficient matrix:
A=[0 sin(phi) v*cos(phi);
2*mu*sin(phi)/r^3    0   -mu*cos(phi)/r^2;
(-v/r^2+2*mu/(v*r^3))*cos(phi) ...
(1/r+mu/(v^2*r^2))*cos(phi) -(v/r-mu/(v*r^2))*sin(phi)];
% Regulator gains by adaptation law:
k=[1 v^(-1/2) 1e-5];
```

Table C.3 (*Continued*)

```
% Observer coefficients follow:
a1=A(1,2);a2=A(1,3);a6=A(2,1);a7=A(2,3);a8=A(3,1);
a9=A(3,2);a10=A(3,3);
F=-0.01;
L2=(a10-F)/a7;
G=F*[0 L2]+[a8 a9]-[L2*a6 0];
H=-L2;
% Trajectory error variables:
dr=x(1,1)-r;
dv=x(2,1)-v;
dphi=x(3,1)-phi;
% Estimated trajectory error variables:
z=x(4,1);
xo2=L2*dv+z;
% Feedback control input:
du=-k(1,2)*dv-k(1,1)*dr-k(1,3)*xo2;
if t>600
    u=0;
else
u=u+du;
end
% Trajectory state equations:
xdot(1,1)=x(2,1)*sin(x(3,1));
xdot(2,1)=u-mu*sin(x(3,1))/x(1,1)^2;
xdot(3,1)=(x(2,1)/x(1,1)-mu/(x(1,1)^2*x(2,1)))*cos(x(3,1));
% Observer's state equation:
xdot(4,1)=F*z+H*du+G*[dr dv]';
```

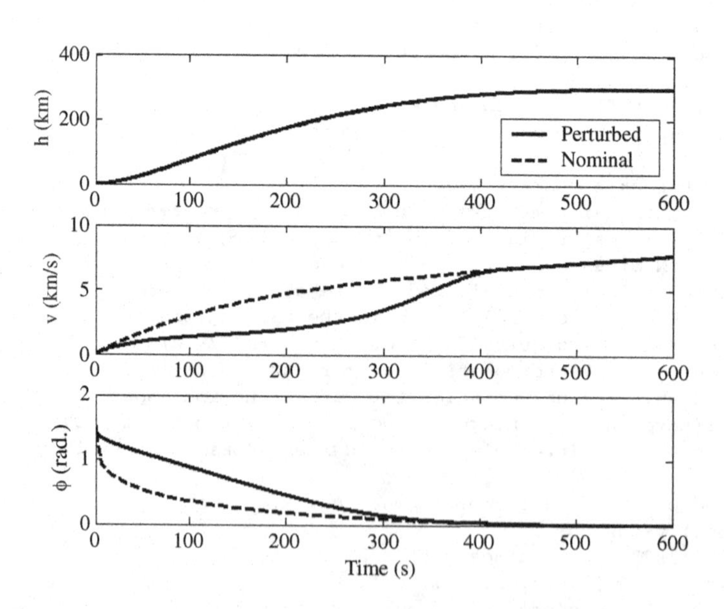

Figure C.17 Closed-loop compensated trajectory of a guided launch vehicle with a bi-output, reduced-order, observer-based adaptive compensator

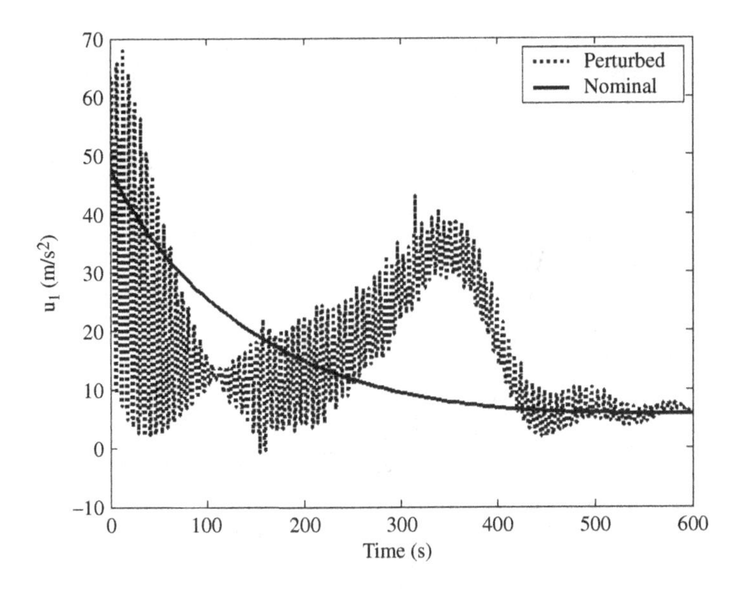

Figure C.18 Closed-loop forward acceleration input of a guided launch vehicle with a bi-output, reduced-order, observer-based adaptive compensator

Example C.5 Compute the drag, mass, and thrust required for the observer-based gravity-turn trajectory guidance system with forward acceleration input of Example C.4 using the two-stage rocket's drag and mass data given in Chapter 4.

The results of the code *rocket_drag_mass.m* run for the closed-loop trajectory simulated in Example C.4 are plotted in Figures C.19 and C.20. Note that the thrust required is always several times greater than the

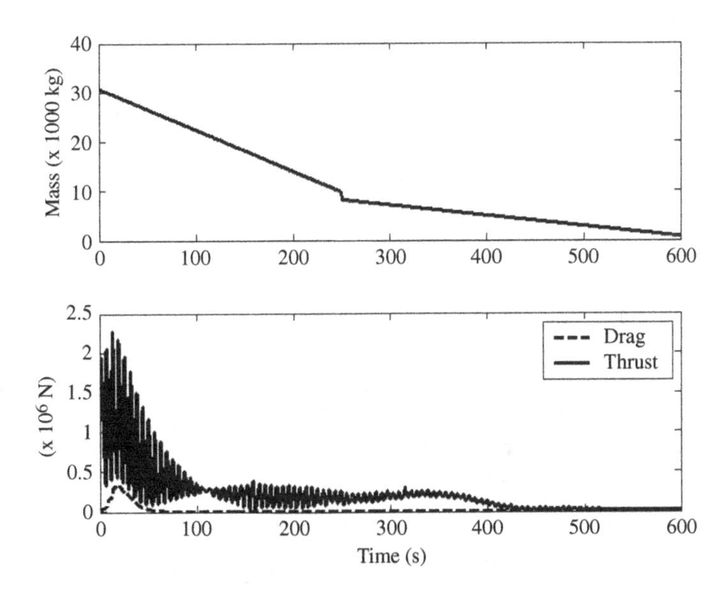

Figure C.19 Mass, atmospheric drag, and required thrust of a guided launch vehicle with a bi-output, reduced-order, observer-based adaptive compensator

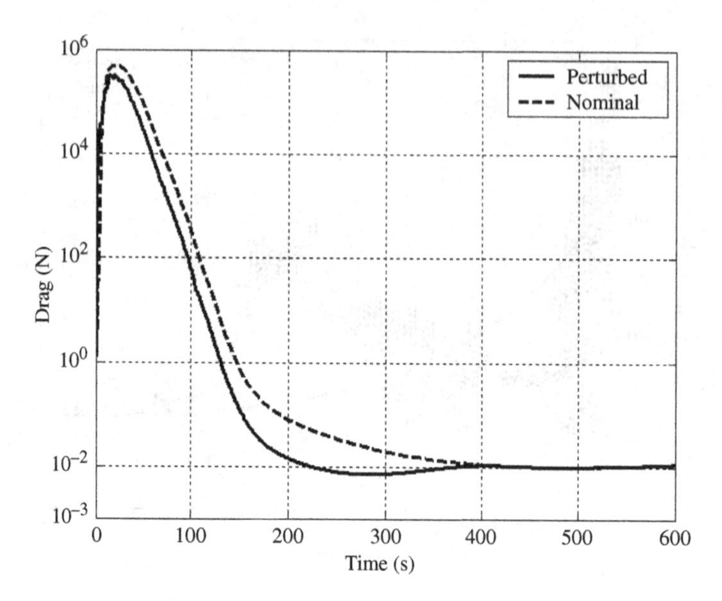

Figure C.20 Atmospheric drag of perturbed and nominal trajectories of a guided launch vehicle with a bi-output, reduced-order, observer-based adaptive compensator

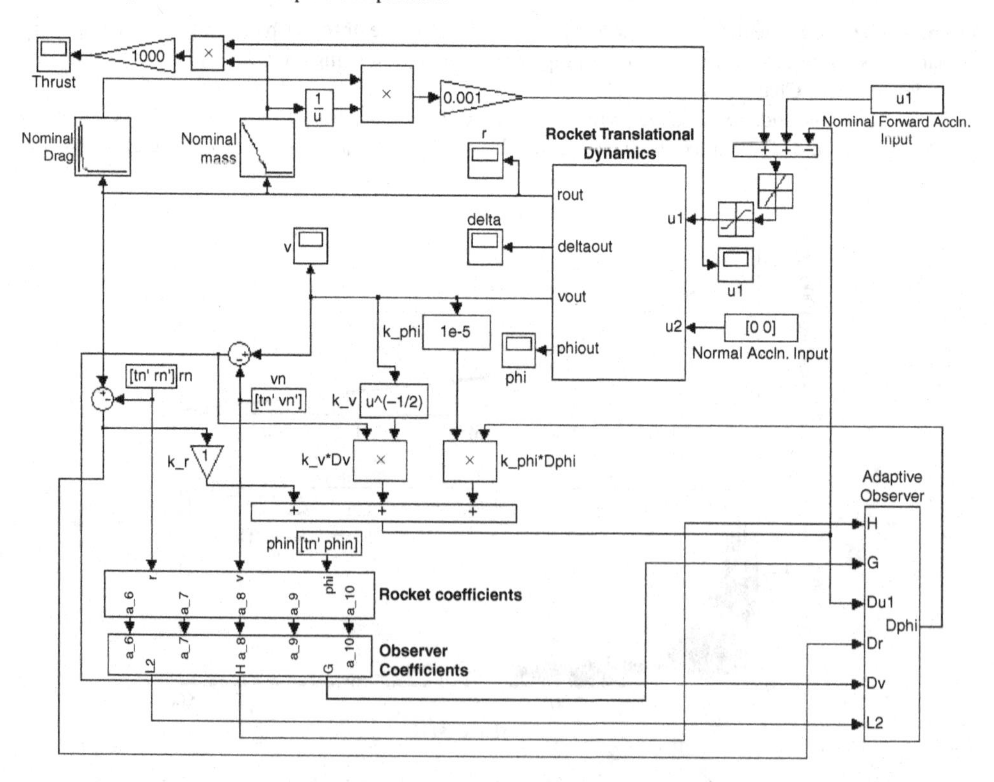

Figure C.21 Simulink block diagram for simulating closed-loop initial response of nonlinear adaptive rocket guidance system with adaptive observer and magnitude and rate limits on thrust acceleration input

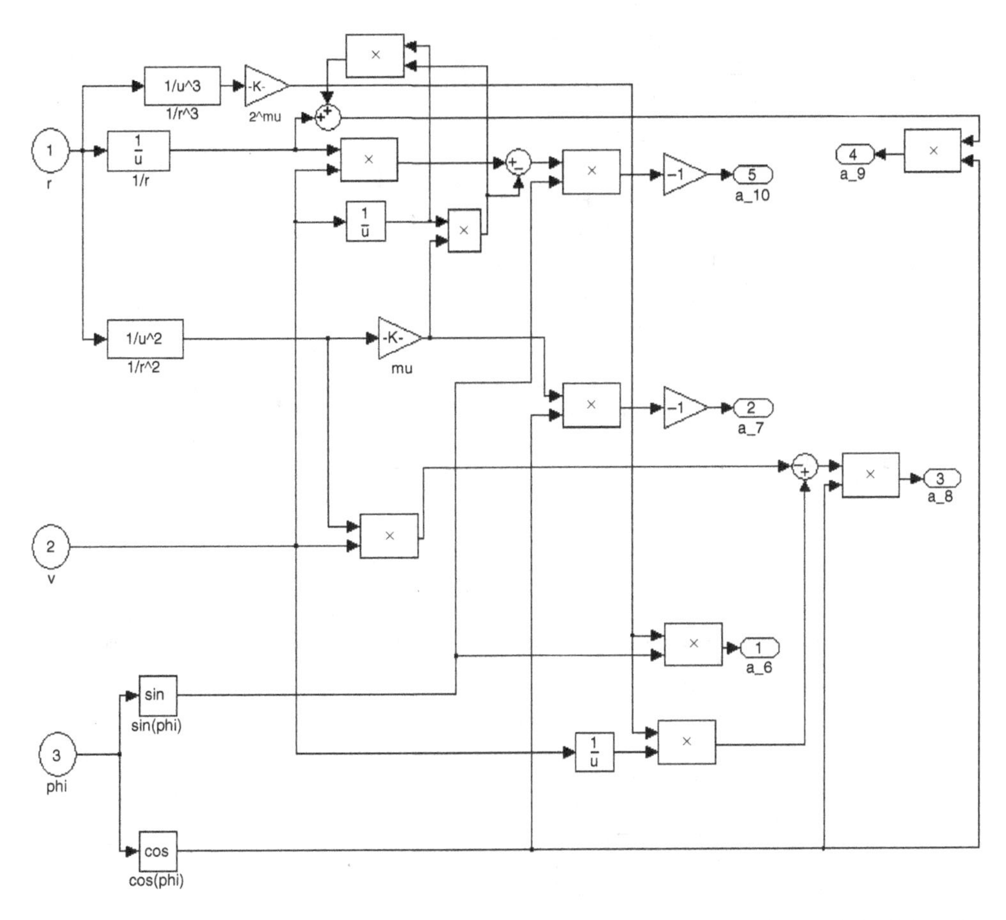

Figure C.22 Simulink *Rocket Coefficients* subsystem block for computing nonlinear rocket translational dynamics coefficients required by the block diagram of Figure C.21

drag. The drag of the perturbed trajectory is seen to be slightly smaller than that of the nominal trajectory in Figure C.20, because of the slightly higher altitude and a smaller speed of the former caused by the initial flight-path error.

Example C.6 Taking into account the atmospheric drag and mass computed in Example C.5, simulate the closed-loop response of the bi-output, reduced-order compensator for forward acceleration input designed in Example C.4 to the following thrust acceleration limits of the rocket engine:

$$6\ \text{m/s}^2 \leq f_T/m \leq 40\ \text{m/s}^2, \quad \left|\frac{\mathrm{d}f_T/m}{\mathrm{d}t}\right| \leq 40\ \text{m/s}^3.$$

Note the increase in the rate limits from $\pm 30\ \text{m/s}^3$ in Example C.2 to $\pm 40\ \text{m/s}^3$ in order to allow a more rapid actuation required in the presence of the observer. We construct a Simulink model, depicted in Figure C.21, which requires separate subsystem blocks called *Rocket Coefficients* (Figure C.22), *Observer Coefficients* (Figure C.23), and *Adaptive Observer* (Figure C.24). The *Rocket Translational Dynamics* subsystem block is the same as that used in Chapter 4. Before beginning the simulation,

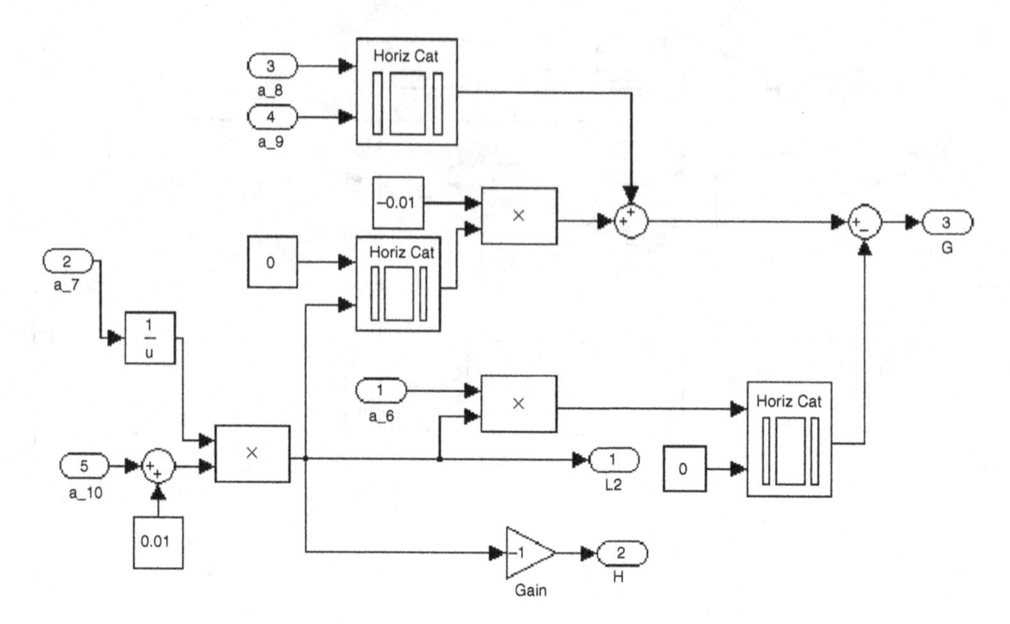

Figure C.23 Simulink *Observer Coefficients* subsystem block for computing adaptive observer gains required by the block diagram of Figure C.21

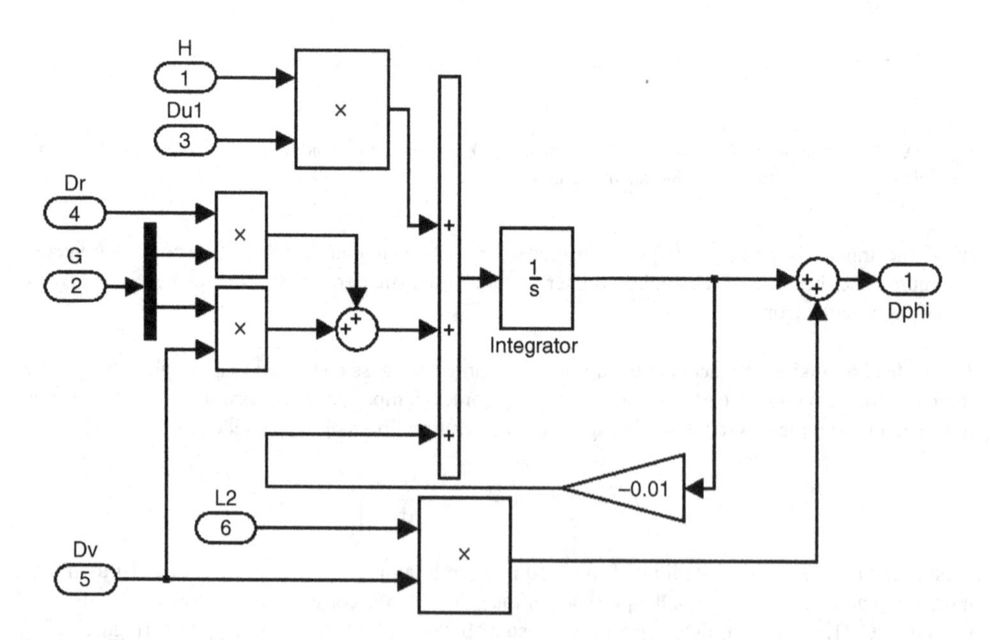

Figure C.24 Simulink *Adaptive Observer* subsystem block for adaptive observer dynamics required by the block diagram of Figure C.21

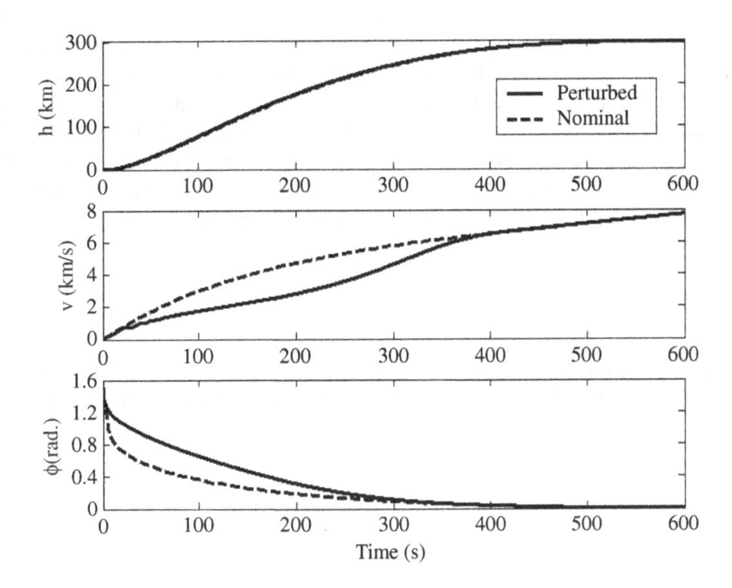

Figure C.25 Closed-loop simulated trajectory of nonlinear adaptive rocket guidance system with adaptive observer and magnitude and rate limits on thrust acceleration input

the nominal trajectory is defined in the MATLAB workspace by row vectors tn, rn, vn, phin, and the nominal control input by the matrix un1 = [tn; u1]. Furthermore, the mass and drag computed by the code *rocket_drag_mass.m* (Chapter 4) along the nominal trajectory are stored as column vectors m and D, respectively. Note the presence of table look-up blocks for nominal mass and drag as well as thrust acceleration saturation and rate limit blocks in Figure C.21. The results of the simulation carried out by the fourth-order Runge–Kutta method with a relative tolerance of 10^{-6} are plotted in Figures C.25 and C.26.

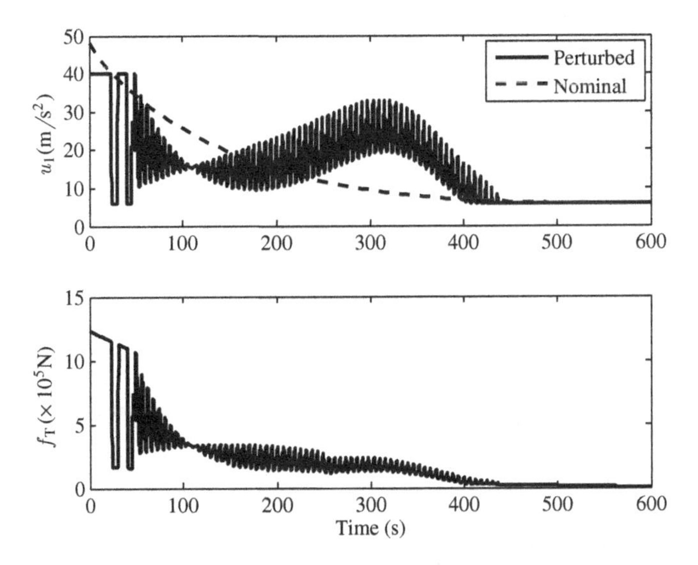

Figure C.26 Closed-loop simulated forward acceleration input and required thrust of nonlinear adaptive rocket guidance system with adaptive observer and magnitude and rate limits on thrust acceleration input

There is no difference in the closed-loop trajectory compared to that observed in Figure C.6 for the case of state feedback regulator (Example C.1), even though an observer and rate and magnitude control limits are being implemented. Figure C.26 shows that thrust saturation occurs only during the initial phase of the flight due to the reduced feedback gains compared to those in Example C.2. Furthermore, the saturation limits cause a massive reduction (almost 50%) in the required peak thrust from 2.3×10^6 N (Figure C.19) to only 1.25×10^6 N (Figure C.26). Clearly, the bi-output, observer-based adaptive compensator is much more efficient than the adaptive full-state feedback regulator for the present example.

References

Gibson, J.E. (1963) *Nonlinear Automatic Control*. McGraw-Hill, New York.

Slotine, J.-J.E. and Li, W. (1991) *Applied Nonlinear Control*. Prentice Hall, Englewood Cliffs, NJ.

Tewari, A. (2002) *Modern Control Design with MATLAB and Simulink*. John Wiley & Sons, Ltd, Chichester.

Index

Printed in the United States
By Bookmasters